Gravity-Superconductors Interactions: Theory and Experiment

Editors

Giovanni Modanese
Free University of Bolzano
Faculty of Science and Technology
Bolzano University
Italy

&

Glen A. Robertson
Institute for Advanced Research in the Space,
Propulsion and Energy Sciences
USA

CONTENTS

FOREWORD

It is our pleasure to write the Foreword for *Gravity-Superconductors Interactions*, the first eBook to further its goal of presenting to the scientific community the state of theoretical and experimental research concerning the latest results in the emerging field of *physics for novel gravity-like fields* that might represent a new paradigm shift regarding the very nature of gravitation.

We were privileged to work together for several years at the European Space Agency (ESA), experiencing first-hand the immense technical difficulties and extreme cost to place relatively small payloads into low earth orbit. Space propulsion is still dealing with the technologies developed in the 50s and 60s of the last century, and the vision portrayed by Werner von Braun in his famous article in Collier's magazine in 1952, entitled *Man on the Moon*, did not become a reality.

The shuttle era is coming to an end soon, and NASA is thus preparing to fly as soon as possible its next generation space vehicle, *Ares I-X*, a two stage rocket, which, to some extent, is looking like a modernized version of the Saturn V rocket from the Apollo moon program developed by von Braun in the 1960s. In his recent article, T. Jones in Aerospace America, May 2009, is describing clearly the enormous development efforts, the vast infrastructure, and the sophisticated technology required to eventually deliver 25 metric tons of payload into low earth orbit, which is somewhat less than the 29 metric tons of the soon to be retired space shuttle. Despite all the engineering ingenuity, undoubtedly displayed in the design of *Ares I-X*, it is a sobering lesson to learn that these extreme technological efforts will result in a space transportation system of relatively modest capability, but being of highest technological complexity.

The problem is not with the engineering, which even goes beyond the present state of the art. It is linked to the underlying propulsion physics that remains unchanged since the days of ancient Chinese rockets. It is the physical principle of classical momentum conservation which stands in the way of producing an efficient and effective propulsion system. It is the *basic physics* itself that prevents progress. The same holds true in the field of energy generation, especially fusion which may be out of reach as was discussed in the recent article by M. Moyer entitled *Fusion's False Dawn* in Scientific American, March 2010. As it seems, only *novel physics* can overcome this barrier.

Therefore, the motivation to further exploring the mysterious nature of gravitation is understandable, aiming beyond Newtonian (Einsteinian) gravity. Gravitation has maintained the interest of researchers at every stage in the history of physics, and it became Einstein's quest to unify gravitation with the other forces since 1916, the year he published his general relativity theory. The still unfinished manuscript on his desk, found after he passed away, clearly showed that he was still elaborating on his lifelong dream, namely to extend the description of the force of gravity as geometry, which had worked so marvellously well in the case of gravitation, to the other physical interactions. The *geometrization of physics*, *i.e.,* associating a metric tensor with each physical interaction, is still an open question, and it remains to be understood, if and how this beautiful principle can be extended to encompass all the other forces.

Hence, it should be no surprise that *new theoretical attempts along with experimental work* are presented in this eBook to continue where Einstein was forced to leave off. The quantization of the gravitational field has been unsuccessful, despite great efforts in this direction. The problem may be that the number of fundamental forces is not known, in other words, there is a *belief that only four forces exist* (strong, weak, electromagnetic, and gravitational force). Perhaps gravity is of a more subtle nature than Newtonian gravity, and an interaction between gravity and electromagnetism might exist? At least, the Maxwell equations of electrodynamics and the linearized Einstein field equations, termed Einstein-Maxwell equations, show surprising structural similarity.

New gravitational experiments have been published since 2006, and geometrical theories from the 1950s (for instance, by Finzi, Heim, Wheeler) were extended and combined with concepts of modern physics (symmetry, symmetry breaking, London equations, Ginzburg-Landau theory, spacetime as a physical field, *etc.*) and have gained some prominence, trying to explain novel experimental results for extreme gravitomagnetic and gravity-like fields. In his monograph on *Quantum Field Theory*, M. Kaku presents a calculation of the Coleman-Weinberg potential that might be employed to calculate the coupling strength for the extreme gravitomagnetic fields. Most recently, as pointed out by A. Zee in *Quantum Field Theory in a Nutshell*, gravity might be the square of two spin 1 fields (it should be noted that particles of spin 1 can be described by Yang-Mills fields), an idea that also might be applicable in the explanation of the experiments on extreme gravitomagnetic fields that are 18 orders of magnitude larger than those predicted by general relativity, and, if confirmed, are outside general relativity. These and other exciting ideas are presented here to the reader, and might shed new light on the nature of gravity as well as the number and type of fundamental forces that exist in Nature.

Novel theories on the geometrization of physics should provide new statements and propositions that unmistakably should lead to recognizable facts, which should, for instance, occur from the existence of extreme gravitomagnetic and gravity-like fields observed at cryogenic temperatures, rather than by speculation or chance. As Einstein felt, the most important objective of any theory is to comprise as few and basic elements as possible without contradicting physical experience in conjunction with practical applications. For example, as presented in this text, a relationship between the different phenomena of electromagnetism and gravitation might have been discovered. Any novel theory must be verifiable by laboratory experiments or astronomical observations. In order to verify a theory, it must provide a procedure on how the measurable information can be extracted. Since experiments do not produce physical principles, any novel theory must produce meaningful forecasts and also be falsifiable.

According to *Dirac's dictum*: a special regulator of a theory that reflects quality is its beauty. Einstein's theory of general relativity is an example of such a theory. The successful geometrization of physics combined with proper symmetries (group theory) would fit this picture as would the experimental generation of gravity-like fields at cryogenic temperatures by symmetry breaking.

In this eBook, these two important topics are addressed and discussed from various points of view. Needless to say, beauty cannot be the sole yardstick for the correctness of a theory or physical phenomenon, and there is always the danger that, for instance, physical models are invented to fit an experimental situation. An example to be remembered are the (non-existing?!) gravitational waves measured by Weber. It is of utmost importance that any discovery is verified by other laboratories before it can be claimed as valid. Verifying gravitational experiments is not an easy endeavour since highly sensitive devices have to be produced and utilized at cryogenic temperatures, often at liquid Helium temperature. Even if experimental findings or theories eventually cannot be verified, one should not denounce the serious experimenter or theorist for failure, since the history of science has shown that every step forward is a complicated venture, needless to say that all programs for novel theoretical models initially contain many unclear points. But this is true even for established theories. The theory of general relativity has unified gravity with inertia. The equation of motion is for material points moving along geodesics. General relativity interprets gravitation in terms of curvature of spacetime that is, the homogeneity and isotropy of spacetime are violated. Energy and momentum conservation are valid only in flat spacetime, Since gravitational waves require the full nonlinear Einstein field equations, the superposition principle does not seem to hold, in contrast to electromagnetic waves. This would be true only in the weak field limit.

Recently, a number of important and interesting experimental results on gravitomagnetic and gravity-like fields, *generated in the laboratory*, have been obtained. Gravitational experiments are notoriously difficult as can be seen from the fact that the physics of gravitational wave astronomy, despite the early efforts of J. Weber starting out in 1969, is still not an established fact.

As was pointed out by the well known theoretical physicist Richard P. Feynman in his now famous lecture, *There's Plenty of Room at the Bottom*, given already in 1960, and published in the journal *Engineering and Science* (February 1960), there occur numerous strange phenomena in the complex situations of solid state physics. He prophetically foresaw an enormous number of technical applications that could arise from such physics. He also mentioned Kammerling Onnes, the pioneer of low temperature physics and superconductivity. Why should it not be possible that a combination of low temperature and solid state physics could lead to strange phenomena and, this is the most important point, to a large number of technical applications, but this time in the field of gravitational engineering? This is what this eBook is all about.

Finally, in order for science to progress, both theorists and experimenters have to be willing to take a certain scientific risk that is, *getting off the trodden path*. If a blind alley is met, the courage to reverse one's direction of research is required. If, however, ideas of novel gravitational fields at cryogenic temperatures turn out to be true, the new scientific age of gravitational engineering might have begun. It also would mean that famous string theory is not the answer or, at least, not the complete answer, and thus will not reveal the secret of the Universe.

Whether or not this eBook is telling the final scientific truth, or even stands for a paradigm shift cannot be decided at this moment. Nevertheless, we are convinced that substantial benefit for the reader does come from it.

Jean-Marie Muylaert
Director, Von Karman Institute
Rhode-Sainte Genèse
Belgium

Jochem H. Hauser
Professor of HPC, Ostfalia
University of Applied Sciences, Germany
Scientific Director, High Performance Computing and
Communications for Space GmbH, Hamburg
Germany

PREFACE

Recent developments in this new subject are summarized by many of the researchers most directly involved. If gravitation has to be eventually reconciled with quantum mechanics, the macroscopic quantum character of superconductors might actually matter. Is it possible to generate gravity-like fields by condensed-matter systems, in conditions accessible in a laboratory? General Relativity and lowest-order Quantum Gravity predict in this case very small emission rates, so these phenomena can only become relevant if some strong quantum effect occurs. The novel experimental evidence of anomalous interactions between gravity and superconductors requires on one hand a pragmatic approach, for the verification of the effects and their detailed characterization in a model-independent way. On the other hand, this evidence is a challenge to our ability to develop creative ideas, new physical and formal concepts for their explanation, in an unknown territory placed between general relativity, quantum mechanics and new physics. A high level of multi-disciplinarity is needed, because the research spans here from superconductors (London equation, Ginzburg-Landau theory…) to gravitation (spacetime geometry, metrices, *etc.*) to quantum field theory and its formalism, to the techniques connected to field generation and detection.

This eBook is unique in its genre and represents a global novelty. It maintains a careful balance between different techniques and approaches. It will be of interest to researchers in General Relativity and gravitation theories, in field theory, in experimental gravitation, in low-temperature and high-temperature superconductivity and in more applied fields like telecommunications and propulsion technology. The subject is quite new (no more than 10-12 years) and there are presently no competitors to this eBook; it is the first attempt to put together several contributions on the subject for a broad audience. Most contributors have a direct experience in the field. R.C. Woods, G.A. Robertson, G. Hathaway, H. Reiss, E. Podkletnov and T. Junker have all run experimental trials. G. Fontana has published several works about high-frequency gravitational waves, and Ning Wu about gauge theories of gravity. R. Lewis is an experienced experimental particle physicist. G. Modanese has been studying quantum gravity for about 20 years, published in the main journals and has analysed several experiments in direct contact with the authors from 1995 onwards. J. Hauser and W. Droescher had a long exposure to B. Heim and his theory; they have extended it and have been closely following M. Tajmar's superconductor gravity experiments since their beginning.

G. Modanese

Free University of Bolzano
Faculty of Science and Technology
Bolzano University
Italy

G.A. Robertson

Institute for Advanced Research in the Space
Propulsion and Energy Sciences
USA

List of Contributors

W. Droescher
Institut fuer Grenzgebiete der Wissenschaft, Maximilianstr. 8, 6010 Innsbruck, Austria

G. Fontana
University of Trento
Via Sommarive 14, 38100 Trento (Italy)
E-mail: giorgio.fontana@unitn.it.

G. Hathaway
Hathaway Consulting Services
1080 19th Sideroad, King City, Ontario, Canada L7B 1K5; Tel: 905-795-9382
E-mail: ghathaway@ieee.org

J. Hauser
Faculty H, Ostfalia Univ. of Applied Sciences, Campus Suderburg, H.-Meyer-Str. 7, 29556 Suderburg, Germany
E-mail: jh@hpcc-space.de

T. Junker
Göde Stiftung – Goede Foundation
Am Heerbach 5, 63857 Waldaschaff, Germany; Tel: +49-6095-999-198
E-mail: timo_junker@gravitation.org

R.A. Lewis
Penn State University, Physics Dept. (retired)
1308 Andover Drive, Boalsburg PA 16827
E-mail: r3l@psu.edu

G. Modanese
Free University of Bolzano, Faculty of Science and Technology
P.zza Università 5, 39100 Bolzano, Italy; Tel. 0471-017003
E-mail: giovanni.modanese@unibz.it

E. Podkletnov
Tampere University
Korkeakoulunkatu 1, FI-33720 Tampere, Finland
E-mail: epodkletnov@hotmail.com

H. Reiss
Department of Physics, University of Wuerzburg
Am Hubland, D-97074 Wuerzburg, Germany
harald.reiss@physik.uni-wuerzburg.de

G.A. Robertson
Institute for Advanced Research in the Space, Propulsion & Energy Sciences
Madison, AL 35757, USA; Tel: 256-694-7941
E-mail: gar@ias-spes.org

R.C. Woods
Department of Electrical and Computer Engineering, Louisiana State University
Baton Rouge, LA 70803-5901, U.S.A.; Tel: 225-578-8961
E-mail: cwoods@lsu.edu

CHAPTER 1

Gravity-Superconductors Interactions: Historical Background

Giovanni Modanese[1,*] and Glen A. Robertson[2,*]

[1]*Free University of Bolzano, Faculty of Science and Technology, P.zza Università 5, 39100 Bolzano, Italy and* [2]*Institute for Advanced Research in the Space, Propulsion & Energy Sciences, Madison, AL 35757, USA*

Abstract: This chapter summarizes from a historical perspective the works on gravity-superconductors interactions from the De Witt 1966 brief paper; to the Anandan, Chiao and Ross 1980s theoretical papers; to the Li, Peng, Torr 1990s theories and superconductor experiments as those of Podkletnov *et al.* in 1992, 1995, & 2001, Chiao *et al.* in 2003 and Tajmar *et al.* in the late 2000s with some discussion on other theories and experiments. More detailed work can be found in the other chapters of this eBook.

Keywords: Ravitational forces, theories of gravitation, superconductors, high-Tc superconductors, low-Tc superconductors, type-II superconductors, general relativity, quantum gravity, vacuum energy density, gravitational waves, gravitomagnetism, gravity-like fields.

1. INTRODUCTION

The idea of a possible special interaction between superconductors and the gravitational field was first expressed in an article by De Witt [1] and further in a series of papers in the early 1980s by Anandan and Chiao [2], who considered the possibility of constructing antennas for gravitational radiation, using Josephson junctions as transducers. In 1983 Ross derived the modified London equations for a superconductor in a gravitational field [3], and showed that these equations are consistent with the results found earlier by DeWitt, *i.e.*, that the phase shifts due to a gravitomagnetic field are accommodated by minuscule supercurrents. In the early 1990s, Li and Torr derived a similar relationship involving gravitomagnetic and gravitoelectric coupling to gravity. In 1990, Peng and Torr used the generalized London equations to treat the interaction of a bulk superconductor with gravitational radiation, and concluded that such a superconducting antenna would be many orders of magnitude more sensitive than a Weber bar [4, 5].

On the experimental side there are a few recent works, which until now have not been independently confirmed and have not reached a clear and definitive outcome. The experiments by Podkletnov *et al.* with rotating YBCO disks are discussed in Section 2, their replication attempts are described in Section 2.1. Experiments on anomalies in the weight of superconductors at T_c are summarized in Section 3. The experiments by Tajmar *et al.* with rotating superconductors are analyzed in Chapter 6 of this eBook. Tests by Chiao *et al.* with transducers between e.m. and gravitational waves are described in Chapter 7. Experiments by Podkletnov with high voltage discharges through superconducting electrodes are discussed in Section 4 and in Chapters 8 and 9.

1.1. Superconductors and Gravity

Superconductivity is a macroscopic quantum effect. Therefore, it is natural to suspect that the gravity-superconductors coupling is different from the standard classical coupling mediated by the energy-momentum tensor, *e.g.*, crossing into quantum gravity and seeming ideal for the development of gravitational quantum interference devices. Differing from atoms and molecules, which have almost irrelevant gravitational interactions [5, 6], although these interactions can be observed in precision experiments of neutron interference [7].

*Address correspondence to Giovanni Modanese:** Free University of Bolzano, Faculty of Science and Technology, P.zza Università 5, 39100 Bolzano, Italy; Tel: 0471-017003; E-mail: giovanni.modanese@unibz.it **and Glen A. Robertson**: Institute for Advanced Research in the Space, Propulsion & Energy Sciences, Madison, AL 35757, USA; Tel: 256-694-7941; E-mail: gar@ias-spes.org

A gravitational quantum interference device was the main motivation of the work by Anandan and Chiao, that is to see if superconductors could act as very sensitive gravitational detector. Superconducting Quantum Interference Devices (SQUIDs) can detect very small amounts of magnetic flux, multiple of the elementary fluxon $h/2e$. The equations of the superconductor wavefunction in an external gravitational field are mathematically similar to those in an external electromagnetic field, exhibiting in principle flux quantization, Meissner effect *etc.* but the coupling constants are very different and the gravitational effects usually very small. For instance, a "gravitational Meissner effect" could only be observed in neutron stars; even for a mass density of the order of 10^{17} kg/m^3, the gravitational London penetration depth is ca. 12 km [8]. The gravitational flux quantization in a superconducting ring is discussed among others in the review by Kiefer and Weber [9].

Another typical property of superconductors which makes them suitable as detectors is their low dissipation level. Braginsky *et al.* [10] in the '70s proposed several schemes for experiments of laboratory generation and detection of time-variable gravitational fields, and some of these schemes involved superconducting oscillators. The Probe B experiment for the measurement of the gravitomagnetic field of the Earth on orbiting satellites uses superconducting gyroscopes. This choice has technical reasons, but it has been discussed [11] whether the superconducting character of the gyroscopes might affect their gravitational coupling.

The theoretical formalism employed in the cited papers varies. Some use the metric formalism for the description of the field, thus allowing a direct inclusion and check of the equivalence principle. Others employ the weak-field approximation of the Einstein equations in the form of the so-called Einstein-Maxwell equations.

From a different perspective, in Chapter 4 of this eBook, Robertson looks inward at the superconductor properties themselves for guidance toward the understanding of what superconductor properties could lead to gravitational phenomena. This is accomplished by first looking at properties for rapid power flow internal to the superconductor, applying these properties to energy radiated in gravitational wave, establishing an experiment to setup this power flow (which happens to look much like the Podkletnov impulse experiment) and evaluating the theoretical calculations to the Podkletnov impulse experiment. The theoretical calculations and the experimental data are shown to be surprising similar both in graphical form and value.

1.2. Gravitomagnetic Fields

It is well known from general relativity that the gravitational field contains weak components of magnetic type, called the gravitomagnetic field. These components have the property to be produced by moving objects and to act on objects in movement. It is possible to write the Einstein field equations in weak field approximation in a form very similar to that of Maxwell equations. The important difference, with respect to the Maxwell equations for electromagnetism, is that gravitomagnetic effects are suppressed by factors $1/c$ or $1/c^2$. So they are always very small with respect to "gravitoelectric" effects - those which have a Newtonian limit.

In the linear Maxwell-Einstein approximation, the relevant components of the gravitational field are reduced simply to two vector fields $\mathbf{E}_g$ and $\mathbf{B}_g$ which behave in close analogy to electromagnetic fields, since they satisfy the equations

$$\nabla \cdot \mathbf{E}_G = -4\pi G \rho_m$$

$$\nabla \times \mathbf{E}_G = -\frac{\partial \mathbf{B}_G}{\partial t}$$

$$\nabla \cdot \mathbf{B}_G = 0$$

$$\nabla \times \mathbf{B}_G = -\frac{4\pi G}{c^2}\mathbf{j}_m + \frac{1}{c^2}\frac{\partial \mathbf{E}_G}{\partial t}$$

where $\mathbf{j}_m$, ρ_m are the mass-energy current and density. These equations are discussed in more detail in Chapter 6. Their application to weak and quasi-static fields is well established since the '50 [12, 13] but their use for the description of gravitational waves requires special care and may give inaccurate results (consider just the fact that the fields $\mathbf{E}_g$ and $\mathbf{B}_g$ carry helicity 1, while gravitational waves described in the usual TT gauge have helicity 2). In their application of the Einstein-Maxwell equations, Li and Torr (see below) find a "divergence" due to the supposed $\mu \to 0$ limit of the vacuum permeability in a superconductor; such limits must also be taken with care, because they could take us beyond the weak-field validity range of the linearized equations.

Finally note that in full General Relativity, forces among masses are included as part of the stress tensor source. This procedure guarantees conservation of energy, momentum and angular momentum. The Einstein-Maxwell equations must be augmented by constraints to ensure conservation (see Chapter 6).

1.3. The Work by Li and Torr

In the '80s Torr and Li [14, 15] made some bold predictions on the effects of a rotating superconductor on the gravitational field. According to their calculations with the Einstein-Maxwell equations, a "gravitational attenuation" was attainable. Although their model has been recently criticized by Kowitt [16] and by Harris [17], Li and Torr calculated the gravitomagnetic field which would be produced by a superconductor containing circulating supercurrents. According to them, in this case also the movement of the ions, which produce a mass current, has particular importance. The fact that spin alignment is present would also be relevant, as the alignment of the spin of the lattice ions would be a source of gravitomagnetic field. The objections to this model are of two types. First, the total spin amount that is possible to obtain in normal condensed matter is always very small, also in presence of alignment. Therefore, given the weak coupling with the gravitomagnetic field, one does not understand how it could reach detectable intensity. Another objection is of more technical nature. Ning Li and Torr use the Maxwell equations for gravity, which hold in weak field approximation. However, when they find that some terms of these equations "explode" in the $\mu \to 0$ limit, they keep this result even if it is inconsistent with the initial linear approximation.

An interesting fact is that the Li and Torr model has relevance to the Podkletnov superconductor rotation experiment (Section 2), though done previously to it. That is, not only the alignment of the spin of the lattice ions would produce a gravitomagnetic field, but in the presence of an outside applied time-dependent vector potential, this would be converted into a perceptible gravitoelectric field. Such a conversion is necessary, if one wants to apply this theory to claims of weak gravitational shielding, because in that case the test bodies are at rest, and so they would not respond to a gravitomagnetic field.

1.4. Confrontation Between General Relativity and Experiment

These efforts should be placed into the wider context of the confrontation between General Relativity and experiment [18]. It is widely recognized that General Relativity has successfully explained all the gravitational phenomenology observed in the laboratory until now. Gravitational waves have escaped direct detection, however, and modifications of the gravitational forces at large distances appear to be necessary to explain the details of galactic rotation (Milgrom theory).

Furthermore, the precision with which the value of G is known is clearly unsatisfactory compared to the precision with which the other fundamental physical constants are known. While the definition of the other fundamental constants relies on microscopic experiments performed with high precision devices, for the G constant it is necessary to use more or less sophisticated versions of the Cavendish experiment. In particular, we do not know the behaviour of gravity at short distances (millimeters or less). According to Gillies [19], in the second half of 1900 more measurements of G were made than ever before. Some discrepancies were observed in recent times with respect to the best official value fixed in 1982. The value of G has been called into question by new measurements from respected research teams in Germany, New Zealand, and Russia [20]. The new values disagree considerably, up to 0.6%. Recent theoretical predictions suggests that gravity penetrates extra, compact dimensions so that the gravitational inverse square law must be modified at short ranges (less than 1 mm). Arkani-Hamed, Dimopoulos, and Dvali [21] have offered this as an explanation of the hierarchy problem.

Surprising results were obtained recently by the observation of the motion of the Pioneer space probes [22]. Direct measurements are possible thanks to the radio signals transmitted by the spacecrafts, which give precise information about their position, speed and acceleration. From the analysis of the data a residual acceleration was found, not explainable through the usual orbit tracking models. Some proposals for a theoretical explanation of the phenomenon were presented. One of these [23] calls for the existence, in addition to Einstein's gravitational field, of a scalar field which, supposed it has certain properties, could lead to the observed effect. This field could also be responsible for the rotation anomalies observed in the galaxies, because it gives a modification of the behaviour of gravity at large distances. The cited model belongs to a wide group of extensions of the standard theory of gravitation [24], which originate from the Brans-Dicke model and in more recent times refer to string theory and to the so-called dilaton field. Another notable extension of General Relativity is the Extended Heim Theory (EHT; see Chapter 11 of this eBook), which offers an explanation for some of the problems mentioned above. EHT could also account for the enhanced gravitomagnetic and gravitolectric effects observed in the Tajmar experiments (Chapter 6).

2. THE FIRST PODKLETNOV EXPERIMENT

The first version of this experiment, published in 1992 [25], consisted of radiating a large YBCO disk below its critical temperature with RF radiation while the superconductor was magnetically levitated. It was serendipitously noticed that a balance placed above the disk behaved strangely. A check then showed that any object above the disc lost some of its weight, and rotating the disc, the effect was increased. Generally, a small test mass suspended above the superconductor apparently showed a 0.05% weight loss under certain conditions. A relatively large size (145 mm diameter × 6 mm thick) sintered single layer disk with specific grain size distribution of relatively small grains was required to be levitated by Meissner magnetic levitation up to 7 mm above a single ''pancake'' coil operated at frequencies of 50–106 Hz. In addition, the disk was spun to high speeds using edge-oriented pseudo-rotating magnetic fields provided by a magnetic stator positioned around the periphery of the disk and operated at similar frequencies. The experiment was performed below 60 K in the vapours of liquid helium.

In 1997, Podkletnov published an updated version of the experiment on the Los Alamos Physics Archiv site [26]. This was essentially the same experiment as reported in 1992 and achieved up to 1–2% weight loss in test objects but with the following experimental differences. A larger superconducting disk in a ring configuration (27 cm outside diameter × 8 cm inside diameter × 1 cm thick) was used with a small–medium grain size distribution in a bi-layer structure. One layer was superconducting and the other was a normal conductor at the experimental operating temperatures. The disk was was placed inside a steel cryostat and levitated above either three or six individual solenoidal coils and two toroidal coils threaded through the center hole of the disk and operated in a 2-phase mode to drive the disk in rotation up to 5000 rpm.

The toroidal YBCO disk was prepared using conventional ceramic sintering technology followed by partial melt-texturing: the upper surface was quickly heated to 1200 °C using a planar high-frequency inductor. The analysis of the cross-section of the toroid revealed the existence of two zones with different crystal structures. The upper part of the disk (6-7 mm thick) had an orthorhombic structure typical of the quench and melt growth process, with T_c=94±2 K. The critical current density J_c exceeded 15000 A/cm^2. The transition temperature of the lower layer was equal to 60±10 K. Samples made of various materials were tested for weight change, including metals, glass, plastic, wood and so on. All the samples were hung over the cryostat on a cotton thread connected to sensitive balances of three different kinds. The distance between the samples and the cryostat varied from 25 to 3000 mm. Samples made from the same material and of comparable size, but with different masses, lost the same fraction of their weight. The best measurement gave a weight loss of 0.5% while the disk was spinning at 5000 rpm, with typical values ranging from 0.3 to 0.5%. Samples placed above the inner edge of the toroid (5-7 mm from the edge) were least affected, losing only 0.1 to 0.25% of their weight. The external boundary of the "shielding cylinder" was quite clear (no more than 2 cm). The maximum values of weight loss were obtained below 70 K. During the time when the rotation speed was being decreased from 5000 to 3500 rpm, using the solenoids as braking tools, the shielding effect reached maximum values: the weight loss of the samples was from 1.9

to 2.1%, depending on the position of the sample with respect to the outer edge of the disk. The samples' maximum weight loss was observed only when the magnetic field was operating at high frequencies, on the order of 3.2 to 3.8 MHz.

Podkletnov's work originated outside the gravitational-physics community and spurred much controversy. As a chemist and an expert of materials science, he was studying the properties of high-Tc ceramic superconductors [27]. His first article was published without any clamour in the superconductivity journal Physica C in 1992. In 1995 Modanese published an electronic preprint (later appeared in [28]) whose main thesis was, that the effect could not be explained by General Relativity, but had a possible explanation in Quantum Gravity, if the superconducting condensate was regarded as a contribution to the vacuum energy density or cosmological term. The preprint drove some attention to Podkletnov's work. A few replication attempts were reported in the following years (compare Section 2.1). Note that although it is possible to measure variations in g much smaller than 0.5%, there could be a production threshold; so it is not guaranteed that an incomplete replication of the experimental conditions gives a weaker but still observable effect.

With its claimed gravitational modification of 0.5% - 2% max, the Podkletnov effect lies in an intermediate range between much smaller gravitational effects, which require very accurate detection, and a number of unlikely pseudo-scientific claims of larger effects, obtained with simpler devices. The effect appears to be too small for applications to propulsion, unless its energetic efficiency is much increased. Although, if confirmed, it is large enough to shake our current view of gravitational physics.

Critics have often required "witnesses" of these effects, or have produced "logical arguments" to disprove them without the need of a replication. Both attitudes are unfair in our opinion, even though similar methods are often, by necessity, employed by those who screen the large number of exotic claims available in the informal literature. The ideal path to a new discovery in science is detailed publication followed by independent confirmation. It should be recognized, however, that in the course of history several discoveries have followed a more tortuous path, and that sometimes the work of researchers not belonging to certain elites, even if valid and exploited by others, did not receive the due credit. One of the aims of this eBook is to bring some order to the subject and collect all the available information.

2.1. Replication Attempts

Several attempts have been made to repeat the original gravity shielding experiment, but to the authors' knowledge, a complete verification has yet to be done. The most noticeable attempts are given in the following subsections.

2.1.1. NASA Experimental Attempt

Beginning in the late 1990s through the early 2000s, a group at NASA-Marshall Space Flight Center (MSFC) in Huntsville, Alabama made two attempts, one in house and one through a NASA-SBIR. Neither attempt was completed as in both cases, the funding did not allow for the development of the rotation system needed for the completion of the test.

The first attempt was a joint effort between the University of Alabama in Huntsville (UAH) and NASA-MSFC in house personnel. Much of the funding was used to develop techniques to produce several large (~12" Diameter) YBCO superconductors, both by UAH and MSFC personnel, and to conduct some static testing. Although several superconductor disks were successfully fabricated [29], no bi-layer SC disk was fabricated. Static tests on a commercially purchased 4 inch single layer YBCO superconductor using a rented industry gravity device was conducted with limited results [30]. Continued development funding was not requested due to conflicting opinions between UAH and NASA personnel as to the direction of the program.

In the second NASA-SBIR attempt, Podkletnov was hired as a consultant by the small business entity. Much of the funding was used in fabricating many superconductors to Podkletnov's specifications. Eventually, a large bi-layer superconductor disk was successfully fabricated to Podkletnov's satisfaction. The small business entity had a simple Helium Dewar fabricated and verified that the large bi-layer superconductor disk could be levitated by an electromagnetic arrangement, but funding ran out before a rotation system could be developed. No NASA or public publication of this effort was made. A SBIR report was delivered to the MSFC, but to the authors' knowledge, no known record still exists. The small business entity reported that some of the YBCO fabrication data was published by Podkletnov in his first impulse gravity generator paper (compare Section 4). Noting that Podkletnov made no reference to the small business entity or NASA-MSFC in this paper.

Although several unverified claims have been made in various media by non-NASA personnel, no verification or disproval of the original gravity shielding experiment could be made by NASA as the experiment was not completed in either attempt.

At this writing, no known attempt has been made by NASA to repeat the second impulse experiment (Section 4).

2.1.2. Woods' Experiment

In 2001, Woods *et al.* [31] reported an attempt at replicating the gravity shielding experiment with 2 inch and 3 inch superconductor disks, which produced no change in the test mass weight. Woods *et al.* argued that the fact that no effect has been observed in tests (to include theirs) is that these tests have not fulfilled the specified conditions for a gravity effect, which are:

- A disk with a diameter greater than 100 mm.
- A disk containing - 30% non-superconducting YBCO, preferably organized into two layers.
- A disk capable of self-levitation, but still containing large numbers of inter-grain junctions.
- An alternating levitation field with a frequency ~10 kHz.
- A second alternating excitation field with a frequency of ~1MHz, for disk rotation.
- Disk rotation speeds of a few thousand rpm for large (>0.05%) gravity effects (the largest effects (>2%) were reported by Podkletnov when the disk was being decelerated at around 3000 rpm).

2.1.3. Hathaway's Experiment

In 2003, Hathaway *et al.* [32] reported an attempt of the gravity shielding experiment, which used a bilayer YBCO disks of 160 mm outside diameter × 40 mm inside diameter × 7–10 mm thick. These disks were fabricated using one of several methods proposed by Podkletnov (in personal communication with the same) or larger grain size distribution was used.

In the experiment, three individual levitation solenoids of similar design to that outlined in the 1997 Podkletnov paper were operated at 105 Hz. Three high-frequency coils in a 3-phase arrangement threaded toroidally through the center hole of the disk were operated at between 2 and 5 MHz. The disk was rotated mechanically by a bearing and gear assembly designed to operate at liquid helium temperatures from a drive motor external to the liquid helium cryostat. The experiment was performed between ~5 and 20 K and the detection of the gravity-like force used a modified analytical balance of sensitivity exceeding that used by Podkletnov in his first experiment.

Some conclusions of the experiments by Hathaway *et al.* [32] are as follows:

1) No weight modification or gravity-like force has been detected to the 0.001% level. The experiments were run using sintered YBCO bi-layer disks with triple grain size and the

maximum disk rotation was no greater than 550 rpm. Therefore, improvements to the experimental apparatus should be made for higher rotation speeds (1000 – 2000 rpm.).

2) The ability to achieve true AC Meissner levitation at 100 kHz was not successful because of the large size and weight of the disks used in the experiments. Apparently, the magnetic field intensity of the solenoid assembly proved to be too weak. Therefore, either greater power levels are required (greater than 1 kW), or the coil design needs to be optimized.

One of the present authors (G.R.) suspects that the influence of the rotation is not a linear function, but squared. Such that even though the sensitivity of the Hathaway *et al.* [32] is quite high, any effect would have been several orders of magnitude smaller at the 550 rpm than suggested by Hathaway *et al.* Therefore, any gravity effect would not have been detectable even at the 0.001% level of sensitivity for the rotation speed used in the experiments. Further, the proper power input into the superconductor seems to be the key as in the Podkletnov impulse experiments, as a smaller superconductor disk was used with greater results (Section 4). This seems to suggest that the input power from levitation was also needed in the Hathaway *et al.* experiment.

Our conclusion is that the Hathaway *et al.* experiment did not satisfy the entire requirement set, as indicated by Woods *et al.* in the previous subsection, needed for verification of the original gravity shielding experiment; even though it was conducted at a much higher level of sensitivity. At best, this experiment sets a lower limit on the experimental criteria.

2.1.4. Other Notable Reports

In a 2002 BBC news item, it was alleged that researchers at Boeing were funding a project called GRASP (Gravity Research for Advanced Space Propulsion), which would attempt to construct a gravity shielding device [33], but a subsequent 2003 Popular Mechanics news item stated that Boeing had denied funding GRASP with company money, although Boeing acknowledged that it could not comment on "black projects". A possible solution of this contradiction has been suggested: it is alleged that the GRASP proposal was presented to Boeing, but that Boeing chose not to fund it [34].

A discussion by one of the authors with the Boeing person supposedly putting this together indicated that pressure from Boeing management would not allow this experiment to be conducted at that time. No other word on a Boeing experiment has been noted in media and it is believed that no experiment was conducted by Boeing.

Finally we note that Tajmar *et al.* (see Section 6 and Chapter 6 of this eBook) reported null results on gravitational anomalies with rotating YBCO; their disk, however, was mechanically rotated and probably sintered and single-layer.

2.2. Incompatibility with Classical General Relativity

A significant gravitational shielding has never been previously observed. Several experiments, starting with the classical measurements of Majorana, have shown that the gravitational force is not influenced by any medium, up to one part in 10^{10} or less (see also[35]). For a "classical" medium the reason for this is essentially the absence of charges of opposite sign which, by shifting or migrating inside the medium, might generate a field which counteracts the applied field. On the other hand, it is easy to check that the probability of a process in which a graviton excites an atom or a molecule of the medium and is absorbed is exceedingly small, essentially due to the smallness of the gravitational coupling at the atomic level.

From the perspective of classical General Relativity, it is readily realized that neither the superconducting disk nor any other object of reasonable density can influence the local geometry so much as to modify the gravitational force by the observed amount. To check this one just needs to write the Einstein equations (or even some generalizations of them) and impose suitable conditions on the source $T_{\mu\nu}$. The Einstein equations allow to find the gravitational field as a response to a source - linear in a first approximation, or non-linear in the presence of strong mass-energy densities. The proportionality constant between field and source is of the order of the Newton constant G for linear responses and even smaller, of the order of G/c^n, for non-linear responses. There exist static fields and fields propagating like waves, but in any case their strength is related to the mass of the source which has generated them. The only sources close to us which are massive enough to generate a detectable field are the earth, the moon, the sun and, to a smaller extent, the other planets of the solar system. Any other object or physical system available on a laboratory scale, irrespective of its chemical composition or microscopic structure, generates gravitational fields of exceedingly small strength. These fields can be detected through very sensitive instruments, but they are typically of the order of 10^{-6} g or less.

These observations are well known and lead to the conclusion, in full agreement with Einstein equations, that the gravitational field generated by a very massive object is in practice unaffected by the presence of any other body whose mass is much smaller. In particular, this does not seem possible that the gravitational acceleration g at the earth surface can be affected, through any human-sized apparatus, by more than approx. 1 part in a billion. The conclusion above rests, as mentioned, upon the hypothesis that the equations of classical General Relativity are appropriate to the situation.

In perturbative quantum gravity the superconductor gives rise to a space-dependent contribution to the vacuum energy density term. This is known to be equivalent in turn to a mass term for the graviton, but this mass turns out to be very small. A realistic estimate for different types of superconductors has been given in [36]: the maximum energy density contribution is of the order of 10^8 J/m^3. According to N. Wu [37], the situation may be different in a gauge theory of gravity.

2.3. Non-Height Dependence in the Weak Shielding Region

Another puzzling point is that in the reported data there is a non-dependence of the shielding effect on the height at which the samples are placed above the superconducting disk. Up to a height of ca. 3 m no difference in the shielding value has been noted, with a precision of one part in 10^3 of the weight [22]. It is also remarkable that during the measurement at 3 m height the sample was placed in the room which lies above the main laboratory, on the next floor; in this way the effect of air flows on the measurements was greatly reduced. For the used 500 g sample the weight loss was ca. 2.5 g. Such an extremely weak height dependence of the shielding is in sharp contrast with the intuitive picture, according to which the gravitational field of the Earth is the vectorial sum of the fields produced by each single "portion" of Earth. (See also [38]). In the absence of any shielding, the sum results in a field which is equivalent to the field of a pointlike mass placed in the center of the Earth; this can be checked elementarily by direct integration or invoking Gauss theorem and the spherical symmetry. But if we admit that the superconducting disk produces a weak shielding, the part of the Earth which is shielded lies behind the projection of the disk as seen from the sample, *i.e.*, within an angle α about the vertical direction, such that $\tan\alpha = h$, where h is the sample height over the disk measured in units of the disk radius. (For simplicity we suppose now the sample to be centered above the disk.)

In order to predict the shielding effect as a function of h, taking into account this geometrical factor, one must integrate the Newtonian contribution $\cos\alpha'/R^2$ over the intersection between the Earth and the cone defined by $\alpha' < \alpha$. We have done this for the values h = 1, 2, 3, 4, 6, 8, 10, through a Montecarlo algorithm [39]. We took into account the higher density of the Earth's core; we also computed analytically the contribution of the tip of the cone, from the Earth's surface to the Earth's core, in order to reduce the fluctuations in the Montecarlo samplings for small R. The resulting values were, for instance, the following: for h=1 the ratio between shielding and maximum-shielding was 0.62; for h=4 it was reduced to 0.102 and for h=10 to 0.018.

This predicted strong height dependence is clearly incompatible with the mentioned experimental data, which instead seem to indicate that in the shielding process all the mass of the Earth behaves effectively as if it would be concentrated in one point. Or, alternatively, it indicates that the "shielding" is actually the consequence of a steady repulsive emission from the disk, similar to that observed in the experiments with high-voltage discharges (Section 4), but much weaker. The latter hypothesis would be validated if further measurements can confirm that a slight tilting of the rotation axis of the disk causes a tilting of the shielding cylinder. On the other hand, if the SC disk would emit an hypothetical secondary field, or some radiation, this would probably appear like a divergent beam. (Also note that in this case one should expect some emission downwards, which does not appear to be the case.)

The modified field pattern observed in the presence of weak shielding is clearly non conservative. If a test mass makes a round trip going up inside the shielding cylinder and coming down outside, the gravitational field exerts a net work on the mass. Formally this amounts to say that the observed field is not the gradient of a gravitational potential. The Einstein equation for a weak static field still holds, namely $\partial_i \Gamma^i_{00} = 0$, with $a^i \propto \Gamma^i_{00}$ (***a*** is the acceleration of the test mass *m*, in the limit when the velocity of *m* is much smaller than light velocity); however, the associated equation $\Gamma^i_{00} = \partial^i h_{00}$ is not true. In fact, the very concept of local "gravity modification" (compare also [40]) is subtle and prone to paradoxes. In the following we discuss the problems of energy conservation and possible violations of the equivalence principle.

2.4. Energetic Balance

In principle, energy must be supplied in order to reduce the weight of an object, because the gravitational energy of the object has negative sign and is smaller, in absolute value, in the presence of shielding. However, since the field is non-conservative in the presence of shielding, we can not compute the difference in the energy of an object between the interior and the exterior of the shielding cylinder by evaluating naively the difference (which turns out to be huge) between an hypothetical "internal potential" $U = -GMM_E(1-\alpha)/R_E = -Mg_ER_E(1-\alpha)$, and an "external potential" $U=-Mg_ER_E$. (Here M is the mass of the object, R_E the Earth radius, M_E the Earth mass and α the "shielding factor, " $\alpha \sim 0.01$.)

Moreover, the gravitational fields with which we are most familiar, being produced by very large masses, are insensitive to the presence of light test bodies; thus, it makes sense in that case to speak of a field in the usual meaning: while a body falls down, we do not usually worry about its reaction on the Earth. But in the present case, the interaction between the shielded object and the "source of the shielding, " is very important. Let us ask a provocative question, suggested by the experimental reality: if the superconducting disk is in a room and the shielding effect extends up to the ceiling, should we expect the disk and all the shielding apparatus to feel a back reaction? (And possibly an even stronger one if the ceiling is quite thick or if there are more floors above?) The most reasonable answer is this: since the ceiling is very rigid, the experimental apparatus is not able to exert any work on it and thus does not feel its presence.

To clarify this point, imagine that we hang over the superconducting disk a spring of elastic constant *k*, holding a body of mass *M* at rest. Then we operate properly on the disk and produce the shielding effect with factor α. Now, the gravity acceleration over the disk becomes $g_E(1-\alpha)$. If the shielding effect is obtained quickly, the mass should begin to oscillate, otherwise it should rise by an height of $\Delta x = \alpha g_E M k^{-1}$, while remaining in equilibrium. In any case, since the kinetic energy and the potential energy of a harmonic oscillator in motion have the same mean value, it is legitimate to conclude that the work done by the shielding apparatus on the system [mass+spring] is of the order of

$$\Delta E \sim k\,(\Delta x)^2 \sim (\alpha g_E\, M)^2\, k^{-1}$$

This example shows that the work exerted by the apparatus on a sample being "shielded" will depend in general on the response of the sample itself. Only if the sample can respond to the effect, will work be done.

In conclusion, we must expect in general an interaction between the partially shielded samples and the shielding apparatus. The energy needed to shield a sample depends on the mass of the sample itself and on the way it is constrained to move. In particular, we deduce from the equation above that if we want to detect the shielding effect by measuring the deformation of a spring (and do this with the minimum influence on the shielding apparatus) we should use, as far as is allowed by the sensitivity of the transducer, a spring with a high rigidity coefficient k. The optimum spring constant also depends on timing. If the source is a short pulse, then a k is desirable, such that $\sqrt{(k/m)}$ matches the pulse width; compare Chapter 8, Section 5.2.

2.5. Other Implications

Will a light ray be deviated when it crosses the shielding region? Will a precision clock show any gravitational slow-down when it is first placed into the shielding region and then pulled out? These phenomena are typically studied using the metric tensor. We saw that in the presence of shielding the relation between the static force Γ_i and the gradient of h_{00} is spoiled, but it could still be possible to use the metric tensor to predict the trajectory of a light ray or the relation between the proper time of a frequency standard and the coordinate time. These effects could also be tested experimentally, at least in principle. Light propagation across the shielding region could be studied, for instance, through interference measurements on a laser beam after several reflections at both sides of the region. It can be estimated that the modification of the metric tensor inside the shielding region may lead to a shift in the frequency of a clock of about 1 part in 10^{11} - approx. 2 orders of magnitude larger than the precision of atomic clocks. But apart from the several practical difficulties connected to these measurements, it is not clear yet on the theoretical side whether the gravity anomaly affects in the same way all the components of the metric, and whether it affects in the same way the static behavior of the field and its variations on a short time scale.

Another important issue is the compatibility between the shielding phenomenon and the equivalence principle. Imagine a box divided in two sections 1 and 2. Suppose that the lower part of the box, with mass m_2, contains a shielding apparatus, complete with power supply generator. Now let the box be in free fall. If "the shielding is OFF, " the acceleration of the box is equal to g. Then we "turn the shielding ON", say with efficiency α; this means that the gravitational force felt by the mass m_1 over the apparatus is multiplied by a factor $(1-\alpha)$. Let us admit that the weight of m_2 itself is not affected. It is easy to see that in this case the acceleration of the box becomes less than g. This means that the gravitational mass and the inertial mass of the box are not equal any more, and this represents a violation of the equivalence principle. If we do not accept the possibility of such a violation, we must admit that the shielding effect can not work like this. We must admit that if the shielding apparatus is rigidly connected to the Earth, then there is effective weight reduction of the samples suspended over the apparatus; but if the whole shielding apparatus is in free fall, then a reaction force from the samples on the apparatus arises, which makes the total weight variation vanish. This means of course that it is impossible to build a flying machine using the gravity shielding effect. It is still possible however, in principle, to build a ground-based "lift" [41].

It is known that the equivalence principle is compatible only with a tensorial gravitational field [42]. Some theoretical models developed to explain anomalous gravity-superconductors interactions involve scalar or vector fields, and this also can lead to violations. (See also Chapter 7 of this eBook.) More radically, the simple fact that superconductors have peculiar gravitational interactions (not just proportional to their mass-energy content) can be seen as a violation of the equivalence principle. Note, however, that several theoretical attempts to connect general relativity to the standard model allow for a violation of the equivalence principle.

2.6. Relation to the London Field

It is known that inside a rotating superconductor the magnetic field does not vanish, but takes a certain value B_L called "London field strength"; this is given, in terms of the electron charge e, the electron mass m and the angular velocity ω of the SC, by $B_L = 2m\omega/e$. This holds for any value of the external magnetic field, including the case $B_{ext} = 0$. The London field has been interpretated by some authors as the field needed to give the superelectrons the same rotation velocity of the crystal lattice. While the normal electrons keep pace with the positive ions due to ohmic friction, the Cooper pairs are mechanically decoupled from the lattice and need the field B_L to be forced to move on circular orbits with frequency ω; otherwise very large bulk "slide" currents would flow, up to 10 or 100 times larger than the critical intrinsic supercurrent, depending on the angular velocity [43]. Now consider cases when, like in Podkletnov's experiment (i) The rotation velocity of the SC is suddenly changed or (ii) The SC has a composite structure, being made of two or more parts with different chemical and crystal properties and different values of the effective electron mass; or both (i) and (ii). In these cases the combined effect of accelerations-decelerations and inhomogeneities in the material can spoil the dynamical equilibrium between superelectrons and lattice ions usually associated with the London field.

The London field is only one example of the possible interplay between a moving SC and the e.m. or gravitational field. This includes the case of rotating SC samples, or samples which oscillate or accelerate along a line. Some of these cases were studied very early. A general formalism, suitable for the description of all these situations, has been given by Peng *et al.* in the cited referecne [4]. They proposed a unified phenomenological approach to study the electrodynamics of both an arbitrary moving SC and a SC under the influence of non-electromagnetic external forces, including the Newtonian gravitation and gravitational waves. This theoretical work has provided a basis for the analysis of several experiments which exploit the London field for precise determinations of the Cooper pair mass and the ratio *h/m*; a similar analysis was applied to the readout systems of the Stanford gyroscope experiment [44]. The theoretical model developed by Tajmar *et al.* to explain the results of his rotating-disk experiments also makes some connection to the London field (see Chapter 6).

3. TRANSIENT WEIGHT MODIFICATION BY SUPERCONDUCTORS

In 1996 the observation was reported of transient anomalous gravitational effects caused by small melt-textured YBCO disks (typical diameter 2 cm, thickness 5 mm). The disks were cooled below T_c and then levitated above a powerful permanent magnet. When the disks returned normal and fell down slowly on the magnet, small test objects hung above them exhibited temporary weight changes. These findings were briefly described at the end of a theoretical article [45].

The observed weight reduction was large enough to be clearly distinguished from the noise (of the order of 1% or more). It was detected only in transient form, lasting up to ca. 3 s. This happened because the weight reduction was coincident with the superconducting transition of the HTC disk, which occurred quite rapidly when the disk warmed up over its critical temperature. More than 400 trials were run, with a suspended proof mass (a glass rods bundle - total weight 63 g). Approximately 10% of the trials gave positive result, *i.e.*, a clear transient reduction in the weight of the proof mass. This apparently random behavior signals that some of the conditions which are necessary to trigger the effect were not well under control. In particular, it was found that the duration of the superconducting transition is very crucial: if the transition was too quick or too slow with respect to certain criteria, no effect was observed. The duration of the transition, in turn, appeared to depend strongly on the thermal conditions of the disk, and the latter can be controlled only with limited accuracy.

In 1997 H. Reiss made experiments with sintered YBCO pellets contained in a plastic capsule and reported weight changes of the pellets themselves [46]. The experiments were difficult, due to liquid nitrogen evaporation, but accurate. Recently Tajmar and co-workers [47] have criticized the results by Reiss (and similar but less accurate claims published by F. Rounds [48]), claiming that they have performed similar measurements with null results. Reiss, however, has replied in [49], maintaining that the precision of the Tajmar experimental trials is insufficient. In 2010, Tajmar and collaborators published a more refined work done with a magnetic suspension balance ([50]; see also Chapter 10 of this eBook [51]).

An accurate replication of the experiments of referecne [45] was done at the Politecnic of Turin, Italy, in 1998-99 by Gonnelli and Binello. This still represents the most convincing demonstration of the existence of the effect. Gonnelli and Binello were respectively head and student of a research group with ample experience in ceramic superconductors. The experiment was witnessed by several physicists. After many visual observations they began to build an automatic detection system based upon a variable capacitor coupled to a metal plate supporting the samples. Recurring problems with this detection system, however, eventually forced the interruption of the experiment without enough precise data to allow for a publication worth the group's reputation. The whole work was only reported in the master thesis of Binello [52].

The weight changes observed by Gonnelli and Binello were of about 1 part in 10000. The observation was reliable, however, since the target was the plastic plate of a precision analytical balance, encased in a glass and placed on an anti-vibration table, about 1 meter above the superconductor. The trials reported in [45], on the other hand, had employed a less precise electronic balance, with unknown reaction and integration time, connected to the target by a rod and a counterweight, which could have affected the measurements. Several configurations were tested by Gonnelli and Binello. Only few of them gave positive results. They report three such series of trials:

Series 1: 80 runs. One-inch disc, melt-textured, "home-made". Magnetic field 0.4 T, proof mass 60 g, transition time ~ 15 s. Results: the balance plate oscillates in 12% of the runs.

Series 2: 60 runs. As above, but with hexagon LevHex by SCI-Columbus (same size, levitation force slightly larger). Results: the balance plate oscillates in 18% of the runs.

Series 3: 90 runs. Like Series 1, but with microwave irradiation (50 mW, 8-12 GHz). Results: the balance plate oscillates in 15% of the runs.

The strength of the pulses was inferred by a simple calculation on the oscillation of the balance plate. This gave the strength $10^{-4}\ g \approx 0.001\ m/s^2$ mentioned above, under the assumption that the duration of the force pulse was 1 s. If the pulse is assumed to be shorter, the inferred strength increases in inverse proportion.

4. EXPERIMENTS ON HIGH VOLTAGE GAS DISCHARGES WITH YBCO CATHODES

After the experience with the rotating YBCO disks, in the years 1997-2002 E. Podkletnov built a device in which a ceramic superconducting cathode and a copper anode caused electrical discharges in low pressure gases, at temperatures between 50 and 70 K [53]. The electrodes were connected to a capacitor array (Marx generator) charged up to 2000 kV; peak currents were of the order of 10^4 A. The cathode had diameter 80-120 mm, thickness 7-15 mm, and was fabricated by OCMTG technology (see below). The apparatus represented the evolution of a series of Van de Graaf generators modified by the application of a YBCO coating to one of the (spherical or toroidal) electrodes [54]. These were used to organize high voltage discharges in rarefied gases, varying the gas composition and pressure, as well as temperature, electrodes distance, containment magnetic field, *etc.*

In discharges at voltage above 500 kV two new phenomena were observed, probably related to each other. First, the discharge did not look like a spark, but was a flat, glowing discharge, which originated from the whole surface of the superconducting electrode. Furthermore, a radiation pulse was emitted at the discharge, which propagated orthogonally to the cathode, towards the anode and beyond it, in a collimated beam, apparently without any attenuation. The radiation pulse carried an energy of 10^{-3} J at least. The features and the nature of this radiation were investigated by several means, still it was not possible to identify it; it could only be excluded that it was electromagnetic radiation or any other radiation with energy-momentum relationship $E=cp$ (see discussion below).

More in detail, the vacuum chamber had the form of a cylinder with the diameter of about 1 m and the length of 1.5 m and was made of 3 sections of tubes of silica quartz, thus allowing observation of the shape, trajectory and colour of the discharge. The superconducting emitter consisted of two layers; the non-superconducting layer was fixed to a copper cryostat using metal indium or Wood's metal, the superconducting layer faced the target electrode. The discharge could be concentrated on a smaller target area using a large solenoid that was wound around the chamber. A small solenoid was also wound around the emitter, so that some magnetic flux could be trapped inside the emitter when it was cooled down below the critical temperature.

The superconducting layer of the emitter was made of YBCO (containing small amounts of Ce and Ag). The other layer was not superconducting in the temperature range of interest and had chemical composition Y-ReBaCuO, where Re represents Ce, Pr, Sm, Pm, Tb or other rare earth elements. The materials of both layers were synthesized using a solid state reaction under low oxygen pressure, then the powder was subjected to a melt texture growth (MTG) procedure. Dense material after MTG was crushed, ground and put through sieves in order to separate the particles with the required size. A bi-layered disk was prepared by powder compaction in a stainless steel die and sintering using seeded oxygen controlled melt texture growth (OCMTG).

The discharge chamber was evacuated to less than 1 Pa using first a rotary pump and then a cryogenic pump. The experiment could be carried out at liquid nitrogen temperatures or at liquid helium temperatures. If lower temperatures were required, the tank was filled with liquid helium and in that case the temperature of the emitter reached 40-50 K. The anomalous radiation emitted was measured along the projection of the axis line which connected the center of the emitter with the center of the target electrode. Laser pointers were used to define the projection of the axis line and impulse sensitive devices were situated at the distance of 6 m and 150 m from the installation (in another building across the area).

Normal pendulums were used to measure the radiation pulses coming from the emitter. The pendulums consisted of spheres of different materials hanging on cotton strings (typically 80 cm long) inside glass cylinders under vacuum. The radiation appeared to penetrate through different shields without any noticeable loss of beam strength. It acted on the pendulums like a repulsive force field, with a force proportional to the mass of the sphere. The maximum deviation was about 10 cm.

In order to assess the effect of the anomalous radiation on light, some preliminary measurements with a laser were also done (see also Chapter 8 of this eBook). The laser beam was brought to cross the anomalous radiation beam with an angle of approx. 0.1°, giving an interaction region ca. 57 m long. The interaction caused the intensity of the initial laser spot to decrease by 7-10% during the discharge and then return quickly to baseline: the sensor (Fluorescence Kinetic Measurement System FOG100) indicated a rise time of 10^{-7} s or less.

The discharge at room temperature in the voltage range from 100 kV to 2000 kV was similar to a discharge between metal electrodes and consisted of a single spark. When the superconductor was cooled down below the transition temperature and the voltage was less than 500 kV, the shape of the discharge changed in such a way that it did not form a direct spark between the two electrodes, but the sparks appeared from many points on the superconducting emitter and then moved to the target (positive) electrode. When the voltage was increased to 500 kV the front of the moving discharge became flat, with diameter corresponding to that of the emitter. This flat glowing discharge separated from the emitter and moved to the target electrode with great speed. For maximum distance between the emitter and the target (about 1 m) it was possible to see the flat glowing sparkle jumping from the emitter to the target. When the distance was reduced to 25 cm, the time of the flash light as defined by the photo-diode was between 10^{-5} and 10^{-4} s. The actual time of the discharge is certainly shorter (compare discussion of circuit parameters in section 4.2 and Chapter 9 of this eBook) and the duration of the anomalous force pulse still shorter (Chapter 8).

4.1. Phenomenological Discussion

Some first "obvious" theoretical remarks were presented in the cited reference [53], based on the description of the experiment given by Podkletnov (see below for some more technical remarks stemming from a replication attempt).

a) The flat discharge could be due to the excellent electrical conductivity of the *ab* crystal planes in YBCO. This makes the surface of the emitter, to a high accuracy, a surface of constant electric potential. In usual spark discharges in a neutral gaseous medium the dielectric is first "broken" at some point near an electrode where the electric field is strong enough to generate cascade ionization. Here, the field at the electrode's surface is extremely uniform and a flat discharge starts only when its strength exceeds everywhere some threshold value.

b) It is known that conduction in YBCO in the *c* direction takes place only through a sort of tunneling between the *ab* planes. It can be estimated that during the current pulse each pair goes through ~1000 *ab* crystal planes. The discharge current crosses two normal-superconducting junctions, the first one between the normal and superconducting part of the emitter, and the second one between the external surface of the emitter and the ionized gas.

c) The observed anomalous non-electromagnetic radiation appears to propagate through brick walls and metal plates without noticeable absorption, but this is not due to a weak coupling with matter, because the radiation acts with significant strength on the test masses that are free to move. The electromagnetic shielding obtained with thick metal plates, the Faraday cage and UHF absorbing panels cannot be regarded as perfect; however, it is quite clear that any residual electromagnetic radiation cannot be responsible for the observed impulse transfer to the targets. In fact, this radiation conveys an impulse which is certainly not related to the carried energy by the usual dispersion relation $E=cp$. One can estimate, for instance, considering the data for the 18.5 grams pendulum, that the kinetic energy associated to the observed displacement is of the order of 10^{-3} J and the momentum is of the order of 10^{-2} kg m/s. If this momentum had to be imparted to the pendulum by radiation pressure, the energy needed in the beam would exceed the total energy available in the discharge (~10^3 J). Moreover, the radiation energy in excess would heat up the pendulums (unless they act as perfect reflectors, which seems very unlikely).

d) The anomalous radiation acts upon targets of any composition, with a force proportional to their mass and apparently independent from their cross-section. The proportionality to mass is confirmed only within the reproducibility of the discharge process; the casual error (standard deviation of the single data) is between 5 and 7%. Further variations originate from inhomogeneities in the beam due to structural defects in the emitter (up to 12%). This implies that the force could be in fact not proportional to mass, but to some other quantity closely enough related to it, like for instance the baryonic number.

e) The term "radiation" is actually unsuitable, and one could possibly envisage an unknown quasi-static force field. In this way one could explain why a momentum is transmitted to the test masses. It is hard, however, to understand how such a field could be so well focused. This configuration cannot satisfy any known static field equation. (A possible exception is the Dirac string field [55]).

f) In order to estimate the strength of the beam's effect, suppose that l is the length of a detection pendulum and g is the local gravitational acceleration. Let d be the half-amplitude of the oscillation. Let t be the duration of the pulse, and a the average acceleration imparted to the targets. One easily computes that if $d \ll l$, then $at \sim d\sqrt{(g/l)}=2\pi d/T$, where T is the period of the pendulum. With the data of [53], taking $t=10^{-7}$ s, one finds $a\sim 10^7$ m/s^2. This is of course a bewildering acceleration for an object of this size.

4.2. Study of a Scaled-Down Replication

In 2004 the first author began a collaboration with the Institute for Gravitational Research (IGF) in Waldaschaff, Germany, aimed at a reduced-scale replication of Podkletnov's discharge experiment. The first task was to assess the possibilities of a scaling-down of the device. This required in turn a circuit analysis and a proper understanding of the discharge mechanism. The capacitance of the circuit was known, but several other points were still unclear, like for instance [56]:

- The effective inductance of the circuit and of the chamber;
- The effective high-frequency impedance of the emitter;
- The relation between breakdown voltage and optimum gap length;
- The gas parameters (electron drift velocity, ionization rate *etc.*).

A 50% scaling in size was considered, along with the possibility of favourable modifications of the other experimental conditions: voltage reduction, increases in pressure and temperature, reduction of the focussing magnetic field. Further questions emerged at each step:

- How does the microscopic structure of the emitter affect beam coherence and collimation?
- Is stimulated emission present?
- What is the role of the normal layer in the emitter?
- What parasitic electromagnetic emissions are present? (Including a dangerous back-radiation reported by Podkletnov in separate accounts).
- Are there any similarities with other devices, like for instance free electron lasers?
- Are solid-state switches a viable alternative to the gas discharge? (Compare the scheme proposed by G. Fontana for an HTSC Gaser in Chapter 3).

When the replication work at 50% size was started (compare Chapter 9), with tests of the vacuum chamber, installation of an insulating high-voltage cabin and acquisition of a commercial Marx generator, standard circuit theory was applied to the analysis of the pre-breakdown and post-breakdown phases. New estimates were done of the electron drift velocity, collision frequency and mean free path in the discharge, leading to the conclusion that the discharge mechanism was not of the Townsend type. Possible limits on the total current and on the current density arising from the Child limit and the Alfven limit were considered.

A puzzle remained as to the origin of the apparent high-frequency resistance of the emitter: according to Podkletnov, the V/I ratio during the gas discharge was approximately 100 V/A, and the discharge was apparently over-dampened. Several possible explanations were analysed: complex impedance of superconductors for high-frequency current, radiation reaction, resistance due to the intrinsic Josephson junctions for current flow in the c-axis of YBCO. None of these could explain the observed resistance.

Much of the uncertainty in the modelling of the device with an equivalent circuit plus discharge comes from the behaviour of the superconducting emitter, for which different microscopic models were proposed, also to account for the anomalous emission. The first model was semiclassical and supposed an acceleration of the Cooper pairs in the emitter due to the strong electric field present. This was later replaced by a model in which the emitter is regarded as a stack of intrinsic Josephson junctions [57]. See also Section 5 of Chapter 5 of this eBook. Our latest analysis of the discharge has evidenced a strong similarity with “vacuum spark discharges” [58] with superconducting cathodes [59]. In Chapter 9 of this eBook T. Junker reports on experimental techniques tested for replication of the impulse gravity generator experiment.

4.3. Partial Replication Reported by C. Poher and D. Poher

Shortly before the publication of this review we learnt of a new paper by C. Poher and D. Poher [60] about a partial replication of Podkletnov’s high-voltage discharge experiment. In their version the gas discharge was replaced by a solid state switch. They applied electric discharges of several megawatts, at 77 K, to propelling devices made of Y123 superconducting layers and thin insulating layers. During the discharges, the devices were strongly pushed in the direction opposed to the electron flow. The layered devices were apparently propelled by their emission of a momentum-bearing flux of an unknown nature. This flux weakly accelerated distant irradiated matter and created several physical effects not yet reported. The emitted beam had no electric charge, and travelled through materials without apparent absorption or dispersion, at a speed much greater than 1% the speed of light. The kinetic energy transferred by the propelling momentum of the devices to external masses were proportional to the square of the electric energy of the discharges. A related patent has been registered by the first author in the US [61]. An informal analysis of Poher’s work, based on an earlier unpublished report, can be found in the cited reference [55].

5. FURTHER WORK

Other chapters of this eBook contain full descriptions of theoretical approaches to gravity-superconductors interaction. In this Section we cite briefly a few more theoretical papers which in some way make reference to anomalous interactions between gravity and superconductors. We have selected papers which either are peer-reviewed, or whose authors also have published in mainstream physics, because the informal publications on this subject are numerous but quite difficult to assess.

In addition to the experimental work described in Chapter 7 of this eBook, Chiao *et al.* have recently published theoretical papers and proposals for experiments [62-64]. In reference [64], the Einstein-Maxwell equations are employed, along with certain hypotheses on the gravitational impedance of superconductors (see also Chapter 7, Section 11, of this eBook). We have briefly discussed the validity of the Einstein-Maxwell equations for the description of gravitational waves in Chapter 5 of this eBook, App. 2. Finally, Chapter 2 of this eBook contains an analysis of the so-called "Chiao-Li-Torr conjecture". Papers by Agop *et al.* [65] combine General Relativity and the London equations. Ummarino [66] employs in his analysis the time-dependent Ginzburg-Landau equation. Licht [67] studies the interaction of a gravitational wave with a superconductor. de Matos *et al.* [68] propose a theoretical interpretation of the Tajmar experiment [69-72] and of the anomaly of the Cooper pairs mass in rotating superconductors.

6. SUMMARY DISCUSSION

The first and second Podkletnov experiments (with rotating disk and high-voltage discharges) appear to be related. In both experiments a composite high-Tc SC under the action of e.m. fields generates repulsive gravity-like forces. In the first experiment the force is weak and stationary, in the second it is much stronger but of short duration. An interpretation of the first experiment as a "shielding" of the Earth's field is theoretically untenable. Both experiments raise serious conceptual problems concerning the Equivalence Principle and can probably only be explained at the quantum level, and taking into account the back-reaction of the target on the field generator. There are presently neither complete confirmations of Podkletnov's results nor complete replications with negative outcome. The replication attempts have met considerable difficulties, for several reasons: the complexity of the device, the non-standard techniques employed, the incomplete description of several details. The version of C. Poher and D. Poher of the high-voltage discharge experiment (Section 4.3) is somewhat different from the original, but the observed effects are similar.

A tentative theoretical explanation of the first experiment has been offered by Ning Wu (Section 2.2). The second experiment can be analyzed following Fontana [73], Robertson [74] or Modanese [75].

The experiments by Tajmar *et al.* share some common ingredients with Podkletnov's experiments (superconducting rotating disks, detection of anomalous gravity-like forces), but are otherwise quite different both materially and conceptually. Tajmar *et al.* did not see the effects claimed by Podkletnov, but other effects, at different temperatures and with different materials. A detailed analysis of Tajmar experiments has been done by Lewis [76]. According to the latest results, it is possible that the observed anomalous forces do not originate from the superconductor but from the liquid Helium. The Extended Heim Theory by Hauser and Droescher [77], which appears to explain well Tajmar's experimental results, makes no statements on Podkletnov's claims.

The Chiao experiment using microwaves needs to be extended to come up with a definitive result on whether the e.m./gravitational waves conversions occur [78]. Again, the experimental situation displays both similarities and crucial differences with the other experiments. The so-called Chiao-Li-Torr conjecture (the postulate that a superconductor acts as an efficient reflector for high-frequency gravitational waves) has remarkable experimental consequences, which have been analyzed by Woods [79].

The weight anomalies at the superconducting transition are in a similar situation: partial confirmations, no clear confutation, different phenomenology examined by different authors (for instance, weight changes reported in the superconductors themselves or in near test masses). No theoretical interpretation has been offered for this class of phenomena yet.

A new generation of experiments is certainly necessary. In Chapter 9, Junker [80] describes a new setup for the replication of Podkletnov's high-voltage experiment. In Chapter 10, Hathaway and Reiss [81] describe some of the requirements that future experiments should satisfy, in order to lead to clear confirmations or refutations of the effects and to supply reliable data for the theoretical analysis.

In Chapter 12 of this eBook [82], Robertson discusses a new gravitational coupling to field densities, where there is a hierarchy of density dependent scalar fields in the Universe down to the laboratory scale experiment (*e.g.*, the Tajmar *et al.* superconductor experiment), with respect to the far Universe field's vacuum mass or minimum dark matter mass. Robertson discusses the similarity between the rotating cold mass in the Tajmar *et al.* superconductor experiment and the appearance of rings about galaxies.

Complementary to Robertson, in Chapter 13 Reiss [83] also looks at the superconductor properties for clues to gravity-superconductors interactions by analyzing the impact of lifetime and decay rates from thermally excited states in superconductors.

An effective theory, in any case, needs truly innovative concepts, since there is no place for the observed anomalies in the standard formal scenarios.

REFERENCES

Note: "arXiv" papers are available from Cornell's electronic preprint depository http://arxiv.org
A version of several published papers is also available there; check with http://scholar.google.com

[1] DeWitt, BS. Superconductors and gravitational drag. Phys Rev Lett 1966; 16: 1092–1093.

[2] Anandan J. Gravitational and inertial effects in quantum fluids. Phys Rev Lett 1981; 47: 463–467. Anandan J, Chiao RY. Gravitational radiation antennas using the Sagnac effect. Gen Rel Grav 1982; 14: 515. Chiao RY. Interference and inertia: A superfluid-helium interferometer using an internally porous powder and its inertial interactions. Phys Rev 1982; B 25: 1655.

[3] Ross DK. The London Equation for Superconductors in a Gravitational Field. J Phys A: Math Gen 1983; 16: 1331.

[4] Peng H, Torr DG. The Electric Field Induced by a Gravitational Wave in a Superconductor: A Principle for a New Gravitational Antenna. Gen Rel Grav 1990; 22: 53.

[5] Peng H, Torr DG, Hu EK, Peng B. Electrodynamics of moving superconductors and superconductors under the influence of external forces. Phys Rev B 1991; 43: 2700.

[6] Weinberg S. Gravitation and Cosmology. Principles and Applications of the General Theory of Relativity. John Wiley and Sons 1985.

[7] Colella R, Overhauser AW, Werner SA. Observation of gravitationally induced quantum interference. Phys Rev Lett 1975; 34: 1472-1474. Speliotopoulos AD, Chiao RY. Coupling of linearized gravity to nonrelativistic test particles: Dynamics in the general laboratory frame. Phys Rev 2004; D 69: 084013.

[8] Lano RP. Gravitational Meissner effect. arXiv 1996: hep-th/9603077.

[9] Kiefer C, Weber C. Annalen Phys 2005; 14: 253-278.

[10] Braginsky VB, Caves CM, Thorne KS. Laboratory experiments to test relativistic gravity, Phys Rev 1977; D 15: 2047.

[11] Droscher W, Hauser J. Gravity-like fields and space propulsion concepts. Paper AIAA-2008-5124. In: Proceedings of the 44th AIAA/ASME/SAE/ASEE Joint Propulsion Conference, 2008. Amer. Inst. of Aeronautics & Astronautics 2008.

[12] Iorio L, Corda C. Gravitomagnetism and gravitational waves. arXiv 2010:1001.3951. Mashhoon B. Gravito-electromagnetism. arXiv 2000: gr-qc/0011014.

[13] Forward RL. General relativity for the experimentalist. Proceedings of the IRE 1961; 49: 892-904.

[14] Ning Li, Torr D. Effects of a gravitomagnetic field on pure superconductors. Phys Rev 1991; D 43: 457-459. Ning Li, Torr D. Gravitational effects on the magnetic attenuation of superconductors. Phys Rev 1992; B 46: 5489-5495.

[15] Ning Li, Torr D. Gravitoelectric-electric coupling *via* superconductivity. Found Phys Lett 1993; 6: 371.

[16] Kowitt M. Gravitomagnetism and magnetic permeability in superconductors. Phys Rev 1993; B 49: 704–708.

[17] Harris EG. Comments on 'Gravitoelectric-electric coupling *via* superconductivity' by Douglas G. Torr and Ning Li'. Found Phys Lett 1999; 12: 201–208.

[18] Will CM. The Confrontation Between General Relativity and Experiment. Space Science Reviews 2009; 148: 3-13.

[19] Gillies GT. The Newtonian gravitational constant: recent measurements and related studies. Rep Prog Phys 1997; 60: 151-225.

[20] Michaelis W, Haars H, Augustin R. A new precise determination of Newton's gravitational constant. Metrologia 1995; 32: 267. Fitzgerald MP, Armstrong TR. Newton's gravitational constant with uncertainty less than 100 ppm. IEEE Trans Inst Meas 1995; 44: 494-497. Izmailov VP, Karagioz OV. Temporal and spatial variations of measured values of the gravitational constant. Measurement Techniques 1993; 36: 1065-1069.

[21] Arkani-Hamed N, Dimopoulos S, Dvali G. The Hierarchy Problem and New Dimensions at a Millimeter. Phys Lett 1998; B429: 263-272.

[22] Turyshev SG, Toth VT. The Pioneer Anomaly. arXiv:1001.3686.

[23] Hoyle DC, Kapner DJ, Heckel BR *et al.* Submillimeter Tests of the Gravitational Inverse-Square Law. Phys Rev 2004; D 70: 042004.

[24] La D, Steinhardt PJ. Phys Rev Lett 1989; 62: 376. Calchi Novati S, Capozziello S, Lambiase G. Newtonian limit of induced gravity. arXiv 2000: astro-ph/0005104.

[25] Podkletnov E, Nieminen R. A Possibility of Gravitational Force Shielding by Bulk $YBa_2Cu_3O_{7-x}$ Superconductor. Physica C 1992; 203: 441.

[26] Podkletnov E. Weak gravitation shielding properties of composite bulk $YBa_2Cu_3O_{7-x}$ superconductor below 70 K under e.m. field. arXiv 1997: cond-mat/9701074.

[27] Hathaway GD. Gravitational experiments with superconductors: history and lessons. In: Frontiers of propulsion science. Amer. Inst. of Aeronautics & Astronautics 2009; pp. 229-246.

[28] Modanese G. Theoretical analysis of a reported weak gravitational shielding effect. Europhys Lett 1996; 35: 413-418.

[29] Koczor R, Noever D. Fabrication of Large Bulk Ceramic Superconductor Disks for Gravity Modification Experiments and Performance of YBCO Disks Under e.m. Field Excitation. Paper AIAA 99-2147. In: 35th AIAA/ASME/SAE/ASEE Joint Propulsion Conference; 1999; Los Angeles, USA. Amer. Inst. of Aeronautics & Astronautics 1999.

[30] Li N, Noever D, Robertson T, Koczor R, Brantley W. Static Test for a Gravitational Force Coupled to Type II YBCO Superconductors. Physica C 1997; 281: 260-267.

[31] Woods RC, Cooke SG, Helme J, Caldwell CH. Gravity Modification by High-Temperature Superconductors. In: Proc. 37th AIAA/ASME/SAE/ASEE Joint Propulsion Conf.; 2001; Salt Lake City, USA. Amer. Inst. of Aeronautics & Astronautics 2001.

[32] Hathaway G, Cleveland B, Bao Y. Gravity modification experiment using a rotating superconducting disk and radio frequency fields. Physica C 2003; 385: 488–500.

[33] Boeing tries to defy gravity. BBC News, July 29, 2002. Available from http://news.bbc.co.uk/2/hi/science/nature/2157975.stm

[34] Banke J. Gravity Shielding Still Science Fiction. Space.com 2002. Available from http://www.space.com/businesstechnology/technology/gravity_research_020731.html

[35] Unnikrishnan CS, Mohapatra AK, Gillies GT. Anomalous gravity data during the 1997 total solar eclipse do not support the hypothesis of gravitational shielding. Phys Rev D 2001; 63: 062002.

[36] Modanese G. Local contribution of a quantum condensate to the vacuum energy density. Mod Phys Lett A 2003; 18: 683-690

[37] Ning Wu. Symmetry breaking and anomalous gravitational effects. In: Modanese G, Robertson GA, Eds. Gravity-Superconductors Interaction: Theory and Experiment. Bentham 2012; Chapter 14.

[38] Unnikrishnan CS. Does a superconductor shield gravity? Physica C 1996; 266: 133-137.

[39] Modanese G. Updating the theoretical analysis of the weak gravitational shielding experiment. Paper IAA-97-4.107. In: Proceedings of the 1997 IAF Congress. 1997; Turin, Italy. arXiv 1996: supr-con/9601001.

[40] Bertolami O, Tajmar M. Gravity control and possible influence on space propulsion: A scientific study. Paper ESA-CR(P) 4365-2003. Available from http://adsabs.harvard.edu

[41] Modanese G, Taylor CY. Evaluation of an Impulse Gravity Generator Based Beamed Propulsion Concept. In: Proceedings of the 38th AIAA/ASME/SAE/ASEE Joint Propulsion Conference, 2002. arXiv 2002: physics/0209023.

[42] Low, FE. Classical Field Theory: Electromagnetism and Gravitation. New York: J. Wiley 1997. Feynman RP. Lectures on Gravitation. New York: Addison-Wesley 1995.

[43] Modanese G. The London field in bulk layered superconductors. arXiv 1999: cond-mat/9909441.

[44] Fairbank JD, Deaver Jr. BS, Everitt CWF, Michelson PF, Eds. Near Zero: New Frontiers of Physics. New York: Freeman 1988. Cerdonio M, Prodi G, Vitale S. Gen Rel Grav 1988; 20: 83.

[45] Modanese G, Schnurer J. Possible quantum gravity effects in a charged Bose condensate under variable e.m. field. Phys Essays 2001; 14: 93-105.

[46] Reiss, H. A Possible Interaction Between Gravity and High Temperature Superconductivity – By a Materials Property? In: Proceedings of the 15th Europ. Conf. ThermoPhys Prop; 1999; Wuerzburg, Germany. Weight anomalies observed during cool-down of high temperature superconductors. Phys Essays 2006; 16.

[47] Tajmar M, Hense K, Marhold K, de Matos C. Weight Measurements of High-Temperature Superconductors during Phase Transition in Stationary, Non-Stationary Condition and under ELF Radiation. In: STAIF 2005 - Space Technology and Applications International Forum. American Institute of Physics 2005.

[48] Rounds FN. Anomalous Weight Behavior in $YBa_2Cu_3O_{7-x}$ Compounds at Low Temperature. arXiv 1997: physics/9705043.

[49] Reiss H, Hathaway G. Minimum Experimental Standards in the Laboratory Search for Gravity Effects, STAIF 2006 - Space Technology and Applications International Forum. American Institute of Physics 2006; pp. 1290.

[50] Tajmar M, Plesescu F, Seifert B. Measuring the Dependence of Weight on Temperature in the Low-Temperature Regime Using a Magnetic Suspension Balance. Meas Sci Tech 2010; 21; 015111.

[51] Hathaway G, Reiss H. Experimental Considerations in Superconductor Gravity Experiments. In: Modanese G, Robertson GA, Eds. Gravity-Superconductors Interaction: Theory and Experiment. Bentham 2012; Chapter 10.

[52] Binello F. Studio di possibili effetti sul campo gravitazionale da parte di superconduttori ad alta T_c in presenza di campi elettromagnetici. Master Thesis. Torino, Italy. Politecnico di Torino, 1999. Supervisors R. Gonnelli, A. De Marchi, G. Genta.

[53] Podkletnov E, Modanese G. Investigation of high voltage discharges in low pressure gases through large ceramic superconducting electrodes. J Low Temp Phys 2003; 132: 239-259.

[54] Podkletnov E, Modanese G. Impulse Gravity Generator Based on Charged $YBa_2Cu_3O_{7-y}$ Superconductor with Composite Crystal Structure. arXiv 2001: physics/0108005.

[55] Lewis R. Review of Poher experiment on fields produced by electric discharges in a superconductor. arXiv 2011: 1101.2419.

[56] Modanese G. Evolution of the IGG concept at IGF from 2004 to 2007. arXiv 2011: 1102:5447.

[57] Modanese G, Junker T. Conditions for stimulated emission in anomalous gravity-superconductors interactions. In: Christiansen MN, Rasmussen TK, Eds. Classical and Quantum Gravity Research. Nova Science 2008; pp. 245-269.

[58] Korop ED, Meierovich BE, Sidel'nikov YuV, Sukhorukov ST. Micropinch in a high-current diode. Sov Phys Usp 1979; 22: 727. Koshelev KN, Pereira NR. Plasma points and radiative collapse in vacuum sparks. J Appl Phys 1991; 69: R21. Erber CK, Koshelev KN, Kunze HJ. Time development of Mg micropinches in a low-inductance vacuum spark discharge. J Quantitative Spectroscopy & Radiative Transfer 2000; 65: 195-206.

[59] Mesyats GA. Pulsed electrical discharges in vacuum at cryogenic electrode temperatures. IEEE Trans Electr Insul 1989; 24: 883-888. Shkuratov SI. Pulsed electrical breakdown from superconducting electrodes in vacuum diodes. IEEE Trans Dielectr Electr Insul 1995; 2: 251-262.

[60] Poher C, Poher D. Physical phenomena observed during strong electric discharges into layered Y123 superconducting devices at 77 K. Appl Phys Res 2011; 3: 51.

[61] Poher C. US Patent Application 0251717, 2010.

[62] Chiao RY. Conceptual tensions between quantum mechanics and general relativity: Are there experimental consequences? arXiv 2003: gr-qc/0303100.

[63] Chiao RY. Generation and detection of gravitational waves at microwave frequencies by means of a superconducting two-body system. arXiv 2007: 0710.1378.

[64] Chiao RY, Mintera SJ, Kirk Wegter-McNelly K. Do mirrors for gravitational waves exist? Physica E 2010; 42: 234-255.

[65] Agop M, Buzea CG. On Gravitational Shielding in Electromagnetic Fields. arXiv 1999: physics/9911011. Agop M, Buzca CG, Ciobanu B. Local gravitoelectromagnetic effects on a superconductor. Physica C 2000; 339: 120-128. Agop M, Buzea CG. A new theoretical approach of the gravitational shielding: Podkletnov's effect. Physica C 2000; 341-348: 307-308.

[66] Ummarino GA. Possible alterations of the gravitational field in a superconductor. arXiv 2000: cond-mat/0010399

[67] Licht AL. Gravitational waves on conductors. arXiv 2004: gr-qc/0402109

[68] Tajmar M, de Matos CJ. Gravitomagnetic Field of a Rotating Superconductor and of a Rotating Superfluid. Physica C 2003; 85: 551-554. de Matos CJ. Gravitoelectromagnetism and Dark Energy in Superconductors. Int J Mod Phys 2008; D16: 2599-2606. de Matos CJ. Physical Vacuum in Superconductors. arXiv 2009: 0908.4495.

[69] Tajmar M, Plesescu F. Fiber-Optic-Gyroscope Measurements Close to Rotating Liquid Helium. Paper CP1208. In: Proceedings of Space, Propulsion & Energy Sciences International Forum-SPESIF, 2010. American Institute of Physics 2010; pp. 220.

[70] Tajmar M, Plesescu F, Seifert B. Anomalous fiber optic gyroscope signals observed above spinning rings at low temperature. J Phys Conf Ser 2009; 150: 032101.

[71] Tajmar M, Plesescu F, Seifert B, Schnitzer R, Vasiljevich I. Investigation of Frame-Dragging-Like Signals from Spinning Superconductors using Laser Gyroscopes. Paper CP969. In: Proceedings of Space Technology and Applications International Forum - STAIF 2008. American Institute of Physics 2008; pp. 1080.

[72] Tajmar M, Plesescu F, Seifert B, Marhold K. Measurement of Gravitomagnetic and Acceleration Fields around Rotating Superconductors. In: Proceedings of Space Technology and Applications International Forum - STAIF 2007. American Institute of Physics 2007; pp. 1071-1082.

[73] Fontana G. High Temperature Superconductors as Quantum Sources of Gravitational Waves: the HTSC GASER. In: Modanese G, Robertson GA, Eds. Gravity-Superconductors Interaction: Theory and Experiment. Bentham 2012; Chapter 3.

[74] Robertson GA. Quantum Effects in the Type II Superconductor that Lead to Power Radiated in Gravitational Waves. In: Modanese G, Robertson GA, Eds. Gravity-Superconductors Interaction: Theory and Experiment. Bentham 2012; Chapter 4.

[75] Modanese G. Quantum Gravity Evaluation of Stimulated Graviton Emission in Superconductors. In: Modanese G, Robertson GA, Eds. Gravity-Superconductors Interaction: Theory and Experiment. Bentham 2012; Chapter 5. Modanese G. Anomalous gravitational vacuum fluctuations which act as virtual oscillating dipoles. In: R Sobreiro R, Ed. Quantum Gravity. InTech 2012.

[76] Lewis RA. Tajmar experiments on "condensing the vacuum". In: Modanese G, Robertson GA, Eds. Gravity-Superconductors Interaction: Theory and Experiment. Bentham 2012; Chapter 6.

[77] Hauser J, Droescher W. Emerging Physics for Gravity-Like Fields. In: Modanese G, Robertson GA, Eds. Gravity-Superconductors Interaction: Theory and Experiment. Bentham 2012; Chapter 11.

[78] Lewis RA. The Chiao Gravity-superconductor EM transducer: An Overview. In: Modanese G, Robertson GA, Eds. Gravity-Superconductors Interaction: Theory and Experiment. Bentham 2012; Chapter 7.

[79] Woods RC. Interactions Between Superconductors and High-Frequency Gravitational Waves. In: Modanese G, Robertson GA, Eds. Gravity-Superconductors Interaction: Theory and Experiment. Bentham 2012; Chapter 2.

[80] Junker T. Setup of the impulse gravity generator experiment replication. In: Modanese G, Robertson GA, Eds. Gravity-Superconductors Interaction: Theory and Experiment. Bentham 2012; Chapter 9.

[81] Hathaway G, Reiss H. Experimental Considerations in Superconductor Gravity Experiments. In: Modanese G, Robertson GA, Eds. Gravity-Superconductors Interaction: Theory and Experiment. Bentham 2012; Chapter 10.

[82] Robertson GA. Gravity-Superconductor Interactions from a Hierarchy of Density Dependent Scalar Fields: A Dark Matter Connection. In: Modanese G, Robertson GA, Eds. Gravity-Superconductors Interaction: Theory and Experiment. Bentham 2012; Chapter 12.

[83] Reiss H. Impact of Lifetime and Decay Rates of Thermally Excited States in Superconductors on a Gravity Experiment. In: Modanese G, Robertson GA, Eds. Gravity-Superconductors Interaction: Theory and Experiment. Bentham 2012; Chapter 13.

CHAPTER 2

Interactions Between Superconductors and High-Frequency Gravitational Waves

R.C. Woods*

Department of Electrical and Computer Engineering, Louisiana State University, Baton Rouge, LA 70803-5901, USA

Abstract: Interactions between superconductors and high-frequency gravitational waves have been predicted by a number of authors, and also challenged by others. The evidence for and against possible sources of interaction is reviewed. Assuming the veracity of the claims allows a number of possible applications to be considered, which may double as experimental tests of the results, and a number of these are considered in detail as possible ways to clarify the current controversial position.

Keywords: Gravitational waves, generation of gravitational waves, detection of gravitational waves, high-frequency gravitational waves, superconductors, high-Tc superconductors, low-Tc superconductors, type-II superconductors, general relativity, gravitomagnetism.

1. INTRODUCTION

Gravitational waves and their applications may soon become an important emerging technology. The initial theoretical work of Einstein [1], Forword and Miller [2], and Romero and Dehnen [3] laid the foundations that have supported subsequent experimentation. The 1994 observations of Hulse [4] and Taylor [5], for which they were awarded the Nobel Prize, confirmed the production of very-low-frequency gravitational waves of astronomical origin, and eliminated any remaining scepticism concerning their existence and nature. Subsequent development has included the large-scale LISA, LIGO, Virgo, and other prototype projects for low-frequency gravitational wave detection. More recently, the properties of high-frequency gravitational waves (HFGWs, defined by Douglass and Braginsky [6] as GWs having a frequency between 100kHz and 100MHz) and very-high-frequency gravitational waves (VHFGWs, defined by Douglass and Braginsky [6] as GWs having a frequency between 100MHz and 100GHz) have been investigated theoretically. (Often the distinction between HFGW and VHFGW is blurred by many writers who refer simply to HFGW for succinctness and to distinguish from low-frequency GWs for which completely different techniques are needed). Typical recently proposed HFGW laboratory generators [7, 8] are predicted to produce a GW signal that has a constant polarization angle, unlike a rotating binary star system in which the polarization angle rotates at twice the orbital frequency. For a binary star system in circular orbit, the GW emitted amplitude is constant whereas the predicted GW detected signal varies at the detector due to the shift in polarization.

Interactions between superconductors and (very-)high-frequency gravitational waves ((V)HFGW) have been claimed or predicted by a number of authors over many years. Such predictions have always been controversial, not least because up to the present time it has not been possible to detect or receive any frequency of GW directly, and so claims of interactions remain claims without evidence proving or disproving such interactions. As long ago as 1972, Weinberg [9] suggested that superfluids might act as efficient mirrors for GW. The general basis for this deduction is that the bosons making up a superfluidic Bose-Einstein condensate are, by definition, in a quantum-mechanical eigenstate of zero momentum (with a very high occupancy which is allowed as bosons do not obey the Fermi exclusion principle). Then, when a GW is incident, the eigenstate cannot change its state by particle rearrangement as does a Fermi gas of electrons; instead, as an extensive eigenstate containing a large number of bosons in any non-microscopic

***Address correspondence to R.C. Woods:** Department of Electrical and Computer Engineering, Louisiana State University, Baton Rouge, LA 70803-5901, USA; Tel: 225-578-8961; E-mail: cwoods@lsu.edu sample, the superfluid has no option but to resist any conformance with the space-time distortion characteristic of the GW. (The alternative is for each and every boson in the superfluid to change its eigenstate simultaneously, which is theoretically possible but is thermodynamically extremely unlikely. This is the same as the reason why it is theoretically possible but thermodynamically extremely unlikely that all the randomly vibrating molecules in a macroscopic mass will simultaneously have a significant random velocity component directed upwards so that the mass spontaneously jumps into the air from a rest position. The smaller the mass, the more likely this is, as demonstrated by Brownian motion of microscopic oil drops in the well-known laboratory demonstration experiment). In resisting the space-time distortion of the GW, the superfluid reflects all GW power and so behaves as a perfect GW mirror (it is argued [9]).

1.1. GW Propagation and Superconductors

It is generally accepted that in a vacuum and in any other materials (apart from superconductors) the GW propagation phase velocity is equal to the speed of light in a vacuum ($c = 3\times10^8$m s^{-1}). This can be shown by casting the Einstein equations of General Relativity into linear approximations suitable for weak gravitational fields, from which a set of four "Gravito-Electromagnetic" equations, analogous to the four well-known Maxwell-Heaviside equations of classical electromagnetism, can be derived [10]:

$$\operatorname{div}\mathbf{B}_G = 0 \tag{1}$$

$$\operatorname{div}\mathbf{D}_G = -\rho \tag{2}$$

$$\operatorname{curl}\mathbf{E}_G = -\dot{\mathbf{B}}_G \tag{3}$$

$$\operatorname{curl}\mathbf{H}_G = -\mathbf{J}_G + \dot{\mathbf{D}}_G \tag{4}$$

where $\mathbf{D}_G$, $\mathbf{B}_G$, $\mathbf{E}_G$, and $\mathbf{H}_G$ are the gravito-electric flux density, gravito-magnetic flux density, gravito-electric field, and gravito-magnetic field respectively (where the "electric" and "magnetic" names are retained only for the purposes of maintaining the analogy with the Maxwell-Heaviside equations), ρ is the mass volume density, and $\mathbf{J}_G$ is the mass flow density, with gravitational permittivity

$$\varepsilon_G = 1/4\pi G \tag{5}$$

and gravitational permeability

$$\mu_G = 4\pi G/c^2 \tag{6}$$

giving the analogous constitutive relations

$$\mathbf{D}_G = \varepsilon_G \mathbf{E}_G \tag{7}$$

and

$$\mathbf{B}_G = \mu_G \mathbf{H}_G \,. \tag{8}$$

(Note that the appearance of negative signs in Eqns. (2) and (4) is a fundamental departure from what would represent a strict electromagnetic analogy; similar charges repel, but two positive masses attract each other). Then it is a simple matter to combine these equations under appropriate conditions to extract the classical wave equation with velocity equal to c.

More recently, Li and Torr [11, 12] have published calculations of the propagation behaviour of gravitational waves inside superconductors. They claimed that the phase velocity of gravitational waves in any superconducting material would be ~10^6m s^{-1}, *i.e.*, ~300 times less than the phase velocity of gravitational waves in other materials. Therefore, using the language of conventional geometric optics, essentially what was claimed is that a superconductor has a gravitational refractive index of $n_G \approx 300$. This also gives a corresponding wavenumber increase by a factor $n_G \approx 300$ times within superconductors.

This result has major consequences for the design of future instruments to generate and detect gravitational waves, which may need to be able to focus, refract, reflect, or manipulate gravitational waves for efficient coupling to detectors, transmitters, generators, resonant chambers, and other sensors. In particular, high-frequency gravitational waves (HFGWs) have wavelengths on the same order as the dimensions of typical superconductor components envisaged for HFGW use and so any suggestion of phase velocity change will affect the design of those components.

Kowitt [13] and Harris [14] independently argued that the Li and Torr [11] results are not credible. Kowitt's objection in 1994 [13] was that Li and Torr [11] assumed a value of the magnetic permeability of zero (or, equivalently, a magnetic susceptibility of –1) inside a superconductor. This is an experimental result, widely accepted for macroscopic samples, but it is not clear that the macroscopic experimental result is true microscopically. It is perfectly true that a value of magnetic permeability different from that of free space may be obtained macroscopically in magnetically active samples in which individual magnetic moments of atoms align with the applied field. Since this enhanced permeability arises from the aggregate effect of localised magnetic moments, it follows that a renormalized permeability is not definable microscopically. However, zero magnetic permeability in a superconductor results from cancellation of the magnetic field **H** within the superconductor by distributed current loops giving a distributed magnetisation **M**. In a typical type-II superconductor (see §3.1), the circulating supercurrents causing this distributed magnetic moment must be external to the magnetic vortices, of course. These circulating supercurrents extend throughout the material, whereas in a conventional magnetic material the notional currents producing the magnetic moments are those caused by orbiting electrons bound to atoms. While it is the case that in a conventional magnetic material it is not possible to define an effective permeability that has local microscopic significance, this argument does not apply to the extended supercurrents found in superconducting materials. For these reasons, Kowitt's objection [13] may not be tenable.

Harris raised a different objection in 1999 [14] that must be taken seriously; he pointed out that Torr and Li [12] assumed an arbitrary value for the observer's distance of 10^{-6}cm at one point, and 10^{-13}cm at another point in their calculation, and that both values are unreasonable. This objection is harder to refute but it applies only to the 1993 calculation of Torr and Li [12].

Fontana (quoted by Baker, Davis and Woods [15]) has confirmed that the phase velocity of HFGWs is slowed in a superconductor, and Modanese and Fontana [16] also imply the same point. It is possible that the HFGW wavelength is changed inside a superconductor but not as much as predicted by Li and Torr [11]. Establishing the precise reduction may require an experimental test. It appears that at the current time, this issue should be classified as unresolved. This is unsatisfactory at the moment but should not prevent attempts to deduce consequences from assuming that it is true, since these may provide further tests of the result.

If the Li and Torr [11] result were correct, it would have fundamental implications for interactions between GW and matter, since there appears to be minimal interaction between GW and all conventional (non-superfluid or superconducting, at least) forms of matter. A confirmed significant gravitational refractive index would therefore represent by far the strongest interaction known between GW and any form of matter or radiation, and would revolutionise the design rules for engineering any HFGW equipment [7].

These consequences may suggest further possible tests of a prediction that is rather difficult to test directly experimentally with current technology. In particular, one of the clearest implications for HFGW engineering is that a large mismatch in HFGW phase velocity, and therefore a large value of gravitational refractive index, inevitably results in a huge mismatch in HFGW propagation wave impedance at superconductor-air interfaces, and consequently requires that the magnitude of the Fresnel reflection coefficient for GW must be very close to 100% [17], giving large Fresnel reflections from those interfaces. Even if Li and Torr are incorrect regarding the actual magnitude of the GW phase velocity reduction, any significant difference of the gravitational refractive index from unity (roughly, 10 or greater) would still result in a substantial GW reflection coefficient. This means that Li and Torr's result has major consequences for the design of GW components able to focus, refract, reflect, and otherwise manipulate gravitational waves for efficient coupling to detectors, transmitters, generators, resonant chambers, and other sensors, and will produce a number of design problems in superconducting lenses proposed for HFGW generation and detection [15].

As indicated above, this issue has in recent years been the source of some controversy. Notwithstanding this, Minter, Wegter-McNelly and Chiao [18] recently published a calculation taking an approach completely different from that of Li and Torr [11], yet which supports the view that a superconductor will efficiently reflect HFGW. The argument of Minter, Wegter-McNelly and Chiao [18] considers a film superconductor that may even be significantly thinner than the London penetration depth (the typical distance to which a magnetic field penetrates into a superconductor, generally of the order of 50 – 500nm). By analogy with the well-known calculation for electromagnetic waves, they solved the Gravito-electromagnetic equations (Eqns. (1) – (4)) for GW incident upon a superconductor, and deduced complete GW reflection in this situation. A potential problem with their analysis is that the group velocity of a Cooper pair is predicted to be significantly greater than *c*, the speed of light in a vacuum. Traditionally forbidden, the authors point out that superluminal velocities have been experimentally measured for photons [19]. However, causality is not violated by this observation regarding Gaussian photons, whereas the absence of causality violation in the case of Cooper pairs is not so clear. Their final conclusion regarding mirrors for GW, although derived in a completely different way from the method of Li and Torr [11], is essentially the same as that derived from their work [17] assuming that Li and Torr are correct. Chiao [20] has extended this analysis by examining the interactions between gravitation and superconductors in more detail.

Because of the controversy surrounding both the Li and Torr [11] result and its corollary [17] that appears to agree with the prediction of Minter, Wegter-McNelly and Chiao [18], this result has been christened "the Li-Torr-Chiao conjecture": the postulate that a superconductor acts as an efficient reflector for HFGW. It seems certain that some consequences of the conjecture would be advantageous in GW engineering and that some would not. However, it should be possible to exploit this conjecture significantly to build novel types of GW device. One important reason for examining consequences of this conjecture is that if a simple experimental test could be devised, the issue of its veracity might be easily settled. Amongst consequences that have been examined recently are novel methods of focusing and concentrating HFGW, and novel transduction methods. These have been combined variously with high-Q mechanical resonators, GW radiation pressure, manipulation of magnetic vortices, and charged spheres.

2. FRESNEL REFLECTION

Assuming the veracity of the Li and Torr [11] results (including $n_G \approx 300$), if a superconductor is a slowing material for HFGW, then in principle many geometrical optical components might be designed by analogy with conventional optics. If HFGW are capable of being focused, then the usual results from geometrical and physical optics would be expected to carry over in virtually unchanged form. This means that the minimum diffraction-limited focused spot diameter is $2\lambda/\pi$ (corresponding to the best case of focusing from a convergence angle of 90°), where λ is the HFGW wavelength in the material at the focusing point [21]. Hence, a diffraction-limited spot has area $n_G^2 \approx 300^2 = 90{,}000$ times smaller inside a superconductor than in free space, and so the corresponding energy density is 90, 000 times ($\cong$ 50dB) larger inside a superconductor than in free space. Clearly this result alone merits further attention in the design of noise-limited signal detectors, since it implies considerably easier signal detection under adverse conditions (potentially a 50dB lowering of the noise floor), as well as the notion that HFGW may be focused by suitably shaped superconductor, though to obtain this 50dB improvement *all* of the HFGW detection must be accomplished inside the superconductor, which may not be so easy to arrange in practice.

For a HFGW having a typical frequency of 4.9GHz [7], the wavelength $\lambda_0 = 6.1$cm in air, corresponding to $\lambda = 200$μm in a superconductor. One immediate consequence is that superconducting components for manipulation of HFGW at this frequency must be machined to a tolerance of ~20μm (~one thousandth of an inch) or better for effective optical performance (using the well-known $\lambda/10$ practical criterion for optical manufacturing). In view of the current difficulty of fabricating high-temperature superconductor ceramic materials to tight tolerances, it may be more cost-effective to use low-temperature superconductor alloys instead, for which conventional machining techniques may be employed. The disadvantage, of course, is that liquid helium refrigerant must then be used.

The relations between the components of waves incident upon a plane boundary between two different optical media are given by the optical Fresnel relations [21]. These relations are derived assuming amplitude matching (no discontinuities of the field quantities) at the interface, and the results depend only upon the phase velocities in the materials. The full statement of the Fresnel relations includes the angle of incidence, θ_i, to the normal vector, as well as the refractive index, n. The value of n is defined as the magnitude of the phase velocity in the original propagation material divided by the phase velocity in the material on the other side of the boundary; $n > 1$ for a wave travelling to a slower material and $n < 1$ for a wave travelling to a faster material. However, for simplicity, generally the results for normal incidence ($\theta_i = 0$) are sufficient for deducing the important behaviours concerned. At normal incidence, the expression for the reflection coefficient simplifies to

$$r = \frac{1-n}{1+n}. \tag{9}$$

and the transmission coefficient for normal incidence simplifies to the expression

$$t = \frac{2}{1+n}. \tag{10}$$

(For comparison, the results for grazing incidence, $\theta_i = 90°$, are that $t = 0$ and $|r| = 1$ independent of n, the phase of r depending on the incident polarization). The transmission coefficient, t, gives the ratio of the amplitude of the transmitted wave to that of the incident wave, but in almost all cases of interest here the transmittance,

$$T = nt^2 = \frac{4n}{(1+n)^2}, \tag{11}$$

is a more useful measure as it is equal to the ratio of the transmitted intensity to the incident intensity. The transmission coefficient, t, is itself defined in terms of the wave amplitude alone and does not take account of the different wave impedance in the second medium, affecting energy transmission.

Since these results are fundamentally independent of the nature of the underlying waves (and, for example, are unchanged in principle for acoustic waves [22]), it appears clear that corresponding results for HFGW will be direct analogues of the conventional optical results summarised in the previous paragraph, but substituting the gravitational refractive index n_G for the optical refractive index n. Using these general results, for HFGW travelling from air to superconductor, $n_G = 300$, and so $r = -299/301 \approx -1$ (*i.e.*, corresponding to 180° phase change on reflection at a boundary of air to superconductor, with almost no amplitude change on reflection). The corresponding transmittance is $T = 300\times2^2/301^2 \approx 0.013 = 1.3\%$.

Conversely, for HFGW travelling from superconductor to air, n is replaced by the inverse value $1/n_G = 1/300$, and so $r = 299/301 \approx +1$ (*i.e.*, corresponding to zero phase change on reflection at a boundary of superconductor to air, with almost no amplitude change on reflection). The corresponding transmittance is $T = 2^2\times300^2/(301^2\times300) \approx 0.013 = 1.3\%$, exactly the same as for HFGWs travelling in the reverse direction from air to superconductor.

A secondary issue concerning the usage of superconductors in GW experiments (and, particularly, the "normal" superconductors rather than the "high-temperature" superconductors) is of some concern here. The "high-temperature" superconductors will typically be cooled using liquid nitrogen at 77K (or possibly liquid helium at 4.2K), but "normal" or "low temperature" superconductors are frequently cooled using a dilution refrigerator to temperatures lower than the helium superfluid transition temperature of 2.17K. Since Weinberg has conjectured that superfluids might act as efficient mirrors for GW [9], surrounding a superconductor of any type with a jacket of refrigerant that is itself superfluid may immediately prevent the GW interactions that are claimed with the superconductor. On the other hand, Minter, Wegter-McNelly and Chiao [18] have suggested that neutral atoms in a superfluid will not interact with GW because they are uncharged. Unless and until this point is finally settled, it seems that any HFGW experiment using helium refrigerant cooled below 2.17K may be called into question as it cannot be assured that the refrigerant itself does not enter into the interactions with the HFGW.

2.1. Simple Superconducting HFGW Refractive Components

These results have importance regarding the feasibility of a HFGW refractive structure using superconducting material. A transmittance of only 1.3% at each air-superconductor interface corresponds to an insertion loss at each surface of 19dB, potentially offset by up to 50dB increase in diffraction-limited power density possible inside the superconductor, or a net improvement in power density of up to 31dB without any steps being taken to reduce reflection loss. However, since there is 19dB insertion loss in either propagation direction, a simple superconductor lens analogue of a simple optical glass lens would have a transmittance of only $0.013^2 \approx 0.017\%$, *i.e.*, the combined transmittance of two air-superconductor boundaries, one each side of the lens. In other words, a simple HFGW lens made of superconductor material will introduce an energy loss of 99.98%, or an insertion loss of 38dB. This is, of course, an intolerable overhead on the use of such a lens. A simple GW superconducting lens analogue thus appears impractical without further refinement.

The well-known technique of "optical blooming" or "optical coating" was developed many years ago to minimise Fresnel loss in conventional lenses, and merits examination in the present context. The basic principle is illustrated in Fig. **1**. (In Fig. **1**, a small non-zero angle of incidence is shown for clarity, but the principle strictly only functions accurately at normal incidence for the coating thickness shown). The coating material is added in a uniform film over a reflective interface between two non-opaque materials. The film thickness is exactly one-quarter of a wavelength of the HFGW travelling in the coating material, and the refractive index of the coating is exactly equal to the geometric mean of the two surrounding materials. If one material is air (refractive index unity) and the other is superconductor, then the coating material must have refractive index $n_c = \sqrt{n_G} = 17.3$. This condition on the value of the refractive index ensures that Fresnel reflections from the air-coating and coating-superconductor interfaces have equal amplitudes, and the coating thickness ensures that at normal incidence the two reflection positions are displaced by $2\times\lambda_c/4$ (where λ_c is the HFGW wavelength inside the coating), equivalent to an optical path difference of $\lambda_o/2$; *i.e.*, the two reflections are out of phase by 180° and so exactly cancel by destructive

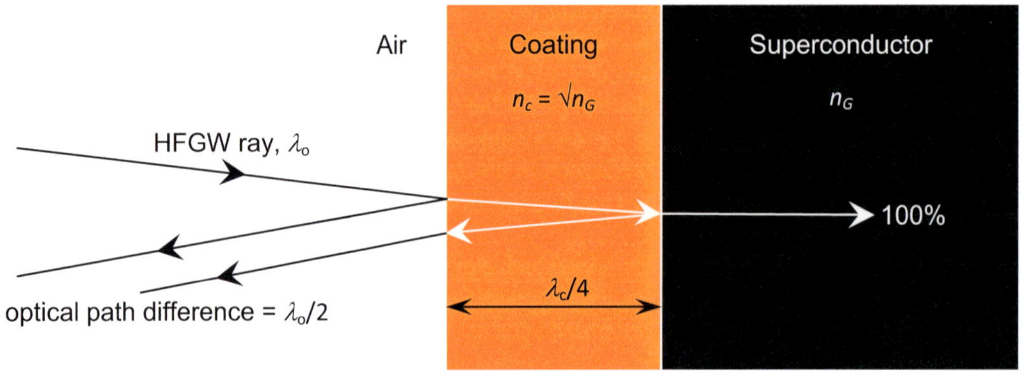

Figure 1: Basic principle of optical blooming.

interference. Just as in the analogous optical case, there is therefore no mechanism for losing energy at the interface, and so the overall transmittance must be 100%.

Although this scheme is very effective for conventional optics and is widely used in all but the very simplest of lens designs, there are a number of major problems with using it for HFGW lenses made from superconductor. Firstly, it is not clear how a coating material having $n_c = 17.3$ for GW can be made. At present, the only known values of the HFGW refractive index are for superconductors ($n_G = 300$, assuming the Li and Torr [11] prediction is correct), and for all other materials ($n_G = 1$). One suggestion has been to make a suspension of superconductor particles held in an inert filler (maybe epoxy) [17]. For this to give an intermediate GW refractive index, the superconductor particle size must be much less than $\lambda_o/(4n_c) = 0.9$mm for HFGW frequency 4.9GHz. Also, for each particle to act substantially as a superconductor, the superconductor particle diameter must be much greater than the superconductive coherence length, ξ, which is typically a few μm. Perhaps the best particle size would be the geometric mean of these values, or around 60μm. To give the precise refractive index value needed, assuming that the overall refractive index is the weighted mean of the contributing components by volume, the superconductor filling factor would be 5.5% by volume. However, herein lies the second difficulty: optical blooming relies on precise cancellation of reflections, and the coating material must be formulated very carefully to ensure that its refractive index is as closely as possible equal to $\sqrt{n_G}$ so that the Fresnel reflections from the two interfaces do indeed cancel precisely. This is particularly important in the present case because the unmodified Fresnel reflection coefficient is so large and so reducing the overall reflection to a tolerable value requires accurate matching of the reflections from the two coating surfaces. The large standard error currently on the value of n_G, lack of secure knowledge about how a suspension of superconducting particles will behave as a GW transmission medium, and the difficulties of manufacturing a precisely uniform particle size and dispersion within the medium, all indicate that this may be very difficult to achieve in practice. Thirdly, the coating thickness must be $\lambda_o/(4n_c)$ within very fine limits. Using precision manufacturing techniques now available, this may be easier to achieve than the allied condition on the required refractive index of the coating. However, it must be noted here that the blooming action only works precisely at one wavelength and hence at one frequency. For traditional optical applications this is not a major problem as the Fresnel reflection from air-glass surfaces is quite small and the improvement obtained from even an inaccurate cancellation at the spectral extremes is worthwhile, but it will be a serious problem for wideband HFGW detectors. Finally, the simplified calculation is only correct for normal incidence, and becomes increasingly inaccurate at non-zero angles of incidence. It is not possible to devise a coating thickness that is optimal for any non-zero range of angles of incidence. Again, because of the large Fresnel reflections being cancelled in the HFGW case, this problem may be difficult to overcome.

An estimate of the difficulties of cancelling the Fresnel reflections in this case may be made as follows. From Equation (9), the reflection coefficient at each edge of the coating is ideally $|r| = 0.89$. Hence, 1% inaccuracy in the reflection amplitude arising from an inaccurate value of n_c, or an equivalent inaccuracy in the coating thickness, results in an overall transmittance reduced by 3.6%. Development of a multiple coating technology analogous to that used in the highest precision optical designs might reduce this sensitivity at the cost of much more expensive fabrication.

As an aside, one important implication of these results is their application to HFGW detectors using a superconductor coating to enhance the electrical resonant Q of an RF cavity [7, 15]. The superconductor thickness used will be critical. The incident HFGW must penetrate this cavity without significant loss (for efficient HFGW detection) and so to avoid reflection of HFGW, the superconductor coating must have a thickness of precisely $\lambda/2$ measured in the superconductor, or $\lambda_o/(2n_G)$. This gives a round-trip optical path difference of λ_o between the reflections from the inside and outside surfaces of the superconductor, but also there is a 180° phase change on reflection at the boundary from air to superconductor (not present on reflection from the superconductor to air boundary) so that normal reflections from the inside and outside surfaces will interfere destructively. For a typical frequency of 4.9GHz, the HFGW wavelength is $\lambda = 200$μm in the superconductor and so the superconductor coating around the corresponding RF resonant cavity must be 100μm thick. This should be achievable using traditional low-temperature superconductor materials where the disadvantage will be the use of liquid helium refrigerant; high-temperature superconductor materials are easier to refrigerate but harder to fabricate to a close tolerance. It is also acceptable to make the superconductor thickness an integer multiple of $\lambda/2$ if this eases fabrication problems (since the superconducting layer will be thicker and therefore mechanically stronger), although the dimensional precision required of this thickness is unchanged.

The design of refractive lenses for HFGW based upon superconductors presents some interesting possibilities because of the large gravitational refractive index. Snell's law of refraction [21] states that when a ray is incident upon a boundary between two optical media, $\sin\theta_i/\sin\theta_r = n$, where θ_r is the angle of refraction and n is the refractive index (replaced by n_G for HFGW travelling from air to superconductor, or replaced by $1/n_G$ for HFGW travelling from superconductor to air). This law is unaffected overall by the presence of any anti-reflection coatings. Therefore, for HFGW travelling from air to superconductor, the maximum possible angle of refraction is $\theta_r = \arcsin(1/300) = 0.2°$. This means that, to a good approximation, all refracted rays inside the superconductor are normal to the surface. As a direct consequence, the critical angle, θ_C, for total internal reflection (*i.e.*, reflection of HFGW rays incident from the inside of the superconductor onto a superconductor-air surface) is also $\arcsin(1/n_G) = 0.2°$. Any HFGW incident on a superconductor-air surface at more than 0.2° to the normal will be reflected with virtually no power loss. Total internal reflection at incident angles greater than the critical angle occurs regardless of the use of anti-reflection coatings.

The consequences of these results have been explored [17]. To summarise, in principle very wide aperture lenses may be made (close to *f*/0.5, a very fast lens) with near-zero spherical and chromatic aberrations. Some other special geometries for refractive lenses were proposed and evaluated [17]. If instead the reduction in wavelength turns out to be modest (a factor of ~1.5 or ~2 only) then existing designs of refractive glass optical systems can be used almost unchanged with HFGWs by replacing glass with superconductor. However, since Fresnel reflections at each side of the lens appear difficult to reduce to reasonable proportions, fixed refractive lenses will probably not represent the most efficient choice for GW optics. It may be easier to exploit the large reflection coefficient of HFGW from superconductor surfaces, by fabricating focusing HFGW mirrors rather than fixed-focus lenses.

2.2. Superconducting HFGW Reflecting Components

In view of the difficulties in designing effective HFGW refractive superconducting components as described in the previous section, it may be more immediately practical to explore the possibilities afforded by superconductors as HFGW reflectors. HFGW reflecting components can closely mimic equivalent conventional optical designs and existing optical mirrors, but in addition heavy supporting structures would be easy to incorporate since any non-superconducting material is expected not to affect the HFGWs in any way. This means that sturdy structural frameworks constructed of normal material will not obstruct or vignette the aperture at all, even directly in the ray paths. This means that the HFGW version of a reflection design can be expected to be made almost without compromise regarding its dimensional stability. In conventional optics, reflecting components are often preferred in precision applications because of the complete absence of chromatic aberration, and superconducting HFGW mirrors would retain this advantage.

Firstly, it should be noted that for normal incidence the HFGW Fresnel reflectance from an uncoated air-superconductor interface is 98.7% (*i.e.*, representing an insertion loss of 0.06dB), and (while not outstanding compared to the best available optical mirrors) this may be adequate for many applications involving a small number of reflections. If not, a reflection-*enhancing* coating might be used. This could be exactly the same material (superconductor particles suspended in an inert matrix) proposed earlier (§2.1) for anti-reflection coating, compounded with the same proportions of constituents and thus again having $n_c = \sqrt{n_G} = 17.3$, but applied now so as to have a thickness of $\lambda_0/(2n_c)$ (for HFGW at normal incidence). At this thickness, the positions of the reflections from the two surfaces are displaced by $\lambda_c/2$, equivalent to an optical path difference of λ_0; *i.e.*, the phase difference between them is 360° and so the two reflections add by constructive interference, ideally giving an overall reflectance of 100%. Since the uncoated reflectance is already very close to 100% it appears entirely feasible to improve the reflectance using a coating in this manner, provided a suitable coating material can be developed. This method potentially has the same disadvantages as using a coating to *reduce* reflections. As in the case of an anti-reflection coating, the wave interference only works exactly for rays at normal incidence. However, in the case of reflection enhancement the basic unenhanced reflection coefficient is so close to unity that small errors due to manufacturing tolerances or non-normal incidence will have little consequence, unlike the case of HFGW anti-reflection coating. The same cautions regarding the unknown behaviour of a superconducting particle suspension, its uniformity and precise manufacture, and introduction of chromatic aberrations, also still apply, but again because the improvement needed is so small they can probably be ignored in the case of reflective components. Whether the use of reflection enhancement coating is cost-effective will depend upon each application, since the projected improvement will be quite small.

Fig. **2** shows an example of a focusing superconductor reflector design. For a small-aperture spherical reflector the focal length will be $-R/2$ approximately, as for an optical mirror, where R is the radius of curvature (convex is positive). For a parabolic reflector, parallel incident rays converge at the parabolic focus. Neither type of reflector produces chromatic aberration; the spherical reflector has some spherical aberration at large apertures but the parabolic reflector eliminates this, so that large apertures will be quite feasible in that case.

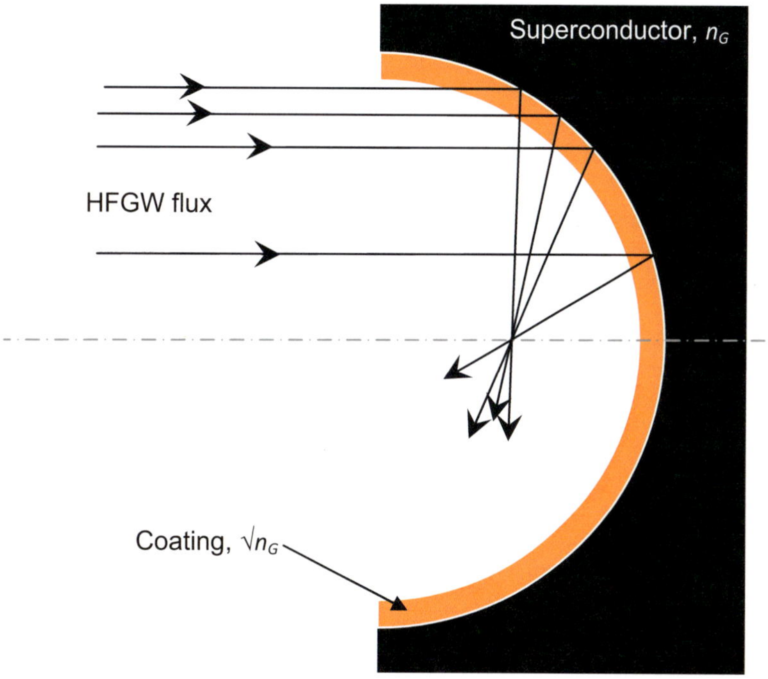

Figure 2: Superconductor used as a HFGW reflector. The superconductor surface may use a reflection-enhancing coating, of thickness $\lambda_c/2$.

A more sophisticated use of superconductors as HFGW reflectors is illustrated in Fig. **3**, showing a Fabry-Perot étalon [21]. An étalon is basically a prototype one-dimensional cavity resonator consisting of two plane parallel reflective walls of (ideally) infinite extent. One application of such a resonator would be to increase the detected amplitude of HFGWs for one of the HFGW detector designs currently proposed [23]. The Fabry-Perot étalon acts as a high-Q resonant cavity for HFGWs and so will magnify the received signal by Q, while significantly reducing the reception bandwidth. Bandwidth reduction at the receiver is usual in all high-frequency applications. In optical applications, the quantity of interest is the "finesse" of the étalon, defined as the frequency separation between adjacent resonant modes of the étalon, divided by the FWHM frequency of a resonant mode. An étalon's finesse may be shown to take the value $|\pi r/(1-r^2)|$ which, in the case where no reflection-enhancing coatings are used, is equal to

$$\frac{\pi}{4}\left(n_G - \frac{1}{n_G}\right) \approx \frac{\pi n_G}{4}. \qquad \textbf{(12)}$$

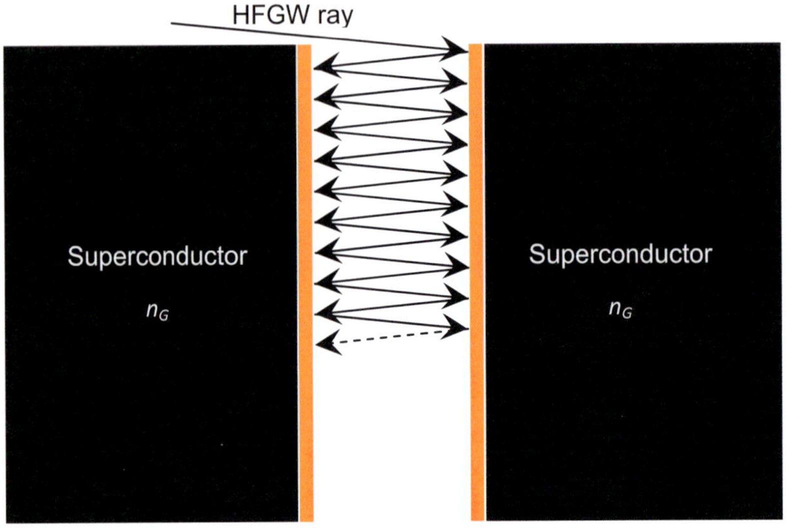

Figure 3: Superconductors used to form a Fabry-Perot Étalon or high-Q resonant cavity using parallel plane walls. The superconductor surfaces may use a reflection-enhancing coating, of thickness $\lambda_c/2$.

Using the previous value $r = -299/301$ for an uncoated superconductor gives a typical finesse of the order of 240 for a superconducting HFGW cavity. This is significantly better than that usually achieved in a conventional optical étalon, for which typical finesse values are ~100. Clearly the finesse can be increased enormously by using even a moderately successful reflection-enhancing coating to bring r closer to unity, so that even greater resonance can be achieved. It is worth noting that as a result of the resonant enhancement, the HFGW intensity inside such an étalon is increased by a factor $1/(1 - r^2)^2$, and so a considerable improvement in detector sensitivity and selectivity could potentially be achieved if this could be directly exploited.

3. NOVEL REFRACTIVE COMPONENTS

In a classic 1962 paper, Gertsenshtein [24] described how the non-linearity of Einstein's field equations meant that electromagnetic and gravitational waves were coupled in such a way that propagation of one type of wave in a vacuum would simultaneously generate a wave of the other type. The amplitude of the generated wave is, however, so small that practical laboratory exploitation of this effect is fraught with apparently insuperable difficulties. Effective manipulation of gravitational waves has therefore generally been thought possible only by events of astronomical scale. In this section, the possibility of a strong interaction between high-frequency gravitational waves and magnetic fields within a superconductor is explored. Zel'dovich pointed out in 1974 [25] that the classical Gertsenshtein effect cannot give appreciable coupling in any medium except a vacuum. This is because an electromagnetic wave in a transparent medium is subject to the electromagnetic refractive index and for a given frequency the electromagnetic wavelength will always be less than its value in a vacuum, whereas a gravitational wave at the same frequency has an unchanged wavelength in any medium (excepting superconductors [11]). Within a superconductor, the electromagnetic refractive index will be considerably less than $n_G \approx 300$ [11] and in any case an electromagnetic wave cannot propagate at all in a superconductor without rapid attenuation [26]. Therefore, except in a vacuum, the two types of wave cannot overlap exactly in phase over an extended distance (the "synchro-resonant" or "phase coherent" condition required to give a detectable Gertsenshtein effect) and so a substantial interaction is impossible. It follows that interaction between GW and a magnetic field within a superconductor must be unrelated to the classical Gertsenshtein effect.

A novel type of HFGW lens design is proposed here using a magnetic field to adjust the focal length in an infinitely-variable manner. This component has no exact analogue in conventional electromagnetic geometrical optics because using current technology it is not possible to vary the refractive index of an optical material by varying an external parameter. This lays the foundations for controlling and manipulating HFGWs in future gravitational wave systems. Therefore, some significant novel applications are potentially opened up that may offset the large potential insertion loss penalty associated with interface reflections as identified in §2.1 above.

3.1. Basic Interaction Between HFGW and Magnetic Field in a Superconductor

In any superconductor, the superconductivity may be eliminated temporarily by applying a sufficiently large magnetic field. In general, all superconductors so far discovered fall into two mutually exclusive groups: "type-I" and "type-II'. In a "type-I" superconductor (generally, most of the pure element superconductors), there is a single critical applied external magnetic field value below which all magnetic field is excluded from the superconductor (a perfect diamagnet, having magnetic susceptibility of –1 and relative permeability of 0), but which, if exceeded, "quenches" (removes) the superconductivity as effectively as raising the temperature higher than the critical temperature T_c. In a "type-II" superconductor (generally, most of the technologically important superconducting alloys and the "high-temperature" superconductors such as $YBa_2Cu_3O_{7-\delta}$ or YBCO), there are two critical field values. Below the "lower critical field", B_c, the superconductor is a perfect diamagnet like a type-I superconductor. Above the "upper critical field" the material is completely "quenched". In applied magnetic fields having magnitudes between B_c and the upper critical field, it is more energetically favourable for some parts of the superconductor (a small proportion of the whole) to "quench", allowing the field to penetrate this locally non-superconducting material. These quenched regions or "magnetic vortices" channel the magnetic field through the material cooled below T_c. In detail, what happens is that the superconducting order parameter, Δ, is identically zero on the vortex axis at its centre, and it transitions to the bulk value smoothly at a distance approximately ξ (the superconducting coherence length) from the vortex axis [27]. In a typical type-II superconductor, the internal radius of a single vortex, ξ, is of the order of 1μm or so. The vortex boundary is an interface between quenched and superconducting material; the local magnetic field within a vortex is higher than the upper critical field. The total energy of the system is lowered in this configuration because the potential energy of the field being completely expelled (as in a type-I superconductor) is higher than the field partially penetrating the material, but offsetting this is the energy required to create the vortices. The field B is still expelled from the material outside the vortices but the density of magnetic vortices depends upon the applied external magnetic field B, since each vortex encloses an amount of flux equal to the lower critical field B_c multiplied by the total perpendicular area occupied by the vortex. (In a type-I superconductor, the energy required to create vortices is higher than that to be gained by partial field penetration, so that no vortices can form at any applied field).

Now consider the basic geometry shown in Fig. **4**, where the applied magnetic field B is everywhere greater than the lower critical field, B_c. Following the prediction of Li and Torr [11], the HFGW phase velocity will be reduced only outside the vortices, or at a distance greater than ξ from the centre of any vortex. There is therefore extremely large Fresnel reflection [15, 17] for any HFGW incident upon the outside (cylindrical curved) surface of a vortex. Although the vortex is itself a "fast" structure compared to its surroundings rather than a "slowing" structure, the extremely large Fresnel reflection means that this cylinder of quenched material will act as an "anti-waveguide". Moreover, its small diameter means that it will be well beyond waveguide cut-off for HFGWs having wavelengths significantly larger than 1μm or so (in non-superconducting material, because the wavelength reduction inside superconducting material does not apply). It follows that there will be essentially no HFGW propagation along the vortices (ignoring evanescent excitations that will decay after a short distance).

At this point recall that for a HFGW having a typical frequency of 4.9GHz [7] perhaps arising from laboratory generation, the wavelength $\lambda_o = 6.1$cm in air, corresponding to $\lambda = 200$μm in a superconductor, so that treatment of HFGW using the ray picture inside a superconductor is justified approximately as it is for long-wavelength optical radiation. (Note that although the wavelength is reduced inside a superconductor, the frequency is unchanged; the wavelength reduction arises entirely from a reduction in GW phase velocity rather than from a frequency increase. The GW amplitude will be different inside a superconductor, but allowing for energy loss at the interfaces the energy density is unchanged, because the amplitude change results from the changed wave impedance related to the different phase velocity). The picture described here is certainly very different from that typically used for describing low-frequency gravitational waves, where wavelengths of many kilometres and detectors similarly sized are required. This difference is due entirely to the vast difference in frequency between low and high frequencies; similar differences in electromagnetic waves dictate corresponding widely different technological approaches to handling audio and microwave frequencies.

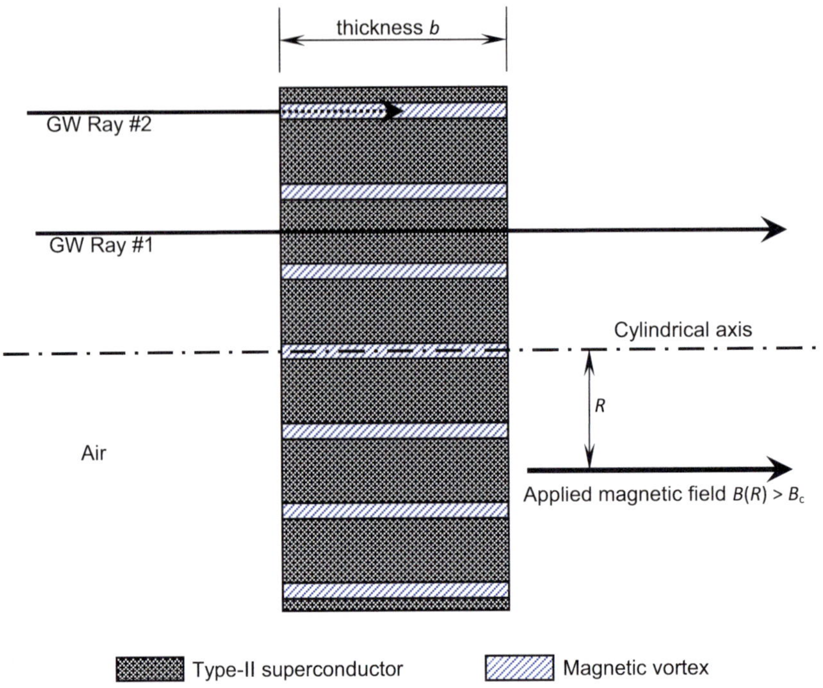

Figure 4: Interaction of HFGW with type-II superconductor.

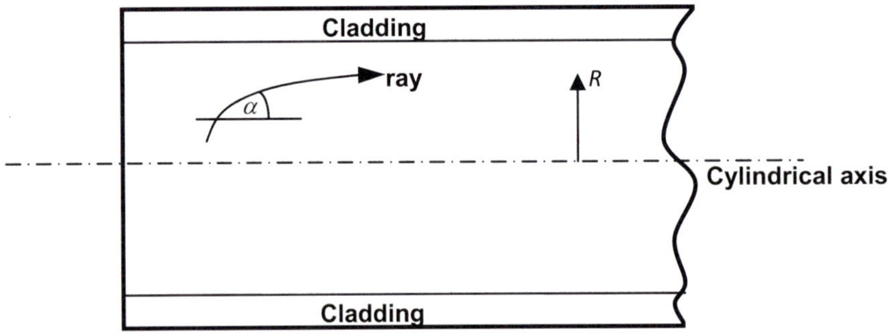

Figure 5: Path of a meridional ray in a graded-index optical fibre.

The ray picture of wave propagation may be used to find the effect, on the HFGW, of a gradual refractive index transition: the result will be analogous to light rays propagating in a graded-index optical fibre, as depicted in Fig. **5** [21]. In such a fibre, the optical refractive index $n(R)$ depends only on the radius R from the fibre cylindrical axis. Using geometric ray optics (*i.e.*, assuming that light travels in rays of zero width measured perpendicular to the velocity) and by considering adjacent layers of material having infinitesimal thickness, it is straightforward to show that at every point on a meridional ray (*i.e.*, a ray crossing the fibre cylindrical axis), the angle α between the ray path and the cylindrical axis always satisfies the equation

$$n(R)\cos\alpha = n_o, \tag{13}$$

where n_o is the optical refractive index at the furthest excursion of that particular ray from the cylindrical axis. So, at a given value of refractive index n at specified radius R, the two possible solutions (corresponding to rising and falling) for the ray angle are

$$\alpha = \pm \arccos [n_0/n(R)]. \tag{14}$$

Therefore, because of the symmetry about a fibre diameter normal to the cylindrical axis, this light ray follows a curved path whose end trajectories are indistinguishable from specular reflection when observed at a distance removed from the region of index grading. Similarly, a HFGW will be effectively subject to specular reflection since the vortex itself is a "fast" structure compared to the bulk material that is a "slowing" structure, and there is therefore no possibility of a transmitted ray component even though the transition between the two regions may not be abrupt.

It follows that HFGWs can logically only propagate through the remaining superconducting matrix surrounding the vortices. To find the effective phase velocity of HFGW through the superconductor we must therefore find the HFGW phase velocity through a material containing impenetrable obstructions parallel to the propagation direction. This is a very difficult problem to solve accurately and here the approximation is made that this material will act as a waveguide having an effective diameter equivalent to the mean distance between adjacent vortices. Of course, the waveguide properties, including the guide propagation constant, depend upon the lateral dimensions of the guide. The mean vortex separation depends upon the applied magnetic field. Therefore, it appears possible to adjust the effective phase velocity through the superconductor by adjusting the applied magnetic field. Increasing the applied field increases the areal density of vortices; decreasing the applied field reduces the areal density of vortices. Hence, increasing the field reduces the effective guiding diameter and increases the HFGW phase velocity; decreasing the field increases the effective guiding diameter and reduces the HFGW phase velocity. This means that the effective GW phase velocity through the material is closer to the value c/n_G at lower applied fields, provided the GW wavelength is sufficiently large that the granularity of the vortex structure is not an issue.

For HFGW wavelengths significantly smaller than the typical separation distance between vortices, the behaviour is best described differently. There are two types of paths through the material, *i.e.,* through a vortex or not through a vortex. The paths through a vortex will either have velocity c or will still be beyond waveguide cut-off, so that the overall continuously variable phase velocity will not occur. Instead, the resultant GW on passing through the material will, in near-field, produce a silhouette (or phase silhouette) of the vortex pattern. In other words, the GW merely images the vortex pattern immediately on passing through the superconductor. This situation is not considered further here, since this seems to be of little technological interest, though it may afford a novel technique for imaging the vortex positions in a superconductor.

By suitably adjusting the applied magnetic field while being always in a cylindrically symmetrical pattern, the GW velocity through the material may be arranged to vary continuously as a function of the radius from the centre. For a uniform thickness of superconductor measured parallel to the GW propagation direction, a varying phase velocity corresponds to a varying GW phase shift as a function of radius. If the GW phase shift varies in the appropriate manner with radius, then a converging or diverging lens will result, whose focal length depends upon the applied magnetic field pattern and so can be controlled electrically. This variable focus lens would have no moving parts and so would be uniquely placed for rapid and reliable auto-focusing of GWs in laboratory instruments. Given laboratory-scale generation and detection of HFGWs [7], this device would offer a novel experimental test of the Li and Torr result, as well as a useful HFGW manipulation component unparalleled in conventional electromagnetic optics.

3.2. GW Propagation Through a Superconductor with Applied Magnetic Field

In the basic structure (Fig. **4**) considered first, GW ray #1 travels through superconducting material, while GW ray #2 is incident upon a vortex acting as a GW waveguide beyond cut-off. Therefore, ray #2 becomes evanescent inside the vortex and its intensity decays exponentially; it will be assumed to play no further part in the operation of the device. However, ray #1 has phase velocity dependent upon the disposition of vortices in the material. The thickness of superconductor, b, is fixed and must be significantly greater than the evanescent decay length of HFGWs within each vortex. (This requirement is easy to satisfy since the typical vortex diameter is so small).

The significance of these magnetic vortices in the present context is that the material within the vortices, having zero or small superconducting order parameter Δ, must behave as normal (non-superconducting) material as far as the propagating GW is concerned. Therefore, the GW is propagating through a piece of material containing a lattice of vortices within each one of which the wave propagation properties are completely different from the rest of the material.

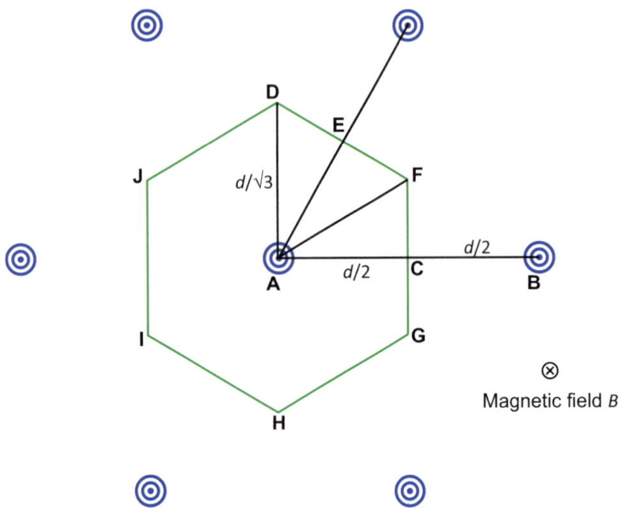

Figure 6: Idealised magnetic vortex geometry in a typical type-II superconductor. The magnetic field is normal to the page, and vortex positions are indicated by ◎.

To estimate the effective phase velocity of ray #1, suppose that the magnetic flux through each vortex is Φ_o. For the coupled-electron wavefunction to be single-valued and continuous, it is well-known that the value of Φ_o must be quantised in units of the fundamental flux quantum, $\pi\hbar/e = 2.08\times10^{-15}$Wb, where $\hbar$ is Planck's constant/2π and e is the fundamental charge. If the value of Φ_o is uniform within vortices located over the entire superconductor, then in an applied field B the local areal density of vortices is given by B/Φ_o and so the average area occupied by a single vortex is Φ_o/B. The vortices generally form a regular triangular lattice (two-dimensional close-packing [28]), and so the relation between the vortex separation, d, and the applied magnetic field may be found simply as follows. In Fig. **6**, d is the distance AB, and the distances AC and AE are both equal to $d/2$. Since angle EAD is 30°, distance AD is $d/(2\cos 30°) = d/\sqrt{3}$. Now, triangle FAD is equilateral, so distance DF is also $d/\sqrt{3}$, and hence the area of triangle FAD is $\tfrac{1}{2}$(AE)(DF) $= d^2/(4\sqrt{3})$. The vortex A is located at the centre of the regular hexagon DFGHIJD whose area is 6 times the area of triangle FAD, or $d^2(\sqrt{3})/2$. This hexagon is the 2-dimensional unit cell of the vortex lattice, so the external applied field B must be given by

$$B = 2\Phi_o/(d^2\sqrt{3}). \tag{15}$$

Solving for d gives the distance between vortices,

$$d = \sqrt{(2/\sqrt{3})}\sqrt{(\Phi_o/B)} \approx 1.075\,\sqrt{(\Phi_o/B)}, \tag{16}$$

which will be much greater than a typical internal vortex diameter.

Accurate calculation of the effective phase velocity of ray #1 in Fig. **4** appears to be a difficult task since simpler related problems, such as optical propagation through cylindrical optical fibres, yield only to some very gross approximations, or to numerical analysis, or have solutions known only empirically [21]. Also, no consistent theory has yet been developed for describing gravitation at the microscopic level of quantum mechanics. It seems safe to assume that precise calculation of the phase shift of a gravitational wave travelling through such a lattice of vortices will be difficult to the point of being impossible with current knowledge. However, empirical approximations similar to those employed in fibre-optics can be employed in this case.

The V-parameter, or normalised frequency, is used to describe wave propagation through cylindrical optical fibres [21] and this approach is used approximately here for estimating the behaviour of a guiding structure having a diameter $d \approx \sqrt{(2/\sqrt{3})}\sqrt{(\Phi_o/B)}$. The definition of V is:

$$V = \frac{\pi d}{\lambda_o}\sqrt{n_g^2 - n_{g2}^2} \approx \frac{\pi d}{\lambda_o}\sqrt{2n_g^2\Delta}, \tag{17}$$

where n_g is the effective gravitational refractive index of the guiding material, n_{g2} is the effective gravitational refractive index of the material surrounding this guide, and Δ is the normalised refractive index difference,

$$\Delta = (n_g - n_{g2}) / n_g. \tag{18}$$

Because of the averaged effect of the presence of vortices consisting of magnetically quenched superconductor, n_{g2} is slightly less than n_g and so Δ is small but positive. In general, finding the wave velocity in a waveguide is a difficult problem, but we can use the approximate result [21]

$$v_{ph} \approx \frac{c / n_g}{1 - \pi^2\Delta / V^2}, \tag{19}$$

so that by using the definition of V, Equation (17),

$$v_{ph} \approx \frac{c / n_g}{1 - \lambda_o^2 / (2n_g^2 d^2)} = \frac{2cn_g}{2n_g^2 - \lambda_o^2 / d^2}, \tag{20}$$

and substituting the value of the effective guiding diameter $d \approx \sqrt{(2/\sqrt{3})}\sqrt{(\Phi_o/B)}$ gives:

$$v_{ph} \approx \frac{4cn_g\Phi_o}{4n_g^2\Phi_o - \sqrt{3}B\lambda_o^2}. \tag{21}$$

(Note that the phase velocity, v_{ph}, is independent of the precise geometry of the vortices). Hence, the phase shift, ϕ = (wavenumber) × (distance), after travelling through a superconductor of thickness b, is approximately given by:

$$\phi = \frac{\omega b}{v_{ph}} \approx \omega\left(\frac{n_g}{c} - \frac{\sqrt{3}\pi^2 Bc}{n_g\omega^2\Phi_o}\right)b = \left(\frac{n_g\omega}{c}\right)b - \left(\frac{\sqrt{3}\pi^2 c}{n_g\omega\Phi_o}\right)Bb. \tag{22}$$

This phase shift can be varied continuously by varying the magnetic field B. This result gives $\phi = n_g\omega b/c$ for $B = 0$, as would be expected of the superconductor without vortices, *i.e.*, either a type-I superconductor, or a type-II superconductor in a field $B < B_c$. When the material is completely quenched, either magnetically or by being at temperature higher than T_c, the phase shift is simply the same as for a non-superconductor, $\phi = \omega b/c$, corresponding to $n_g = 1$.

The specific additional phase shift (*i.e.*, caused solely by the applied field B) per unit applied magnetic field and per unit propagation length is given by $-(\sqrt{3})\pi^2 c/(n_g\omega\Phi_o)$. Assuming one flux quantum per vortex, $\Phi_o = \pi\hbar/e$, the specific phase shift becomes $-(\sqrt{3})\pi ce/(n_g\omega\hbar)$. At GW frequency 3GHz, taking $n_g = 300$ [11], and other fundamental physical constants taking their usual values, this expression gives a maximum specific excess phase shift of –0.4rad μT^{-1} μm^{-1}. Although this would represent a large interaction between a magnetic field and a GW, strictly it is not related either to the Gertsenshtein effect [24] or to the Li effect [29] as the magnetic field is not periodic or part of a phase-coherent electromagnetic wave. By far the dominant error source in the result is likely to be the current lack of precision established in the Li and Torr [11] result. A very conservative estimate of the systematic error likely in this result is that the slowing factor may be uncertain by as much as ±50%, and this will lead to a corresponding uncertainty in the final result for the field-induced phase shift, Equation (22).

At first sight it appears that the magnitude of the interaction between HFGW and vortices (indicated by the specific additional phase shift) diverges at low frequencies, since it is inversely proportional to ω. However, in the low frequency limit, the approximations used in deriving Equation (22) are no longer valid. This is because the GW wavelength is then much greater than the typical distance between vortices. Accordingly, on physical grounds, there can no longer be any waveguiding effect, and the GW phase velocity through the superconductor is given simply by c/n_g. There is then no interaction with the vortices, and so the GW phase shift is $\phi = n_g \omega b/c$ for $B = 0$, identical to the case of the superconductor without vortices. Accordingly, there must be a minimum frequency $\nu_{\min}$ for which the approximations in Equation (22) are reasonably accurate. This minimum can be estimated by setting the GW wavelength equal to the vortex separation, giving

$$\nu_{\min} \approx \sqrt{\frac{\sqrt{3}}{2}} \left(\frac{c}{2n_g} \right) \sqrt{\frac{B}{\Phi_o}} \quad \textbf{(23)}$$

In addition, of course, at increasingly high frequencies, the "waveguides" formed by the vortices are heavily overmoded [21] giving an effective change in wave velocity reducing to zero. Under these conditions, the GW phase shift from Equation (22) reverts to the result $\phi = n_g \omega b/c$, the same as for a uniform material without vortices. The HFGW frequency at which this occurs depends upon both the applied field B and the superconductor thickness b, so that rather than evaluating this frequency it is more useful to analyse the performance expected, as in §3.4 below.

3.3. Theoretical Design of GW Variable-Focus Lens

In conventional geometric ray optics, as shown in Fig. **7**, a lens of focal length f produces an optical path difference of $\sqrt{(R^2 + f^2)} - f$ for rays at radius R from the optical axis. Expanding using the binomial theorem, this is approximately equivalent to an optical phase shift varying according to

$$\phi = \phi_o - \frac{\pi R^2}{\lambda_o f} \quad \textbf{(24)}$$

for a small radius $R \ll f$, where ϕ_o is a constant. This result is exploited in conventional ray optics by the well-known Fresnel lens and zone plate optical components [21].

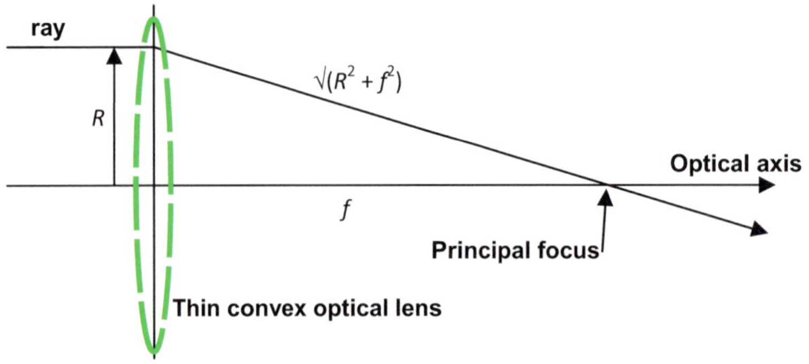

Figure 7: Optical lens interpreted as a phase shifter.

For a uniform value of applied magnetic field $B(R)$, the arrangement of Fig. **4** acts only as a variable GW phase shifter, but lens action results when the magnetic field $B(R)$ has appropriate non-uniformity. Suppose that the applied field $B(R)$ in Fig. **4** has cylindrical symmetry around the indicated cylindrical axis (*i.e.*, no dependence upon the azimuthal (rotational) angle). Then, equating the phase shift formulae in Eqns. (22) and (24), the required variation of applied magnetic field for producing a lens of focal length f is given by:

$$B(R) = B(0) + \frac{n_g \omega^2 \Phi_o}{2\sqrt{3}\pi^2 c^2 bf} R^2, \quad \textbf{(25)}$$

where $B(0)$ is the applied field on the cylindrical axis, and the phase shift on the cylindrical axis is given by

$$\phi_o = \left(\frac{n_g \omega}{c}\right) b - \left(\frac{\sqrt{3}\pi^2 c}{n_g \omega \Phi_o}\right) B(0) b. \quad \textbf{(26)}$$

A negative focal length (*i.e.*, a diverging lens) is obtained if B has its maximal value on the cylindrical axis; a convex lens is produced if B has its minimal value on the cylindrical axis.

In order to obtain a magnetic field that is cylindrically symmetric and also depends quadratically on R, there are several possibilities. Two of these are examined further below in this section.

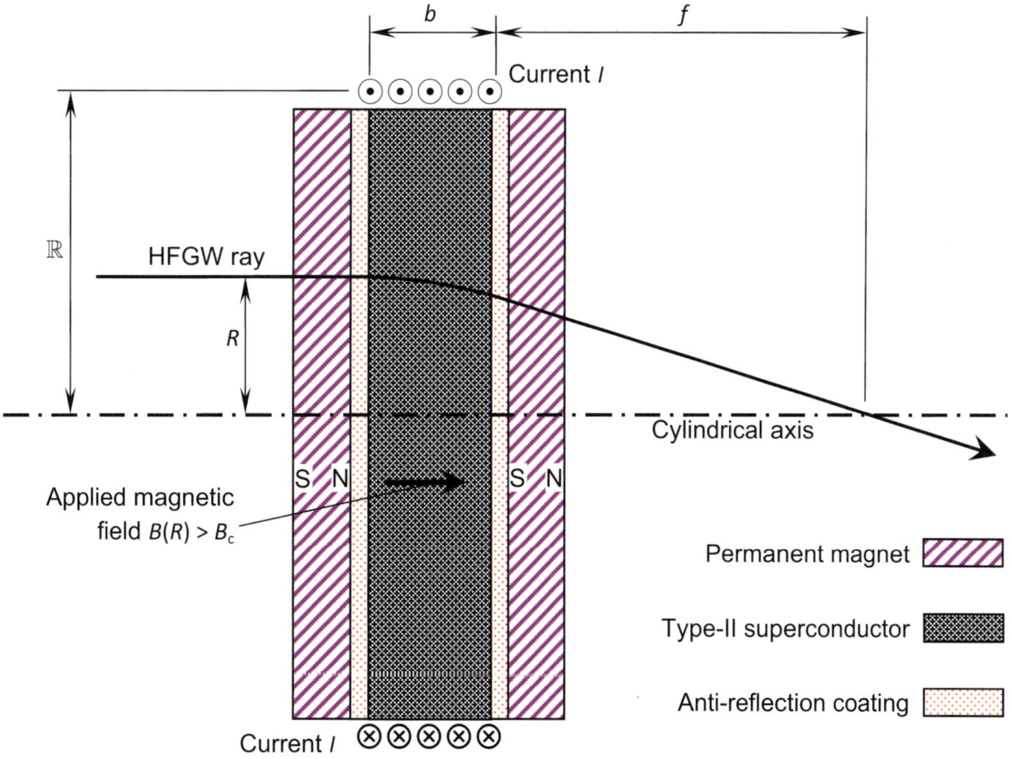

Figure 8: Basic HFGW controllable lens structure using encircling coil.

The first possibility is that the required magnetic field variation is produced by a coil of wire encircling the superconductor and concentric with the cylindrical axis, as shown in Fig. **8**. The total current surrounding the superconductor is I, equal to the current through each turn multiplied by the number of turns (all assumed to have radius **R**). The field variation must be superimposed upon a uniform background field slightly larger than the lower critical field B_c, which is usually around 0.1T for practical superconductors [30]. This value of background field is conveniently within the range available from permanent magnets which are fixed to each side of the superconducting disk. (Two permanent magnets produce a more uniform field within the superconductor than just one magnet).

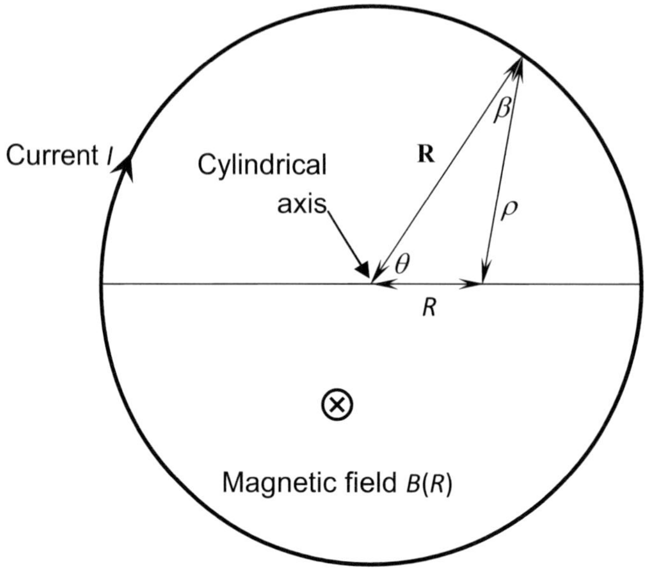

Figure 9: Calculation of the magnetic field in the plane of a thin circular current loop.

To find the focal length of the lens produced, it is necessary to find the field from the current loop in the loop plane at radius R from the cylindrical axis. This can be calculated as follows. The geometry of the loop appears as in Fig. **9** viewed directly along the cylindrical axis. The magnetic field at a point R from the centre of the loop is given by the Biot-Savart law [31]:

$$B(R) = \frac{\mu_o I}{4\pi} \int_{\theta=0}^{\theta=2\pi} \frac{\mathbf{R}\,\mathrm{d}\theta \cos\beta}{\rho^2}. \tag{27}$$

This integral cannot be evaluated exactly analytically, so an approximation is used here. Firstly, the integrand must be explicitly expressed in terms of **R**, R, and θ. Since **R** and R are constants within the integration, the integration may then proceed by integrating a power series in R. To achieve this, the sine law is applied to the triangle **R**, R, ρ, giving:

$$\sin\beta = R\frac{\sin\theta}{\rho}. \tag{28}$$

Then, applying the cosine law to side ρ gives:

$$\begin{aligned}\frac{\cos\beta}{\rho^2} &= \frac{\sqrt{1-\sin^2\beta}}{\rho^2} = \frac{\sqrt{\mathbf{R}^2 + R^2\cos^2\theta - 2\mathbf{R}R\cos\theta}}{\left(\mathbf{R}^2 + R^2 - 2\mathbf{R}R\cos\theta\right)^{\frac{3}{2}}} \\ &= \frac{1}{\mathbf{R}^2}\left(1 - 2\frac{R}{\mathbf{R}}\cos\theta + \frac{R^2}{\mathbf{R}^2}\cos^2\theta\right)^{\frac{1}{2}}\left(1 - 2\frac{R}{\mathbf{R}}\cos\theta + \frac{R^2}{\mathbf{R}^2}\right)^{-\frac{3}{2}}.\end{aligned} \tag{29}$$

For small values of $R \ll \mathbf{R}$, the parentheses in Equation (29) can be expanded using the binomial theorem to give:

$$\frac{\cos\beta}{\rho^2} = \frac{1}{\mathbf{R}^2}\left(1 + 2\frac{R}{\mathbf{R}}\cos\theta - \frac{R^2}{\mathbf{R}^2} - \tfrac{1}{2}\frac{R^2}{\mathbf{R}^2}\sin^2\theta\right) \tag{30}$$

as far as terms in R^2. With this result, Equation (27) becomes:

$$B(R) = \frac{\mu_o I}{4\pi} \int_{\theta=0}^{\theta=2\pi} \frac{d\theta}{\mathbf{R}} \left(1 + 2\frac{R}{\mathbf{R}}\cos\theta - \frac{R^2}{\mathbf{R}^2} - \frac{1}{2}\frac{R^2}{\mathbf{R}^2}\sin^2\theta \right) = \frac{\mu_o I}{2\mathbf{R}} \left(1 - \frac{5R^2}{4\mathbf{R}^2} \right), \tag{31}$$

which is valid for $R << \mathbf{R}$. (By symmetry, there can be no odd powers of R in the power-series expansion for the field). Therefore, equating the coefficients of R^2 in Eqns. (25) and (31), and assuming one flux quantum per vortex ($\Phi_o = \pi\hbar/e$), gives the GW focal length:

$$f = -\frac{4n_g \hbar\omega^2 \mathbf{R}^3}{5\sqrt{3}\pi\mu_o ec^2 Ib}. \tag{32}$$

The focal length is inversely proportional to the total circulating current I. At zero coil current, the focal length is infinite (*i.e.*, the structure has no diverging or converging properties), so by varying the coil current the lens power (= $1/f$) is continuously variable between zero and the value given by Equation (32). A short focal length (*i.e.*, a high power lens) is obtained by using a small value of the coil radius **R** and a thick slab of superconductor (a large value of b). In Equation (32), the sign of the focal length depends upon the direction of the current I, as this determines whether there will be a maximum or minimum value of magnetic field on the cylindrical axis. If the sign of the focal length requires that the field from the coil on the cylindrical axis opposes the static field from the permanent magnets, then the designer must ensure that the minimum field is always greater than the lower critical field B_c.

The maximum optical power (= $1/f$) of this lens design is limited because shorter focal length magnitudes, corresponding to greater lens power magnitudes, are obtained for lower GW frequencies $\nu = \omega/2\pi$. Hence, there is a minimum value of $\omega = 2\pi\nu_{\min}$ below which the analysis fails (see Equation (23)) and the structure no longer behaves as a lens as a result of the applied field (though it may act as a fixed-focus lens by virtue of its geometry). This maximum lens power is obtained by operating the variable-focus lens at the lowest GW frequency allowed by Equation (23), corresponding to the maximum waveguiding effect while the waveguiding approximation retains at least some semblance of validity. This hypothesis allows an estimate of the most powerful focal length possible by substituting the waveguiding condition, Equation (23), into the focal length formula, Equation (32). In addition, substituting the expressions for magnetic field (Equation (31)) and the vortex flux gives

$$f_{\min} \approx -\frac{\mathbf{R}^2}{5n_g b} \tag{33}$$

for the minimum magnitude of focal length possible for this design, in the case where there is no additional magnetic field applied through permanent magnets (Fig. **8**). If permanent magnets are used in addition to set up the static background magnetic field, then the minimum focal length will have larger magnitude.

The second design possibility examined here is illustrated in Fig. **10**. This eliminates the large surrounding large current-carrying coil and replaces it by a small coil of wire carrying a current and located on the cylindrical axis, which may be more convenient in certain applications. If the coil is small compared to the other dimensions in the experimental geometry, the coil may be approximated as a perfect magnetic dipole. The usual expression for the field from a perfect (small) magnetic dipole is [31]:

$$\mathbf{B} = \frac{\mu_o m}{4\pi\Re^3}\left(\mathbf{a}_\Re 2\cos\theta + \mathbf{a}_\theta \sin\theta\right), \tag{34}$$

where the dipole moment is $m = IA$ for a total current I (as before, measured as the current in each turn multiplied by the number of turns) enclosing an area A, the radial distance from the dipole is $\Re$, the vectors $\mathbf{a}_\Re$ and $\mathbf{a}_\theta$ are unit-length basis vectors for the coordinates $\Re$ and θ respectively, and θ is the angle that $\mathbf{a}_\Re$ makes to the dipole axis (the same as the cylindrical axis in the arrangement of Fig. **10**). In this application, the dipole field formula must be re-written in terms of the magnitude of the field at radius R from the dipole or cylindrical axis on a thin planar plate of superconductor situated at axial distance s from the dipole, since for small angles the GW phase shift is dependent upon the field total magnitude rather than on one resolved component. From Equation (34), using the small-angle trigonometric approximations (for $s >> b$ and $s >> R$) shows that for small R the total field magnitude is given by:

$$B(R) = \frac{\mu_0 m}{4\pi s^3}\left[2 - \frac{15}{4}\left(\frac{R}{s}\right)^2\right] = B(0) - \frac{15\mu_0 m R^2}{16\pi s^5}, \tag{35}$$

taking the expansion only as far as the term in R^2 (and there is no term in R^3 since by symmetry no odd powers of R can exist in the expansion).

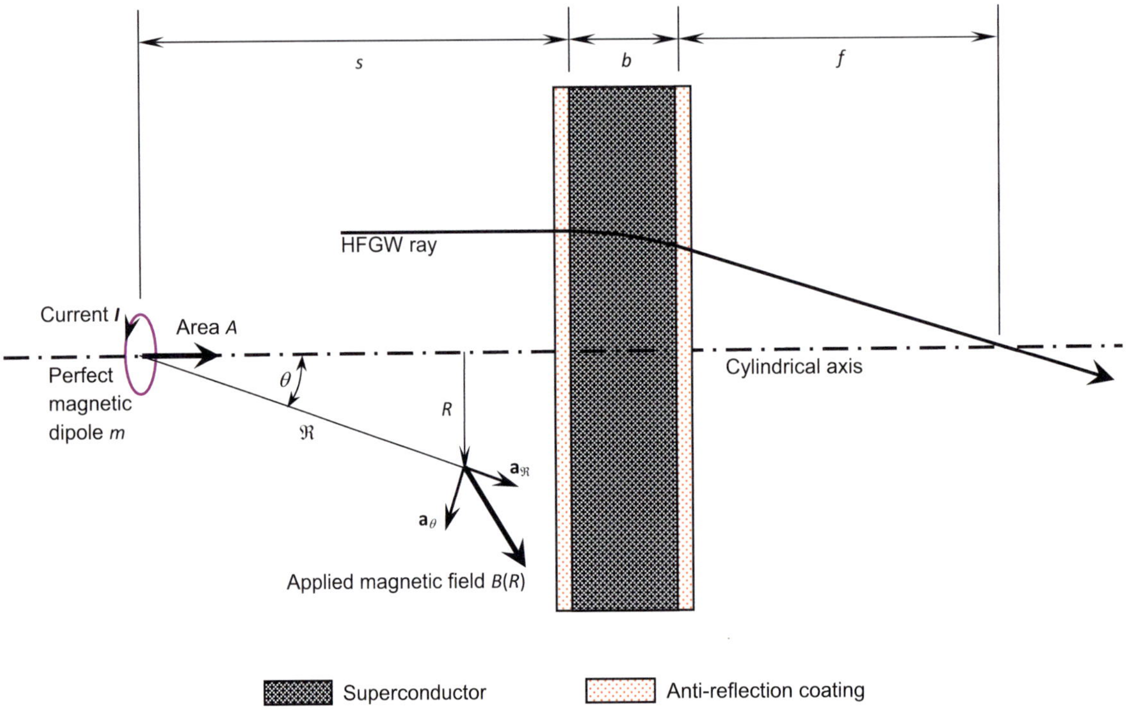

Figure 10: Basic HFGW variable-focus lens using magnetic dipole for producing the required magnetic field.

Therefore, equating Eqns. (25) and (35) gives the expression for the GW focal length f for this arrangement, provided $f >> b$:

$$f = -\frac{8n_g\omega^2\Phi_0 s^5}{15\sqrt{3}\pi\mu_0 c^2 bm} = -\frac{8n_g\omega^2\hbar s^5}{15\sqrt{3}\mu_0 c^2 ebIA}. \tag{36}$$

In this case, the focal length is still inversely proportional to the total current flow, *via* the magnetic dipole moment m; a short focal length is obtained by using a magnetic dipole close to the superconductor and a thick slab of superconductor. In this expression, if permanent magnets are *not* used to provide a static background field as shown in Fig. **8**, then the focal length will always be negative, giving a diverging lens for all parameter values. However, a variable focal-length converging lens can easily be constructed by making one of the superconductor faces convex rather than planar as assumed here for simplicity (or by adding a static field from permanent magnets). The usual formulae for combining optical lenses will apply. As previously for the first design, there is a limit to the optical power possible with this design,

$$f_{\min} \approx -\frac{s^2}{30 n_g b}. \quad \textbf{(37)}$$

Clearly, the most powerful focal length is an important parameter for the designer to consider, since it depends crucially upon the geometry used in each particular design.

3.4. Practical Design of GW Variable-Focus Lens

There are currently very large error limits upon the value of n_g predicted by Li and Torr [11], at least ±50% due to their use of linearised gravitational equations as theoretical approximations. The most common simple practical requirement will be for a lens to focus generated HFGW onto the sensitive aperture of (for example) a HFGW detector [7]. A variable-focus lens enables compensation for any manufacturing error upon commissioning the instrument. To allow for this uncertainty in a practical design, one strategy will be to combine a variable lens with a fixed-focus GW reflector so that the variable lens may be adjusted until the focal length of the combination is precisely what is required.

In both designs of variable lens, it is likely that there will be large Fresnel reflection effects from the front and back surfaces of the superconductor slab producing the focusing effect. Therefore, effective suppression of these reflections on both front and back surfaces will probably be essential, if these devices are to find practical utility. In practice, as examined above (§2.1), this may be difficult to achieve (even if an anti-reflection coating can be used), and the advantages of using a variable-focus GW lens need to be balanced against either the very significant insertion loss or the difficulties associated with reducing this loss to a more manageable magnitude. It is assumed here that these losses may be suppressed to an acceptable level by some means.

Suppose that a HFGW at frequency 4.9GHz [7] is to be manipulated by a lens of the first design, Fig. **8**. The gravitational refractive index n_g is the least accurately known design parameter at the present time, and in this section the value $n_g = 300$ is taken as previously. The other fundamental physical constants take their usual values; the remaining variable parameters available to the lens designer are the coil current I, the coil radius **R**, and the superconductor thickness b. The background uniform field from the permanent magnets serves only to create magnetic vortices in the superconductor; provided it is greater than B_c and less than the upper critical field, its exact value plays no further part in the lens operation. Table **1** shows some realistic selected values of these parameters, corresponding to typical design decisions that might be made. In practice, the magnetic drive coil can be made using many turns of wire so that the current drive actually necessary will be reduced from the values shown in Table **1** by a factor equal to the number of turns in the coil. Also shown in Table **1** are the resulting shortest value of the focal length calculated according to Equation (32), and the maximum aperture possible for each design, calculated from the maximum usable active radius arbitrarily chosen as **R**/10, so that the small-value approximations used in Equation (30) remain accurate. Notwithstanding these values, in practice the accuracy of Equation (24) (the description of a lens in terms of a phase shifter with quadratic radial variation of phase) may limit the maximum usable aperture to around *f*/5, depending upon the precision required. There is, of course, no need to fabricate either superconductor or magnets larger than the usable lens aperture, except for the purposes of optimising the background field uniformity.

Table 1: For typical parameters of the first lens design (Fig. **8**) and using $n_G = 300$, values of calculated focal length at 4.9GHz from Equation (32), maximum aperture for which Equation (30) is accurate, minimum GW frequency from Equation (23), and minimum focal length from Equation (33).

Coil current, I/A	Coil radius, R/m	Superconductor thickness, b/cm	Focal length, f/m	Maximum aperture	Minimum GW frequency, $\nu_{\min}$/GHz	Minimum focal length, $f_{\min}$/m
0.1	1	2	–0.12	*f*/0.6	2.6	–0.03
0.2	1.5	1.5	–0.27	*f*/0.9	3.0	–0.1
0.3	2	1	–0.65	*f*/1.6	3.1	–0.27
0.4	2.5	1	–0.95	*f*/1.9	3.2	–0.42
0.5	3	1	–1.3	*f*/2.2	3.3	–0.6

The design parameters used in Table **1** are reasonable and easy to obtain experimentally without presaging future developments necessary to enable this technology, and the resultant GW focal lengths are within the range of experimental values that will be useful to a designer of instruments and detectors of HFGW. Very large apertures are possible, which will have particular value in the design of instruments to detect HFGW of very low intensity, where both the lens speed and the availability of infinitely adjustable focusing are important.

For each design case in Table **1**, the field produced by the magnetic coil is around $1\mu T \approx 2\%$ of the earth's magnetic field. Although the earth's uniform field merely contributes an extra term to the value of $B(0)$ in Equation (29), because it is the field variations that are significant in this device then the non-uniformity of the earth's magnetic field may be a concern. This may require the lens to be surrounded by magnetic screening. Any fixed-focus superconducting HFGW lenses [17] should not need screening from the earth's magnetic field because this field is considerably smaller than B_c and so no magnetic vortices will be present.

For a lens of the second design, shown in Fig. **10**, the parameters available to the lens designer are the dipole coil current I, the dipole-superconductor distance s, the superconductor thickness b, and the dipole coil area A. Table **2** shows the corresponding results of calculating the focal length from Equation (36) using some realistic parameter values. Once again, reasonable performance is obtainable from reasonable design parameters.

Table 2: For typical parameters of the second lens design (Fig. **10**) and using $n_G = 300$, values of calculated focal length at 4.9GHz from Equation (36), maximum aperture for which Equation (35) is accurate, minimum GW frequency from Equation (23), and minimum focal length from Equation (37).

Coil current, I/A	Coil-superconductor distance, s/m	Superconductor thickness, b/m	Coil area, A/cm^2	Focal length, f/m	Maximum aperture	Minimum GW frequency, ν_{min}/GHz	Minimum focal length, f_{min}/m
0.05	0.1	0.01	1	–0.10	f/5.1	0.32	–0.00011
1	0.2	0.01	1	–0.16	f/4.1	0.51	–0.00044
10	0.5	0.003	1	–5.3	f/53	0.41	–0.0093
1	0.3	0.003	0.5	–8.3	f/138	0.20	–0.0033
2	0.4	0.003	0.5	–17	f/218	0.18	–0.0059

It will also be possible to utilise these designs in the case of much longer-wavelength GWs. The only limitation is the practical one of obtaining suitable sized superconducting materials, but this problem may be solved by using an array of a number of smaller superconducting tiles. If necessary, a superconducting tessellation may be supported by a non-superconducting framework as non-superconducting materials are completely transparent to GWs. In this case the second design (Fig. **10**) would have a significant practical advantage over the first design (Fig. **8**) because it eliminates the need for a very large circumscribed electric coil. Other advantages of the arrangement of Fig. **10** over that of Fig. **8** are that in Fig **10** the focusing is controlled remotely, and there is no need to make the superconductor precisely circular as in Fig. **8**. If remote focusing is not required, and if the cost of manufacture of a large and precisely circular superconductor and coil are not prohibitive, then the arrangement shown in Fig. **8** may be more suitable as the arrangement of Fig. **10** has the disadvantage that at long focal lengths the available aperture is limited and therefore only optically slow lenses are possible using that arrangement. In principle, the results are also applicable to GWs having frequencies greater than 100GHz [8], limited only by the fabrication precision needed to manufacture the superconducting components and to make a close approximation to a perfect magnetic dipole.

3.5. Reduced Value of Gravitational Refractive Index

The gravitational refractive index is not a parameter that can be varied by the designer; it appears to be a fundamental constant for superconductors. The current best-available value of the gravitational refractive index is $n_G \approx 300$ [11], assuming that there is an effect at all. However, at the present time, the controversy surrounding this value indicates that an interaction considerably smaller than this remains a distinct possibility. In other words, there are very large error limits upon the value of n_G. This would result in significantly reduced Fresnel reflection loss than predicted earlier. Some authors would claim that there is no effect, and so would simply set $n_G = 1$ [13, 14]. (For completeness, it must also be pointed out that it is also conceivable that n_G might be found to take a value considerably greater than 300, in which case a much greater effect than previously envisaged would be possible. This scenario appears to be extremely unlikely from the present perspective). Hence, it is prudent to examine the effects of a significantly reduced value of n_G (though not as low as unity, in which case there would be no possibility of any interactions). At the very least, such an examination will indicate the sensitivity of the predictions to the actual value of the material parameter that seems least controllable and least accurately known at the moment.

Table 3: For typical parameters of the first lens design (Fig. **8**) and using $n_G = 6$, values of calculated focal length at 200GHz (see text) from Equation (32), maximum aperture for which Equation (30) is accurate, minimum GW frequency from Equation (23), and minimum focal length from Equation (33).

Coil current, I/A	Coil radius, R/m	Superconductor thickness, b/cm	Focal length, f/m	Maximum aperture	Minimum GW frequency, ν_{min}/GHz	Minimum focal length, f_{min}/m
0.1	1	2	–4.1	f/20	130	–1.7
0.2	1.5	1.5	–9.1	f/30	150	–5.0
0.3	2	1	–22	f/54	160	–13
0.4	2.5	1	–32	f/63	160	–21
0.5	3	1	–44	f/73	170	–30

Since the value of n_G has extremely large error limits, it is a moot point exactly what alternative lower value of n_G might be used instead. As a purely arbitrary example, a value $n_G = 6$ gives Fresnel loss at each uncoated interface of 50% or 3dB and may afford a useful comparison with the previously assumed value $n_G = 300$. This example is chosen purely to illustrate how much the performance might change with a drastically reduced value of n_G that is nevertheless just within the bounds of possibility. Assuming this arbitrary lower value, to illustrate the dependence of the variable-focus lens performance upon the value of n_G used, the same parameters of Table **1** were used in the re-calculation of the lens performance shown in Table **3**. The reduced Fresnel loss in this case would make the lens insertion loss considerably smaller, so that this loss might even be tolerable in some lower performance applications without using any extra means to reduce it. On the other hand, as indicated previously, with such a small value of n_G then the manufacture and fabrication of an effective anti-reflection coating would be considerably eased in any case. Resulting from the reduced value of n_G, the variable-focal-length lens performance is significantly less, as would be expected, and the minimum GW frequency value is also much higher, so the parameters in Table **3** were calculated using a GW frequency of 200GHz. Although it has markedly reduced performance, obviously the device still operates as a feasible GW variable-focus lens offering useful potential in a typical receiver application [7].

4. RESONANT ENHANCEMENT OF GW-SUPERCONDUCTOR INTERACTIONS

The essence of the Li-Torr-Chiao conjecture is that superconductors hold a unique position in interacting with GW to a considerably greater degree than any other type of matter. If correct, this would be a remarkable conclusion that would surely hold the key to further scientific, technological, and ultimately economic development of this field. Nevertheless, at the present moment, this field is severely hampered by the absence of direct methods of generating and detecting GW of any frequency. In principle it should be possible to use mechanical resonance to enhance the GW interaction further. Mechanical resonances can now have extremely high Q values, sufficiently high to justify a rather different type of experiment from those suggested so far. This section examines the likely response of a high-Q mechanical resonator that is coated with a thin superconductor layer. Regardless of the veracity of the Li-Torr-Chiao conjecture, the resonator will be subjected to a periodic force, analogous to electromagnetic radiation pressure. If the Li-Torr-Chiao conjecture is untrue, the GW radiation pressure will certainly be minimal as it originates from the GW impedance mismatch at the interface between vacuum and resonator. It is widely accepted that the magnitude of this interaction involving ordinary masses (and, by extension, superconductors, if the Li-Torr-

Chiao conjecture is false) is so close to zero that it can safely be ignored except in the case of astronomical masses (or possibly large masses in a laboratory setting subjected to very sensitive instrumentation). However, if the Li-Torr-Chiao conjecture is correct, the magnitude of the GW radiation pressure is increased very significantly. Measuring the amplitude of the induced mechanical vibrations will indicate the power of the incident HFGW. If the resonator is a piezoelectric device then the output will be a direct electrical signal at double the HFGW frequency. The present problem, therefore, is to calculate the response of a high-Q mechanical resonator coated with a thin superconductor layer. The analysis is presented as an order-of-magnitude estimate of the likely observed effect.

4.1. Radiation Pressure from Incident Gravitational Wave

Disregarding the effects of GW polarization, which will act as a perturbation on the results and will therefore not be considered further here, the power (P) carried by a gravitational plane wave in the propagation direction through cross-sectional area A is given by the formula for the GW Poynting vector [32]:

$$P = \frac{c^3 \omega^2 A h^2}{32\pi G}, \tag{38}$$

where ω is the angular frequency of the gravitational wave, G is the Newtonian universal gravitational constant, and h is the effective dimensionless gravitational wave amplitude. This value may be divided by the energy carried by each graviton, $\hbar\omega$, to give the number of gravitons passing through unit area per unit time. Therefore, the mean time between graviton hits is (in an approximation not taking account of the statistical randomness of graviton arrival):

$$\tau = \frac{32\pi G \hbar}{c^3 \omega A h^2}. \tag{39}$$

Under perfect reflection conditions, such as would be the case if the Li-Torr-Chiao conjecture were correct, each graviton changes its momentum by $2\hbar\omega(\cos\theta_i)/c$ (where ω/c is the usual quantum-mechanical wavenumber) on hitting a perfect specularly reflecting mirror at angle of incidence θ_i to the normal (because the normal component of the motion of the quantum particle's is exactly reversed under these circumstances, whereas the parallel component is unchanged).

Even if the Li-Torr-Chiao conjecture were incorrect, formally any graviton reflected at the resonator surface would still impart a momentum change of $2\hbar\omega(\cos\theta_i)/c$ but only a very small proportion of the incident gravitons would be reflected under this hypothesis. To retain generality, the power reflection coefficient (or the quantum-mechanical probability of graviton reflection) will be denoted r^2, where $r^2 \approx 1$ if the Li-Torr-Chiao conjecture is correct giving almost perfect Fresnel refection, and $r^2 \approx 0$ if the Li-Torr-Chiao conjecture is incorrect. (Note that the conclusion $r^2 \approx 1$ if the Li-Torr-Chiao conjecture is correct is merely an extreme case of Fresnel reflection at large gravitational refractive index, and is not a result of total internal reflection).

Therefore, the average force produced by the gravitational wave is:

$$F = \frac{2\hbar\omega(\cos\theta_i)/c}{\tau} r^2 = \frac{c^2 \omega^2 A h^2 \cos\theta_i}{16\pi G} r^2. \tag{40}$$

Since this force value is a time average, the time-variation of the force will have an amplitude equal to F and will oscillate in time about the average value at angular frequency 2ω. A highly-resonant mechanical system tuned to angular frequency 2ω will therefore respond with enhanced amplitude to this particular stimulus. The basic concept is illustrated in Fig. **11**. The screen is included so that HFGW are incident only from one side of the resonator, making the drive unidirectional. The resonator is surrounded by superconductors acting as GW screens (which they must do if they behave as near-perfect reflectors of

GW) so that there is no HFGW path to the reverse side of the resonator. (Further exploitation of the GW-blocking properties of a superconductor is discussed in §4.5 below). Thermal (Boltzmann) excitation of HFGW cavity modes inside the superconducting chamber will need to be eliminated at the resonator frequency by adjusting the internal dimensions of the screening box appropriately.

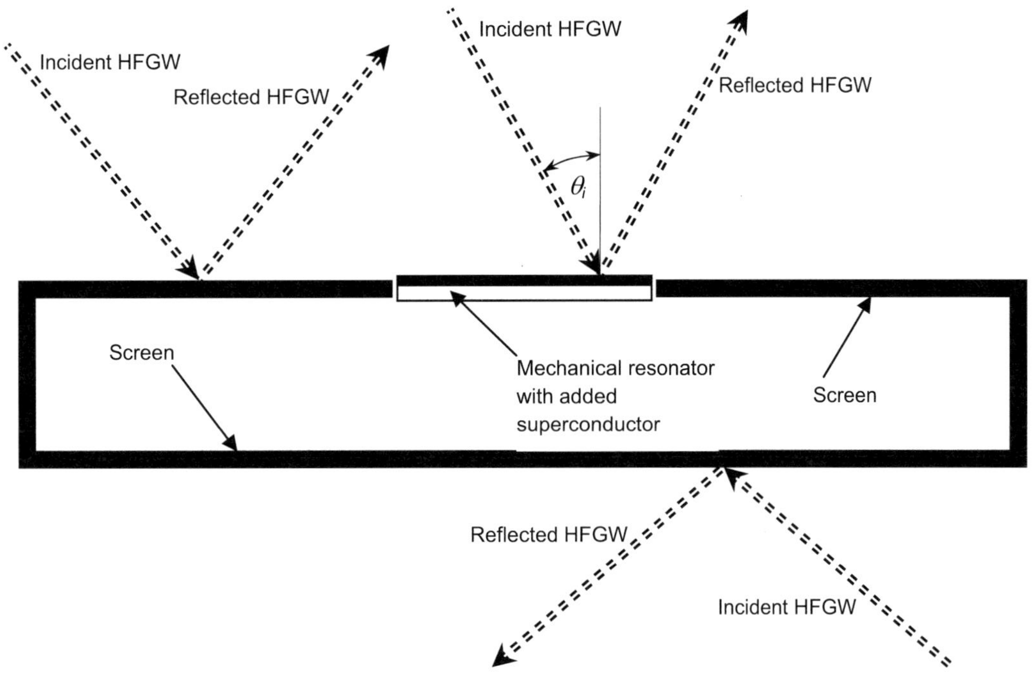

Figure 11: Concept of interaction between superconductor (= ▬) and HFGWs incident at random angles, enhanced by a mechanical resonator.

4.2. Acoustic Longitudinal Resonators

At the frequencies of interest here (HFGW in the low microwave band around several GHz) it seems likely that the most suitable resonators would be piezoelectric mechanical resonators, particularly Film Bulk Acoustic Resonators (FBARs). Exploratory FBAR devices have been fabricated operating at resonant frequencies up to 7.5GHz [33, 34], and represent the highest frequency mechanical resonators available other than atomic and molecular structures [8]. An FBAR (see Fig. **12**) is a mechanical (acoustic) resonator consisting of a vibrating membrane (typically around $100\times100\mu m^2$ in planform, and around $1\mu m$ thickness), fabricated using well-established integrated-circuit microfabrication technology. The motion of the vibrating membrane is sensed piezoelectrically, typically using aluminium nitride (AlN) as the piezoelectric material. These devices have in recent years been highly developed as cost-effective RF filters mainly for use in commercial cellular telephones, in which high-Q resonators are needed for the typical carrier frequency band around 1.9GHz. The mechanical resonator in a typical FBAR is an ideal resonator for the present application.

The important part of this mechanical resonator consists of a material that undergoes acoustic oscillations at a certain frequency. Ignoring damping, the material of the resonator obeys the mechanical wave equation [22] which may be written in simplified form in one dimension as:

$$C\frac{\partial^2 \dot{u}_x}{\partial x^2} = \rho\frac{\partial^2 \dot{u}_x}{\partial t^2}, \qquad \textbf{(41)}$$

where C is the elastic constant of the material, ρ is its mass density, and $\dot{u}_x$ is the particle velocity in the material. The x-axis is defined as in Fig. **12**. Only longitudinal vibrations are assumed, as is the case for practical resonators in the frequency range of interest here, so that the analysis can be kept within reasonable bounds. (This implies an assumption that no shear stresses or strains are present). This relation corresponds to a bulk acoustic phase velocity in the resonator material of

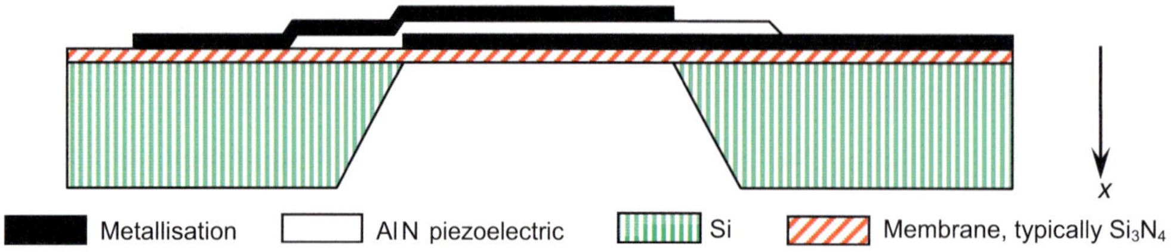

Figure 12: Basic FBAR construction (side view, not to scale).

$$V = \surd(C/\rho). \qquad \textbf{(42)}$$

Auld [22] showed that damping may be introduced by renormalizing the elastic constant as $C + j\omega_a\eta$, where η is the viscosity constant of the material and ω_a is the acoustic angular frequency. This is equivalent to modifying the wave equation to be

$$C\frac{\partial^2 \dot{u}_x}{\partial x^2} + \eta\frac{\partial}{\partial t}\frac{\partial^2 \dot{u}_x}{\partial x^2} = \rho\frac{\partial^2 \dot{u}_x}{\partial t^2}. \qquad \textbf{(43)}$$

The solutions of this equation are space- and time-harmonic, so that the formal equivalencies

$$\frac{\partial}{\partial x} = \mp jk \text{ and } \frac{\partial}{\partial t} = j\omega_a \qquad \textbf{(44)}$$

may be used. This gives the dispersion relation:

$$Ck^2 + j\eta\omega_a k^2 = \rho\omega_a^2. \qquad \textbf{(45)}$$

For small viscosity η (*i.e.*, a high-Q resonance) and expanding to first order only in η, this can be written as

$$k = \omega_a\sqrt{\frac{\rho}{C}}\left(1 + \frac{j\eta\omega_a}{C}\right)^{-\frac{1}{2}} = \omega_a\sqrt{\frac{\rho}{C}}\left(1 + \frac{j}{Q}\right)^{-\frac{1}{2}} \approx \omega_a\sqrt{\frac{\rho}{C}}\left(1 - \frac{j}{2Q}\right), \qquad \textbf{(46)}$$

where the resonator quality factor Q is defined by Auld [22] as

$$Q = \frac{C}{\omega_a\eta}. \qquad \textbf{(47)}$$

For the undamped case, *i.e.*, $\eta \to 0$, this gives $Q \to \infty$ and Equation (46) collapses to Equation (42).

Since the solutions of Equation (43) are harmonic, the general solution may be written as

$$\dot{u}_x = X\exp(j\omega_a t - jkx) + Y\exp(j\omega_a t + jkx), \qquad \textbf{(48)}$$

where X and Y are amplitude coefficients that are determined by the boundary conditions. As an approximation to find an order-of-magnitude response, the boundary conditions used here are that the oscillatory forcing term $F\exp(2j\omega t)$ (from Equation (40)) is applied at $x = 0$, and that zero force is applied at $x = L$, where L is the thickness of the resonator material. The x-stress in the resonator material, T_x, obeys the relation [22]

$$\frac{\partial T_x}{\partial x} = \rho \frac{\partial \dot{u}_x}{\partial t}, \tag{49}$$

so that at $x = L$ it is required that

$$0 = \frac{j\omega_a \rho X}{-jk} \exp(j\omega_a t - jkL) + \frac{j\omega_a \rho Y}{jk} \exp(j\omega_a t + jkL), \tag{50}$$

from which the relation between the two amplitude coefficients is obtained as

$$X = Y \exp(j2kL). \tag{51}$$

At $x = 0$ it is required that the x-stress in the resonator material corresponds to the external forcing term, giving

$$\frac{F}{A} = \frac{j\omega_a \rho X}{-jk} + \frac{j\omega_a \rho Y}{jk} = \frac{\omega_a \rho}{k}(Y - X) = \frac{\omega_a \rho Y}{k}\left[1 - \exp(j2kL)\right], \tag{52}$$

so that from Equation (48) the solution at $x = 0$ is

$$\dot{u}_x(0) = \frac{F}{A} \frac{k \exp(j\omega_a t)}{\omega_a \rho\left[1 - \exp(j2kL)\right]}\left[\exp(j2kL) + 1\right] = -\frac{F}{A} \frac{k}{j\omega_a \rho \tan(kL)} \exp(j\omega_a t), \tag{53}$$

and the solution at $x = L$ is

$$\begin{aligned} \dot{u}_x(L) &= \frac{F}{A} \frac{k \exp(j\omega_a t)}{\omega_a \rho\left[1 - \exp(j2kL)\right]}\left[\exp(j2kL)\exp(-jkL) + \exp(+jkL)\right] \\ &= -\frac{F}{A} \frac{k}{j\omega_a \rho \sin(kL)} \exp(j\omega_a t). \end{aligned} \tag{54}$$

It follows that the forced displacement at $x = 0$ and at $x = L$ may be written as

$$u_x(0) = \frac{F}{A} \frac{k}{\omega_a^2 \rho \tan(kL)} \exp(j\omega_a t) \text{ and } u_x(L) = \frac{F}{A} \frac{k}{\omega_a^2 \rho \sin(kL)} \exp(j\omega_a t) \tag{55}$$

respectively. For the structure to resonate, the thickness (L) must be a half-integral number of acoustic wavelengths (the same criterion that applies to the length of a piano string), or

$$L = i\pi|V|/\omega_a, \tag{56}$$

where i is an integer giving the number of half-wavelengths in the vibration mode. This condition allows (kL) to be close to ($i\pi$) so that the denominators in Equation (55) are small and the mechanical amplitude can be correspondingly large, governed by the value of Q. In this case, from Equation (46), and upon expanding to first order in ($1/Q$), the value of $\tan(kL)$ is given by

$$\tan(kL) = \tan\left[\omega_a \sqrt{\frac{\rho}{C}}\left(1 - \frac{j}{2Q}\right)\frac{i\pi|V|}{\omega_a}\right] = \tan\left[i\pi\left(1 - \frac{j}{2Q}\right)\right] \approx -\frac{ji\pi}{2Q} \tag{57}$$

and the value of $\sin(kL)$ is given by

$$\sin(kL) = \sin\left[i\pi\left(1-\frac{j}{2Q}\right)\right] \approx -\frac{ji\pi}{2Q}(-1)^i, \tag{58}$$

where $(-1)^i$ indicates an oscillating sign depending upon the value of i.

4.3. Acoustic Resonator Forced by HFGW Radiation Pressure

Combining the results from Eqns. (40), (55), (57), and (58), the vibration amplitude at $x = 0$ produced by an incident HFGW at precise acoustic resonance is approximately given by

$$|u_x(0)| = \frac{c^2\omega^2 h^2 \cos\theta_i}{16\pi G} r^2 \times \frac{2Qk}{i\pi\omega_a^2\rho}, \tag{59}$$

and the amplitude $|u_x(L)|$ at $x = L$ has the same magnitude but the vibration is in exact antiphase to that at $x = 0$ for odd values of the mode integer i, and is exactly in phase for even values of i. This is similar to the well-known behaviour of transverse acoustic resonators such as piano strings. Since this mechanical resonator is excited so as to oscillate at angular frequency $\omega_a = 2\omega$, the mechanical vibration amplitude at the two ends of the longitudinal resonator, $x = 0$ and $x = L$, becomes

$$|u_x| = \frac{c^2h^2(Q\omega_a)\cos\theta_i}{32i\pi^2G\rho V} r^2. \tag{60}$$

To maximise this amplitude, clearly the largest possible value of the product $(Q\omega_a)$ at the fundamental resonance $(i = 1)$ is best. From Eqns. (55), (57) and (58), the value $i = 1$ signifies a mode where the mechanical oscillations are in antiphase at the two ends of the resonator. This corresponds to the case where the material thickness is equal to half an acoustic wavelength in that material, and of an oscillation at its fundamental longitudinal mode parallel to its thickness dimension. Note that this is not the same as a diaphragmatic vibration where the restoring force arises from the bending of the cantilevered diaphragm, resulting in a much lower resonant frequency. Fig. **13** shows schematically the difference between these modes. Hence, the total thickness of the resonator oscillates with an amplitude of twice the value given by Equation (60). To sense this oscillation, the resonator is coated with a piezoelectric material whose strain is governed by the resonator material. The overall mechanical strain, S_x, conveyed to the piezoelectric material oscillates with amplitude given by

$$|\Delta S_x| = \frac{c^2h^2Q\omega_a^2\cos\theta_i}{16\pi^3G\rho V^2} r^2 = \frac{c^2h^2\omega_a\cos\theta_i}{16\pi^3G\eta} r^2. \tag{61}$$

Minimising the viscosity constant η, or alternatively maximising the device parameter $Q/(\rho V^2)$, will produce the largest strain amplitude; in practice, the Q of a particular resonator technology is frequently more often recorded than η. The superconducting HFGW mirror might be used as one electrode of the two needed to sense the voltage generated by the piezoelectric material. The other electrode might also be superconductor, or could be an ordinary evaporated or sputtered metal thin-film as used in conventional FBAR fabrication. To estimate the piezoelectric potential generated, a simple model will be used here. In practice, the shunt capacitance of the sensing electrodes will also be important depending upon the load impedance of the sensing amplifier and the area A of the resonator, both of which are unknown design parameters at this stage. However, since the resonator area does not appear in Eqns. (59), (60), or (61), the smallest practical resonator area can be used (just large enough to render unwanted edge effects negligible) to minimise the shunt capacitance.

The piezoelectric material conforms to the mechanical strain given by Equation (61) and the mechanical stress (**T**) is assumed to be zero, corresponding to a thin piezoelectric film with low elastic constant. This approximation simplifies the estimation of the resulting piezoelectric voltage generated. The constitutive relation for the mechanical strain, essentially defining the piezoelectric coefficient **d**, is (using abbreviated subscript tensor notation) [22]:

$$\mathbf{S} = \mathbf{d} \cdot \mathbf{E} + \mathbf{C}^{-1}\big|_E : \mathbf{T} \approx \mathbf{d} \cdot \mathbf{E}\,, \qquad (62)$$

where **E** is the electric field. To find the oscillatory voltage amplitude φ generated due to the strain, the piezoelectric film thickness is needed. This is probably of the same order as the membrane thickness (L) and if the two are assumed equal then

$$\varphi = |L\mathbf{E}| = \left|\frac{\pi V \mathbf{S}}{\omega_a d}\right| = \frac{c^2 h^2 (Q\omega_a)\cos\theta_i}{16\pi^2 dG\rho V} r^2\,. \qquad (63)$$

Different thicknesses of the piezoelectric material will result in different values of generated voltage, but this is a design parameter fixed at the resonator design stage.

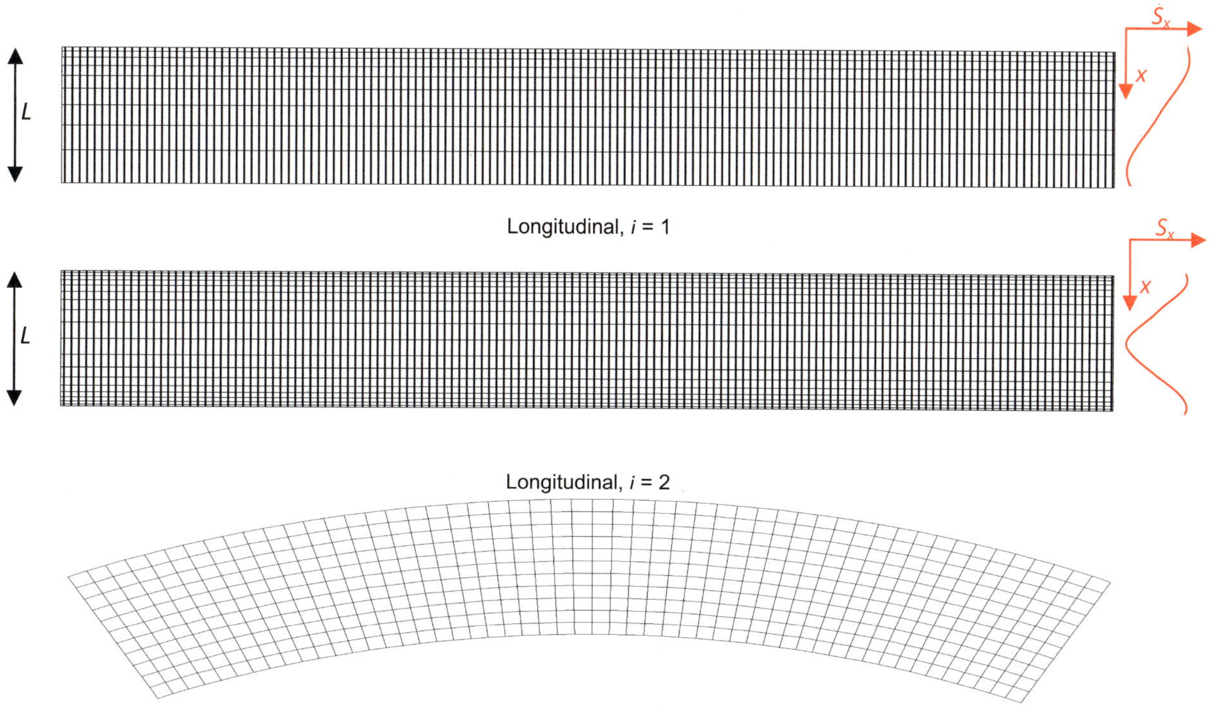

Figure 13: Comparison of resonator modes. (Curves on the right indicate schematically the instantaneous material strain as a function of x).

4.4. Numerical Estimates of Voltage Generated

Using Equation (63), values of the piezoelectric voltage produced can be estimated for materials that might be used for fabricating resonators intended for this application. Using the approximate estimate outlined above (§4.3), the only important design parameter relevant to the piezoelectric material is its piezoelectric coefficient d, which is inversely proportional to the output voltage. Of the piezoelectric materials commonly used in MEMS applications currently, the Group III nitrides have the smallest values, around $d \approx 4\text{pm V}^{-1}$ for both AlN and GaN [35]. This value is used in the numerical estimates below in this section; this is an approximation because the effective (renormalized) value of the piezoelectric coefficient also has a dependence upon the substrate material as a result of the mechanical constraints imposed [36], but detailed data on this small correction is not yet available for all likely membrane materials. The results for a number of resonator materials are summarised in Table **4**. In all cases, $r = 1$ was assumed, of course.

From Equation (47), the product ($Q\omega_a$) is a constant for any given resonator material, so that the detection frequency is actually immaterial as far as estimated sensitivity is concerned but it is included in Equation (63) because typical available data lists the material Q at a specified frequency. The dimensionless HFGW amplitude $h = 10^{-26}$m/m was chosen arbitrarily; this value is slightly greater than the cosmological relic HFGW amplitude expected at around 10GHz [37].

Table 4: Values of generated piezoelectric voltage calculated from Equation (63) for various resonator membrane materials, at incident HFGW amplitude $h = 10^{-26}$m/m.

Material	Data source	Density, ρ/kg m^{-3}	Velocity, V/m s^{-1}	Measured Q...	...at frequency/GHz	Voltage, φ/pV
Si_3N_4/AlN	[34, 38]	3290	9707	100	1.9	7.98
silicon	[39]	2330	8540	2800	1.0	189
sapphire	[39]	3980	11500	48000	1.0	1410
diamond	[40]	3520	18000	700	1.23	18.2
YAG	[22]	4550	7675	49200	1.0	1890

Even with 100% reflection of gravitons at normal incidence ($\theta_i = 0$), giving the largest possible elastic interaction with the HFGW, the results are clearly not large. (Incidentally, this observation indicates that without a significant breakthrough in generating HFGW at high power there could be no significant exploitable propulsion effects arising from HFGW reflection). Resonator operation would need to be in a vacuum to maximise Q, to reduce Brownian motion contributing to the measured noise floor, and also to prevent condensation forming on the resonator at the reduced temperature necessary for superconductor operation. The membrane materials providing the largest resonant effect are clearly sapphire and YAG (yttrium aluminium garnet). This results from their very large acoustic Q factors (*i.e.*, low loss and low viscosity constant η) giving significant magnification of the effects of radiation pressure on a mechanical structure.

The largest predicted voltages generated are in the nV range and may be detectable with suitable low-noise receivers operating under narrow bandwidth. The usual Johnson noise formula $4k_BT_AR_L$ (where k_B is Boltzmann's constant, T_A is absolute temperature, and R_L is the load resistance) for the mean square thermal noise voltage generated in unit bandwidth shows that in a nominal 1Hz receiver bandwidth, and for a receiver cooled to 77K (readily obtainable using liquid nitrogen refrigerant, and an adequate temperature for operating the "high-temperature" superconductors), the receiving amplifier's Thévenin equivalent input load resistance R_L must be less than ~250Ω to detect the voltages produced from HFGW in sapphire or YAG resonators without being hopelessly overwhelmed by thermal noise. On cooling to 4.2K (significantly more difficult than 77K though routine in low-temperature laboratories), the input load resistance R_L must be less than ~4500Ω to avoid thermal noise swamping the HFGW signal.

4.5. Use of Superconducting GW Chopping

The basis of the results in the previous section (§4.4) is the assumption that superconductors reflect HFGW with ~100% efficiency. This requires that a superconductor must block GW transmission equally efficiently. If this is true (with all of the caveats that apply to the assumption that superconductors act as efficient GW reflectors) then this effect can also be exploited. An elementary example has already been shown in Fig. **11** where superconductors are to be used as stationary GW shields around the GW-sensing resonator. A more sophisticated example is described in this section: reduction of the GW detector bandwidth (potentially to as low as a fraction of 1Hz) using a mechanical superconducting chopper wheel, to reduce random noise assuming that wide bandwidth is not needed. This would be similar to the technique used in conventional optical signal retrieval systems but using a rotating sectored superconducting disc rather than a rotating optical shutter.

This principle is illustrated in Fig. **14** where, rather than true radial sectors, the wheel is shown with a number of equally-spaced high-T_c superconducting discs fixed on it. (Small superconducting discs may be easier to fabricate in practice than sectors, though this has the practical disadvantage that the mark/space ratio of the chopped signal then depends on the GW beam position. Careful control of the GW beam collimation may mitigate this though possibly at the expense of reduced detector aperture). GW travelling on the required path are unambiguously "labelled" by the rotating chopper wheel. When the chopper wheel rotates, and if the superconductor behaves as a perfect mirror thus blocking the GW, the GW is pulse-modulated at a constant repetition frequency determined by the rotation speed and the number of superconducting sectors fixed on the wheel. Ideally, the diameters of the individual superconducting discs or sectors are large compared to the GW wavelength in free space, so that there is little or no diffraction around the vanes of the chopper. If the superconducting vanes are too small, GW diffraction will reduce the depth of modulation. This could be improved to some extent by placing a superconducting collimation aperture for the GW beam adjacent to the chopper wheel.

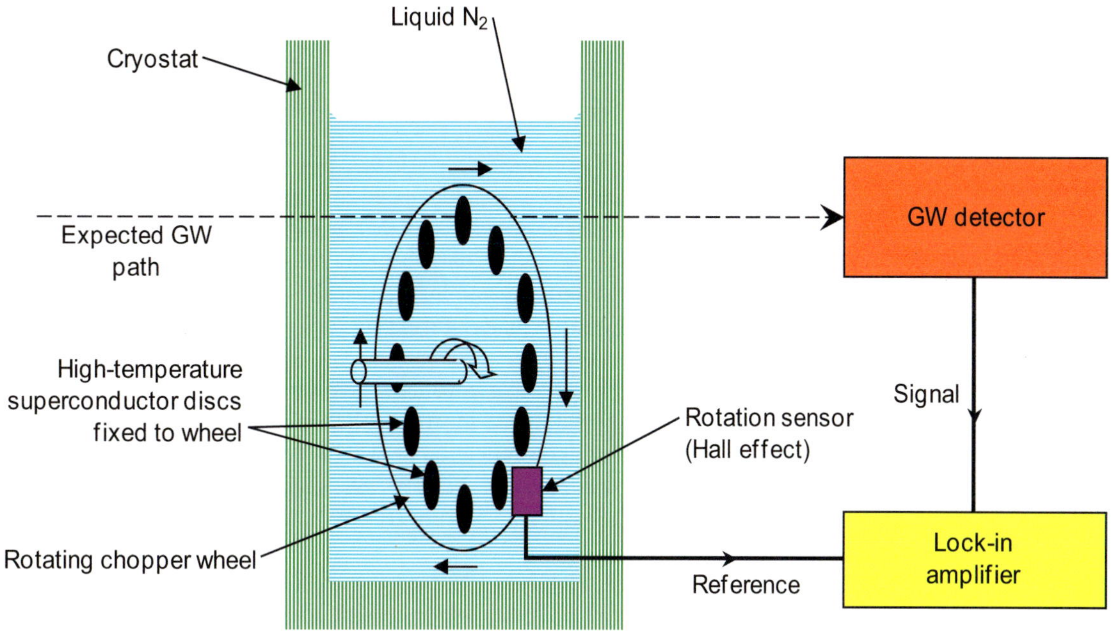

Figure 14: Basic idea of using a superconducting GW chopper to improve sensitivity of a GW detector.

A standard lock-in (synchronous) amplifier is referenced to the rotation of the sectored superconducting disc; chopping frequencies of up to several 10s of Hz are typical in optical systems. The reference signal could conveniently be obtained by using a Hall sensor, since the superconducting sectors are perfect diamagnets and so will produce a significant disturbance in a magnetic field in their immediate vicinity. (A Hall sensor would be preferred to optical pickup of the reference signal under cryogenic conditions). In the GW case there is the extra complication of having to operate the superconducting chopper cryogenically, but in principle this is feasible at 77K using high-temperature superconductors, especially considering that the superconducting chopper disc can be strengthened mechanically using ordinary material without compromising its GW performance. High-T_c superconducting material is favoured in this application because of the much greater difficulties of constructing such a moving device at 4.2K in liquid helium as would be needed for low-T_c material. The wheel should be rotated by a motor connected through a chain drive or gear train (of course no elastic driving belt will retain its flexibility when submerged in liquid nitrogen) or possibly even a direct-drive motor.

If the Li-Torr-Chiao conjecture is true, this technique would allow signal recovery with very much reduced bandwidth and hence a very low noise floor using synchronous detection, and if successful would be a direct demonstration of the superconducting vanes serving to block GW. In fact, assuming the veracity of the Li-Torr-Chiao conjecture, this approach would drastically reduce the noise floor of any other GW detector as well, not just the arrangement described in the previous section.

5. CONCLUSIONS

Li and Torr [11] predicted a substantial reduction in phase velocity for HFGWs travelling within a superconductor material. This result has been disputed but currently has not been conclusively confirmed or disproved, and so is open to confirmation or refutation. One way to shed light upon this controversy is to deduce some consequences that may be tested rather more easily than the basic result concerning phase velocity may be tested with current technology. This chapter has examined some of those consequences.

Fresnel reflection at the air-superconductor surfaces will be almost 100% unless an effective anti-reflective coating material can be developed; this material would need a precisely defined HFGW refractive index and also a precisely controlled thickness of one quarter of the HFGW wavelength inside the coating. The tolerances on fabricating such a coating may be prohibitive even if a suitable material (such as a suspension of superconductor particles in an inert carrier) could be found and developed. If such a coating were feasible, the unique refractive properties of the superconductor indicate that all incident HFGW rays will be refracted very close to the normal, and the critical angle for total internal reflection will be 0.2°.

Such a small critical angle leads to unique lens geometries for HFGW. High-performance HFGW lenses may be possible using superconductor, but only if a suitable material for "optical blooming" can be developed. Given all of these difficulties, it appears to be more practical for HFGW optics to use reflection techniques universally, as even without using a reflection-enhancing coating the Fresnel reflection coefficient is very close to unity. For use in multiple-reflection instruments it may be useful to develop such a reflection-enhancing coating, consisting of the same material used for anti-reflection coating but applied at a uniform thickness of one half the HFGW wavelength inside the coating. Since the reflection coefficient is already so close to unity, the requirements of a reflection-enhancing coating are not nearly so stringent as those of an anti-reflection coating. Both high-performance mirrors (spherical and parabolic) and resonant cavities appear to be feasible components relying upon reflection of HFGW from superconductor surfaces. In particular, a HFGW Fabry-Perot étalon designed for HFGW use seems likely to outperform an equivalent optical étalon and may be used in applications requiring HFGW tuning.

Two concept designs for an infinitely-variable-focus lens component suitable for use with HFGWs have been developed. One allows remote control of the lens power, which may be advantageous in some instrument configurations. The power of the lens is adjustable by varying the current drive to a magnetic coil producing a non-uniform magnetic field over the face of a plane disk of type-II superconductor such as YBCO. The operation of this component relies upon the result of Li and Torr [11] that HFGWs are slowed within a superconductor. Although the precise magnitude of the Li and Torr effect is currently unconfirmed, if this effect occurs at all then fabrication and demonstration of this novel component will be possible. Therefore, in this sense, this component may also be regarded as a possible test of the Li and Torr result, as an alternative or in addition to simply measuring the HFGW velocity directly through a superconductor. Accurate analysis is difficult and an approximate treatment based upon the well-known analysis of fibre-optic waveguides has been used. This should be accurate to within the limits of the Li and Torr result. Numerical simulation may provide a more sophisticated treatment, though probably such a detailed analysis will not be productive until the magnitude of the Li and Torr effect has been established more precisely than at the moment. The underlying Li and Torr effect implies performance limitations that have been estimated and shown to present some practical concerns for the designer, but these concerns are not serious enough to prevent operation of the device and to give useful performance in a typical application.

In these types of variable-focus GW lenses, GW frequencies below the minimum frequency identified here will not be phase-shifted controllable by a magnetic field as needed for technological exploitation. GW at frequencies below this minimum will still be subject to the Li and Torr effect [11] (if it exists at all, of course), but for frequencies less than a few GHz the wavelengths become too long for comfortable handling by superconductors (necessitating huge installations). Low frequency GW might still be a useful technology if an effective method of controlling them could be conceived. However, the ν^6 dependence of the total quadrupole GW power radiated by a rotating rod [41] means that the sensitivity at low frequencies will be minute compared even to the likely low system sensitivity at high frequencies [7], for which currently a complete terrestrial transmission-detection system may be realisable with known technology only with considerable outlay and component development.

The consequences of a significantly reduced magnitude of the Li and Torr effect have also been explored and shown to reduce the performance of the device significantly as would be expected since the devices rely heavily upon this effect for their operation, but significant performance is still predicted. The importance of this work is to establish the sensitivity of the design objectives on the various material parameters that cannot be readily controlled by the designer.

Such lens concepts may be combined with other fixed (or variable) focusing arrangements to produce versatile GW lens components. Technologically useful values of focal length are readily achievable using design parameters that would be typical for a designer having superconductors and electrical components readily available. This presupposes that Fresnel reflection effects can be reduced to insignificance using some artifice.

This development allows possibilities unparalleled in conventional optics because the focal length may be varied completely electrically, with no mechanical moving parts. Since there are no moving parts, the focal length may be controlled with much greater reliability than could be obtained by making a GW analogue of an equivalent optical glass lens design. Control can also be considerably faster, since its maximum speed is limited only by electronic circuit design, whereas the speed of conventional moving-optics designs is determined by the limited speeds of mechanical actuators and motors. Therefore, fast auto-focusing, zooming, and imaging tomography using electronic servos may become possible following development of the necessary HFGW detectors.

As a test of the Li-Torr-Chiao conjecture, radiation pressure from a HFGW could be used to excite a mechanical resonator at double the HFGW frequency. The natural resonant properties of the resonator material give magnification of up to several orders of magnitude of the force applied to the resonator material. A brief survey of a small number of common resonator materials indicates that a resonant membrane fabricated from sapphire or YAG, combined with a piezoelectric AlN or GaN layer, may be the most suitable. Even so, the effects are not large, and their unambiguous observation would need careful attention to noise levels such as the use of lock-in techniques to reduce the effective receiver bandwidth. Used as a HFGW detector for HFGW amplitude $h = 10^{-26}$m/m, this arrangement is projected to give a voltage output greater than 1nV. Since these effects are very small it seems unlikely that significant propulsion effects could be obtained using HFGW without a transformational improvement in high-power HFGW generation techniques.

REFERENCES

[1] Einstein A. Näherungsweise Integration der Feldgleichungen der Gravitation, Sitzungsberichte der Königlich Preußischen Akademie der Wissenschaften, Berlin, 1916; 688-696.

[2] Forward RL and Miller LR. Generation and Detection of Dynamic Gravitational-Gradient Fields. J Appl Phys 1967; 38: 512-518.

[3] Romero BF and Dehnen H. Generation of Gravitational Radiation in the Laboratory. Z Naturforsch 1981; A 36: 948-955.

[4] Hulse RA. The Discovery of the Binary Pulsar. Rev Modern Phys 1994; 66: 699-710.

[5] Taylor JH. Binary Pulsars and Relativistic Gravity. Rev Modern Phys 1994; 66: 711-719.

[6] Douglass DH and Braginsky B. Gravitational-Radiation Experiments. In: General relativity: an Einstein centenary survey. Eds. Hawking SW and Israel W, CUP 1979; pp. 90-137.

[7] Woods RC and Baker RML. Gravitational Wave Generation and Detection Using Acoustic Resonators and Coupled Resonance Chambers. In: Proc. AIP Space Technology and Applications International Forum, Albuquerque, NM, 2005; Am Inst Phys 2005; 746: 1298-1305.

[8] Woods RC and Baker RML. Generalized Generators of Very-High-Frequency Gravitational Waves. In: Proc. 1st AIP Space, Propulsion and Energy Sciences International Forum, Huntsville, Al, 2009; American Inst Phys. 2009; 1103: 515-523.

[9] Weinberg S. Gravitation and Cosmology: Principles and Applications of the General Theory of Relativity. Wiley 1972.

[10] Wald RM. General Relativity. University of Chicago Press 1984; section 4.4.

[11] Li N and Torr DG. Gravitational Effects on the Magnetic Attenuation of Superconductors. Phys Rev 1992; B 46: 5489-5495.

[12] Torr DG and Li N. Gravitoelectric-Electric Coupling *via* Superconductivity. Found Phys Lett 1993; 6: 371-383.

[13] Kowitt M. Gravitomagnetism and Magnetic Permeability in Superconductors. Phys Rev 1994; B 49: 704-708.

[14] Harris EG. Comments on 'Gravitoelectric-Electric Coupling *via* Superconductivity' by Douglas G. Torr and Ning Li. Found Phys Lett 1999; 12: 201-208.

[15] Baker RML, Davis EW and Woods RC. Gravitational Wave (GW) Radiation Pattern at the Focus of a High-Frequency GW (HFGW) Generator and Aerospace Applications. In: Proc. AIP Space Technology and Applications International Forum, Albuquerque, NM, 2005; Am Inst Phys 2005; 746: 1315-1322.

[16] Modanese G and Fontana G. Effect of the Vacuum Energy Density on Graviton Propagation. In: Proc. AIP Space Technology and Applications International Forum, Albuquerque, NM, 2004; Am Inst Phys 2004; 699: 1198-1205.

[17] Woods RC. High-Frequency Gravitational Wave Optics. In: Proc. AIP Space Technology and Applications International Forum, Albuquerque, NM, 2006; American Inst Phys 2006; 813: 1305-1312.

[18] Minter SJ, Wegter-McNelly K and Chiao RY. Do Mirrors for Gravitational Waves Exist? Physica E 2010; 42: 234-255.

[19] Chiao RY and Steinberg AM. Progress in Optics. Ed. E. Wolf. North-Holland 1997; 37: 345.

[20] Chiao RY. Quantum Incompressibility of a Falling Rydberg Atom, and a Gravitationally-Induced Charge Separation Effect in Superconducting Systems. arXiv 2010: 1003.1149. Available from http://arxiv.org

[21] Saleh BEA and Teich MC. Fundamentals of Photonics. Wiley-Interscience 1991; pp. 6, 18-26, 59, 95, 116, 205, 279-285, 312.

[22] Auld BA. Acoustic Fields and Waves in Solids, Vol. I 2nd. Ed. Krieger 1990; pp. 43, 66, 92, 115, 124-131, 170, 171, 279, 367.

[23] Baker RML, Woods RC and Li F-Y. Piezoelectric-Crystal-Resonator High-Frequency Gravitational Wave Generation and Synchro-Resonance Detection. Proc. AIP Space Technology and Applications International Forum, Albuquerque, NM, 2006; Am Inst Phys 2006; 813: 1280-1289.

[24] Gertsenshtein ME. Wave Resonance of Light and Gravitational Waves. Sov Phys JETP 1962; 14: 84-85.

[25] Zel'dovich YaB. Electromagnetic and Gravitational Waves in a Stationary Magnetic Field. Sov Phys-JETP 1974; 38: 652-654.

[26] Schmüser P. Basic Principles of RF Superconductivity and Superconducting Cavities. In: CAS-CERN Accelerator School: Intermediate Course on Accelerator Physics. Zeuthen, Germany, 15-26 Sep 2003; pp. 183-202. Available from http://www.aps.anl.gov/News/Conferences/2004/RFSC-Limits/schmeuser.pdf

[27] Arovas DP, Berlinsky AJ, Kallin C and Zhang S-C. Superconducting Vortex with Antiferromagnetic Core. Phys Rev Lett 1997; 79: 2871-2874.

[28] Crabtree GW, Leaf GK, Kaper HG, Vinokur VM, Koshelev AE, Braun DW, Levine DM, Kwok WK and Fendrich JA. Time-Dependent Ginzburg-Landau Simulations of Vortex Guidance by Twin Boundaries. Physica C 1996; 263: 401-408.

[29] Li FY, Tang M and Zhao P. Interaction Between Narrow Wave Beam-Type High Frequency Gravitational Radiation and Electromagnetic Fields. Acta Physica Sinica 1992; 41: 1919-1928.

[30] Liang R, Dosanjh P, Bonn DA, Hardy WN, and Berlinsky AJ. Lower Critical Fields in an Ellipsoid-Shaped $YBa_2Cu_3O_{6.95}$ Single Crystal. Phys Rev B 1994; 50: 4212–4215.

[31] Cheng DK. Field and Wave Electromagnetics, 2nd. Ed. Addison-Wesley 1989; pp. 235, 240.

[32] Hobson MP, Efstathiou G and Lasenby AN. General Relativity. CUP 2006; p. 513.

[33] Ruby R and Merchant P. Micromachined Thin Film Bulk Acoustic Resonators. Proc. 48th IEEE Int. Frequency Control Symposium 1994; 135-138.

[34] Lakin KM, Belsick J, McDonald JF and McCarron KT. High Performance Stacked Crystal Filters for GPS and Wide Bandwidth Applications. Proc IEEE Ultrasonics Symposium 2001; 833-838.

[35] Bernardini F and Fiorentini V. First-Principles Calculation of the Piezoelectric Tensor d of III–V Nitrides. Appl Phys Lett 2002; 80: 4145-4147.

[36] Tonisch K, Cimalla V, Foerster C, Romanus H, Ambacher O and Dontsov D. Piezoelectric Properties of Polycrystalline AlN Thin Films for MEMS Application. Sensors and Actuators 2006; A132: 658-663.

[37] Grishchuk LP. High-Frequency Relic Gravitational Waves and Their Detection. Proc. 2nd HFGW Workshop, IASA Austin, Texas, 2007. Available from http://www.gravwave.com/docs/Grishchuk%20%20HFGW%20Lect.pdf

[38] Accuratus. Silicon Nitride, Si_3N_4, available online at http://www.accuratus.com/silinit.html

[39] Zhang H, Pang W, Yu H and Kim ES. High-Tone Bulk Acoustic Resonators on Sapphire, Crystal Quartz, Fused Silica, and Silicon Substrates. J Appl Phys 2006; 99: 124911-1-5.

[40] Abdolvand R, Ho GK, Butler J and Ayazi F. ZnO-on-Nanocrystalline Diamond Lateral Bulk Acoustic Resonators. Proc IEEE MEMS 2007; 795-798.

[41] Baker RML. Novel Formulation of the Quadrupole Equation for Potential Stellar Gravitational-Wave Power Estimation. Astronomische Nachrichten 2006; 327: 710-713.

CHAPTER 3

High Temperature Superconductors as Quantum Sources of Gravitational Waves: The HTSC GASER

Giorgio Fontana*

University of Trento, Via Sommarive 14, 38100 Trento, Italy

Abstract: Orthorhombic Cuprate High Temperature Superconductors (HTSC) are characterized by the presence of both *s-wave* and *d-wave* Cooper pairs in their electronic structure, and it is known from theory that electronic transitions between *s* and *d* states may produce the emission of gravitational waves. In normal materials, *s-d* transitions are suppressed by competing electric and magnetic transitions that have much higher probabilities; this is not the case of Cooper pairs in HTSC, which cannot support states other than *s-wave* or *d-wave*. To give theoretical foundations to a potential technology capable of constructing a quantum source of gravitational waves, this paper will also discuss means for creating coherence and population inversion and means to increase the emission probability, with close resemblance to the conceptual developments that led to the experimentation of the first laser. The expected performances of the device are derived from quantum gravitational theories. Additional properties of the active materials are considered in order to enforce the theoretical foundations of the device. A proof-of-concept device, operating at about 1 THz, is described.

Keywords: Gravitational waves, gravitational waves from quantum systems, quantum coherence, laser emission, theories of gravitation, superconductors, high-Tc superconductors, low-Tc superconductors, s-wave and d-wave pairs in superconductors, general relativity, gravity-like fields .

1. INTRODUCTION

It is commonly believed that only astrophysical objects with large masses and high density can produce gravitational waves with amplitude compatible with experimental physics [1]. This is not an accurate statement.

Subatomic particles do indeed have nuclear mass density and can be driven to the required dynamics capable of producing gravitational waves. This is well known and has been studied since 1960 by some authors.

Gravitational wave "lasers" have been proposed and theoretically studied [2], but the development of the proposals was suspended because stimulated emission was deemed not feasible [2-4]. In fact it is not possible to build a resonant cavity for gravitational waves; moreover the electromagnetic channel usually offers much higher emission probabilities.

After the initial studies, few attempts have been made to overcome the above mentioned limitations by invoking super-radiance and by adopting superconducting cavities [5, 6].

The concepts presented in this paper will show that the functions of the resonant cavity can be substituted by synchronizing the elementary oscillators with Bose condensation; moreover a class of quantum systems is shown to exist for which the electromagnetic channel can be effectively controlled and minimized, thus providing a viable path to the development of a quantum source of high frequency gravitational waves (HFGW).

This paper summarizes results obtained by the author in past years [7-13].

2. GRAVITATIONAL WAVES FROM QUANTUM SYSTEMS

The emission of gravitational waves by quantum systems has been thoroughly analyzed by Halpern and Laurent [2] and Ford [3]. Transitions for which the orbital quantum number L changes by ±2 and for which

***Address correspondence to Giorgio Fontana:** University of Trento, *Via* Sommarive 14, 38100 Trento, Italy; E-mail: giorgio.fontana@unitn.it

Giovanni Modanese and Glen A. Robertson (Eds)

the total quantum number J changes by 0 or ±2 are gravitational quadrupolar transitions, for which the emission of photons is forbidden and the emission of quanta of gravitational waves is allowed. For comparison, the selection rules for photons are those for which the orbital quantum number L changes by ±1 and for which the total quantum number J changes by 0 or ±1, *etc.* For instance, for an atom, gravitational transitions are those between orbitals 3d and another orbital characterized by a lower energy among 3s, 2s and 1s, (Fig. **1**).

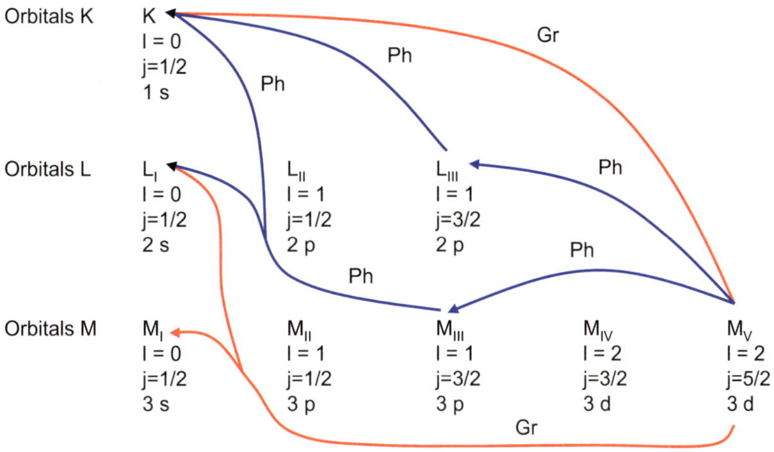

Figure 1: Examples of photon and graviton transitions.

In typical atomic systems gravitational transitions compete with multiple photon transitions, therefore methods for counteracting photon transitions have to be developed.

Detailed analysis comparing the emission of gravitons to the more favorable emission of photons has shown that the ratio R of the transition probabilities for matrix elements of equal structures is [2]:

$$R = 8\pi G m^2/e^2 , \tag{1}$$

Turning up to be of the order of $1.6\text{x}10^{-36}$ for the proton [2] and of the order of $4.8\ \text{x}10^{-43}$ for the electron. An intermediate value can be found for the muon.

With $J=2$, the transition probability per unit of time for the emission of gravitational waves is proportional to:

$$M^2(d{\cdot}f/c)^4 f \tag{2}$$

This quantity increases with the square of the moment of inertia Md^2 where d is the radius of gyration.

The frequency f is inversely proportional to the moment of inertia; thus in the quantum theory, the smaller the moment of inertia is, the higher the emission probability of gravitational waves becomes. Therefore, quantum systems composed by couples of closely interacting electrons seem favored for the emission of gravitational waves respect to alternative approaches.

A known method to further increase the emission of gravitational waves consists in using induced emission at resonance *i.e.,* Gravitational wave Amplification by Stimulated Emission of Radiation (GASER or GRASER).

Because of the lack of mirrors or resonant cavities suitable for gravitational waves, and taking into account the physical quantities involved, the characteristic length, l, of the single pass, one way, molecular GASER has been found to be [2]:

$$l \approx 10^{38} \frac{\delta\omega}{\omega} \quad \text{cm}, \tag{3}$$

where $\frac{\delta\omega}{\omega} = \Delta$ is the line-width of the transition. $\Delta=10^{-14}$ is a typical value for molecular transitions, therefore GASER action is apparently impossible to achieve with normal atomic systems and adopting acceptable lengths for the devices.

This same conclusion is confirmed by adopting a semi-classical simplified approach, in which gravitational waves are treated classically while quantum systems are treated with quantum mechanics [4]. Using the *Fermi Golden Rule* to calculate the transition probability per unit of time for a transition from an initial state |i> to a final state |f> we have:

$$(Transition\ prob./\ time) = (2\pi / h)\left|\langle f|H_I|i\rangle\right|^2 \delta(E_f - E_I - h\omega) \tag{4}$$

where H_I is the interaction Hamiltonian, the delta function results from neglecting the decay of the state. For the ideal system the probability tends to infinity, while for an atom the delta distribution should be replaced by a Lorentzian function with peak value:

$$Lp = \frac{2}{\gamma\pi} = \frac{3mc^3}{\pi\omega^2 e^2} \tag{5}$$

At the frequency of 1 THz the line-width γ is about $3 \cdot 10^{-8}$ and $\Delta \approx 3 \cdot 10^{-17}$. The cross section σ, for the gravitational transition is:

$$\sigma = 96.5(Gm^2/e^2)(h^2/m\,e^2)^2 = 10^{-57} cm^2 \tag{6}$$

where h is Planck's constant and when multiplied by typical solid state densities of 10^{24} cm^3 gives the e-folding length for the gravitational wave stimulated emission $L \approx 10^{33}$ cm, about 100,000 times the estimated radius of the universe.

Again, the gravitational transitions have been found to be smaller than the corresponding electric and magnetic transition by a factor that agrees with Equation (1)

Table **1** reports the results for cross section and inverse lifetimes for electromagnetic and gravitational transitions. $\alpha=e^2/hc$ =1/137 is the fine structure constant, *a,* is the size of the emitting system, which has been assumed to be the Bohr radius (h^2/me^2).

Table 1: Cross-sections and inverse lifetimes for electromagnetic and gravitational transitions.

Transition	σ	1/τ
Electric dipole	$0.396\ \lambda^2$	$(32/81)^2\ \alpha^3\omega$
Electric quadrupole	$1.68\ a^2$	$(1/16)^2\ \alpha^5\omega$
Gravitational quadrupole	$96.5\ a^2\ Gm^2/e^2$	$(19/40)^2\ (Gm^2/e^2)\alpha^5\omega$

Fortunately, in the solid state the electron wave-functions of couples of electrons may collapse to a single macroscopic wave-function, originating the well-known phenomenon of superconductivity [14]. If superconducting electrons may be involved in gravitational wave emission, the delta function in Equation (4) is no longer too rough of an approximation because the state of the condensate is only marginally modified by the emission. If the state emits a quantum of gravitational waves, it looses a member at the same time. In addition, the size of the quantum system is no longer the Bohr radius, but can be the size of a big macroscopic object.

A quite general theoretical approach to High-Frequency Gravitational Waves emitted by normal modes in crystals has been developed by Sacchetti and Trevese [15]. After a detailed calculation, the power spectrum of the emitted gravitational waves has been found to be (Equation 18 and 19 of [15]) :

$$P_{GW}(\omega) = 2\gamma_{TO} \frac{\chi \omega_{TO}^{2}}{4r^{2}c} \frac{\sum_{q}\left(E_{TOq}^{2} + E_{TO'q}^{2}\right)}{\left(2\omega_{TO} - \omega\right)^{2} + \left(4\gamma_{TO}\right)^{2}}; \tag{7}$$

where *TO* means Transverse Optical and the remaining symbols are defined in [15].

For resonant transitions between two bosonic states we have $2\omega_{TO}=\omega$ and $\gamma_{TO} \rightarrow 0$, therefore $P(\omega) \rightarrow \infty$, with the limitation that the output power in gravitational waves cannot exceed the input power. Equation 21 of [15] is clearly not applicable to the present case because it is intended only for dense fermionic systems excited with monochromatic electromagnetic radiation.

The device that adopts superconductivity, electron pairing and spin-2 transitions for the generation of Gravitational Waves is the HTSC GASER, that was first described in [7].

3. INDIRECT EVIDENCE OF SPIN-2 TRANSITIONS IN HTSC

The occurrence of gravitational transition in HTSC seems to be indirectly shown in already published measurement data, if the analysis takes into account the possibility of gravitational *s–d* transitions, which is normally ignored in material science.

Superconductivity is characterized by a macroscopic quantum phase-coherent state. In Low-T_c superconductors (Pb, Nb, Nb_3Sn, *etc.*) the phonon-mediated electron-electron interaction produces spin-singlet pairing with *s-wave* symmetry according to Bardeen, Cooper, and Schrieffer (BCS) [14].

In High-T_c cuprate superconductors the structure of Cooper pairs was extensively studied in recent years [16-20] with the conclusion that there is a predominant, but not exclusive, *d-wave* pairing. Tunneling experiments using Josephson junctions between HTSC and Low-T_c superconductors have consistently shown the presence of an admixture of *s-wave* and *d-wave* pairing in the order parameter of orthorhombic cuprate HTSCs.

In the phenomenological *Ginzburg-Landau* theory of superconductivity, the order parameter is a wave function and after the formalization of the microscopic theory of superconductivity [14], Gor'kov [21] established the link between the old phenomenological theory and the new microscopic theory.

Near T_c the *Ginzburg-Landau* equations can be derived from the BCS theory and the order parameter is identified with the pair wave-function, more precisely it has been defined as the superposition of the wave-functions of all the pairs. Therefore, the admixture of *s-wave* and *d-wave* pairing in the order parameter seems to agree with the presence of a superposition of wave-functions of two populations of Cooper pairs, one associated to the *s-wave* component of the order parameter and the other associated to the *d-wave* component.

In [17] a best fit data analysis of a BSCCO sample named *mb232* is in agreement with the existence of these two populations with critical temperatures of 28K for $T_{c,d}$ and 10 K for $T_{c,s}$. Because of Gor'kov's result, the *s-wave* population is composed of Cooper pair with total spin S=0 in agreement with BCS theory.

BCS theory suggests the possible existence of states with S≠0 with no energy gap respect to states with S=0, but measurements [17] show the existence of an additional state with S=2 and a measurable energy gap respect to the state with S=0 because of the different critical temperatures. The two states could be attributed to pairs confined in the CuO layers and to pairs not confined in the layers.

This is not in contradiction with BCS. In fact, in anisotropic HTSC two pairing states have been shown to exist, but only the state with lower energy is stable, the other state must be empty and it manifests its existence only in experiments that create a direct coupling to it with wave-function selective Josephson-junctions.

Because pairs in the two states differ for the total energy and the quantized angular momentum, we now relate the measured properties of mixed order parameter in HTSCs to the possible gravitational wave transitions. The difference of critical temperatures is a measure of the energy gap.

For the energy gap Δ, BCS predicts:

$$\Delta(T) \approx 1.76 k_B T_c \sqrt{\cos\left[\frac{\pi}{2}\left(\frac{T}{T_c}\right)^2\right]} \qquad \textbf{(8)}$$

For T< $T_{c,d}$ /2 and T< $T_{c,s}$ /2, the term under the square root can be neglected, giving for the cited sample a differential *s-wave/d-wave* energy gap of 2.74 10^{-3} eV and a line frequency for the gravitational transition of 660 GHz. On the other hand the coefficient 1.76 in Equation (8) has been measured to be as large as 3.6 for BSCCO, this may increase the frequency of the HFGW up to about 1.3 THz.

The power emitted in the form of gravitational waves is formally related to the recombination rate, corrected for the momentum conserving reaction on the crystal lattice, which will be converted to heat.

The condensation of both the *s-wave* component and the *d-wave* component ensures that the relative energies of the pairs are almost not affected by thermal fluctuations and impurities except for distances less than the coherence length. On the other hand, the line width is critical for achieving amplification, but no experimental data is available on the residual relative fluctuations of the pair energies.

The strong correlation among pairs is the main difference between superconductivity and other forms of Bose-condensation such as superfluidity and Bose-condensates of trapped atoms [22]. Because of the strong correlation, fluctuations are averaged over the pairs. In fact, the Josephson effect [23] and the related large literature on superconducting quantum interference devices (SQUIDs) indicate that the coherence of the wave-function is experimentally maintained over very large distances, with no apparent local splitting of the ground state into fine structures or the appearance of drifts, which is also prohibited by BCS theory itself. Being the *d-wave* component the one with the highest binding energy, it represents the state with the highest population.

To achieve GASER action it is therefore necessary to increase the density of the metastable *s-wave* component. A possible mechanism for achieving population inversion could be the electrical injection of *s-wave* Cooper pairs with a Josephson junction made with Low-T_c materials, which couple only to the *s-wave* component of the HTSC.

The *s+d* component can be removed by a normal metal contact. The transport is performed in the bulk superconductor and it requires no work; power is required for the removal of *s+d* Cooper pairs and for the creation of *s-wave* Cooper pairs in the Low-T_c material. Reference [17] presents detailed experimental data. It has been found that in a Pb/BSCCO Josephson junction, $T_{c,Pb}$ is different from, $T_{c,s}$ in BSCCO, with a strong indication that the *s-wave* component in BSCCO is not induced by proximity to Pb.

In addition, detailed discussion of experimental results [17] indicates that second order tunneling [24] may not play a substantial role in the tunneling process: the *s-wave* component is present in BSCCO, but only in the vicinity of the junction. This behavior might be interpreted as a very short diffusion length and consequently a highly efficient *spin-2* recombination, which is a great advantage for the emission of gravitational waves.

So far, the observed Josephson tunneling between BSCCO and Pb has suggested two interpretations, the first one is the homogeneous presence of a mixed order parameter in orthorhombic HTSCs, the second one is the homogeneous presence of a mixed order parameter in Pb [25].

A third possibility is the one presented here: a coherent gravitational waves field allows *s-wave* tunneling to an available energy level of the Cooper pair condensate in BSCCO, followed by a rapid transition of the pairs to the dominant *d-wave* symmetry; being the low probability of the process compensated by the smallness of the linewidth, which is directly related to coherence.

The emission of gravitational waves could explain the observed behavior of Pb/BSCCO Josephson junctions [17], especially the fact that the degree of admixture of the order parameter can dynamically depend on the mentioned processes.

As reported in the introduction, multiple photon transitions and multipole electric/magnetic transitions may compete with gravitational transitions. Therefore it must be established if electromagnetic waves can escape or, on the contrary, they are trapped in the bulk superconductor as required by the basic GASER theory [2].

Reported values of the surface resistance of HTSCs at THz frequencies [26, 27] seems to indicate that if multipole electric/magnetic transitions could take place in the bulk superconductor, then the associated electromagnetic waves could not escape the material except for transitions localized not more than few hundred nanometers below the surface.

Literature reports a surface resistance of $YBa_2Cu_3O_{7-\delta}$ (YBCO) thin films at 77K – 1.3Thz of 1Ω, that is about an order of magnitude higher than that of gold at the same temperature, penetration depths of the order of 200nm have been also measured for YBCO.

A more specific theory relates Superconductivity in cuprate HTSC to 4s to 3d electron exchange amplitude in Cu atoms [28]. This is unequivocally associated to gravitational *spin-2* transitions, according to Halpern and Laurent theory.

4. STRUCTURE OF THE HTSC GASER AND PREDICTED PERFORMANCES

The structure of the HTSC GASER is shown in (Fig. **2**). The structure is similar to Pb/BSCCO Josephson junctions prepared for investigating the nature of pair wave-function in BSCCO [17].

In the orthorhombic HTSC side, *s-wave* and *d-wave* pairs build up their respective wave-functions, characterized by different critical temperatures which define the energy gap. The dominant wave-function is *d-wave*. The injection of *s-wave* pairs is possible by letting a super current to circulate through a *s-wave* Josephson junction with high critical current, *i.e.,* with an insulating barrier of negligible thickness.

The effect of Cooper pair condensation in two bosonic quantum states is equivalent to the fact that some relevant parameters, for instance phase, energy and radiation pattern of all the pairs are exactly the same or are linked by simple functions. Therefore assuming Cooper pair density of $N=10^{20}/cm^3$, with 10 cm^3 of material, wave-function coherence amplifies the probability of gravitational wave emission to the level of typical spontaneous photon emission in a single atom, see Equation (1). Cooper-pair Bose condensate acts as a much heavier elementary pair, the gain being N^2. This is not the only useful gain, a beaming effects due to radiation coherence also appears.

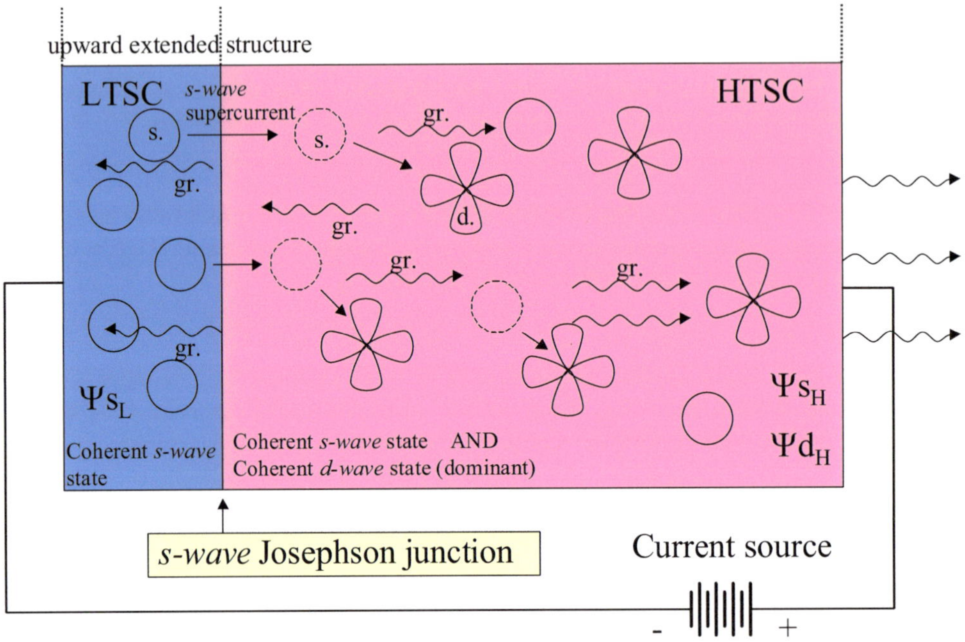

Figure 2: Structure of the proposed HTSC GASER.

At large distance from the source, the overall coherence gain respect to non-condensed and random emitting sources is N (wave function coherence) * N (coherent superposition of emitted gravitons - beaming effect) = N^2. The gain saturates when the whole super-current participates to gravitational wave emission and directivity is dominated by geometry, which depends on emission wavelength and the size of the GASER.

Coherence gain is due to super-radiant states in electron gases and is rigorously derived in [29]. More specifically, according to Ford [3], for gravitational waves the total energy-momentum radiated by the system for the semi-classical theory is:

$$S^{\mu\nu}_{s.c.} = -8\pi \int d^3x d^4x' d^4x'' G_r^{;\mu}(x,x') G_r^{;\nu}(x,x'') \times \left[\langle T_{\alpha\beta}(x') \rangle \langle T^{\alpha\beta}(x'') \rangle - \frac{1}{2} \langle T(x') \rangle \langle T(x'') \rangle \right], \quad (9)$$

whereas for the quantum theory it is:

$$S^{\mu\nu}_{q} = -8\pi \int d^3x d^4x' d^4x'' G_r^{;\mu}(x,x') G_r^{;\nu}(x,x'') \left\langle T_{\alpha\beta}(x') T^{\alpha\beta}(x'') - \frac{1}{2} T(x') T(x'') \right\rangle, \quad (10)$$

where $T_{\mu\nu}$ is the energy momentum tensor intended as quantum operator, whose expectation $<T_{\mu\nu}>$ enters the semi-classical Einstein equation $G_{\mu\nu} = 8\pi<T_{\mu\nu}>$. In the semi-classical theory the total energy-momentum radiated by the system depends upon a product of expectation values; differently for the quantum theory it depends on expectation values of products.

For calculating the emission of gravitational waves in the quantum theory, the semi-classical theory is a good approximation for states that can be considered coherent states, like those produced by superconductivity. In addition, on very general basis, Ford suggested that if an excited boson field emits gravitational waves, they must be considered of quantum origin, *i.e.*, gravitons.

The superposition of a large number of coherent sources is expressed by the interference pattern that they project on a detector at a large distance with respect to the wavelength λ. For instance, the maximum emission for a linear array is for $a \sin\vartheta = n \lambda$, where a is the spacing of the point sources, and n is an integer to which the maximum is associated.

The amplitude of the maxima is N^2 times the amplitude produced by each source, where N is the number of sources. More generally a circular planar surface distribution of point sources will produce a beam of waves orthogonal to the plane of the point sources provided that the wavelength is much shorter than the diameter of the distribution. The emission is bi-directional.

Interestingly it is possible to conceive a distribution of coherent point sources on a spherical surface, as shown in Fig **3**.

For outgoing waves they emulate a point source located at the center of the sphere with power equal to the coherent summation of the waves emitted by all the point sources. The radiation directed towards the center is focused to the center of the sphere, therefore a large power density can be obtained.

The order of magnitude of the power density gain for a given total input power is A/λ^2, where A is the total area covered by the point sources.

A curved crystal (Fig. **3**), can produce a curved wave-front of high-frequency gravitational waves, even if no refractive or reflective material is available and capable of producing a focused beam of gravitational waves starting from plane waves. Focused gravitational waves may produce non linear effects, discussed in the detection section.

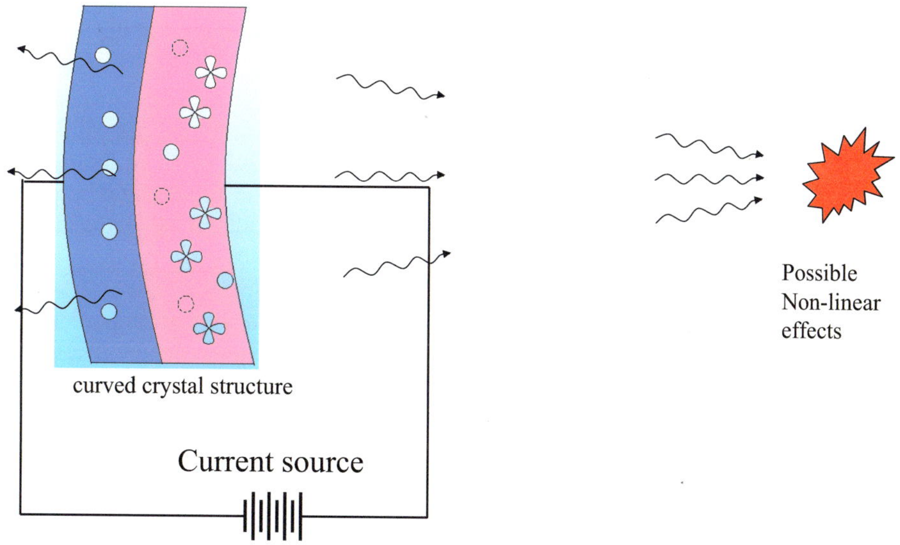

Figure 3: Curved HTSC GASER.

For quantum transitions the frequency is given by *E/h* and the power emitted is also related to the gap energy *E* times the recombination rate, which is the number of electron pairs that traverse the junction each second, reduced by spurious effects. We therefore apply ideal conditions, where spurious effects are negligible.

The energy gap is the one reported by measurements on Pb/BSCCO Josephson junctions [17]. The maximum current is the typical critical current of HTSCs. The expected performances of the HTSC GASER are summarized in Table **2**.

Table 2: Expected performances of BSCCO HTSC GASER.

Parameter	Value	Units
Energy of s/d wave gap in BSCCO	~1.3	THz
Power output with 10 kA/cm^2	~10	W/ cm^2
Power output for multi-layer	~10^3	W/ cm^2
Volumetric power output	~10	MW/m^3

About 10 giga-Ampere of electric current should be required to produce ~10 MW of HFGW. A multi-layered structure is possible if wave-function coherence is preserved in the HTSC side of the GASER (electrical parallel connection).

To allow a higher working temperature, the LTSC Magnesium Diboride (MgB_2) might be used as a source of *s-wave* pairs. MgB_2 superconducts at 40 K, and has the remarkable property of having two energy gaps [30], clearly due to anisotropy.

Therefore MgB_2 is not only a possible *s-wave* pair injector, but also represents an additional experimental proof that cooper pairs with different energy may share the same material.

5. FABRICATION OF A PROTOTYPE HTSC GASER

5.1. Crystallographic Structure

Generally, the fabrication of prototype devices is a complex procedure. It starts with the design of the device, structure and size, goes through the definition of the materials, nature and purity, the definition of the fabrication process, techniques and steps.

Finally, the properties of the produced device will be measured to recognize if the results agree with expectation. The procedure is usually repeated many times.

The structure of the proposed HTSC GASER is quite simple; an orthorombic cuprate HTSC mono-crystal should be produced along with a Josephson junction with a LTSC. Both sides of the device should be metalized for connecting the GASER to a current generator. To define the preferable geometry of the HTSC mono-crystal we make some considerations on directivity patterns in wavefunction coherent systems.

Directivity of emission is generally determined by a combination of geometrical properties of the emitter, for instance the shape of the resonant cavity of a LASER. For a wavefunction coherent system in the solid state, composed of microscopic oscillators emitting waves, the directivity function is the product of two components: the directivity function of the single emitter, and the superposition function (interference) of many such emitters.

The collective wavefunction is the main mechanism that synchronizes the oscillators.

5.2. Space and Time Synchronization of the Microscopic Sources

In [2] it has been shown that there exists a directivity pattern for transitions emitting gravitons, the directivity here depends on the *d-wave* wavefunction orientation. In HTSC, the macroscopic wavefunction has been observed by interference effects in Josephson junctions. The macroscopic wavefunction is oriented according to the crystallographic orientation of the HTSC crystal.

Even if this factor has a minima and maxima, its dependence on the angles is smooth, and it appears to be not critical, nevertheless crystals with different crystallographic orientation should be built to test the theory.

The angular distribution of GWs for a quantum system is derived in [2]. For J=M=2 it takes the form:

$$\cos^8\left(\frac{\vartheta}{2}\right)+\sin^8\left(\frac{\vartheta}{2}\right). \qquad \textbf{(11)}$$

It has been found that the maximum amount of gravitational waves occurs in a direction for which the corresponding electromagnetic radiation is excluded [2].

The d-wave wavefunction in HTSCs develops in the a-b crystallographic axes [31], making a maximum of the gravitational waves emitted in the direction of the c axis. This emission is bi-directional.

The directivity factor related to the superposition function of many emitters is what mainly determines the directivity function and the directivity gain.

The collective behavior of the microscopic sources can be analyzed adopting Gor'kov (1959) results. He proved that near *Tc* the Ginzburg-Landau equations can be derived from the BCS theory and the complex order parameter ψ is identified with the pair wave-function, more precisely it has been defined as the superposition of the wave-functions of all the pairs.

The local phase of the order parameter averages the phase of the microscopic wave-functions of the Cooper pairs. Therefore the relative phase of the order parameter can control the spatial distribution of the phase of the coherent microscopic sources involved.

The directivity pattern of the HTSC GASER depends on the directivity patterns of all microscopic sources, that are aligned with the crystal structure, and on the phase relationship among them, which depends on the phase of the wave-function.

$$\frac{\hbar^2}{2m}\left(\nabla-\frac{iq}{\hbar c}\mathrm{A}\right)^2\psi=u\,|\psi|^2\,\psi-v\psi \qquad \textbf{(12)}$$

Because of the presence of flux vortices, the phase of the wave-function "inside" a type II superconductor (HTSC are type II) can be controlled with a properly designed magnetic field applied before cooling the HTSC, this method can freeze the magnetic field inside the HTSC in the form of supercurrent vortices. Moreover, under high current regimes the supercurrent acquires a filamentary structure immersed in non superconducting regions; therefore the injection current itself can induce the phase shift that gives directionality to the emission.

The relationship between the phase of ψ and the internal magnetic field is obtained by solving the above Ginzburg-Landau equation, where **A** is the vector potential. Introducing a gradient in the phase of the wave-function ψ is necessary for the proper operation of the HTSC GASER. In fact without a gradient in the phase of the wave-function, all the microscopic sources would be perfectly synchronized in time, and the phase of the emission would not depend on space. Because the microscopic sources are distributed on a tri-dimensional crystal lattice, the overall emission would generally average to zero, except when the spacing of the sources would be an integer number of full wavelengths at the emission frequency, which is not the case of the HTSC GASER, for which the spacing of the sources is orders of magnitude smaller than the emission wavelength. Therefore the trapped magnetic field is required to phase shift the emissions of the individual sources, allowing constructive interference along a defined direction, which must be compatible with the radiation pattern of the individual sources.

5.3. Production Techniques and Materials

Orthorombic cuprate HTSC mono-crystals can be produced using various techniques. For bulk material, which appears suitable to planar devices, the melt-textured technique is the most promising.

This technique requires the sintering of the precursor powders to produce unaligned, grainy HTSC. Afterward, a thermal process termed incongruent melting is performed. The grainy HTSC melts and reacts to form another solid plus a liquid phase. An oriented solidification is then performed, using a $SmBa_2Cu_3O_x$ seed to facilitate the crystallization with the required orientation. Very precise thermal gradients and time-temperature processing are involved.

The melt-textured techniques and the related different methods have been reviewed by Desgardin [32].

Thin film techniques seem better suited for making curved GASERs, unfortunately there is no need in current technological applications of curved HTSC films similar to those required here, and therefore a new production technology should be developed for this application.

Thin film depositions have been reviewed by Narlikar [31]. Silicon or Aluminum substrates, buffered by thin layers of CeO_2 and Y-stabilized ZrO_2 can be coated with HTSCs.

Thermal coevaporation may offer 20-30 nm/min growth rates. Pulsed laser deposition (PLD) offers similar rates. Magnetron sputtering is widely used for the deposition of layers of metals with high melting temperatures, it easily adapts to the deposition of HTSC with rates of about 10 nm/min.

Metalorganic chemical vapor deposition (MOCVD) may offer 1 μm/min growth rates. Sol-gel techniques have also been successfully employed for the preparation of thin films of HTSCs. Because the coherence gain is proportional to the volume of the HTSC, the melt-textured HTSC seems the best choice for a possible first prototype.

The production of HTSC/LTSC Josephson junction is a standard technique, see for instance the article by Moessle [17]. The Pb LTSC can be evaporated on the HTSC surface directly or with the interposition of an Au or Ag interlayer having thickness between 1 and 6 nm.

The HTSC GASER must operate at temperatures below the lowest transition temperature of both HTSC and LTSC, *i.e.,* liquid helium for Pb LTSC.

The recently discovered superconductor MgB_2 is an extremely interesting option for the LTSC layer because of its high transition temperature of 39 K and low cost [30].

Because the HTSC GASER is based on a very precise transition between two energy levels (macroscopically related to critical temperatures $Tc \propto M^{-\alpha}$, with M mass of the element and α isotope effect), the isotopic purity of the raw materials used for sintering the HTSC layer appears a key factor for successfully manufacturing the HTSC GASER.

Superconductors in which phonons play a role in inducing electron pairing exhibit a well known isotope effect [14, 33].

The situation in HTSCs is not completely understood, with experiments giving controversial results, most confirming the isotope effect [34].

The isotope effect is confirmed on copper (Abrikosov, 2000) [35]. The isotope effect of Oxygen is also investigated. The role of Barium isotopic effect is apparently not investigated, while Yttrium is mono-isotopic in nature.

Barium and Copper may require isotopic selection by a stable isotope enrichment facility to provide the highest uniformity of the energy gap down to the microscopic scale.

5.4. Experimental Setup and Driving Electronics

A possible setup is based on a planar, disk shaped, GASER with area of about 100 cm^2 and thickness of the order of 1 cm. A planar thyristor (SCR – semiconductor controlled rectifier) or a gas thyratron is connected in series with the HTSC GASER and a large set of parallel connected capacitors (C) close the circuit.

The capacitors are distributed all around the sandwich. The configuration is intended for pulsed operation with high current. It offers low circuit inductance (L) and low series resistance (R), and is electrically similar to the configuration used for pulsed laser diodes.

If the Q of the series equivalent RLC circuit is higher than 1, with a single discharge the GASER will emit a series of exponentially damped pulses of HFGWs with modulation frequency of $1/2\pi\sqrt{(LC)}$.

To permit the complete discharge of the capacitors, the thyristor must conduct for the full discharge time and a recirculation diode should be connected in parallel with it. Alternatively a single pulse can be obtained without the recirculation diode.

GaAs-based semiconductors can operate at liquid helium temperatures, instead Si-based semiconductors can operate at liquid nitrogen temperatures.

The process requires charging the capacitors with an external electric energy source, after the required energy is accumulated in the capacitor, the discharge can be initiated by a gate pulse sent to the thyristor.

The electric energy in the capacitor is $E=.5\ CV^2$, therefore high voltages are required for a high energy discharge.

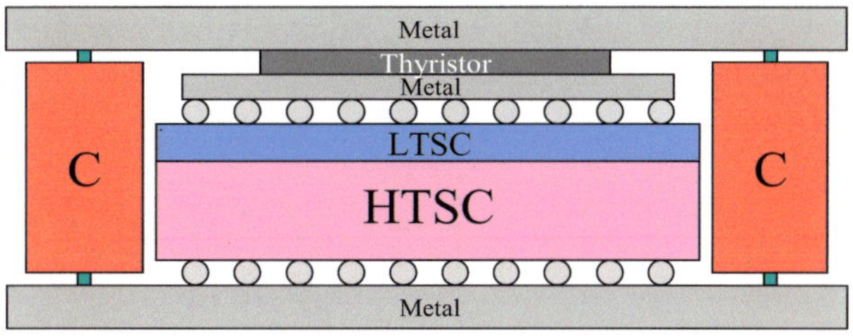

Figure 4: The proposed experimental setup (vertical section).

Fig. **4** shows a vertical section of the experimental setup for pulsed operations. Differently from standard Josephson junctions used in research instrumentation, the HTSC GASER is a high power device with current densities up to 10 kA/cm^2.

To achieve such a high critical current in the Josephson junction, the thickness of the insulating barrier must be minimized; alternatively the barrier can be removed at all, in fact the concept of the barrier is functional to the analysis, we clearly take the limit of its thickness to a single atomic layer in the practical device. The critical current can be also increased using junctions with fractal surface in order to maximize the interface area.

Indium or Bismuth soft metal collapsible ball contacts may facilitate the connection of the device to the GASER cage and driving electronics. A current limited voltage generator should be connected to the top and bottom electrodes for charging the capacitors, the current limit determines the pulse repetition rate.

Assuming $Q=R^{-1}\sqrt{(L/C)}>>1$, the peak current is $I_P=V_C\sqrt{(C/L)}$. With capacitance of the order of 10 μF and inductance of the order of 0.1μH, peak currents of the order of 100 kA can be obtained with about 10000 V of charging voltage.

6. DETECTION OF HFGWS AND EXPECTED RESULTS

Gravitational waves, possibly emitted by the HTSC GASER, can be detected by specialized detectors for HFGW in the GHz to THz range. It has been shown that the detection of 1.3 THz HFGW with $h \geq 10^{-30}$, as it is expected from the proposed planar HTSC GASER, could be possible with the synchroresonance converter described by Li [36].

Many different detector are also described in the literature as reported by Baker [37] and in the eBook by Misner [1] and many other articles [38, 39], unfortunately many detectors have been developed for low frequency GWs and it is unclear if they can be adapted to the THz range. Additionally, it should be emphasized that the "classical" theory of gravitational wave propagation and detection is based on the solution of the linearized Einstein equations, intended to treat small perturbations on flat spacetime.

The expected instanton solutions [40, 41], memory effects [42], massive graviton solutions in spacetime with a cosmological constant [43] and self-interaction effects [44-47] might indeed offer the possibility of a simplified detection of GASER radiation. Standard gravimeters and accelerometers could be the detection instruments. Invoking self-interaction effects, the curved GASER with the parameters already described might produce E≈1mJ of gravitational energy, which is able to accelerate test masses in the vicinity of the focus. The possibility of controlling the operations of the source allows the adoption of lock-in techniques for synchronous detection, thus improving the signal to noise ratio with long integration time.

The nonlinear approach to gravitational waves has recently received attention and some authors have determined that gravitational waves might be fully nonlinear longitudinal waves [48], the conditional is required because of the novelty of the approach. The argument is that for the gravitational wave the capability of carrying energy and momentum, quantities that coincide with its own sources, requires a nonlinear wave. The authors conclude that gravitational waves are indeed emitted according to the quadrupole formula, but they propagate and can be detected as nonlinear longitudinal waves. For waves traveling along the *z* direction we have:

$$h_{(2)}^{\mu\nu} = -\frac{A_{(1)\rho\sigma}A_{(1)}^{\rho\sigma}}{16}\begin{pmatrix}1&0&0&0\\0&-1&0&0\\0&0&-1&0\\0&0&0&-1\end{pmatrix}\cos\left[2k_\rho x^\rho\right]-\frac{A_{(1)\rho\sigma}A_{(1)}^{\rho\sigma}K_\theta x^\theta \omega^2}{8K_\alpha k^\alpha c^2}\begin{pmatrix}1&0&0&1\\0&0&0&0\\0&0&0&0\\1&0&0&1\end{pmatrix}\sin\left[2k_\rho x^\rho\right]$$

(13)

where $A_{(1)\mu\nu}$ is defined by the solution of the linear wave equation:

$$h_{(1)\mu\nu} = A_{(1)\mu\nu}\exp\left[ik_\rho x^\rho\right] \quad \textbf{(14)}$$

Nonlinear waves are neither transverse nor traceless. Following the arguments of ref [48] only the second term of equation 13 will dominate the solution leading to a form of longitudinal wave whose amplitude is inversely proportional to the distance from a point source as expected for energy conservation in wave phenomena. For ideal beamed waves the amplitude does not depend on distance.

The authors of reference [48] discussed the effect of a passing nonlinear gravitational wave on a pair of test particles along the *z* direction, and the change in particle separation has, for wavelength $\lambda >> z$, the following expression:

$$\xi_{(2)}^{z} = \frac{\xi_{(0)}^{z}A_{(1)\rho\sigma}A_{(1)}^{\rho\sigma}z\omega}{16c}\left[2\omega t-\sin\left(2\omega t\right)\right] \quad \textbf{(15)}$$

where $\xi_{(0)}^{z}$ is the initial separation and $\xi_{(2)}^{z}$ is the change in separation during the passage of the nonlinear wave. This is a memory effect like those discussed in [42].

For $\lambda << z$ the direction of the motion depends on $cos(2z/\lambda)$, therefore it is expected that the non-linear wave is capable of inducing vibrations in macroscopic objects, like conventional linear waves; the memory effect is still present but the observable effect depends on boundary conditions of the object and the wave packet. The fully non linear wave is an alternative analytical approach to the conventional linear wave plus the memory (a propagating energy background that supports the wave packet).

As an additional possibility, specific for the oscillating component of the wave, this same device, here proposed as generator, could be operated as detector, provided that the energy gap at the receiving side is exactly the same as the one at the transmitting side.

It is well known that the transition temperature in HTSC can be changed (increased) by applying pressure to the crystal [49]. Therefore, pressure can be applied to the detector to tune the energy gap and obtain that the required transition energy gap becomes identical to that of the generator. At the receiving side *s-wave* pairs are excited to *d-wave* pairs in the HTSC, thus *s-wave* pairs diffuse by density gradient through the Josephson junction from the LTSC to the HTSC to recover the equilibrium condition, this causes a current in the external circuit, that comprises an electrical current detector. The excitation process is proportional to the graviton/pair_condensate cross-section which is inversely proportional to the transition line-width, therefore replicating the approach followed in this paper in the reverse direction.

It must be observed that reciprocity will permit the design and construction of HTSC GASERS pumped by HFGW instead of DC electric current. They will be characterized by identical operating requirements except the *s-wave* injecting junction and related electronics, which need to be removed. If the HFGW pumped HTSC GASER is pumped by a conventional junction HTSC GASER (or another future source of HFGW with the required frequency), it might behave like a HFGW diffractive optics or beam-forming device according to mechanisms discussed in other contributions to this eBook.

Claims of observation of gravitational anomalies near HTSC excited by electromagnetic fields have already appeared in the scientific literature [50, 51] and reported in this eBook, the theoretical prediction of this paper may be related to the claimed experimental results and future experiments are required to substantiate the claims.

7. CONCLUSIONS AND DISCUSSION

This article has discussed the possibility that a class of high temperature superconductors could be used as active material for the construction of a quantum source of High Frequency Gravitational Waves termed HTSC GASER. The concept is similar to the laser diode, with the difference that the material is chosen to enhance *spin-2* gravitational transitions, and suppress electric and magnetic transitions. Coherence is provided by Bose condensation instead of electromagnetic resonance in a cavity.

Research based on astrophysical sources of Gravitational Waves has not yet produced any convincing and repeatable direct evidence of the mere existence of this elusive radiation. Indirect evidence and reasonable modeling do indeed suggest that the search for Gravitational Waves must continue with broader efforts.

The alternative to the detection of Gravitational Waves from astrophysical sources is the Hertz like experiment based on generators and detectors, for which this paper offers a potential and promising approach.

Practical generation and detection of gravitational waves could become the most important discovery of the beginning of this millennium.

This discovery might have already happened according to other reports in this eBook. Efforts must be directed towards quick replications in agreement with the formal analysis reported in this and other papers devoted to analysis and modelization.

REFERENCES

Note: "arXiv" papers are available from Cornell's electronic preprint depository http://arxiv.org

[1] Misner CW, KS Thorne, and JA Wheeler. Gravitation. San Francisco: Freeman 1973.

[2] Halpern L, Laurent B. On the Gravitational Radiation of Microscopic Systems. Il Nuovo Cimento 1964; XXXIII No 3: 728-51
[3] Ford LH. Gravitational Radiation by Quantum Systems. Annals of Physics 1982; 144: 238-48.
[4] Harris EG. Personal communication 1999.
[5] Nagibarov VR, Kopvillem UH, Samartsev VV. The Macroscopic High-Frequency Quantum Generator and Detector of the Gravitational Radiation. Phys Lett 1973; 42A: 399-400.
[6] Musha T, Sekine M. Induced Emission of Gravitational Waves. Lettere al Nuovo Cimento 1976; 16: 14-6.
[7] Fontana G. A possibility of emission of high frequency gravitational radiation from junctions between d-wave and s-wave superconductors. ArXiv: cond-mat/9812070.
[8] Fontana G. Gravitational Radiation and its Application to Space Travel. In: Proc. of Space Technology and Applications International Forum STAIF-2000; 2000: Albuquerque, New Mexico, USA. AIP Conference Proceedings 2000; 504: 1085-92.
[9] Fontana G. Gravitational Radiation and its Application to Space Travel, Principles and Required Scientific Developments. ArXiv: physics/0110042.
[10] Fontana G. Possible Graviton Transitions and Gaser Action in High-Tc Superconductors. ArXiv: cond-mat/0208276.
[11] Fontana G. Design of a Quantum Source of High-Frequency Gravitational Waves (HFGW) and Test Methodology. In: Proc. of Space Technology and Applications International Forum STAIF-2004; 2004: Albuquerque, New Mexico, USA. AIP Conf Proc 2004; 699: 1114-22.
[12] Fontana G. Design of a Quantum Source of High-Frequency Gravitational Waves (HFGW) and Test Methodology. ArXiv: physics/040022.
[13] Fontana G, Baker RML Jr. The High-Temperature Superconductor (HTSC) Gravitational Laser (GASER). Paper HFGW-03-107. The Gravitational-Wave Conference; 2003: Mclean, Virginia, USA. The MITRE Corporation 2003.
[14] Bardeen J, Cooper LN, Schrieffer JR. Theory of Superconductivity. Physical Review 1957; 108: 1175-204.
[15] Sacchetti F, Trevese D. High-Frequency Gravitational Radiation by Normal Modes of Crystalline Solids. Il Nuovo Cimento 1976; 35B: 61-9.
[16] Tsuei CC, Kirtley JR. Pairing symmetry in cuprate superconductor. Rev Mod Phys 2000; 72: 969-1016.
[17] Moessle M, Kleiner R. c-axis Josephson tunneling between $Bi_2Sr_2CaCu_2O_{8+x}$ and Pb. Phys Rev B 1999; 59: 4486-96.
[18] Kouznetsov KA, Sun AG, Chen B, *et al.* c-axis Josephson Tunneling between $YBa_2Cu_3O_{7-x}$ and Pb: Direct Evidence for Mixed Order Parameter Symmetry in a High Tc Superconductor. Phys Rev Lett 1997; 79: 3050-53.
[19] Van Harlingen DJ. Phase-sensitive tests of the symmetry of the pairing state in the high-temperature superconductors-Evidence for d symmetry. Rev Mod Phys 1995; 67: 525-35.
[20] Sun AG, Gajewski DA, Maple MB, Dynes RC. Observation of Josephson Pair Tunneling between a High-Tc Cuprate (YBCO) and a Conventional Superconductor (Pb). Phys Rev Lett 1994; 72: 2267-70.
[21] Gor'kov LP. Microscopic derivation of the Ginzburg-Landau equations in the theory of superconductivity. Soviet Physics JETP 1959; 36(9): 1364-67.
[22] Putterman SJ. Superfluid Hydrodynamics. Amsterdam: North-Holland 1974.
[23] Josephson BD. Possible new effect in superconductive tunneling. Phys Lett 1962; 1: 251-3.
[24] Tanaka Y. Josephson Effect between s Wave and d_{x2-y2} Wave Superconductors. Phys Rev Lett 1994; 72: 3871-74.
[25] Rae AIM. Role of Anisotropy of Lead in BSCCO-Pb Josephson Tunneling. Phys Rev Lett 2000; 84: 2234-38.
[26] Wilke I, Khazan M, Rieck CT. Terahertz surface resistance of high temperature superconducting thin films. J Appl Phys 2000; 87: 2984-88.
[27] Khazan M, Wilke I, Stevens C. Surface impedance of Tl-2212 thin films at THz-frequencies. IEEE Trans Appl Supercond 2001; 11: 3537-40.
[28] Mishonov TM, Indeku JO, Penev ES. Superconductivity of overdoped cuprates: the modern face of the ancestral two-electron exchange. J Phys Condens Matter 2003; 15: 4429-56.
[29] Dicke RH. Coherence in Spontaneous Radiation Processes. Phys Rev 1954; 93: 99-110.
[30] Canfield PC, Crabtree GW. Magnesium Diboride: Better Late than Never. Physics Today 2003; March: 34-40.
[31] Narlikar AV. High Temperature Superconductivity–1 Materials. Berlin: Springer 2003.
[32] Desgardin G, Monot I, Raveau B. Texturing of high-Tc superconductors. Supercond Sci Technol 1999; 12: R115–R133.
[33] McMillan WL. Transition Temperature of Strong-Coupled Superconductors. Phys Rev 1968; 167: 331-44.
[34] Bernhard C, Holden T, Boris AV *et al.* Anomalous oxygen isotope effect on the in-plane FIR conductivity of detwinned YBa2Cu316,18O6.9. ArXiv: cond-mat/0306097.
[35] Abrikosov AA. Theory of the Negative Isotope Effect on Copper in Underdoped YBCO. Phys C 2000; 336: 163-69.

[36] Li Fang-Yu, Tang Meng-Xi, Shi Dong-Ping. Electromagnetic response for High-Frequency Gravitational Waves in the GHz to THz band. Paper HFGW-03-108. The Gravitational-Wave Conference; 2003: Mclean, Virginia, USA. The MITRE Corporation 2003.

[37] Baker RML Jr. Precursor Proof-of-Concept Experiments for Various Categories of High-Frequency Gravitational Wave (HFGW) Generators. In: Proc. of Space Technology and Applications International Forum STAIF-2004; 2004: Albuquerque, New Mexico, USA. AIP Conference Proceedings 2004; 699: 1093-97.

[38] Thorne KS. Gravitational-wave Research: Current Status and Future Prospects. Rev Mod Phys 1980; 52: 285-97.

[39] Thorne KS. Multipole Expansion of Gravitational Radiation. Rev Mod Phys 1980; 52: 299-339.

[40] Gibbons GW, Hawking SW. Euclidean Quantum Gravity. Singapore, New Jersey, London, Hong Kong: World Scientific 1993.

[41] Sanchez N. Connection between the nonlinear σ model and the Einstein equations of general relativity. Phys Rev D 1982; 26: 2589-97.

[42] Fontana G. Gravitational Wave Propulsion. In: Proc. of Space Technology and Applications International Forum (STAIF-2005); 2005: Albuquerque, New Mexico, USA. AIP Conference Proceedings 2004; 746: 1331-9.

[43] Novello M, Neves RP. The mass of the graviton and the cosmological constant. Classical and Quantum Gravity 2003; 20: L67-L73.

[44] Ferrari V, Pendenza P, Veneziano G. Beam-like Gravitational Waves and Their Geodesics. General Relativity and Gravitation 1988; 20: 1185-91.

[45] Ferrari V. Focusing Process in the Collision of Gravitational Plane Waves. Phys Rev D 1988; 37: 3061-4.

[46] Veneziano G. Mutual Focusing of Graviton Beams. Modern Physics Letters A 1987; 2: 899-903.

[47] Alekseev GA, Griffiths JB. Exact solutions for gravitational waves with cylindrical, spherical and toroidal wavefronts. Classical and Quantum Gravity 1996; 13: 2191-209.

[48] Aldrovandi R, Pereira JG, da Rocha R, Vu KH. Nonlinear Gravitational Waves: Their Form and Effects. Int J Theor Phys 2010; 49: 549-63.

[49] Raphael MP, Reeves ME, Skelton EF. Nonlinear Response of Type II Superconductors: A New Method of Measuring the Pressure Dependence of the Transition Temperature, Tc(P). Rev Sci Instrum 1998; 69: 1451-5.

[50] Podkletnov E, Nieminen R. A Possibility of Gravitational Force Shielding by Bulk $YBa_2Cu_3O_{7-x}$ Superconductor. Physica C 1992; 203: 441-4.

[51] Podkletnov E, Modanese G. Investigation of High Voltage Discharges in Low Pressure Gases Through Large Ceramic Superconducting Electrodes. J Low Temp Phys 2003; 132: 239-59.

CHAPTER 4

Quantum Effects in the Type II Superconductor that Lead to Power Radiated in Gravitational Waves

Glen A. Robertson*

Institute for Advanced Research in the Space, Propulsion & Energy Sciences, Madison, AL 35757, USA

Abstract: Quantum effects on the Planck scale in the Type II superconductor have the potential for new phenomena stemming from high internal energies associated with symmetry breaking phase transitions, which can give Planck time energy fluctuations of $\sim 6x10^{24}$ $[J/s]$ per coherent electron pair in association with the Ginzburg-Landau (GL) scalar field across Josephson junctions; whereby, the Ginzburg-Landau (GL) scalar field provides a medium for producing interactions among quantum energy fluctuations and quantum gravity fields during rapid phase transitions. It is calculated that laboratory scale apparatus could produce radiated Planck scale energies or quantum gravitational waves on the order of 10^{-5} $[J]$ per coherent electron pair with gravitational energy absorption on the human scale (*i.e.*, measurable effect) of 10^{-4} $[J]$.

Keywords: Gravitational forces, theories of gravitation, superconductors, high-Tc superconductors, low-Tc superconductors, type-II superconductors, ginzburg-landau field, josephson junctions, general relativity, symmetry breaking, planck scale, vacuum energy density, gravitational waves.

1. INTRODUCTION

It is well known that quantum and classical domains are distinct; the frontier is marked by the Planck constant $\hbar$, which separates microscopic from macroscopic scales. If the lengths, the energies, the times become much larger than $\hbar$, we are in the classical physics regime with likely little hope of recovering quantum signatures. However, despite this apparent sharp division of the classical and quantum worlds, macroscopic quantum phenomena exist and some behaviors of classical systems can be explained only in the framework of quantum mechanics. High temperature superconductors (HTSC) could provide a link to gravity in that quantum gravity events may affect quantum systems that operate at the macroscopic scale, in effect creating a macroscopic "memory" of the event.

The Planck time $T_{pl} \sim 10^{-44}$ $[s]$ is so short, the Planck length $l_p \sim 10^{-35}$ $[m]$ so minuscule, and the Planck energy $E_p \sim 10^9$ $[J]$ so high (in elementary particle physics) that these units are not used casually. So it takes a certain amount of foolhardiness to even mention Planck-scale physics [1]. However, it is generally believed that at Planck scale, the quantum aspects of gravity become manifest [2].

Bojowald and Morales-Tecotl indicate that most of the time, quantum gravity is regarded as being far away from observational tests as different approaches have to be judged purely on grounds of internal consistency and their ability to solve conceptual problems from a confrontation with observations of the real world [3]. These requirements are already very restrictive for the quantization of a complicated theory such as general relativity. To the extent that in all the decades of intense research, not a single completely convincing quantum theory of gravity has emerged yet, though some investigators see promising approaches in loops, strings, and information theory [4].

1.1. Symmetry and Coherence

To produce gravitational effects from superconductors in terms of curvature distortion of spacetime-a macroscopic scale effect, symmetric and coherent energy fluctuations must be produced. That is, gravitational effects are a symmetric and coherent process, where coherence implies that all the particulates

***Address correspondence to Glen A. Robertson:** Institute for Advanced Research in the Space, Propulsion & Energy Sciences, Madison, AL 35757, USA; Tel: 256-694-7941; E-mail: gar@ias-spes.org

Giovanni Modanese and Glen A. Robertson (Eds)

on the macroscopic scale influence the behavior of gravity in similar manner from a point like source, *i.e.*, center of gravity having symmetry related to a definable mass and radius. On the microscopic scale, matter becomes easily asymmetric under various forces and is only coherent under special circumstances. For example, a group of electron can be symmetric and coherent on the surface of a mass, but under high voltage leave the mass in an asymmetric and unpredictable (non-coherent) manner. In such a case, superconductivity offer both symmetry and coherence at the microscopic scale, allowing for possible quantum gravity phenomena to be observed at the laboratory scale.

1.2. High Power Flow

Due to the small nature of quantum events, high energy flows must accompany quantum gravity events. This could be accomplished through Planck time tunneling across Josephson junctions, which could infer "ultra high" electron energy fluctuations or tunneling times across the junctions. High tunneling times are referred to as the Hartman effect, which suggests the existence of supraliminal and arbitrary large group velocities across the junctions [5]. This condition could exist if superconductor electrons quench or change state only on one side of the junction, allowing for near instantaneous relaxation of the non-quenching superconductor electrons on the other side. This could be accomplished by applying high electric potentials to a properly constructed superconductor while maintaining coherence and symmetry of the superconductor electrons on at least one side of the junction.

1.3. Chapter Overview

In this chapter, quantum gravity is explored for the type II superconductor with particular interest in the Yttrium Barium Copper Oxide (YBCO) superconductor, by first reviewing coherence in the GL scalar field and a unique characteristics of the type II superconductor, which allows for coherence through a quasi-independent pancake layering phenomena to form multiple stacked weakly coupled Josephson junctions through the bulk superconductor [6].

Second, dipole-dipole coupling internal to the superconductor is discussed as the energy transfer mechanism by considering the pancake layers as 1D stacks of dielectric films [7].

Third, the availability of free energy under the application and discharge of an external voltage through the superconductor is discussed in terms of the Landau-Ginzburg free energy potential with consideration to the quantum effects associated with small capacitive single Josephson junctions [8-11].

Fourth, mechanisms for high internal power flow are discussed to introduce the Hartman effect, quenching symmetry models, and Planck scale fluctuation time [5, 12]. In the section on Planck scale fluctuation time, the concept of human artificial units is discussed to establish the energy fluctuation effects between the Planck scale and human observable phenomena.

Fifth, symmetry breaking phase transition is discussed to introduce the freeze-out time between the transition from adiabatic (Planck time fluctuations) and impulse (relaxation time) regimes [13]. The freeze-out time is used to develop the decay time of the electron motion through the layered superconductor, which is used to develop the instantaneous power flow of the symmetry breaking model.

Sixth, the energy radiated in a gravitational wave is determined from the power radiated approximation and corrected for mass coupling, which allows for the development of a model of the energy from a gravitational wave absorbed by a mass [14].

Finally, a comparison of the absorbed energy in a gravitational wave model to the experimental data presented by Podkletnov and Modanese is conducted by calculating superconductor characteristics from the experimental data and comparing them to known superconductor characteristics [15, 16].

2. BACKGROUND

DeWitt in 1966 originally considered interactions between a superconductor and gravitational fields [17]. In a series of papers in the early 1980s, Anandan and Chiao considered the possibility of constructing antennas for gravitational radiation, using Josephson junctions as transducers [18]. In 1983, Ross derived the modified London equations for a superconductor in a gravitational field, and showed that these equations are consistent with the results found earlier by DeWitt [19]. In the early 1990s, Li and Torr derived a similar relationship involving gravitomagnetic and gravitoelectric coupling to gravity [20]. In 1990, Peng and Torr used the generalized London equations to treat the interaction of a bulk superconductor with gravitational radiation, and concluded that such a superconducting antenna would be many orders of magnitude more sensitive than a Weber bar [21].

In the nearly forty years since DeWitt first proposed the superconductor-gravity relationship, the experimental data needed to realize these claims are practically non-existence. The exception is the highly disputed claims of Eugene Podkletnov [15, 16, 22, 23].

Another relevant experiment was reported by Simmonds, *et al.* in 2001, which showed that a super-junctions could detect the earth's rotation by inducing a phase difference across the junctions [24]. Although not addressed as a gravity device, this experiment shows that superconductor junctions are very sensitive even to the global environment, which could possibly lead to a gravitational radiation detector as proposed by Anandan and Chiao [18] or the opposite as reported by Podkletnov and Nieminen [15, 16, 22, 23].

Although few, these experiments provide examples that the superconductor junction is an ideal system for studying the interplay between quantum-mechanically conjugate variables as the Josephson phase difference across the junction and the charge on the junction electrodes. In such systems, their behavior is influenced by the Landau-Ginzburg free energy potential $V(\psi)$ dissipation, such as discrete tunneling of quasiparticles and coupling to the electro-magnetic environment and possibly the gravitomagnetic or gravitoelectric environments as suggested by current string theories [8, 9].

3. COHERENCE IN THE LANDAU-GINZBURG FIELD

The Landau-Ginzburg (GL) field is described as a scalar function $\psi \approx \sqrt{n}\, e^{j\phi}$. Electrons in a room temperature superconductor material or normal conductor with no applied external fields are either confined to an atom or move about the composite molecules with random phases; $\sum\phi \approx 0$ (disorder state). However, they are generally thought to be confined to the vicinity of background ions and are positionally fixed. When the superconductor material is cooled to its critical temperature at which time a phase transition occurs, the electrons suddenly agree on a common or coherence phase; $\sum\phi > 0$ (ordered state). Again they are generally thought to be confined to the vicinity of background ions and are localized as opposed to gathering in some region creating a large space charge potential.

In a type I superconductor and as the bulk superconductor material cools down (or warms up), various size domains (depending on the cool-down (or warm up) profile) of superconductive material can form surrounded by normal conductive material. When two or more domains are in close proximity, a superconductor-normal conductor-superconductor Josephson junction is formed.

In a typical bulk type I superconductor, composed of small randomly arranged crystals or grains, proximity effects would cause the electrons of a single grain to go superconductive (or normal) as a group. In the type II YBCO superconductor, weakly coupled Josephson junctions can also form between individual molecules across the copper oxide planes and across grain boundaries typically composed of an oxide layer [6]. These are referred to as superconductor-insulator-superconductor junctions. The insulation planes degrade the time for proximity effects to cause the electrons of a single grain to go superconductive (or normal) as a group.

Therefore in the type II superconductor, a superconductor domain can be as small as one molecule of superconductor material or composed of a multitude of molecules (*i.e*, grains). Whereby from Clem [6], equation (50) with an average value of the 2-D electron density, which gives rise to an average 2-D thin-film screening length Λ, Clem [6] equation (5), in both the type I & II superconductor at respective temperatures below $T_g = 44\ K$, coherence encompasses all domains, in effect producing one single domain with coherent phase ϕ.

Gradients between domains of differing phases are accompanied by currents that tunnel between the domains as the spaces between the domains form Josephson junctions. The possible current patterns are restricted by the requirement that the GL function ψ must be single-valued. Josephson [25] tells us that current density J is caused by a phase gradient ∇ of the scalar field with a coherence phase ϕ between any two domains (neglecting contributions from external magnetic fields) is given by $\vec{\nabla}\arg\psi = \vec{\nabla}\phi = \left(m/2\hbar e\left|\psi^2\right|\right)\vec{J}$.

3.1. Layered Coherence in the Type II Superconductor

In the superconductor, the superconductor electrons throughout the material are in a condensate state and are by nature coupled or entangled within the bulk superconductor. This provides a link to the quenching or phase transition event of superconductor electrons, which under certain conditions, could produce Planck time tunneling effects across superconductor junctions. The problem is to quench enough superconductor electrons in a manner to transfer the resulting released free energy to the remaining superconductor electrons. This must occur in a coherent fashion without quenching the bulk superconductor material. It is proposed here that such coherence could be achieved through vortex interactions across multiple weakly coupled Josephson junction in the type II superconductor.

In type II superconductors, the penetration depth λ is larger than the coherence length ξ. This makes it energetically favorable for domain walls to form between the superconductor and normal regions. That is, when placed in a magnetic field, the superconductor free (*i.e.*, low resistance) energy can be lowered by causing domains of normal material containing trapped magnetic flux to form with low-energy boundaries created between the normal core and the surrounding superconducting material.

Whereby, in an applied magnetic field exceeding a value referred to as the lower critical field, magnetic flux is able to penetrate in quantized units by forming cylindrically symmetric domains called vortices. For higher applied fields, the vortices overlap and the field inside the superconductor becomes strong everywhere. When the applied field reaches a value called the upper critical field, the material becomes normal. The upper critical field can be in the tens of Tesla range [8].

In the vortices themselves, the highest field is in the vortex core, which has a radius on the order of the coherence length ξ. A region having a radius on the order of the penetration depth λ surrounds the core within which magnetic flux and screening currents flowing around the core are both present. As the magnetic field increases, additional vortices form, accompanied by the breakup of the superconductor electrons, and more and more quasiparticle electrons become localized in the vortex cores.

The vortices arrange themselves in a hexadic pattern when their density is so high as to make the repulsive interactions between them appreciable in magnitude. Each vortex will then occupy the area

$$A_V = \frac{\sqrt{3}}{2}R^2 \quad \textbf{(1)}$$

of the unit cell, where R is the average separation of the vortices.

The average field B inside the super-conductor is given by

$$B = N_m\frac{\Phi_0}{A_S} = \frac{\Phi_0}{A_V} \quad \textbf{(2)}$$

where $\Phi_0 = \pi\hbar/e$ is the fluxoid quantum, $A_S = \pi r_s^2$ is the surface area of the superconductor of radius r_s, and N_m is the maximum number of vortices given by

$$N_m = \frac{A_S}{A_V} = \frac{2\pi r_s^2}{\sqrt{3}R^2} \tag{3}$$

For the YBCO superconductor with a low applied magnetic field (lower critical field case), the separation distance

$$R_L \approx 1.77\lambda \tag{4}$$

and with a high applied magnetic field (upper critical field case)

$$R_U \approx 2.69\xi \tag{5}$$

In this chapter, only the lower critical field case is considered, where for the YBCO Type II superconductor, the value for $R_L \sim R \sim 700\ [nm]$ (or $\sim 7x10^{-5}\ [cm]$) [8, 9].

3.2. Pancake Vortices in the Type II Superconductor

A typical vortex is much longer than its core radius and can be pinned in several places. In the YBCO, such a vortex can twist and turn in the material, and sections of it might move while others remain pinned; it could also undergo a perpendicular excursion along the *ab*-plane of the YBCO superconductor. Also, the coherence length ξ along the *c*-direction is less than the average spacing between the copper-oxide planes, and hence the coupling between *ab*-planes tends to be weak.

Using the notion of parallel superconducting layers that are weakly Josephson coupled to each other, a vortex perpendicular to these layers would conventionally be considered as a uniform cylinder of confined flux surrounded by circulating currents [6]. The model is then a stacking of two-dimensional pancake-shaped vortices, one pancake vortex per layer with surrounding, nearly circular current patterns confined to the layer as shown in Fig. **1**, where s_p is the lattice parameter along the *c*-direction. For the YBCO, the lattice parameter $s_p = 1.19\ [nm]$.

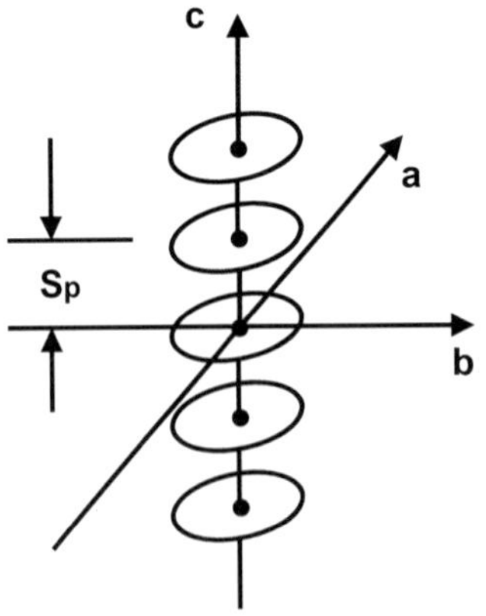

Figure 1: Pancake vortices.

3.2.1. Electron Pair Density in the Pancake Layers

In the pancake model, the interactions across layers can be ignored when no electric field is applied. A state of quasi-independent layering would exist and the average velocity about a vortex in the *ab*-plane of each layer is equivalent.

Then from Poole [8] equations (2.51), (1.18), and (2.28) and noting that $\lambda \sim R_L \sim R$, the average radial electron vortex velocity v_r $[m/s]$ can be given by

$$v_r \approx \left[\frac{e}{m_e}\right][BR] \tag{6}$$

and the current density J $\left[A/m^2\right]$ of each vortex can be approximated from Poole [8] equation (2.34) as

$$J \approx \frac{B}{\mu_0 R}. \tag{7}$$

Now by approximating the charge density D $\left[C/m^3\right]$ about each vortex as

$$D \approx \frac{J}{2\pi v_r} \tag{8}$$

and combining equations (6) and (7) with (8), the charge density D is approximately constant and given by

$$D \approx \left(\frac{m_e}{e}\right)\left(\frac{1}{2\pi\mu_0 R^2}\right) \tag{9}$$

or in terms of the electron number density per volume

$$n_V = \frac{D}{e} \approx \left(9.17x10^{24}\right)\ [m^{-3}], \tag{10}$$

which is approximately three orders of magnitude less than is indicated by the 3-D electron number per unit volume n_{3D}, which for the Type II YBCO superconductor is $1.69x10^{28}$ $[m^{-3}]$ [8, 9].

The lower number density per volume $(n_V < n_{3D})$ can be attributed to the superconductor electron pair breakdown in the vortex cores as previously mentioned and noting that $R \sim R_L > R_U$.

The number of superconductor electrons per layer n_L can then be approximated by

$$n_L \approx n_V\ s_p, \tag{11}$$

which for the YBCO superconductor $n_L \sim 1.09x10^{16}$ $\left[m^{-2}\right]$.

4. ELECTRICAL INTERACTIONS ACROSS LAYERS

It is proposed here that a coupling process between normal and superconductor electrons could be achieved along the vortices passing through a weakly coupled Josephson junction between type II superconductors with different critical temperatures when a high electric potential is placed across the bulk superconductor.

Under special circumstances interatomic couplings can be created on demand through external fields, which excite two-atom resonances, such as those arising from electric dipole-dipole interactions, ground state collisions, or real photon exchange [26-29].

Although, talking about condensate clouds, Lushnikov reports that dipole-dipole forces are long-range and essentially anisotropic. Net contribution of dipole-dipole interactions can be either repulsive (positive dipole-dipole interaction energy) or attractive (negative dipole-dipole interaction energy) depending on the form of condensate, its orientation relative to dipole polarization axes and trap geometry. Dipole-dipole interactions can dominate provided polar molecules are oriented by a strong *enough electric field.* Where, respectively, stability and collapse of the condensate cloud strongly depends on the cloud's anisotropy, which opens a whole bunch of *new phenomena* to be observed and makes task of achieving and control of condensates especially challenging [28].

In the bulk superconductor, the coupling can be thought of as dipole-dipole interactions by considering the pancake layers as 1D stack of dielectric films similar to that proposed by Hemmerich [7]. Hemmerich suggested using lasers to create a quantum entangled state between atoms in a conventional dilute optical lattice for possible quantum communication. The idea is that the laser induced voltages, produced through photon-like excitations, because forward and reverse propagation across the 1D stack of dielectric films as the electric potentials oscillate. Here the direct application of a voltage potential is suggested instead.

In this paper, a dipole-dipole stack is considered to exist between any two pancake layers (*i.e.*, c-axis layers). A properly prepared superconductor can then be composed of a multitude of weakly coupled dipole-dipole stacks or junctions.

4.1. Multi-Layered Superconductor Test Model

Though proximity affects an even weaker coupling between normal and superconductor electrons can be accomplished by adjoining two similar Type II superconductors with slightly different critical temperatures into a dipole-dipole stack. Here the lower critical temperature (T_{c2}) type II superconductor will be referred to as a LTII SC and the higher critical temperature (T_{c1}) type II superconductor will be referred to as a HTII SC. This weaker coupling will then exist between the HTII SC and a thin film layer (TL) into the LTII SC adjacent to the HTII SC. In this paper, the thickness of the TL is taken to be on the order of the lattice parameter s_p.

The establishment of an electrical potential across the two adjoining superconductor could then possibly cause further coupling into the LTII SC through superconductor electron migration from the HTII SC to the HTII SC. Whereby, the coupling process established between the HTII SC and LTII SC can then be reversed by independent quenching of the electrons in this TL, which can be accomplished either by passing the proper current through the bulk superconductor, which could quench the thin layer through heating, or by relieving applied electrostatic forces that may have produced the coupling. In either case, if the coupling process induces electron mass motion back into the HTII SC, the notion of pancake vortices would produce coherence through the body of the bulk multi-layer superconductor along the pinned magnetic flux, in effect, multiplying the electron mass by the number of layers.

Although other designs may be possible, a simple test model under an applied electrical potential was devised and is schematically shown in Fig. **2**. The model consists of a multi-layered superconductor with the LTII SC against an anode (+) and the HTII SC layer against the cathode (-). The magnetic field B from the magnet coils is directed through the multi-layered superconductor and along the same *c*-axis as the applied voltage V_A. To insure current flow through the multi-layered superconductor and a long charge time, a switch (shown in high voltage oil to prevent premature arcing) is placed across the cathode to ground.

It is noted that a uniform current flow through the multi-layered superconductor will be required and methods to insure this, such as uniform density and characteristic of the superconductor, should be taken into account during the fabrication of the superconductor, as well as, proper integration into the test apparatus.

As shown in Fig. **2**, the application of the voltage V_A to the anode (+) and capacitor bank (C_B) would then cause electron and superconductor electron pair migration from the HTII SC to the LTII SC during charge buildup on C_B. Once the desired voltage is reached and the charging voltage is turned off, the switch is closed to produce a high current flow through the superconductor to C_B, which would result in charge neutralization and allow backward motion or relaxation of the superconductor electrons from the LTII SC and back into the HTII SC as long as the HTII SC stays in a superconductive state.

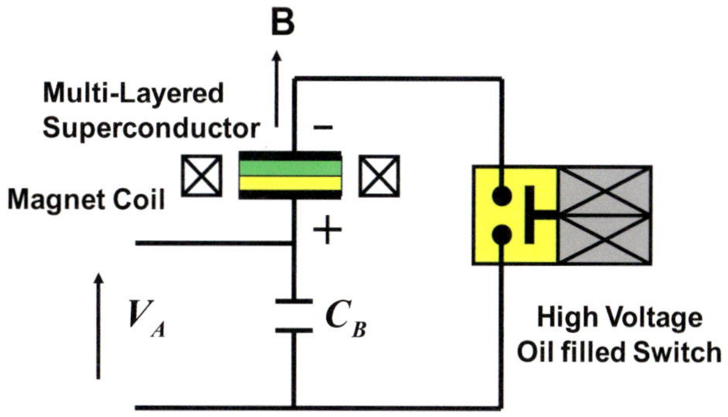

Figure 2: Multi-layered superconductor test model.

5. THE LANDAU-GINZBURG FREE ENERGY POTENTIAL

The potential contribution to the free energy across the interface or junction between the LTII SC and the HTII SC of Fig. **2**, neglecting the contribution from the magnetic field for the temperature dependent case, can be given as

$$V(\psi) = \alpha\,|\psi|^2 + \frac{1}{2}\beta\,|\psi|^4 \qquad \textbf{(12)}$$

where $V(\psi)$ is a potential density $\left[V/m^2\right]$ and ψ is the order parameter or field and where equation (12) can be viewed as a series expansion in powers of ψ^2.

Figure 3: Landau-Ginzburg free-energy functions with (a) $\alpha > 0$ and (b) $\alpha < 0$.

Two cases arise as illustrated in Fig. **3a** and **3b**, depending whether α is positive or negative. If α is positive (Fig. **3a**), the minimum free energy occurs at $|\psi|^2 = 0$, corresponding to the normal state. On the other hand, if $\alpha < 0$ (Fig. **3b**), the minimum occurs when

$$|\psi|^2 = |\psi_\infty|^2 \approx \frac{n_L}{2} \qquad \textbf{(13)}$$

where the notation $|\psi_\infty|^2$ is conventionally used because ψ approaches this value infinitely deep in the interior of the superconductor, where it is screened from any surface fields or currents [9].

Now, if the two materials are similar in crystalline structure, the two adjoining superconductors can be viewed as a single bulk superconductor. The Landau-Ginzburg free energy potential case for $\alpha < 0$ would then apply. For this case and neglecting the temperature dependence

$$V(\psi) = \frac{-\alpha^2}{2\beta} \tag{14}$$

and

$$-\frac{\alpha}{\beta} = |\psi|^2 = |\psi_\infty|^2, \tag{15}$$

which gives

$$\alpha = \frac{2}{|\psi_\infty|^2} V(\psi_\infty) \; [V] \tag{16}$$

and

$$\beta = \frac{2}{|\psi_\infty|^4} V(\psi_\infty) \; [Vm^2] \tag{17}$$

using equations (14) and (15).

Inputting equations (16) and (17) into equation (12) then gives $V(\psi) = V(\psi_\infty)$, where the average free energy across the layer-junction-layer with $\psi = \psi_\infty$, is given as

$$\begin{aligned} |E_J| &= \frac{1}{2} Q \; V(\psi_\infty) A_{sc} \\ &\approx \frac{1}{2} V_j C_J \; V(\psi_\infty) \; A_{sc} \\ &= |\psi_\infty|^2 \, e \, A_{sc}^2 \; V(\psi_\infty) \end{aligned} \tag{18}$$

by noting

$$\frac{n_L}{2} = \frac{Q}{2eA_{sc}} \approx \frac{V_j C_J}{2eA_{sc}} \tag{19}$$

where V_j is the junction voltage and C_J is the junction capacitance and A_{sc} is the surface cross-sectional area of the superconductor. $|E_J|$ is used to account for the negative direction of the energy potential $V(\psi)$ as noted in Fig. **3b**.

5.1. Landau-Ginzburg Energy Potential in the Dipole-Dipole Model

The dipole-dipole model of the junction between the two superconductor layers, with respect to the test model of Fig. **2**, is simply represented in Fig. **4**, where the junction interface is represented by the dotted line, the light gray dots represent superconductor electrons about the electrons (represented by the dark gray dots) in the vortex core due to the applied magnetic field, the black dots represent electrons in the LTII SC layer, and the light dots represent holes left by migrating electrons and superconductor electrons. Each column of three dots represents a pancake layer.

5.1.1. Thermal Quench

During cool-down to the phase locking temperature T_g between the critical temperatures of the two superconductors and with no applied electric potential $(V_A = 0)$, superconductor electrons (light grey dots) are formed in the HTII SC due to normal state change effects and in the thin layer (*TL*) due to proximity effects with the HTII SC as shown in Fig. **4(b)**. The superconductor electrons in the *TL* are coupled with the superconductor electrons in the HTII SC, which gives the Landau-Ginzburg free energy potential as

$$V(\psi_\infty) = \left(\frac{2\Delta}{e}\right) n_L \tag{20}$$

across the junction, where $2\Delta \sim 3.528\, k_B T_c$ is the BCS gap energy between two superconductor electrons. For the YBCO superconductor, $2\Delta \sim 0.028\,[eV]$ and $n_L \sim 1.09x10^{16}\,\left[m^{-2}\right]$, which gives $V(\psi_\infty) \sim 3.05x10^{14}\,\left[V/m^2\right]$ [8, 9].

Now, when a voltage potential (V_A) is applied, as shown in Fig. **4(a)**, some of the HTII SC's electrons in the vortex core (dark grey dots) and superconductor electrons (light grey dots) migrate across the junction with the number of migrating electrons plus superconductor electrons being proportional to the applied voltage potential. The electrons (black dots) in the TL are assumed to migrate toward the anode (+), which allows for the formation of holes (light dots) in the HTII SC when the electrons and superconductor electrons migrated across the junction. This produces a voltage gradient V_j across the junction, which adds into the Landau-Ginzburg free energy potential of equation (20) to give

$$V(\psi_\infty) = \frac{V_j}{A_{sc}} + \left(\frac{2\Delta}{e}\right) n_L \tag{21}$$

where the junction voltage V_j is typically in the range of a few mV or less.

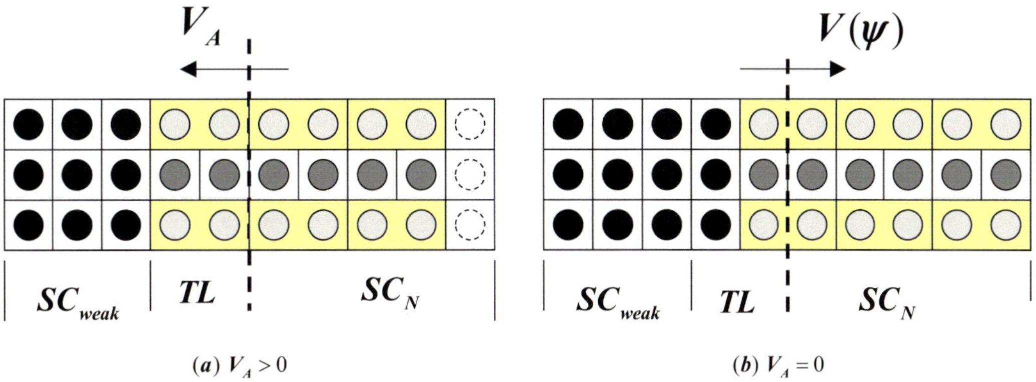

Figure 4: Electron motion between a HTII SC and a LTII SC under external voltage.

5.1.2. Non-Thermal, Time Dependent Quench

If the current flow through the bulk superconductor, during the neutralization of the voltage on the charging capacitor C_B (see Fig. **2**), is less than the critical current of the *TL*, no thermal quenching would result. However, the relaxation of the electrostatic forces within the bulk superconductor would produce a time-dependent relaxation of the superconductor electrons across the junction. This is so, because the superconductor electrons are a characteristic property of the superconductor and, therefore, would not flow in the current to the capacitor.

This would happen by closing the switch in Fig. **2** with the charging voltage off, where the Landau-Ginzburg free energy potential $V(\psi)$ would then force the migrated superconductor electrons from the *TL* and back into the HTII SC. The holes left by the migrating electrons from the vortex core are assumed to be filled by electrons from the current, whereby, there is no backward electron flow in the vortex core. Therefore, one would only expect the junction voltage V_j to add to the Landau-Ginzburg free energy potential, unless *TL* is thermally quenched.

In such a case, the time-dependent Landau-Ginzburg (TDGL) free energy potential $V(\psi,t)$ applies. If $\nabla \cdot J \neq 0$, as is the case with pinned flux, charge would build up to generate a non-uniform potential across the bulk superconductor until a new steady-state solution is established [9]. A central feature of this is a back-flow current, which goes through the vortex cores and returns around their sides. This current is lossless, although it flows directly through the vortex cores in the superconductor.

Here a similar view is applied. That is, during the application of the applied voltage V_A, both electrons and superconductor electron pairs cross into the *TL* due to electrostatic forces. When the switch (see Fig. **2**) is closed, the discharge current flows through the vortex cores and neutralizes the potential across the bulk superconductor. The electrostatic forces, established across the dipole (pancake) layers prior to the discharge, force the superconductor electrons (*i.e.*, super-current) out of the *TL* and back into the HTII SC. The discharge and backward superconductor current flow is modeled as shown in Fig. **5** with respect to the test model of Fig. **2**.

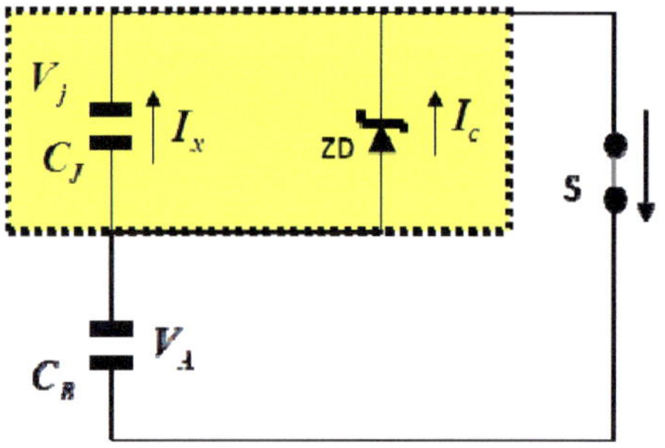

Figure 5: Discharge and backward superconductor current flow during switch closure.

In the model of Fig. **5**, the vortex cores are modeled as a zener diode *ZD* with a reverse bias voltage equal to the junction voltage V_j. This prevents the normal electrons from flowing through the vortex cores during charging of the capacitor C_B to the junction voltage V_A until the junction capacitors are charged to the junction voltage V_j.

When switch *S* is closed and since the C_J carries the equivalent zener diode *ZD* bias voltage, the current flows is trough the vortex cores until the bias voltage on the charging capacitor is reached. The junction now sits with the Landau-Ginzburg free energy potential of equation (21), which quickly quenches the bound between the normal and superconductor electron in the junction between the LTII SC and HTII SC.

This non-thermal quenching scenario means that there is no temperature dependence on α in equation (12) and that the electrostatically coupled superconductor electrons across the junction separate for a short time during the superconductor electron motion back into the HTII SC to reduce the V_j to zero. At which time they again form pair bonds with the superconductor electrons in the *TL* with gap energy 2Δ per electron due to standard proximity effects.

Because the superconductor electrons in the *TL* and in the HTII SC reform pair bonds, the gap energy 2Δ of the superconductor electrons would enter into the time-dependent Landau-Ginzburg free energy potential $V(\psi,t)$ only during the motion of the superconductor electrons across the junction. That is $(\Delta/\Delta t = 0)$, where none of the gap energy is use to for electron motion through the superconductor material. Such that at the time of discharge $(\Delta t = 0)$, the time-dependent Landau-Ginzburg free energy potential is

$$V(\psi_\infty,\ \Delta t = 0)\ A_{sc} \equiv V_J / A_{sc}\ , \qquad \textbf{(22)}$$

where V_J is defined as the time-dependent junction voltage, which gives the time-dependent Landau-Ginzburg free energy $|E_J|$ over some instantaneous time *t* through a single layer-junction-layer as given in equation (18) is then modified as

$$\begin{aligned}|E_J| &= \frac{1}{2}\, V_j\, C_J\, V(\psi_\infty,\, t)\, A_{sc} \\ &\approx \frac{1}{2}\, V_j\, C_J\, V_J\end{aligned} \quad . \tag{23}$$

5.2. Total Landau-Ginzburg Free Energy of the HTII SC

Now taking the assumption of quasi-independent layering, each pancake layer of size s_p acts independently and carries an equivalent junction voltage, which gives the total time-dependent Landau-Ginzburg free energy $|E_T|$ in the HTII SC as

$$|E_T| = \left(\frac{T_h}{s_p}\right)|E_J| \approx \frac{1}{2} C_J\, V_J \cdot \left(\frac{T_h}{s_p}\right) V_j\ , \tag{24}$$

where T_h is the thickness of the HTII SC.

This allows the applied potential V_A to be broken into small pieces across the quasi-independent pancake layers as long as the bulk superconductor is well insulated and internal breakdown does not occur over the short times scales proposed.

Given identical layers, and noting that the maximum time-dependent junction voltage $(V_J)_{\max} = V_j$, where

$$(V_J)_{\max} = V_j \approx \left(\frac{s_p}{T_h}\right) V_A\,, \tag{25}$$

which when combined with equation (24) and noting the time dependency on the discharge voltage V_A gives the final form of the total time-dependent Landau-Ginzburg free energy $|E_T|$ in the HTII SC as

$$\begin{aligned}|E_T| &\approx \frac{1}{2} C_J\, V_j^2 \left(\frac{T_h}{s_p}\right) \\ &= \frac{1}{2}\, C_J\, V_A^2 \cdot \left(\frac{s_p}{T_h}\right)\end{aligned} \quad . \tag{26}$$

6. QUANTUM EFFECTS IN SMALL-CAPACITIVE SINGLE JOSEPHSON JUNCTIONS

For the brief moment during the relaxation of the electrons across each thin pancake layer on either side of the junction, the layer-junction-layer can be viewed as a small-capacitor single Josephson junction. This leads to a time-dependent dimensionless current profile I_p, which must be accounted for in the time dependent junction voltage V_J in equation (26).

Current-voltage $(I-V)$ characteristics of single junctions have been extensively studied for small $I-V$ relationships. The small capacitance C_J per junction gives a large charging energy $e^2/2C_J$, and thus the theory predicts that at low enough temperatures a superconductive small-capacitance tunnel junction behaves like an insulator so long as the junction is isolated from the electromagnetic environment. Experimentally, however, the observation of this charging effect in a single junction is not straightforward, because the measurement leads attached to the single junction should have a high enough impedance to isolate the junction [10, 11].

The charge q and $\hbar/2e$ are quantum mechanically conjugate variables, and a set of the eigenfunctions are Bloch waves. When the energy eigenvalue E is plotted as a function of q, it has a band structure, and in all the allowed bands, it is $2e$ periodic. An example of the energy diagram for $E_J/E_{CJ} = 0.3$, where E_J is the junction energy and $E_{CJ} \approx e^2/2C_J$ is the charging energy, is shown in Fig. **6**.

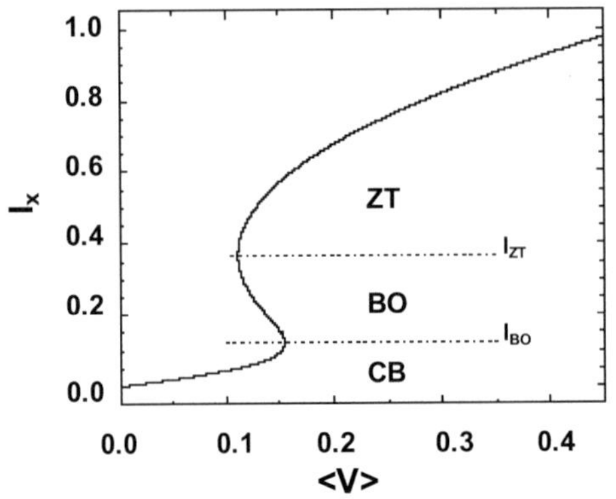

Figure 6: Energy diagram for $E_J/E_{CJ} = 0.3$ [10, 11].

Region *CB*: For sufficiently small I_x, the dominant process is stochastic quasiparticle tunneling, where q changes by e. This tunneling always occurs along the energy parabola $q^2/2C_J$. At sufficiently low temperatures, the tunneling with $\Delta E > 0$ is extremely unfavorable, and the $I-V$ curve is highly resistive.

Region *BO*: For larger I_x, the quasi-charge is frequently driven to the boundary of the Brillouin zone $q = e$, then taken to $-e$ as a superconductor electron tunnels (Bloch oscillation). This process decreases the junction voltage $\langle V \rangle$, and as a result the $I-V$ curve has a region of negative differential resistance, or "back bending," in the low-current part.

Region *ZT*: For still larger I_x, Zener tunneling becomes important, and the junction voltage $\langle V \rangle$ increases again. In Zener tunneling, no quasi-charge is transferred but the state of the system jumps from one band to another as it passes by the gap between the bands.

The current profile I_p can then be applied to the time dependent junction voltage by noting

$$V_J = (I_x)_{\max} R_{qp} I_p, \tag{27}$$

where R_{qp} is the quasiparticle resistance, which is normally on the order or $10^6\ \Omega$, which from equation (25) gives

$$V_J = V_j I_p \approx \left(\frac{s_P}{T_h}\right) V_A I_p, \tag{28}$$

such that the time dependent Landau-Ginzburg free energy of equation (26) takes the form

$$|E_T| \approx \frac{1}{2} V_j C_J V_A I_p. \tag{29}$$

7. MECHANISMS FOR HIGH INTERNAL POWER FLOW

Gravitational energy is generally thought to be quadrupolar in nature, such that they produce a perpendicular and alternating stretching and contraction of a mass. As such, low frequency gravity wave detectors look for this change. Therefore, one would expect phenomenon producing gravity waves to be quadrupolar. The linear motion of the electrons across the pancake layers in itself should not produce a gravitational effect. However, the magnetic field used to create the pancake vortices in each layer induces electron angular motion. As a result, motion along each pinned flux or vortex stack is quadrupolar. To illustrate this, the 3-D Landau-Ginzburg free energy function with $\alpha < 0$ is shown in the "Mexican Hat" form as shown in Fig. **7**. Where the height of the hat represents the total additive junction voltages (the applied voltage V_A) along a pancake stack and the balls represent superconductor electron pairs rotating about the pinned flux.

Note that the superconductor electrons would remain rotating as they move from the bottom to the top of the hat during external voltage relaxation to produce quadrupolar motion. Three are shown to note that there may be phase coupling among the typical granular structure of the YBCO superconductor.

Using the general formulation for calculating gravitational radiation from quadrupolar motion, the power radiated L_{rad} in gravitational waves is roughly approximated from the ratio of the square of the internal power flow $\Delta E/\Delta t$ by

$$L_{rad} \approx \left(\frac{G}{c^5}\right)\left(\frac{\Delta E}{\Delta t}\right)^2 \qquad \textbf{(30)}$$

where $G/c^5 \sim 10^{-52}$ $[W]$ [14].

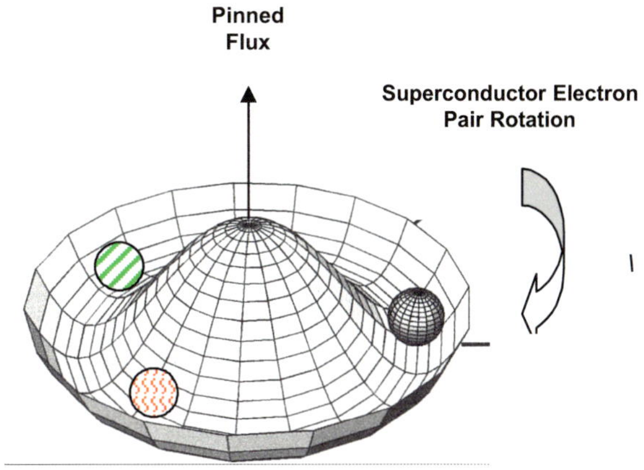

Figure 7: 3-D Landau-Ginzburg free-energy function with $\alpha < 0$.

For moderate values of the applied voltage $V_A \sim 10^5$ $[V]$, junction voltage $V_j \sim 10^{-3}$ $[V]$, junction capacitance $C_J \sim 10^{-7}$ $[F]$, and $\Delta t \sim 10^{-31}$ $[s]$, where $\Delta E \sim |E_T|$, equations (29) and (30) tells us that the internal power flow of the superconductor electrons must be very high even for a few watts of radiated gravitational power.

Since the maximum Landau-Ginzburg free energy is fixed by the superconductor and the applied voltage, the time rate of change Δt of the superconductor electrons must be very high. This requires high group velocities, which necessitates conditions that force the electrons outside their normal relaxation times, which are already relativity short. Two phenomena associated with superconductors could achieve this requirement, **1**) the dissipation of the Landau-Ginzburg free energy potential during a rapid superconductor quench referred to as symmetry breaking phase transition, which implies state changes on a very short time scale and **2**) the Hartman effect, which implies that the effective group velocity of the electrons across a superconductor junction can become arbitrarily large [5, 12].

Therefore, parallels between the dipole-dipole model as previously described and these two phenomena must then be accomplished in relationship to the internal power flow $(\Delta E/\Delta t)$ of the superconductor electrons. These parallels must rest on the behavior of free energy fluctuations during superconductor electron state transitions, especially in the vicinity of the junction between the two superconductor layers.

In order to explain superconductor electron state transitions, quenching models were derived and are presented here to illustrate quenching symmetry between two superconductor electrons and across a junction. These graphical interpretations are not intended to be exact models of the physical effects.

7.1. The Hartman Effect

Generally, one needs to associate gravitational effects using small masses (electrons) with the speed of light and short distances in order to drive high mass power flows. In nature, supraliminal electron velocities have been associated with cosmological events, lasers and electrostatic acceleration [30-32]. In the superconductor another phenomena associated with the junction tunneling process exists. Referred to as the Hartman effect, it indicates that for sufficiently large barrier widths, the effective group velocity of the electrons across a superconductor junction can become arbitrarily large [5, 33, 34].

Defining "tunneling times" has produced controversial discussion. Some of the definitions proposed lead to tunneling conditions with very short times, which can even become negative in some cases. This may seem to contradict simple concepts of causality. The classical causality principle states that the particle cannot exit a region before entering it. Thus the traversal time must be positive. However, when trying to extend this principle to the quantum case, one encounters the difficulty that the traversal time concept does not have a straightforward and unique translation in quantum theory. In fact for some of the definitions proposed, in particular for the so called "extrapolated phase time," the naive extension of the classical causality principle does not apply for an arbitrary potential, even though it does work in the absence of bound states [35, 36].

Generally, the Hartman effect occurs when the time of passage of the transmitted wave packet in a tunneling collision of a quantum particle with an opaque square barrier or junction becomes essentially independent of the barrier width [5, 35]. Since the velocity, when defined by comparing the instants of the incoming and outgoing peaks, may exceed arbitrarily large numbers, this "fast tunneling" has been frequently interpreted as, or related to, a "supraliminal effect" (see *e.g.,* [37-42]).

7.2. Voltage Induced Group Velocity

In the dipole-dipole model, the motion of the superconductor electrons back into the HTII SC would generally be expected to be on the order of the relaxation time τ_{sc} of the superconductor used, which for the YBCO superconductor is $\sim 200\ [fs]$ [8, 9]. Since this motion is assumed to occur over the thickness of the thin layer, which is one pancake layer or one lattice parameter s_p, as shown in Fig. **1**, then the group velocity v of the superconductor electrons can be approximated by

$$v \approx \frac{s_p}{\tau_{sc}}. \qquad \textbf{(31)}$$

For the YBCO superconductor, $v \sim 6x10^7\ [m/s]$, which is ~20% of light speed and implies that arbitrarily large electron velocities v could easily be present within a superconductor.

Also in the superconductor, a Bose-Einstein condensate (BEC)-like property of the electrons has been observed. That is, to an observer, all the electrons appear to be entangled in a multiple paired mode with BCS gap energy 2Δ [8, 9]. From the uncertainty principle, the superconductor electron pair fluctuation occurs on a time scale according to,

$$\Delta t = \frac{\hbar}{\Delta E}, \qquad \textbf{(32)}$$

which gives the fluctuation time $\Delta t = 2.35x10^{-14}\ [s]$ for the YBCO superconductor.

The electron pair fluctuation time scale of equation (32) is still nearly impossible to measure directly. However, using the proximity effects associated with Josephson junctions, the current frequency [8, 9] across the junction between two like superconductors can be measured according to

$$f = \frac{1}{\pi \Delta t} = \frac{\Delta E}{\pi \hbar}. \tag{33}$$

Experimentally equation (33) provides a connection between the time resolved, characteristic Josephson frequency v_f to the junction voltage V_j according to

$$k_f = \frac{v_f}{V_j} = \frac{2\,e}{h} = \frac{e}{\pi \hbar} = 4.84x10^{14}\ [Hz / V]. \tag{34}$$

where the constant k_f is normally given as $483.6\ [MHz/mV]$ [8].

The Hartman effect applied to equation (34) allows for speculation that time resolutions within the HTSC superconductor are subject to voltage variations across the junctions. For example, an instantaneous or a freeze-out time $\hat{t}$ of the motion through a junction and adjacent layer can be defined as

$$\hat{t} \approx \sqrt{\frac{\tau_{sc}}{v_f}} = \sqrt{\frac{\tau_{sc}}{k_f V_j}}, \tag{35}$$

which defines the instantaneous group velocity $\hat{v}$ over the spatial distance s_p as

$$\hat{v} \approx s_p \sqrt{\frac{k_f V_j}{\tau_{sc}}}. \tag{36}$$

However by letting $\hat{v} \to c$, it is found that $V_j \to 3\ [kV]$, which infers that another phenomena must be invoked as this voltage is much higher that normally seen with superconductor junctions.

7.3. Quenching Symmetry Models

In the BCS theory, a weak attractive interaction between electrons, in second order by the electron-phonon interaction, causes an instability of the ordinary Fermi-sea ground state of the electrons with respect to the formation of bound pairs of superconductor electrons occupying states with equal and opposite momentum and spin [8, 9]. These bound superconductor electrons have a spatial extension on the order of some characteristic length and, crudely speaking, comprise the superconductor charge carriers anticipated in the theories. Therefore, these bound superconductor electrons form the basic symmetry surrounding the quenching or electron state change process.

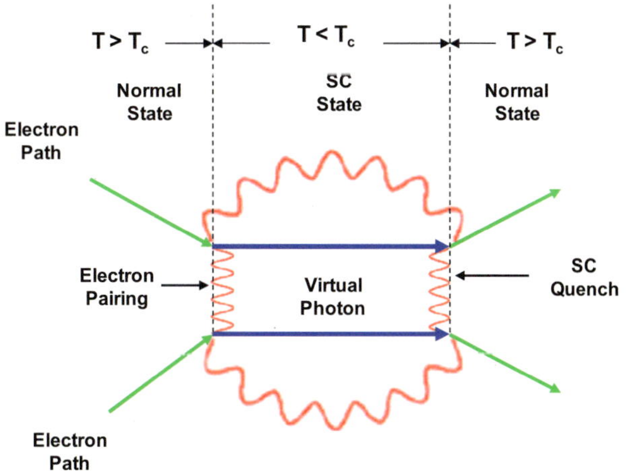

Figure 8: Quenching model of a superconductor electron pair in the bulk superconductor.

In this paper, the state transition symmetry from single electrons to bonded superconductor electrons and *vice versa* is modeled after the photoelectric effect as shown in Fig. **8**. In this model, normal state ($T > T_c$) electrons at different locations within the bulk superconductor agree on a common phase at the critical temperature T_c, which changes the way phonons within the bulk superconductor interact with the superconductor electrons. This phase transition is represented by the wavy-arch during the pair bond or superconductor (SC) state ($T < T_c$). To represent the phonon change in the bulk superconductor, a single virtual photon, which pops out of the vacuum between the superconductor electrons, is used to indicate the energy in the bonding process. The existence time scale of the virtual photon is assumed to be very short. Therefore, multiple virtual photons must fluctuate in and out of the vacuum, one at a time to maintain the bond between the electrons. The virtual photon fluctuations are represented by the wavy line between the two electrons.

When the critical temperature T_c is again reached during warmup ($T > T_c$), the electron-phonon interaction reverts to the normal state. The phonon reversal to the normal state in the bulk superconductor, then forces the virtual photons to return to the vacuum and prevent further ones from coming out.

In Fig. **8**, the cool down state-change is referred to as "Electron Pairing," whereas, the warmup state-change is referred to as "SC Quench." This same terminology is used throughout the paper.

When two superconductor materials with critical temperatures $T_{c1} > T_{c2}$ compose a junction, proximity effects induce a phase-locking temperature T_g between them, at which, the junction network exhibits coherent properties. This is the case for a junction interface between the HTII SC with critical temperature T_{c1} and a TL into the LTII SC having a critical temperature T_{c2}. This is modeled in Fig. **9**.

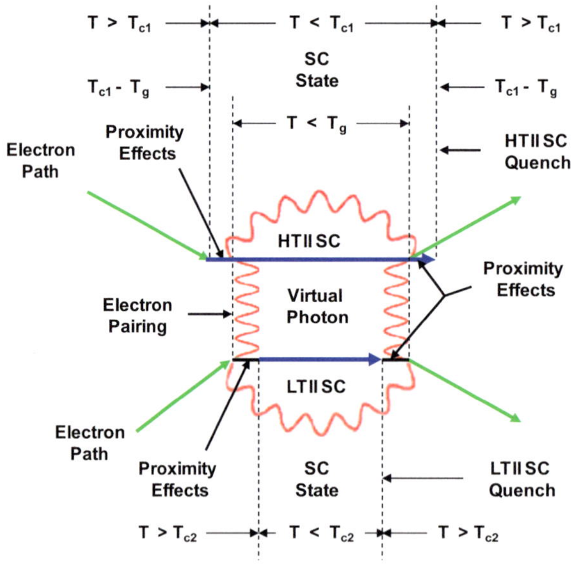

Figure 9: Quenching model of an superconductor electron pair across a junction between the HTII SC and LTII SC.

In the model of Fig. **9**, proximity effects during cool-down prevent the superconductor electrons in the LTII SC from pairing with electrons in the HTII SC until the phase locking temperature T_g is reached. When $T < T_g$, the quenching model of Fig. **9** retains the symmetry to Fig. **8**.

7.4. Planck Scale Fluctuation Time

When dealing with interactions described by massless vector particles (photons) within a relativistic local quantum field theory, Binder indicates that Planck units are assigned to the background fluctuation level and provide for a common base [43]. Therefore, the choice for the fluctuation time of the virtual photon is taken to be the Planck time,

$$T_{pl} = \sqrt{\frac{\hbar G}{c^5}} = 5x10^{-44}\ [s]. \tag{37}$$

However, the Planck time is too fast to be observed, which infers that human units must be artificially imposed when measuring superconductor electron fluctuations (or any other Planck time phenomena). To explain this, it is noted that just before electron phase transition and superconductor pair bonding, each electron had an energy deficit $\Delta E\ \sim 1\ [eV]$. From the uncertainty principle, the electron (actually the electron-atom-photon system) can maintain this deficit before pairing for a time Δt according to equation (32), which for $\Delta E\ \sim 2\ [eV]$, $\Delta t \sim 3.3x10^{-16}\ [s]$.

Human artificial units are imposed on superconductor electron fluctuations by the uncertainty principle and can be determined by combining equations (32) and (37), which yields

$$\Delta t = \left(\left[\frac{c^5}{G} \right] \frac{T_{pl}}{\Delta E} \right) T_{pl} \tag{38}$$

From equation (38) the time related human artificial units $(H.A.U.)_T$ are determine to be

$$(H.A.U.)_T = \left[\frac{c^5}{G} \right] \frac{T_{pl}}{\Delta E} = \frac{E_p}{\Delta E} \tag{39}$$

or the ratio of the Planck energy $E_p = \hbar / \Delta\ E \approx 2\ x10^9\ [J]$ to the system energy ΔE .

Equations (38) and (39) then tell us that Planck scale energy fluctuations $[J/s]$ occur for quantum systems according to

$$\frac{\Delta E}{T_{pl}} = \left[\frac{c^5}{G} \right] \frac{T_{pl}}{\Delta t} = (H.A.U.)_T \frac{\Delta E}{\Delta t} = \frac{E_p}{\Delta t}, \tag{40}$$

which for $\Delta t \sim 3.3x10^{-16}\ [s]$ gives a Planck time energy fluctuation $E_p \sim 6x10^{24}\ [J/s]$ per superconductor electron pair. However, the actual human observable system energy $\Delta E = E_p T_{PL} \sim (6x10^{24})(5x10^{-44}) = 3x10^{-19}\ [J]$ per superconductor electron pair, which still implies that many superconductor electron pair interactions much occur coherently before current laboratory scale equipment, can detect a signal.

8. SYMMETRY BREAKING PHASE TRANSITION

Equation (40) leads us back to the uncertainty principle by noting

$$\hbar = \Delta t \Delta E = T_{pl} E_p, \tag{41}$$

which does not give us much hope in separating Planck scale energy fluctuations from system energy fluctuations.

However, it might be possible through symmetry breaking phase transitions by noting that at the instant of the release of the free energy there is a freeze out time [12] $\Delta t = \hat{t}$ between the transition from the adiabatic (Planck time fluctuations) and impulse (relaxation time) regimes [13], given by

$$\hat{t} = \sqrt{T_{pl} \tau_{sc}}\ . \tag{42}$$

8.1. Symmetry Breaking Quenching Model

Phase transitions between the high and low temperature phases of a superconductor involve spontaneous symmetry breaking between states, which could have total energies that differ by many *TeV* of energy as shown from equation (20). Kibble explains that symmetry-breaking phase transitions are ubiquitous in condensed matter systems and in quantum field theories [44]. There is also good reason to believe that they were prominent in the very early history of the Universe as many such transitions topological defects of one kind or another are formed. Because of their inherent stability, they can have important effects on the subsequent behavior of the system.

Theoretically and experimentally, spontaneous defects in superconductors show up as vortices during rapid quenching, which is a phase transition between the normal and superconductive states. Experiments with thin film superconductors have been conducted and show evidence that the Kibble-Zurek mechanism of vortex formation exists during phase transition [13, 45]. Some differences are noted because in Type II superconductors one must consider a local gauge theory rather than a global symmetry breaking [46-49]. These differences arise from the flux trapping mechanism, which stems from the fluctuations in the magnetic field that get frozen out during the quench [45].

The freeze-out time $\hat{t}$ of equation (42) is on the order of $\sim 10^{-30}\ [s]$, which for *ideal conditions* could produce instantaneous Planck time energy fluctuations $\Delta E/T_{pl} \sim 10^{38}\ [J/s]$ per superconductor electron pair with energy release on the order of *200 [TeV]*. Given the large electron volume density, $n_{3D} \sim 10^{28}$ electrons per cubic meter, coherent and directional quenching could release extremely large energies associated with extremely high electron motion across the junction, which would give instantaneous power radiated in gravitational waves of

$$\hat{L}_{rad} = \left(\frac{G}{c^5}\right)\left(\frac{E_p}{\hat{t}}\right)^2 \approx 10^{25}\ [J/s] \tag{43}$$

and if propagated on the order of the freeze-out time of equation(42) gives the radiated energy in gravitational wave $\sim 10^{-5}\ [J]$ per superconductor electron pair.

A symmetry breaking, quenching model (SB-QM) of two superconductor electrons across a junction between the HTII SC and LTII SC was devised and is shown in Fig. **10** where the symmetry of Fig. **9** is broken at T_q. (In a thermal quench, T_q would be the phase locking temperature T_g of the *TL*.) The back to back waving line between the two superconductor electrons indicate a virtual photon being forced back into the vacuum and reappearing on the Planck time scale. It is during this time that the superconductor electrons can produce internal power flows on the Planck time.

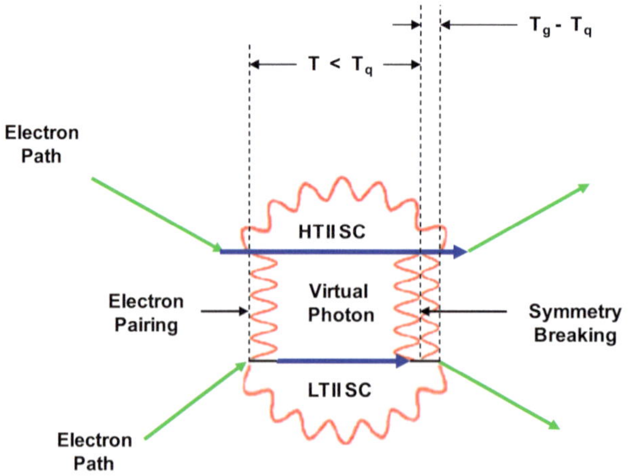

Figure 10: SB-QM of two superconductor electron across a junction between the HTII SC and LTII SC.

8.2. Internal Power Flow Model

Unidirectional and coherent symmetry breaking is however required for the gravitational effect, that is multiple electron phase transitions with electron motion in a single direction. It is proposed here that such a condition can be obtained through junction phenomena such as described by the Hartman effect.

Muga *et al.* notes that, when bound states are present, *e.g.*, for square wells, the advancements may be much more important (even though also bounded by a causal principle) at low energies than those characteristic of the Hartman effect for square barriers. Thus, it seems appropriate to rename the time advancement due to the bound states as an "ultra Hartman effect". this enhanced effect may be readily observable since the low energy time advancement is accompanied by high transmission probabilities close to the onset of a new bound state [36].

Non-thermal quenching between the electrostatically coupled superconductor electrons in the *TL* provides a means to invoke the "ultra Hartman effect" as the mechanism for producing high internal power flow.

A Hartman effect model, shown in Fig. **11** to illustrate the time scale relationship of the fluctuating superconductor electrons across a junction was devised and is based on previous models. In the model, superconductor electrons are shown about the junction potential barrier V_j. The potential ramp and superconductor electrons (green and clear circles) are shown only for clarity.

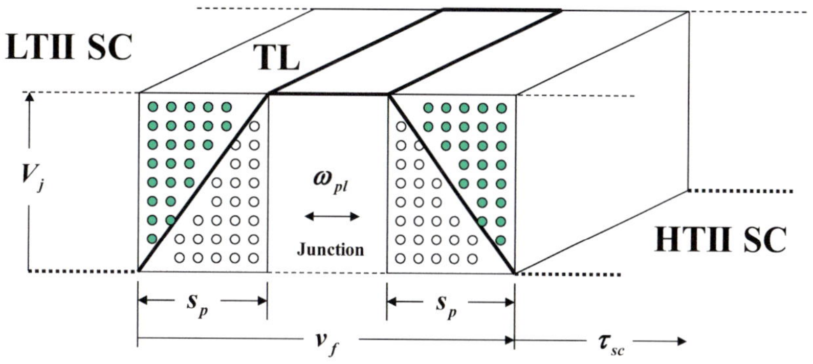

Figure 11: Time scale relationship of fluctuating superconductor electrons across a junction between two pancake layers.

Under steady state conditions (no current flow) with or without an applied electric potential, superconductor electrons (clear circles) appear to fluctuate back and forth across the gap at the Planck frequency

$$\omega_{pl} = \frac{1}{\pi\, T_{pl}} \tag{44}$$

to maintain coherence with the superconductor electrons (green circles) in the LTII and HTII SC.

Under this model and during the applied potential as shown in Fig. **2**, a forward junction current (along with the vortex core current) flows through the HTII SC and the *TL* of the LTII SC with the characteristic junction frequency v_f until the voltage drop across the bulk superconductor equals that on the capacitor. Closing of the switch results in the current I_c until the bias voltage is reached. At which time, superconductor electrons jump backward across the junction at frequency w_{pl} and pair with superconductor electrons in the first layer into the HTII SC. These paired superconductor electrons move across the first layer in the HTII SC at the freeze-out time $\hat{t}$ given in equation (42) due to the residual potential across the pancake layers in the HTII SC and cause a cascading effect of superconductor electron motion across all the junctions in the HTII SC.

However, the voltage across the thickness HTII SC decays as each layer is transverse by the superconductor electrons, such that, a decaying voltage V_n can be defined using equation (25) and given as

$$V_n \approx \left[\frac{T_h}{s_p} - n\right] V_j , \tag{45}$$

where $n = 0,\ 1,\ 2,\ 3\ ....,m$ and $m = T_h / s_p$ is the number of pancake layers in the HTII SC.

Equation (45) indicates a possible step-like decay time $\Delta t = t_d$ of the form

$$\begin{aligned} t_d &= \hat{t} + \hat{t}\left(\frac{V_{m\text{-}1}}{V_A}\right) + \hat{t}\left(\frac{V_{m\text{-}2}}{V_A}\right) + \\ &\dots + \hat{t}\left(\frac{V_{m-n}}{V_A}\right) + \tau_{sc} \end{aligned} , \tag{46}$$

where the voltages $V_m = V_A$ to V_i are the incremental voltage decays due to superconductor electron relaxation in each layer. The change from the freeze-out time $\hat{t}$ to the superconductor relaxation time τ_{sc} indicates that the superconductor electron motion decays to the characteristic frequency v_f at some voltage $V_{m-(n-1)}$.

Equation like that of equation (46) are complex to analyze. Since this is a quantum system, the energy flow time scale is driven by the Planck time scale as given in equation (40). Then noting that the junctions subjects the decay time t_d to the current profile I_p, the decay time t_d is approximated as

$$t_d \approx T_{pl}\ I_p\ . \tag{47}$$

Using equation (47) and (29), the instantaneous power flow for the symmetry breaking model is then given as

$$\frac{\Delta E}{\Delta t} = \frac{|E_T|}{t_d} \approx \frac{|E_T|}{T_{pl} I_p} . \tag{48}$$

9. ENERGY RADIATED IN GRAVITATIONAL WAVE

The energy radiated Q_{rad} in gravitational waves is given from equation (30) by noting

$$Q_{rad} \approx \frac{L_{rad}}{v_f I_p} , \tag{49}$$

where the characteristic junction frequency v_f is used for the time integral as this would be the expected average frequency of energy propagation through the bulk superconductor and I_p indicates a time correction to the characteristic junction frequency v_f due to the junction current-voltage $(I - V)$ characteristic.

Combining equation (49) with equations (30) and (48) gives the total energy in gravitational wave as

$$Q_{rad} \approx \frac{|E_T|^2}{\hbar v_f I_p^3} \tag{50}$$

by noting from equation (37) that

$$\frac{1}{\hbar}=\frac{G}{c^5 T_{pl}^2}. \qquad \textbf{(51)}$$

9.1. Mass Coupling

An indication of the power radiated L_{rad} in a gravitational wave comes from the measurement of the absorbed energy Q by a mass and is given according to

$$Q \approx Q_{rad}\alpha_G \qquad \textbf{(52)}$$

where α_G is the mass coupling constant.

The coupling constant can be quite complex, but is typically related to the total energy available from the source times the characteristics of the mass radiated [50].

Here, the quantum-quadrupole nature of the radiated energy Q_{rad} establishes the pancake model of the type II superconductor as a self-consistent systems of mutually interacting gravitational quadrupoles, where a characteristic number N of quantum masses μ are related to a characteristic velocity scaling through the pancake layers of the bulk superconductor. For this purpose a critical reference is defined by the (gravitational) flux and flux number of mass quanta constituting a confining unit field generating mass

$$m_G = N\mu \qquad \textbf{(53)}$$

with the unit flux composed of N flux quanta the characteristic velocity scaling becomes a number scaling where the local and global aspects are coupled with only one real degree of freedom: the value of N [51].

Applying this concept to the independent layers in the pancake model, the coupling constant α_G with respect to the superconductor (local aspect) and the mass radiated (global aspect) is approximated here as

$$\alpha_G \approx \left(c^2\mathrm{R}\right)\left[N\ t_p^2\right], \qquad \textbf{(54)}$$

where $\mathrm{R}\left[cm^{-2}\right]$ is the three-dimensional scalar curvature invariant associated with the mass radiated, t_p is the average propagation time of the electrons through the bulk superconductor.

The characteristic junction frequency v_f is again chosen as the time integral, where $t_p \approx 1/v_f$, which gives equation (54) as

$$\alpha_G \approx N\left(\frac{c^2\mathrm{R}}{v_f^2}\right). \qquad \textbf{(55)}$$

The junction current profile I_p is not used here as the coupling constant is a property external to the superconductor. That is, the exiting gravitational wave is no longer affected by the electron propagation properties in the bulk superconductor.

Now, in order to establish a direct connection between the coupling constant and the mass radiated, the three-dimensional scalar curvature invariant $\boldsymbol{R}$ of the mass radiated must be converted to the mass density form R_{conv}, which has units of $\left[g/cm^3\right]$ [14]. This is done by noting that $R_{conv}=\mathrm{R}\left(c^2/G\right)$, which gives equation (55) as

$$\alpha_G \approx N\left(\frac{16\pi\rho_m G}{v_f^2}\right), \tag{56}$$

where $R_{conv} \sim 16\pi\rho_m$ and ρ_m is the density of the mass radiated.

The flux quanta N is determined by first approximating the field generating mass m_G from the total number of pancake layers in the HTII SC by

$$m_G \approx \frac{Q_{rad}}{c^2}, \tag{57}$$

which from equation (53) gives the flux quanta as

$$N \approx \frac{Q_{rad}}{\mu c^2}. \tag{58}$$

Then by noting that the quantum mass μ is proportional to the field generating mass m_G per the total number of pancake layers in the HTII SC, the quantum mass μ per layer is given by

$$\mu \approx m_G \div \left(\frac{T_h}{s_p}\right), \tag{59}$$

which from equations (53) and (58) sets the value of the flux quanta N as

$$N \approx \frac{T_h}{s_p} \tag{60}$$

or the number of pancake layers in the HTII SC, which gives the final form of the gravitational coupling constant α_G as

$$\alpha_G \approx \left(\frac{16\pi\ \rho_m G}{v_f^2}\right)\left(\frac{T_h}{s_p}\right). \tag{61}$$

9.2. Absorbed Energy from Gravitational Waves

Combining equations (50) and (61) with equation (52), yields the absorbed energy Q by a mass radiated of density ρ_m as

$$Q \sim (16\ \pi\ \rho_m)\left(\frac{G}{\hbar}\right)\left(\frac{T_h}{s_p}\right)\left(\frac{|E_T|^2}{I_p^3\ v_f^3}\right). \tag{62}$$

Now, by noting equation (34) where $V_j = v_f / k_f$, the Landau-Ginzburg free energy of equation (29) can be given as

$$|\,E_T\,| = \frac{1}{2}\left(\frac{T_h}{s_p}\right)\left(\frac{v_f}{k_f}\right)V_A\ C_J\ I_p, \tag{63}$$

which when combined with equation (62) and noting equation (25) where $V_j = (s_p/T_h)V_A$ gives the absorbed energy Q for a given superconductor and mass radiated as a function of the applied voltage with respect to the junction current profile I_p by

$$Q \sim (4\,\pi\,\rho_m)\left(\frac{G}{\hbar}\right)\left(\frac{T_h}{s_p}\right)^2\left(\frac{C_J^2}{k_f^3}\right)\left[\frac{V_A}{I_p}\right]. \tag{64}$$

10. COMPARISON TO THE IMPULSE FORCE EXPERIMENT

An experiment referred to as the impulse force experiment was reported by Podkletnov and Modanese and bears similarity to the test model of Fig. **2** [15, 16]. The primary differences are in the voltage delivery system and in the electrode-capacitive arrangement. The electrode-capacitive arrangement is shown in Fig. **12**, where the voltage delivery system is a capacitor Marx Bank and the bulk superconductor was placed in a ~10 Torr vacuum with an assumed Helium environment due to leakage in the cryo-system used to cool the superconductor.

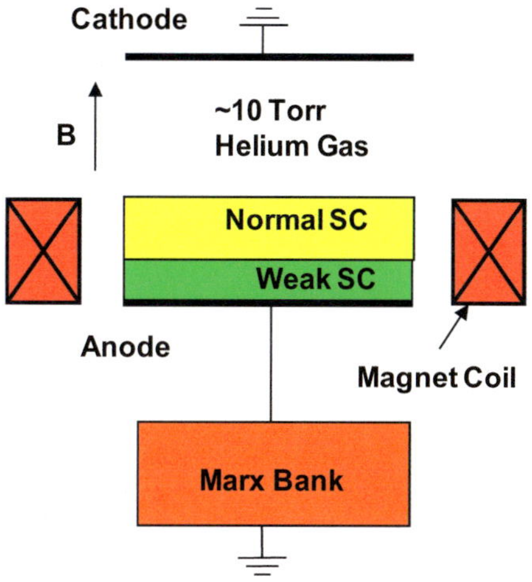

Figure 12: Impulse experiment electrode and marx bank arrangement.

The report indicated that anomalous gravity-like impulse forces were detected on pendulum masses during electrical discharges across the vacuum at voltages $\geq 500\ [kV]$ when a superconductor was attached to the anode and in a superconductive state at temperatures between 40^0 [K] and 70^0 [K]. These temperature are well below the YBCO critical temperature of $\sim 93^{o}$ [K], making it unlikely that the HTII SC layer could be quenched by a current less than the critical current, which implies that a non thermal, time dependent quench would exist during discharge.

10.1. Maximum Values *vs.* Experimental Data

The surface area A_{sc} of the bulk superconductor was not given, but was reported to have a diameter between $80\ [mm]$ and $120\ [mm]$.

Given that the capacitance of the thin layer with thickness s_p is given by

$$C_J = \frac{k\ \varepsilon_0\ A_{sc}}{s_p} \tag{65}$$

and letting $A_{sc} = 5x10^{-3}\ [m^2]$, where the capacitance C_J of the thin layer is then $3.74x10^{-5}\ [F]$ (for $k=1$) and letting

$$V_j = \frac{Q}{C_J} \sim \frac{s_p}{T_h} V_A, \tag{66}$$

where the charge

$$Q = n_L A_{sc} e \,. \tag{67}$$

The maximum junction frequency v_f using equation (34), related maximum junction voltage V_j using equations (66) and (67) and the maximum applied voltage $V_A(4)_{\max}$ and $V_A(8)_{max}$ using equations (66) and (67) (4 and 8 refer to the $[mm]$ thicknesses of the SC samples used) are tabulated in Table **1.**

Table 1: Calculated maximum values.

$V_j = 235\ [mV]$
$v_f = 1.14x10^{14}\ [Hz]$
$V_A(4)_{\max} = 8\ [MV]$
$V_A(8)_{\max} = 16\ [MV]$

It would then be expected that the values in the experiment would be less than or equal to the maximum values listed in Table **1**.

To verify this, it is first noted that the calculated values for the *TL* are the same regardless of the thickness of the HTII SC layer. A tabulation versus the applied voltages was done and is shown in Table **2** for $I_p = 1$, using equation (34) for the junction frequency v_f, equations (66) and (67) for the junction voltage V_j, equation (29) for the total free energy $|E_T|$, and equation (50) for the radiated gravitational energy Q_{rad}.

Table 2: Calculated values for the thin layer.

V_A	v_f	V_j	$\|E_T\|$	Q_{rad}
	(10^{13})	(10^{20})		
MV	*Hz*	*mV*	*J*	*J*
0.50	0.72	15	0.14	0.26
0.75	1.08	22	0.31	0.86
1.00	1.44	30	0.56	2.04
1.25	1.80	37	0.87	3.98
1.50	2.16	45	1.25	6.88
1.75	2.52	52	1.70	10.90
2.00	2.88	60	2.23	16.30

The applied voltages V_A are taken from the experimental data as shown in Table **3** and are noted to be less than the maximum values of Table **1**, which leads to values for the junction voltage V_j, and the propagation frequency v_f, from which the total free $|E_T|$ and the radiated gravitational energy Q_{rad} are calculated, below the maximum values as given in Table **1**.

10.2. Predicted Height Model

By noting that the energy applied to the pendulum mass m can be calculated using the formula

$$Q = m\, g\, h\,, \tag{68}$$

where $g = 9.8\ [m/s]$, and $h\ [m]$ is the upward height moved by the pendulum during the discharge and combining equations (64) and (68), the predicted height $(h = h_p)$ can be given as

$$h_p \sim \left(\frac{4\ \pi\ \rho_m}{m\ g\ h}\right)\left(\frac{G}{\hbar}\right)\left(\frac{T_h}{s_p}\right)^2\left(\frac{C_J^2}{k_f^3}\right)\left[\frac{V_A}{I_p}\right]. \tag{69}$$

The experimental data from reference [15] (which was only plotted in referecne [16]) for the two superconductors reported is given in Table **3** and plotted in Fig. **13**. The thickness of the HTII SC was 8 $[mm]$ for $h_{exp(8)}$ and 4 $[mm]$ for $h_{exp(4)}$.

Table 3: Experimental data from Reference [15, 16].

Voltage	$h_{exp(8)}$	$h_{exp(4)}$
MV	*m*	*m*
0.50	0.0020	0.0010
0.75	0.0052	0.0032
1.00	0.0077	0.0046
1.25	0.0095	0.0056
1.50	0.0109	0.0064
1.75	0.0119	0.0069
2.00	0.0127	0.0072

10.3. Current-Voltage $(I-V)$ Relationship

The curvature in the plot of the experimental data (Fig. **13**) can be explained by examining the current profile I_p through the superconductor junctions with respect to the voltage. Using the experimental values for the height as given in Table **3**, the current profile-voltage $I_p - V$ relationship was calculated and is plotted in Fig. **14**.

The current profile-voltage $I_p - V$ relationship approaches a turning point between the junction voltages of 0.15 and 0.20, which can be explained from the two phases of the decay time of equation (46). Due to the exponential like decay of Fig. **14**, each point can then be expressed as an exponential of the freeze-out $\hat{t}$ and relaxation τ_{sc} times by

$$I_p \approx \gamma_{sc}\left(e^{-t/\hat{t}}\right) \tag{70}$$

for $t \le \hat{t}$ and

$$I_p \approx \gamma_{sc}\left(e^{-t/\tau_{sc}}\right) \tag{71}$$

for $\hat{t} < t \le \tau_{sc}$, where γ_{sc} is a constant related to the superconductor.

The calculated current profile I_p values using equation (69) and the times t for the 4 $[mm]$ and 8 $[mm]$ thicknesses, where $h_p = h_{exp}$ from Table **3**, are tabulated in Table **4**. The superconductor constants $\gamma_{sc}(4) = 4$ and $\gamma_{sc}(8) = 8$ were arbitrarily chosen to fit the current profiles I_p of Fig. **14** and need further study before a thickness connection can be assumed.

As shown in Table **4**, the times $t(4)$ and $t(8)$ at 0.5 $[MV]$ are $> \hat{t}$, but as is shown in the next section, these points are in the **Region** *BO* (see Fig. **6**) and would follow a different current profile than the **Region** *ZT* assumption taken here. At 0.75 $[MV]$ the time value is $\hat{t} \sim 3.28 \times 10^{-30}\ [s]$ and for the rest

$\hat{t} < t \leq \tau_{sc} \sim 2 \, x \, 10^{-16} \, [s]$ as they should be, which indicates that the experimental data with respect to equations (70) and (71) fit well with the model from which equation (69) was developed.

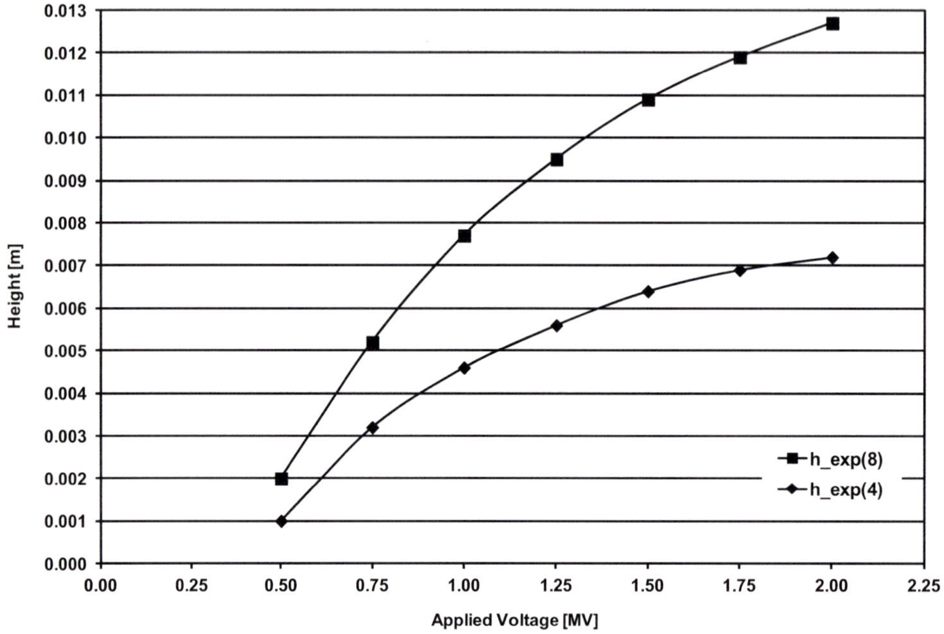

Figure 13: Plot of experimental data ($h_{\exp(8)}$ and $h_{\exp(4)}$) from Table **3.**

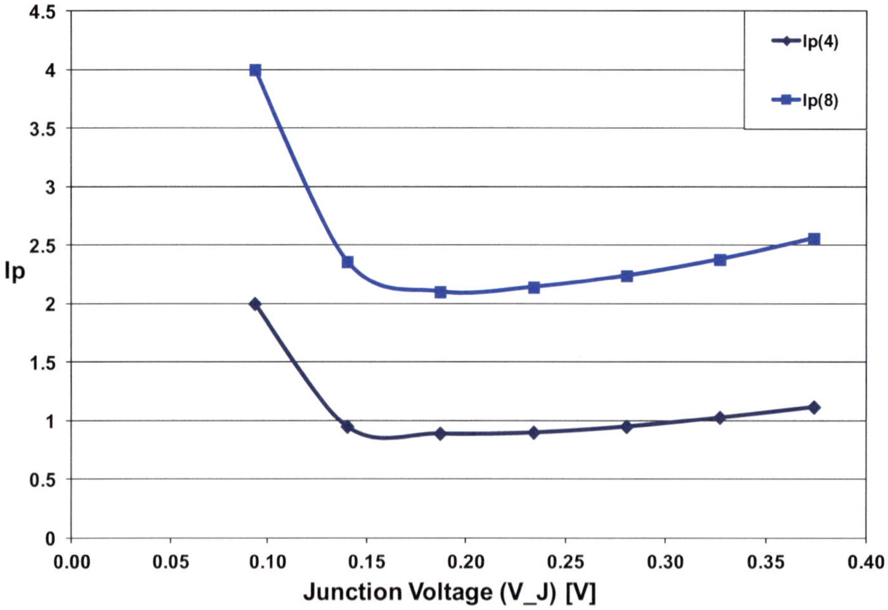

Figure 14: Experimental current profiles I_p of the junction current I_x .

The proper current-voltage $I-V$ relationship for the backward charge motion can now be determined using equations (27) and (28), the junction voltage V_j from Table **2**, the current profile values I_p from Table **4**, and letting $R_{qp} = 10^6 W$. This relationship is plotted in Fig. **15** and Fig. **16** along with the respective applied voltage V_A .

Table 4: Calculated profile and time values for the HTII SC Layer with $\gamma_{sc}(4)=2$ and $\gamma_{sc}(8)=4$.

V_A	$I_p(4)$	$t(4)$	$I_p(8)$	$t(8)$
MV	s	s		
0.50	2.00	1.00x10^{-33}	4.00	1.00x10^{-33}
0.75	0.95	2.45x10^{-30}	2.36	1.73x10^{-30}
1.00	0.89	1.63x10^{-16}	2.10	1.29x10^{-16}
1.25	0.90	1.60x10^{-16}	2.14	1.25x10^{-16}
1.50	0.95	1.48x10^{-16}	2.24	1.16x10^{-16}
1.75	1.03	1.33x10^{-16}	2.38	1.04x10^{-16}
2.00	1.12	1.16x10^{-16}	2.56	8.90x10^{-17}

The $I-V$ profile of Fig. **15** and Fig. **16** are now very similar to the typical $I-V$ relationship of Fig. **6** and represent a value for I_{ZT} near 0.75 $[MV]$, which is $\sim\left[\left(\frac{3}{4}\sqrt{3}\right)\frac{m_e c^2}{e}\right]$. Although it may be just a coincident, the $\frac{1}{2}\sqrt{3}$ may have some connection to the vortex area of equation (1), again further investigation is needed.

The current-voltage $I-V$ relationship of the experimental data as shown in Fig. **15** and Fig. **16** further validates that the experimentally measured energy in the gravitational wave follows the internal characteristics of the type II superconductor.

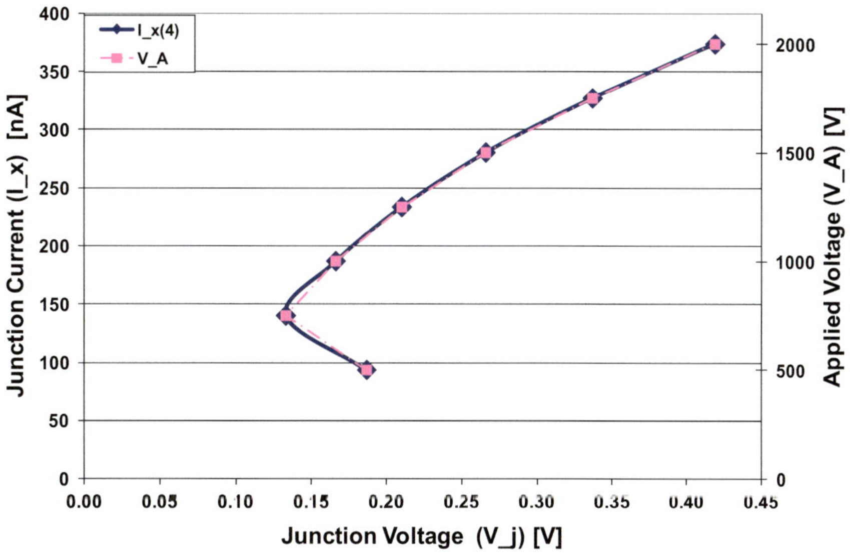

Figure 15: Current-voltage $I-V$ relationship of the 4 $[mm]$ experimental data.

11. SUMMARY

Quantum gravity was explored for the type II superconductor with particular interest in the Yttrium Barium Copper Oxide (YBCO) superconductor, by first reviewing some unique characteristics of the type II superconductor, which allows for coherence through a quasi-independent pancake layering phenomena to form multiple stacked weakly coupled Josephson junctions through the bulk superconductor. Dipole-dipole coupling internal to the superconductor was discussed as the energy transfer mechanism by considering the pancake layers as 1D stacks of dielectric films. The availability of free energy under the application and discharge of an external voltage through the superconductor was discussed in terms of the Landau-Ginzburg free energy potential with consideration to the quantum effects associated with small capacitive single Josephson junctions.

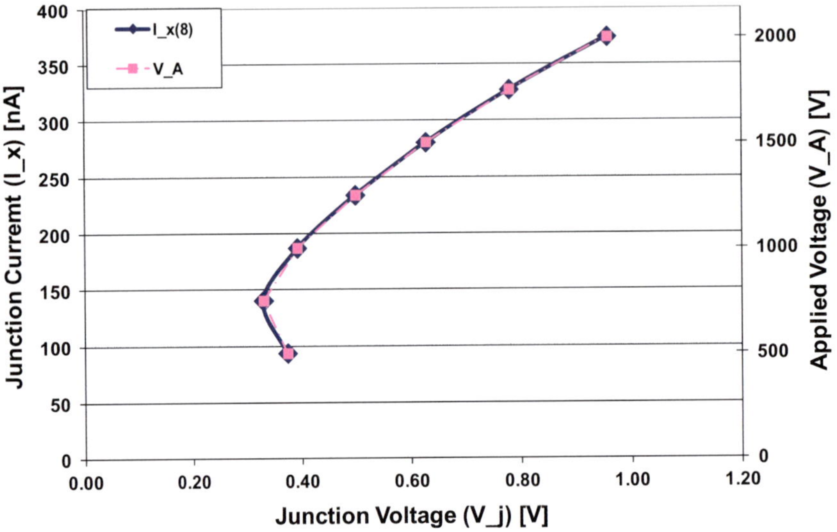

Figure 16: Current-voltage $I-V$ relationship of the $8\ [mm]$ experimental data.

Mechanisms for high internal power flow are discussed and Planck scale fluctuations were introduced. Symmetry breaking phase transition was discussed to introduce the freeze-out time between the transition from adiabatic (Planck time fluctuations) and impulse (relaxation time) regimes. The freeze-out time was used to develop the decay time of the electron motion through the layered superconductor, which is used to develop the instantaneous power flow available during a spontaneous symmetry breaking phase transition. The possible energy radiated in a gravitational wave was determined from the power radiated approximation and corrected for mass coupling, which allowed for the development of a model of the energy from a gravitational wave absorbed by a mass. A comparison of the absorbed energy in a gravitational wave model to the experimental data was conducted by calculating superconductor characteristics from the experimental data and comparing them to known superconductor characteristics.

12. CONCLUSIONS AND DISCUSSION

The high temperature superconductor was analyzed to provide a link to gravity in that quantum gravity events may affect quantum systems that operate at the macroscopic scale, in effect creating a macroscopic "memory" of the event. It was shown that arbitrarily large electron group velocities associated with the Hartman effect and induced by spontaneous symmetry breaking phase transitions could conceivably result in a gravitational disturbance.

Further evidence is given by Pospelov and Romalis who cite that unknown dynamics at the Planck scale could possible lead to such things as Lorentz violation [52]. They tell us that the breaking of Lorentz symmetry enables the CPT symmetry, which combines charge conjugation (C), parity (P), and time-reversal (T) symmetries, to be violated. In conventional field theories, the Lorentz and CPT symmetries are automatically preserved. But in quantum gravity, certain restrictive conditions such as locality may no longer hold, and symmetries may be broken.

Still further, the Hartman effect illustrates Hawking's discussion about ambiguities in defining time in relation to quantum mechanics and cannot be ruled out during spontaneous symmetry breaking phase transition [53]. Therefore, large power flows across the junctions between domains are allowed and provides a base to understand quantum gravity effects at the laboratory level.

REFERENCES

Note: "arXiv" papers are available from Cornell's electronic preprint depository http://arxiv.org

[1] Ng YJ. Selected topics in Planck-scale physics. Mod Phys Lett. A 2003; 18: 1073-1098.

[2] Markopoulou F. Planck-scale models of the Universe. arXiv: gr-qc/0210086.

[3] Bojowald M, Morales-Tecotl HA. Cosmological applications of loop quantum gravity. Lect Notes Phys 2004; 646: 421-462.

[4] Smolin L. Three Roads to Quantum Gravity: A New Understanding of Space, Time and the Universe. Weidenfeld & Nicolson General, Orion Publishing Co. 2000.

[5] Hartman TE. Tunneling of a wave packet. J Appl Phys 1962; 33: 3427.

[6] Clem JR. Two-dimensional vortices in a stack of thin superconducting films: A model for high-temperature superconducting multilayers. Phys Rev B 1991; 43: 1.

[7] Hemmerich A. Quantum entanglement in dilute optical latties. Phys Rev 1999; 60: 943.

[8] Poole CP, Farach HA, and Creswick RJ. Superconductivity. Academic Press 1995.

[9] Tinkham M. Introduction to Superconductivity. McGraw-Hill, Inc. 1996.

[10] Watanabe M. Quantum effects in small-capacitance single Josephson junctions. Phys Rev B 2003; 67: 094505.

[11] Watanabe M. Small-Capacitance Josephson Junctions: One-Dimensional Arrays and Single Junctions. arXiv: cond-mat/0301340

[12] Kibble TWB. Topology of cosmic domains and strings. J Phis A 1976; 8: 1387.

[13] Zurek WH. Cosmological experiments in condensed matter systems. Physics Reports 1996; 276: 177-221.

[14] Misner W, Thorne KS, Wheeler JA. Gravitation. Freeman 1973; Chapter 36.

[15] Podkletnov E, Modanese G. Impulse Gravity Generation Based on Charged $YBa_2Cu_3O_{7-X}$ Superconductor with Composite Crystal Structure. arXiv: physics/0108005.

[16] Podkletnov E, Modanese G. Investigation of High Voltage Discharges in Low Pressure Gases Through Large Ceramic Superconducting Electrodes. J Low Temp Phys 2003; 132: 239.

[17] DeWitt BS. Superconductors and gravitational drag. Phys Rev Lett 1966; 16: 1092.

[18] Anandan J. Gravitational and Inertial Effects in Quantum fluids. Phys Rev Lett 1981; 47: 463. Chiao RY. Interference and inertia: A superfluid-helium interferometer using an internally porous powder and its inertial interactions. Phys Rev B 1982; 25: 1655. Anandan J, Chiao RY. Gravitational radiation antennas using the Sagnac effect. Gen Rel and Grav 1982; 14: 515. Anandan J. Gravitational And Inertial Effects In Quantum Fluids. Phys Rev Lett 1984; 52: 401.

[19] Ross DK. The London Equation for Superconductors in a Gravitational Field. J Phys A: Math Gen 1983; 16: 1331.

[20] Li N, Torr DG. Effects of a Gravitomagnetic Field on Pure Superconductors. Phys Rev D 1990; 43: 457. Li N, Torr DG. Gravitational effects on the magnetic attenuation of superconductors. Phys Rev B 1992; 46: 5489. Li N, Torr DG. Gravito-Electric Coupling *Via* Superconductivity. Found Phys Lett 1993; 6: 371.

[21] Peng H, Torr DG. The Electric Field Induced by a Gravitational Wave in a Superconductor: A Principle for a New Gravitational Antenna. Gen Rel Grav 1990; 22: 53. Peng H, Torr DG, Hu EK, Peng B. Electrodynamics of moving superconductors and superconductors under the influence of external forces. Phys Rev B 1991; 43: 2700.

[22] Podkletnov E, Nieminen R. A Possibility of Gravitational Force Shielding by Bulk $YBa_2Cu_3O_{7-X}$ Superconductor. Physica C 1992; 203: 441-444.

[23] Podkletnov E. Weak Gravitation Shielding Properties of Composite Bulk $YBa_2Cu_3O_{7-X}$ Superconductor Below 70K Under E.M. Field. arXiv: cond-mat/9701074.

[24] Simmonds RW, Marchenkov A., Hoskinson E., Davis JC, Packard RE. Quantum interference of superfluid 3He. Naturc 2001; 412: 5.

[25] Josephson BD. Supercurrents through barriers. Adv Phys 1965; 14: 419.

[26] Brennen GK, Caves CM, Jessen PS, Deutsch IH. Quantum Logic Gates in Optical Lattices. Phys Rev Lett 1999; 82: 1060.

[27] Jaksch D, Briegel HJ, Cirac JI, Gardiner CW, Zoller P. Entanglement of atoms *via* Cold Controlled Collisions. Phys Rev Lett 1999; 82: 1975.

[28] Turchette QA, Woods CS, King BE, Myatt CJ, Leibfried D, Itano WM, Monroe C, Wineland DJ. Deterministic entanglement of two trapped ions. Phys Rev Lett 1998; 81: 3631-3634.

[29] Lushnikov PM. Collapse of Bose-Einstein condensate with dipole-dipole interaction. Phys Rev A 2002; 66: 051601(R).

[30] Kerlick GS, Dymoke-Bradshaw AKL, Dangor AE. Observations of ion acceleration by relativistic electron Beam. J Phys D: Appl. Phys 1983; 16: 613-629.

[31] Esser R, Edgar R. Differential Flow Speeds of Ions of the Same Element: Effects on Solar Wind Ionization Fractions. Appl Phys Lett 2001; 563: 1055-1062.

[32] Andreev AA, Limpouch J. Ion acceleration in short-pulse laser-target interactions. J Plasma Physics 1999; 62 : 179-193.

[33] Olkhovsky VS, Recami E. Recent developments in time analysis of tunneling processes. Phys Reports 1992; 214: 340. Olkhovsky VS, Recami E, Jakiel J. Unified Time Analysis of Photon and (Norelativistic) Particle Tunnelling, and Superluminal group-velocity problem. Physics Reports 2004; 398: 133-178.

[34] Winful HG. Delay Time and the Hartman Effect in Quantum Tunneling. Phys Rev Lett 2003; 91(26): 260401.

[35] Hauge EH, Stovneng JA. Tunneling times: a critical review. Rev Mod Phys 1989; 61: 917.

[36] Muga JG, Egusquiza IL, Damborenea JA, Delgado F. Bounds and enhancements for the Hartman effect. arXiv: quant-ph/0206181.

[37] Fletcher JR. Time delay in tunnelling through a potential barrier. J Phys C: Solid State Phys 1985; 18: 3427.

[38] Enders A, Nimtz G. Evanescent-mode propagation and quantum tunneling. Phys Rev E 1993; 48: 632.

[39] Mugnai D, Ranfagni A, Ruggeri R, Agresti A, Recami E. Superluminal processes and signal velocity in tunneling simulation. Phys Lett A 1995; 209: 227.

[40] Jakiel J. On Superluminal motions in photon and particle tunnellings. Phys Lett A 1998; 248: 156-160.

[41] Chiao RY. Tunneling Times and Superluminality: a Tutorial. arXiv: quant-ph/9811019.

[42] Nimtz G. Superluminal signal velocity. Ann Phys (Leipzig) 1998; 7: 618.

[43] Binder B. Iterative Coupling and Balancing Currents. Available from http://www.quanics.com

[44] Kibble TWB. Symmetry Breaking and Defects. arXiv: cond-mat/0211110.

[45] Kibble TWB. Some Implications Of A Cosmological Phase-Transition. Phys Rep 1980; 67: 183-199.

[46] Hindmarsh MB, Rajantie A. Defect formation and local gauge invariance. Phys Rev Lett 2000; 85: 4660-63.

[47] Rudaz S, Srivastava AM. On the production of flux vortices and magnetic monopoles in phase transitions. Mod Phys Lett A 1993; 8: 1443-50.

[48] Copeland EJ, Saffin P. Bubble collisions in Abelian gauge theories and the geodesic rule. Phys Rev D 1996; 54: 6088-94.

[49] Kibble TWB, Rajantie A. Estimation of vortex density after superconducting film quench. Phys Rev B 2003; 68: 174512.

[50] Shapiro IL, Sola J. Scaling behavior of the cosmological constant and the possible existence of new forces and new light degrees of freedom. Phys Lett B 2000; 475: 236-246.

[51] Binder B. Self-Consistent Quantum-Gravitational Quadrupole Fluctuations. Available from http://philsci-archive.pitt.edu/archive/00001043

[52] Pospelov M, Romalis M. Lorentz Invariance on Trial. Physics Today, 2004.

[53] Hawking S. A Brief History of Time. Bantam Press 1988.

CHAPTER 5

Quantum Gravity Evaluation of Stimulated Graviton Emission in Superconductors

Giovanni Modanese*

Free University of Bolzano, Faculty of Science and Technology, P.zza Università 5, 39100 Bolzano, Italy

Abstract: It can be proven, using the Feynman path integral, that the vacuum state of Quantum Gravity contains localized fluctuations with large mass. Their interaction gives rise to a complex pattern of excited states. In particular, couples of equal virtual masses can be in a symmetrical superposition ψ^+ (ground state) or in an anti-symmetrical superposition ψ^-. Transitions $\psi^+ \rightarrow \psi^-$ can be efficiently "pumped" only by a time-variable vacuum-energy-like term $\Lambda(t)$ associated with coherent matter. The decay $\psi^- \rightarrow \psi^+$ generates off-shell virtual spin 1 gravitons, which can also produce a cascade of stimulated emission, ending with absorption in a target. We compute to leading order the Einstein *A* and *B* coefficients of the transition and give magnitude order estimates for the amplification factor in YBCO crystals or melt-textured samples pumped by bulk supercurrents at frequency $\approx$10 MHz.

Keywords: Gravitational forces, theories of gravitation, superconductors, high-Tc superconductors, type-II superconductors, general relativity, path integrals, quantum gravity, ground state of quantum gravity, vacuum energy density, vacuum fluctuations, gravitomagnetism, einstein A and B coefficients.

1. GRAVITATIONAL VACUUM AND EXCITED STATES

In their pioneering work in 1964, Halpern and Laurent [1] showed that the probability of spontaneous or stimulated gravitational emission from atomic systems is so small that laser action can not be obtained in any realistic conditions. This conclusion is inevitable in the weak-field approximation, and if one considers only incoherent matter and real radiation with energy-momentum ratio $E/p=c$. Recently G. Fontana ([2]; see Chapter 3) has studied the possibility of generating gravitational laser action in certain special circumstances, which can only occur at the junctions of two different superconductors, namely (a) transitions of Cooper pairs between *s* and *d* states, (b) super-radiance due to the macroscopic coherence of the superconductors, (c) suppression of the competing electromagnetic transitions due to the peculiar electrodynamics of superconductors.

In this work we consider a different possibility, based on the existence in Quantum Gravity of strong vacuum fluctuations, called "zero-modes of the Einstein action". The existence of these fluctuations was partly proved in an earlier paper [3] and is here confirmed with explicit non-perturbative calculations of the Feynman path integral (App. 4). We further consider the mutual interaction of the zero-modes and their symmetric and anti-symmetric quantum superpositions ψ^+ and ψ^- (Section 1). We show that the interaction of the gravitational vacuum with a superconductor leads to transitions $\psi^+ \leftrightarrow \psi^-$ and to stimulated graviton emission (Sections 2 and 3). These gravitons, however, are virtual, or "off-shell", having an energy-momentum ratio E/p of the order of 1 m/s. They therefore carry, for the same energy, a much larger momentum than real gravitons, but can only propagate over finite distances, until they are absorbed by a target (Section 4). In Section 5 we give some first estimates of the parameters of the "virtual laser action" which can so be obtained: pumping rate, population inversion condition, spatial amplification factor.

Let us now begin to construct the new picture of the gravitational vacuum which emerges when one takes into account the zero-modes and their interactions. In our previous work [3] we have shown that in the vacuum state of Quantum Gravity virtual negative localized masses appear randomly everywhere, as strong static vacuum fluctuations. In order to avoid a large global curvature, the total average effect of these

***Address correspondence to Giovanni Modanese:** Free University of Bolzano, Faculty of Science and Technology, P.zza Università 5, 39100 Bolzano, Italy; Tel: 0471-017003; E-mail: giovanni.modanese@unibz.it virtual masses must inevitably be renormalized to zero. This is, in our view, guaranteed by the "cosmological constant paradox" [4]: nature appears to be endowed with a dynamical mechanism (probably intrinsic to gravitation, at a very fundamental level) which relaxes to zero any constant positive or negative contributions to the vacuum energy density coming from particle physics or even from gravity itself. So, even though such contributions are formally infinite, in the end they do not affect the curvature of spacetime. The full explanation of this mechanism can only be achieved within a complete non-perturbative theory of Quantum Gravity. Some partial evidence of the dynamical emergence of flat spacetime has been obtained in the lattice theory and in effective field theory approaches [5].

Therefore we shall not be concerned with the global effect of our massive vacuum fluctuations on spacetime. We shall instead consider their interactions, which result in a novel pattern of purely gravitational excited states, above a ground state in which all fluctuations pairs with equal mass are in a symmetrical superposition.

The equations of isolated and non-interacting zero-modes (Appendix 1) are based on a spherically symmetric Ansatz and express the condition $\int d^4x\sqrt{g}R = 0$. Their regular solutions with mass M depend on an arbitrary function with null integral, and outside a certain radius look exactly like Schwarzschild metrics. They are "classical" fields, not in the sense of being solutions of the Einstein equations in vacuum, but in the sense of being classical configurations (functions of spacetime coordinates) which are weighed in the functional integral with non-vanishing measure (Appendices 3, 4).

The local density of gravitational energy and the general gravitational Hamiltonian H are notoriously ill-defined, but the total energy of an asymptotically flat field configuration is known (ADM energy). This is computed as a surface integral of the metric at spatial infinity and takes into account the features of the field source, giving its total mass-energy. This total mass-energy is equal to lowest order to the sum of the masses of the sources, and to higher orders it includes the gravitational binding energy, like for instance $\frac{-2GM^2}{5R}$ for an homogeneous sphere or $\frac{-Gm_1m_2}{r}$ for a couple of pointlike masses. A first crucial application of this concept concerns the interaction energy of two zero-modes. The field of two zero-modes seen at spatial infinity is the superposition of two Schwarzschild metrics with certain parameters; therefore, even if the sources of these metrics do not really exist, the ADM interaction energy is the same as for real particles.

The second application of the ADM energy concerns the energy of a single zero-mode. We can say that each zero-mode corresponds to a quantum state $|i\rangle$ and that $\langle i|H|i\rangle = c^2 m_i$. The states $|i\rangle$ are localized and mutually orthogonal. They correspond to field configurations centered at different points. In the following we shall also suppose for simplicity that their Schwarzschild radii are always much smaller than their distance. We could call such states "purely gravitational, long-lived virtual particles". Note that fixing $\langle i|H|i\rangle$ amounts to a much weaker statement than giving the gravitational quantum Hamiltonian operator H or its density, because $\langle i|H|i\rangle$ is a matrix element and it is a classical limit of the total energy for an asymptotically flat configuration. So whenever we write here the full gravitational Hamiltonian H, in fact we only exploit the formal properties of its matrix elements, like in a Heisenberg representation of quantum mechanics. This is coherent with our path integral approach to the full-interacting case. Stronger statements, like for instance assuming the validity of operator eigenvalue equations like $H|i\rangle = c^2 m_i |i\rangle$, would lead to contradictions. On the other hand, gravitational interaction Hamiltonians on a background metric like that employed in Section 3 do not suffer from these limitations.

According to this line of thought, the "true ground state" of the gravitational vacuum would be obtained in principle as the limit of an infinite uncoherent superposition of flat spacetime plus single zero-mode wavefunctions, after introducing interactions and UV cut-offs for the interaction energies of virtual masses which are very large and very close to each other. This definition of the ground state is clearly difficult to put on a rigorous basis. We are mainly interested, however, into the *excitations* with respect to this ground state, and the most relevant among these excitations are those resulting from interactions of couples.

Consider a couple of states $|1\rangle$ and $|2\rangle$ with masses m_1 and m_2. We have

$$\langle 1|H|1\rangle = c^2 m_1 |1\rangle, \quad \langle 2|H|2\rangle = c^2 m_2 |2\rangle, \quad \langle 1|2\rangle = 0 \qquad \textbf{(1.1)}$$

Putting now $m_1=m_2=m$ and taking the interaction into account, the degenerate non-interacting levels are splitted. Define the symmetrical and anti-symmetrical superpositions ψ^+ and ψ^-:

$$|\psi^+\rangle = \frac{1}{\sqrt{2}}(|1\rangle + |2\rangle) \quad |\psi^-\rangle = \frac{1}{\sqrt{2}}(|1\rangle - |2\rangle) \qquad \textbf{(1.2)}$$

The energy splitting ΔE is given, as known, by

$$\Delta E = E_- - E_+ = \langle\psi^-|H|\psi^-\rangle - \langle\psi^+|H|\psi^+\rangle = 2\langle 1|H|2\rangle. \qquad \textbf{(1.3)}$$

Note that the matrix element $\langle 1|H|2\rangle$ can be taken to be real without loss of generality. Suppose that $\langle 1|H|2\rangle$ can be computed to a first approximation through its classical limit. We then obtain

$$\Delta E = 2\frac{Gm^2}{r}, \qquad \textbf{(1.4)}$$

being r the distance between the symmetry centers of the states $|1\rangle$ and $|2\rangle$.

Let us now consider transitions between the ground state and the excited states which can be induced by coupling the vacuum to an external source. First we look for a relation between the frequency of the external source and the virtual mass of the excited states. In order to fix the ideas, we choose a finite volume V. In the ground state, all couples with equal mass will be in their symmetric superposition state. Any transition of one couple from its symmetric to its antisymmetric state gives an excited state with energy (1.4). Since there exist zero-modes with any (negative) mass, at any distance, there is actually a continuum of excited states. Suppose to act on the system with an external perturbation of frequency ω. Anticipating some points of Section 3 ("Pumping process"), take for instance ω in the range 1-10 MHz. Consider virtual couples having equal masses, at distance r. From equation (1.4) we see that the perturbation can cause transitions to excited levels involving different masses at different distances, provided the ratio m^2/r is the same. Since in our example the maximum distance scale where the external perturbation acts is of the order of the superconductor coherence length 10^{-9} m (see Section 3), the largest virtual mass excited in this way is of the order of 10^{-13} kg. This is small, but definitely much larger than any atomic-scale mass, and we shall see that the matrix element $\langle\psi^+|\hat{\mathbf{d}}|\psi^-\rangle$ of the mass dipole is also much larger than atomic mass dipoles.

2. DIPOLE-LIKE EMISSION AND LIFETIME OF EXCITED STATES

We have introduced the concept of ground state in effective theory of Quantum Gravity as given by the Fock vacuum plus a superposition of random zero-modes. Non-interacting zero-modes with equal mass are coupled in degenerate symmetric and anti-symmetric wavefunctions. The introduction of interaction removes the degeneration. The excited states form a continuum and the interaction of the vacuum with an external oscillating source leads to transitions, with a probability which we shall compute in Section 3.

In this Section we compute the lifetime of an excited state. The decay of the excited state occurs with the emission of an off-shell graviton which is then absorbed by a material target (the complete process is described in Section 4). As we shall see, the typical energy-momentum relation for this kind of gravitons is $f\lambda \approx 1$ m/s, therefore they are completely off-shell and have components with spin 0 and 1. It follows that the dominant graviton emission process in the decay of an excited zero-mode is oscillating-dipole emission. Quadrupolar emission, which is the only process ensuring conservation of energy, momentum and spin in the emission of on-shell gravitons, can in this case be disregarded. Since we are only interested into a lowest-order perturbative estimate (tree diagrams) we can use the linearized Einstein theory in the form of the "Maxwell-Einstein" equations (see Appendix 2)

$$\begin{aligned} &\nabla\cdot\mathbf{E}_G = -4\pi G\rho_m \\ &\nabla\times\mathbf{E}_G = -\frac{\partial\mathbf{B}_G}{\partial t} \\ &\nabla\cdot\mathbf{B}_G = 0 \\ &\nabla\times\mathbf{B}_G = -\frac{4\pi G}{c^2}\mathbf{j}_m + \frac{1}{c^2}\frac{\partial\mathbf{E}_G}{\partial t} \end{aligned} \tag{2.1}$$

Here $\mathbf{E}_G$ is the gravito-electric (Newtonian) field, $\mathbf{B}_G$ is the gravito-magnetic field, and $\mathbf{j}_m$, ρ_m are the mass-energy current and density. The elementary quantization of the field modes in a finite volume V follows the familiar scheme used for the computation of spontaneous and stimulated electromagnetic emission of atoms in a cavity [6]. The Einstein A-coefficient of spontaneous emission turns out to be related to the B-coefficient and to the mass dipole moment by the relation

$$A = \left(B\frac{8\pi\hbar}{\lambda^3}\right) = \left(\frac{G}{\hbar^2}|\langle\hat{\mathbf{d}}\rangle|^2\frac{8\pi\hbar}{\lambda^3}\right) \tag{2.2}$$

where the electromagnetic coupling constants have been replaced, up to an irrelevant adimensional factor, by the gravitational constants, according to equations (2.1). The operator $\hat{\mathbf{d}}$ is the mass-dipole moment and the matrix element is taken between the initial and final state of interest.

It is straightforward to check that there is an oscillating mass dipole between the states ψ^- and ψ^+:

$$\begin{aligned} &\langle\psi^+|\hat{\mathbf{d}}|\psi^-\rangle = \frac{1}{\sqrt{2}}(\langle 1| + \langle 2|)\,\hat{\mathbf{d}}\,\frac{1}{\sqrt{2}}(|1\rangle - |2\rangle) = \\ &= \frac{1}{2}(\langle 1| + \langle 2|)(m_1\mathbf{r}_1|1\rangle - m_2\mathbf{r}_2|2\rangle) = \frac{1}{2}m\mathbf{r} \end{aligned} \tag{2.3}$$

where $\mathbf{r}_1 = \frac{1}{2}\mathbf{r}$, $\mathbf{r}_2 = -\frac{1}{2}\mathbf{r}$; here $\mathbf{r}$ is the displacement between the masses m_1 and m_2, which in the end are taken to be equal. The origin of the coordinate system is in the center of mass.

This mass dipole moment has purely quantum origin, because in our system there are no masses of different signs, and it is known that in this case the classical mass dipole moment computed with respect to the center of mass is zero. We could say that the non-zero matrix element (2.3) is due to the quantum tunnelling between the states $|1\rangle$ and $|2\rangle$. This corresponds to a mass oscillation. Note that the sign of the matrix element is undefined, because we could have chosen as ψ^- state also the linear combination $(|2\rangle - |1\rangle)$. The sign of the matrix element has no influence on the transition probability and on the A-coefficient computed through equation (2.2).

One might wonder if there exist other microscopic systems where, for similar reasons, the mass dipole moment operator has non-zero matrix elements. This certainly does not happen in atoms. Consider for instance the hydrogen atom. The total wavefunction is given by the product of the wavefunctions of the electron and of the proton:

$$\psi(\mathbf{r},\mathbf{R}) = \chi(\mathbf{r})\phi(\mathbf{R}) \tag{2.4}$$

where $\mathbf{r}$ is the electron coordinate and $\mathbf{R}$ the proton coordinate; both coordinates are referred to the center of mass. When we write the system's Hamiltonian and the Schroedinger equation we use the reduced coordinate ρ, therefore the two functions χ and ϕ are obtained from the solution of the Schroedinger equation by expressing $\mathbf{r}$ and $\mathbf{R}$ in function of ρ. This implies that if in a certain solution the electron is localised in a region near $\mathbf{r}_0 = \frac{m_p}{m_e + m_p}\boldsymbol{\rho}_0$, then the proton is localised near $\mathbf{R}_0 = \frac{-m_e}{m_e + m_p}\boldsymbol{\rho}_0$, otherwise the total wavefunction is zero. In this case the situation is similar to the classical case and the mass dipole moment operator is zero when applied to a state:

$$\begin{aligned}&\hat{\mathbf{d}}\psi(\mathbf{r},\mathbf{R}) = \left(m_e\hat{\mathbf{r}} + m_p\hat{\mathbf{R}}\right)\chi(\mathbf{r})\phi(\mathbf{R}) = \\ &= m_e\hat{\mathbf{r}}\chi(\mathbf{r})\phi(\mathbf{R}) + m_p\hat{\mathbf{R}}\chi(\mathbf{r})\phi(\mathbf{R}) = \left(m_e\mathbf{r}_0 + m_p\mathbf{R}_0\right)\chi(\mathbf{r})\phi(\mathbf{R}) = 0\end{aligned} \tag{2.5}$$

In other words, things still go as if the proton recoils when the electron oscillates. For non-localized states one has to consider the full matrix elements, but the conclusion is the same.

We thus remark once again that the existence of the mass dipole moment is a purely quantum effect, due to the interference between the two "mass eigenstates" $|1\rangle$ and $|2\rangle$. The interference is caused by the gravitational interaction between the two states. This interaction is, on one hand, anomalously strong in comparison to the usual gravitational interactions at the atomic scale, because the mass of the involved zero-modes is much larger than the atomic masses. On the other hand, this interaction is much smaller than the interaction in a hypothetical gravitational bound state formed by two masses of this size. This can be easily checked, for instance, by computing the corresponding Bohr radius: this is of the order of $\hbar^2 / Gm^3 \approx 10^{-19}$ m, while the zero-modes in the states ψ^+ and ψ^- are separated by a distance of the order of 10^{-9} m. One can also easily check that the acceleration of each zero-mode due to the presence of the other is very small if compared to those due to atomic or molecular forces, and that the supposed localization of the zero-modes is well compatible with the Heisenberg principle. The picture of the states $|i\rangle$ as localized, static almost-eigenstates which interact only through their gravitational interference is therefore consistent. The lifetime of the states $|i\rangle$ in a zero-mode is much larger than the inverse of their interaction frequency $\Delta E / \hbar$, and yet much smaller than the time it would take them to "collapse" on each other. We shall give a more detailed discussion of this point in another paper.

Equation (2.2) gives the lifetime τ of the excited level ψ^- by spontaneous emission. With the values of m and r found in the previous Section supposing an excitation frequency of the order of 1 MHz, one finds $B \approx 10^{12}$ m^3/Js^2 for the stimulated emission coefficient and $\tau = A^{-1} \approx 1$ s for the lifetime for spontaneous emission (taking $\lambda f \approx 1$ m/s: compare Section 4). The general dependence of B on the frequency ω and on the length r of the dipoles is easily obtained from equations (1.4), (2.2) and (2.3):

$$B \approx \frac{1}{\hbar}\omega r^3 \tag{2.6}$$

Note that B is independent from the Newton constant G.

3. PUMPING PROCESS

We shall now compute the probability of a transition from a state ψ^+ to a state ψ^- under the action of an external perturbation with oscillation frequency ω in resonance with the energy difference (1.4). Given the large number of available states, the resonance occurs for any frequency, and also if the perturbation is not exactly monochromatic.

We want to use the Fermi rule

$$\frac{dP_{In,Out}}{dt} = \frac{1}{\hbar}\left|\langle In|H'|Out\rangle\right|^2 \rho\left(E_{Out}\right) \tag{3.1}$$

where $P_{In,Out}$ is the transition probability between the states $|In\rangle$ and $|Out\rangle$ and $\rho(E_{Out})$ is the density of final states. In our case $|In\rangle$ is the ground state, with all couples of zero modes with equal mass in their symmetric state, and $|Out\rangle$ is the state in which one of the couples has passed to the antisymmetric state. The interaction Hamiltonian H' can be written as the space integral of an Hamiltonian density $H'_{\mathbf{x}}$, computed over the interaction region:

$$H'(t) = \int d^3x H'_{\mathbf{x}}(t,\mathbf{x}) \tag{3.2}$$

We shall consider, as external perturbation, a local "cosmological" term, or vacuum energy density term

$$H_{\mathbf{x},\Lambda}(t,\mathbf{x}) = \frac{1}{8\pi G}\sqrt{g(t,\mathbf{x})}\Lambda(t) \tag{3.3}$$

due to the presence of coherent matter described by a macroscopic wavefunction equivalent to a classical field. We have discussed this in [7, 8]. The Λ term in superconductors turns out to be much larger than the cosmological background: one typically has $\Lambda c^4/G$=10^6-10^8 J/m^3 in superconductors, depending on the type, while the currently accepted value for the cosmological background is of the order of $\Lambda c^4/G$=10^{-9} J/m^3. The value above for superconductors is the result of a complex evaluation of the relativistic limit of the Ginzburg-Landau Lagrangian, which yields the following expression for Λ in terms of the pairs density ρ [7]:

$$\Lambda(t,\mathbf{x}) = -\frac{1}{2m}\left[\hbar^2(\nabla\rho)^2 + \hbar^2\rho\nabla^2\rho - m\beta\rho^4\right], \tag{3.4}$$

where β is the second Ginzburg-Landau coefficient and *m* is the Cooper pair mass. This energy density has strong variations in space and time, following the behaviour of the macroscopic wavefunction. In order to obtain high-frequency oscillations in time, one can typically induce in the material Josephson currents [8]. Spatial variations have a typical scale of 1 nm, so we take this as the size of the volume *V* where the perturbation is spatially constant and the transition probability is computed. For this reason we left only the time dependence in equation (3.3).

3.1. Comparison Between the Perturbation H_Λ and the Usual Field/Particle Interaction

We would like to stress that although the perturbation H_Λ has the form of a vacuum energy density, because it couples with the gravitational field through a tensor proportional to $g_{\mu\nu}$, it is actually caused by the presence of matter; it is a peculiar matter coupling, however, which in the matrix element (3.1) is much more effective than the standard particle/field coupling, as we now show.

Consider for simplicity the case of a non-relativistic particle (the generalization is straightforward). The field/particle interaction Hamiltonian density is in that case

$$H_{\mathbf{x},particle} = \frac{1}{2m}h_{ij}(t,\mathbf{x})p^i p^j \tag{3.5}$$

where *m* is the particle mass, $h_{ij} = g_{ij} - \eta_{ij}$ and *i*, *j* are spatial indices. This Hamiltonian can be used to compute semi-classically the probability that a classical gravitational wave causes a transition between two atomic states; in that case, $h_{ij}(t,\mathbf{x})$ is regarded as a numerical function (the field of the wave), while p^i and p^j are operators, acting on the electron wavefunction. A standard procedure to obtain it is to write the free particle action $S = m\int d\tau = m\int\sqrt{dx^\mu dx_\mu}$ in general-covariant form, *i.e.*, $S = m\int\sqrt{g_{\mu\nu}(x)dx^\mu dx^\nu}$. Writing $g_{\mu\nu}$ as $\eta_{\mu\nu} + h_{\mu\nu}$, one finds that $\eta_{\mu\nu}(dx^\mu dx^\nu / d\tau^2) = 1$ and the square root is expanded as $1 + (1/2)h_{\mu\nu}(dx^\mu dx^\nu / d\tau^2)$ plus higher order terms.

One must also make the assumption that the components h^{00} and h^{0i} of $h^{\mu\nu} = g^{\mu\nu} - \eta^{\mu\nu}$ can be set to zero for a wave-like field. This comes from the Feynman gauge condition, plus reduction to helicity components; the reasoning can be generalized to off-shell waves along the z-direction with $k_0 \neq k_3$, and is essentially the following. One writes a field mode with wave vector k as a polarization tensor $e_{\mu\nu}$ multiplied by e^{ikx}. The polarization tensor satisfies the gauge condition $k_\mu e^\mu_\nu = 1/2 k_\nu e^\mu_\mu$. A residual coordinate transformation acts on this polarization tensor as

$$e'_{\mu\nu} = e_{\mu\nu} + k_\mu \varepsilon_\nu + k_\nu \varepsilon_\mu$$

where ε_μ are arbitrary vectors. Consider a wave propagating in the z-direction, *i.e.*, $k_1 = k_2 = 0$, $k_3 \neq 0$, $k_0 \neq 0$. The transformation above is rewritten as

$$\begin{aligned} e'_{13} &= e_{13} + k_3 \varepsilon_1 \\ e'_{23} &= e_{23} + k_3 \varepsilon_2 \\ e'_{33} &= e_{33} + 2k_3 \varepsilon_3 \\ e'_{00} &= e_{00} - 2k_0 \varepsilon_0 \end{aligned}$$

In this way, one can set to zero all the components e_{0i} and e_{00} .

Suppose now to apply equation (3.5) to our case, *i.e.,* to compute a transition probability $\psi^+ \rightarrow \psi^-$ due to the coupling of gravitation to single particles in ordinary matter. In this case, the particle momentum is a given numerical function of time, while $h_{ij}(t, \mathbf{x})$ is a quantum operator which acts on the Fock vacuum creating or destroying a graviton. (In the following we shall often denote the field operator as $\hat{h}$ and omit the indices.) The numerical factor $\frac{p^i p^j}{2m}$ is of the order of the kinetic energy of the particle $\frac{p^2}{2m}$, *i.e.,* of the order of 10^{-19} J for an atomic system. We shall see below that this is also the magnitude order of the $\Lambda(t)$ term. But while the interaction Hamiltonian $H_{\mathbf{x},\Lambda}$ can have non-vanishing matrix elements also when acting linearly between the states ψ^+ and ψ^-, because it is proportional to the non-linear function $\sqrt{g} = 1 + \mathrm{Tr}(h) + \ldots$, the Hamiltonian $H_{\mathbf{x},particle}$ has non-vanishing matrix elements only to second order.

Namely, we can write a matrix element of the form $\langle In|\hat{h}|Out\rangle$ as

$$\langle In|\hat{h}|Out\rangle = \langle 0, In_{z-m}|\hat{h}|0, Out_{z-m}\rangle,$$

where $|In_{z-m}\rangle$ and $|Out_{z-m}\rangle$ denote the zero-mode components of the initial and final states, and $|0\rangle$ denotes the Fock vacuum, without gravitons. The matrix element is clearly zero, because it contains a single field acting between two Fock vacuum states. In other words, we can say that since neither in the initial state nor in the final state there are gravitons, the standard vertex (3.5) can have non-zero matrix element only when it is taken twice and is therefore proportional to $\left(\frac{p^2}{2m}\right)^2$; but this is of magnitude order 10^{-38} in S.I. units, as seen, and gives a factor 10^{-76} after insertion in (3.1). On the other hand, the $\hat{h}\hat{h}$, $\hat{h}\hat{h}\hat{h}$, … terms in the expansion of H_Λ can give non-zero matrix elements already to first order in Λ.

We are not able to compute these matrix elements without a complete theory, because inside the Schwarzschild radius of the zero-modes the weak field expansion is not valid. The situation resembles that of early nuclear physics, where the nuclear matrix elements were largely unknown, apart from some general properties or magnitude orders; this did not prevent researchers from obtaining important data on the processes, based on the available information and on the crucial knowledge of the final states density (see below).

3.2. Evaluation of the Transition Probability

In order to determine the density of final states, we need to find the number $N(E)$ of zero-modes pairs with energy E in the excited antisymmetric state in the volume V. We are interested into the derivative of $N(E)$: $\rho(E)=\frac{dN(E)}{dE}$. Given any energy E, there are several pairs with that energy. Those with the largest virtual mass, say M, have the maximum size $R=V^{1/3}\approx 10^{-9}$ m (remember that the energy of the first excited state with resonance frequency ω is $E=\hbar\omega=Gm^2/r$).

The number of states depends also on the mass m, so let us denote it by $N_m(E)$. It is given by the third power of the ratio between the volume size R and the size r of the pairs:

$$N_m(E)=\left(\frac{R}{r}\right)^3=\left(\frac{RE}{Gm^2}\right)^3=\left(\frac{R}{Gm^2}\right)^3 E^3$$

hence

$$\rho_m(E)=\frac{dN_m(E)}{dE}=\left(\frac{R}{Gm^2}\right)^3 3E^2$$

Next define $m=\alpha M$ (M is the mass of the largest pair having energy E). It follows $r=\beta R$, with $\beta=\alpha^2$ (r, here and above, is the size of the pairs with mass m and energy E). We can re-write

$$\rho_m(E)=\rho_\alpha(E)=\frac{1}{\alpha^6}\frac{3}{E}$$

or

$$\rho_m(E)=\rho_\beta(E)=\frac{1}{\beta^3}\frac{3}{E}$$

There appears to be an "infrared divergence" for small β. Consider, however, the matrix element $\left|\langle\psi^+|H_\Lambda|\psi^-\rangle\right|^2$. After the action of H_Λ on ψ^-, in the product between the states one takes as volume in the space integral the volume of ψ^+ and ψ^-, which is proportional to r^3; so in the end there is a factor β^6, which cancels the infrared divergence. In other words, only large pairs have a large transition probability.

On evaluating the magnitude order of ρ with $\beta\approx 1$, $E\approx 10^{-27}$ J (Section 2), we obtain $\rho(E)\approx 10^{27}\,\mathrm{J}^{-1}$. The volume integral in equation (3.2) is taken on a region of volume $V\approx 10^{-27}$ m^3, therefore the squared matrix element is of order 10^{-38} in S.I. units. Dividing by $\hbar$ in accordance with the Fermi rule (3.1), we obtain the transition probability

$$\frac{dP}{dt}=\frac{1}{\hbar}\left|\langle\psi^+|H_\Lambda|\psi^-\rangle\right|^2\rho(E)\approx 10^{34}10^{-38}10^{27}\approx 10^{23}\,\mathrm{s}^{-1} \qquad \textbf{(3.6)}$$

This implies that the excitation time is very short (10^{-23} s) and the pumping acts very quickly. In fact, the efficiency of the pumping is limited by a global energetic balance rather than by the transition probability (3.6). As we have seen in [8], the number of transitions per unit time and volume which can be energetically sustained in a layered superconductor with layer thickness z, crossed by a supercurrent with density j, is $N=\frac{j}{2ez}$. We shall evaluate this number in Section 5 inserting reasonable superconductor parameters. After integrating N over the duration of the current pulse one can compare it to the number of available excited states per unit volume (used above in the derivation of the final states density). This comparison is crucial to determine the occurrence of population inversion between the states ψ^+ and ψ^-.

In conclusion, also note that the large transition probability (3.6) implies a strong interaction between the source $\Lambda(t)$ and the vacuum fluctuations with the right frequency. This should make possible a collapse of the vacuum wavefunction (which is in general a superposition) on the particular states ψ^+ involved, before the transition $\psi^+ \rightarrow \psi^-$ occurs. Alternatively, the probability of a state ψ^+ can be thought as unitary, since it involves only virtual particles. We shall give a more detailed discussion of this point in another paper.

4. INTERACTION OF THE EMITTED VIRTUAL GRAVITONS WITH A TARGET

In this Section we shall analyze the balance of the energy-momentum exchange which takes place when the virtual gravitons emitted in the $\psi^+ \rightarrow \psi^-$ transitions are absorbed by a target. We shall also make a comparison with the exchange of virtual gravitons which occurs in the familiar gravitational interactions, like for instance the free fall of a body in the Earth field. We have seen in the previous Sections that

1. The gravitational vacuum has a non-trivial structure, with massive localized fluctuations which interact with each other, giving rise to a continuous pattern of excited levels.

2. When the gravitational field is in one of these excited states, there are no gravitons; but decay to the ground state causes the emission of a virtual, off-shell graviton. We have computed the decay probability and the corresponding A and B Einstein coefficients.

3. Transitions to an excited state can be induced by the local coupling of the gravitational vacuum to a coherent source $\Lambda(t)$. We have called this the "pumping process" (Section 3). The transition rate is quite large (equation (3.6)) for coherent energy densities of the order of 10^8 J/m^3, as can be found in high-Tc superconductors subjected to high-frequency currents. On the contrary, standard interactions through the linear field-particle coupling (3.5) give exceedingly small excitation rates.

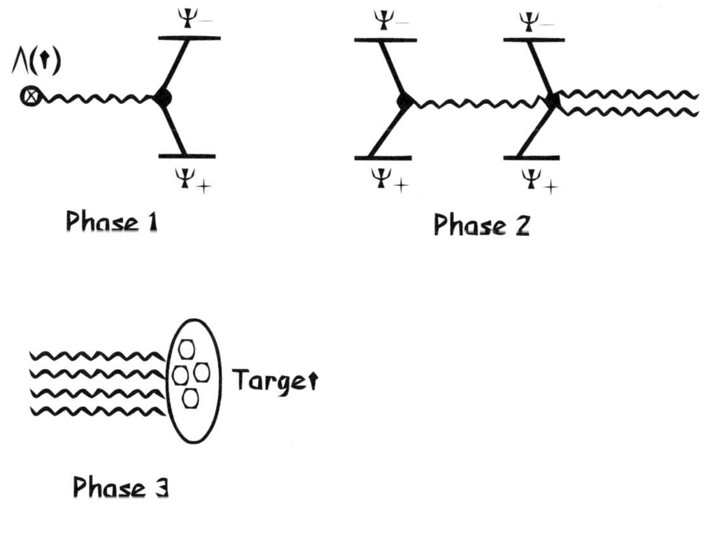

Figure 1: Excitation of the gravitational vacuum by coupling to a coherent source $\Lambda(t)$, followed by virtual graviton emission, stimulated emission and absorption of the virtual gravitons in a target.

The process described above can be represented by the diagrams in Fig. **1** (Phase 1 and Phase 2). Note that in Phase 2 we have added the possible occurrence of stimulated emission, besides spontaneous emission. Phase 3 includes the propagation of the virtual gravitons and their absorption in a target. This will also allow us to justify the assumption, made in Section 2 for the computation of the A-coefficient, that the energy-momentum relation for the virtual gravitons is $E/p \approx f\lambda \approx 1$ m/s.

In Section 2 we have computed the graviton emission probability in the decay from an excited state ψ^- (couple of zero-modes in antisymmetric state) to a state ψ^+ (couple in symmetric state). The probability is large only for virtual gravitons, because

- The mass dipole moment between the two states is relatively large, but not the quadrupole momentum. Therefore on-shell gravitons with spin 2 can not be emitted.
- The spontaneous emission probability is proportional to the factor λ^{-3}, and this can be large enough only if $f\lambda << c$.

We must now justify the assumption $f\lambda \approx 1$ m/s made in Section 2. Since the gravitons are supposed to be virtual, the complete process must include their complete absorption in a target. Suppose that in the absorption process each graviton cedes to the target its entire energy and momentum. The E/p ratio in the target must be the same as in the gravitons. Take for instance, as target, a proton initially at rest. After the graviton absorption we have $\frac{E}{p} = \frac{1}{2}v$. The proton mass is $\approx 10^{-27}$ kg, the graviton energy is $E = hf \approx 10^{-27}$ J (for frequency of the order of 10^7 Hz – see Section 2), hence $\frac{E}{p} \approx 1$ m/s, as supposed. If a graviton beam is absorbed in a target containing protons and light atoms, the balance can be applied to the entire target. A similar balance holds for masses which are multiple of the proton mass; also consider that we did not suppose that the gravitons are monochromatic, but there is an entire emission frequency band corresponding to the spectrum of the pumping $\Lambda(t)$ (the excited gravitational states have a continuum spectrum).

4.1. Formal Quantum-Mechanical Representation of the Process

Since the gravitons appearing in Phase 2 and 3 of the process are virtual, we can not consider the three phases depicted in Fig. **1** as a succession of real events. We can not imagine, for instance, that the gravitons emitted in Phase 2 can propagate indefinitely like a beam of real particles. In fact we are considering a quantum process and we are supposing to write the total amplitude as a sum of amplitudes over intermediate states. In general, given an initial state $|In\rangle$ of a system, a final state $|Out\rangle$ and an evolution operator $U(0, t)$, we can choose an intermediate time t_m and write the transition amplitude as

$$\langle In|U(0,t)|Out\rangle = \langle In|U(0,t_m)U(t_m,t)|Out\rangle =$$
$$\langle In|U(0,t_m)\sum_n |n\rangle\langle n|U(t_m,t)|Out\rangle =$$
$$\sum_n \langle In|U(0,t_m)|n\rangle\langle n|U(t_m,t)|Out\rangle = \sum_n U_{In,n}U_{n,Out}$$

where $\{|n\rangle\}$ is a complete system of intermediate states. When there are many intermediate states compatible with the conservation laws associated with U, the transition probability $|U_{\mathrm{In,\,Out}}|^2$ is given by the sum of several terms, with possible interference:

$$|U_{In,Out}|^2 = |U_{In,1}U_{1,Out} + U_{In,2}U_{2,Out} + ...|^2$$

On the other hand, if in the sum there is a dominant term n^*, the probability is factorized:

$$\left|U_{In,Out}\right|^2 \approx \left|U_{In,n*}\right|^2 \left|U_{n*,Out}\right|^2$$

We have seen that in our case there is essentially only one intermediate state n^* compatible with the conservation of energy and momentum, and for this reason we have computed separately the probabilities (*not the amplitudes*) of Phase 1 and Phase 2 using the Fermi golden rule. The absorption probability of gravitons in the target is supposed to be 1, as it is also implied by the quantum formula for the usual static gravitational potential (see Section 4.2).

The various states appearing in the process can be represented as follows (without taking into account, for simplicity, the stimulated emission):

$$\left|In\right\rangle = \left|0;\psi^+,T\right\rangle$$

where $|0\rangle$ denotes the Fock vacuum, ψ^+ the zero-mode ground state and T the target. There are two dominant intermediate states, namely

$$\left|n_1^*\right\rangle = \left|0;\psi^-,T\right\rangle$$
$$\left|n_2^*\right\rangle = \left|E,p;\psi^+,T\right\rangle, \text{ with } E/p \approx 1$$

and the final state is

$$\left|Out\right\rangle = \left|0;\psi^+,T^*\right\rangle$$

where T^* denotes the excited target. The probabilities which we have computed in the previous Section are respectively:

- $\left|U_{In,n_1^*}\right|^2$ in equation (3.6): probability of the pumping $\psi^+ \rightarrow \psi^-$ under the action of the external source $\Lambda(t)$;
- $\left|U_{n_1^*,n_2^*}\right|^2$ in equations (2.2), (2.3), (2.6): decay probability $\psi^- \rightarrow \psi^+$ with emission of a virtual graviton.

4.2. Comparison with the Virtual Graviton Exchange in Classical Gravitational Interactions

It is interesting to compare the energy-momentum transfer occurring in this process with the transfer which takes place in the familiar gravitational interaction between two masses. It is known [9] that in quantum gravity the static interaction potential of two masses m_1 and m_2 placed a distance r apart can not be expressed, like in quantum electrodynamics or QCD, with the Wilson loop, but with the quantum formula

$$U(r) = m_1 m_2 \lim_{T\to\infty} \frac{1}{T} \int_{-T/2}^{T/2} dt_1 \int_{-T/2}^{T/2} dt_2 \left\langle 0 \right| h_{00}(t_1,\mathbf{r}_1) h_{00}(t_2,\mathbf{r}_2) \left| 0 \right\rangle$$

This equation describes the exchange of gravitons, for an ideally infinite time, between two static masses ($\mathbf{r} = \mathbf{r}_1 - \mathbf{r}_2$; see Fig. 2).

The flux of gravitons is both proportional to m_1 and m_2 and the propagator gives the amplitude of their propagation from $\mathbf{r}_1$ to $\mathbf{r}_2$, but note that their emission and absorption probabilities are equal to 1. If we expand the Feynman propagator in four-momentum space, we can see which energies and momenta are exchanged. One first finds

$$U(r) = Gm_1m_2 \lim_{T\to\infty} \frac{1}{T} \int_{-T/2}^{T/2} dt_1 \int_{-T/2}^{T/2} dt_2 \int dE \int d^3p \frac{e^{i\mathbf{pr} - iE(t_1 - t_2)}}{E^2 - \vec{p}^2 - i\varepsilon} \quad \textbf{(4.1)}$$

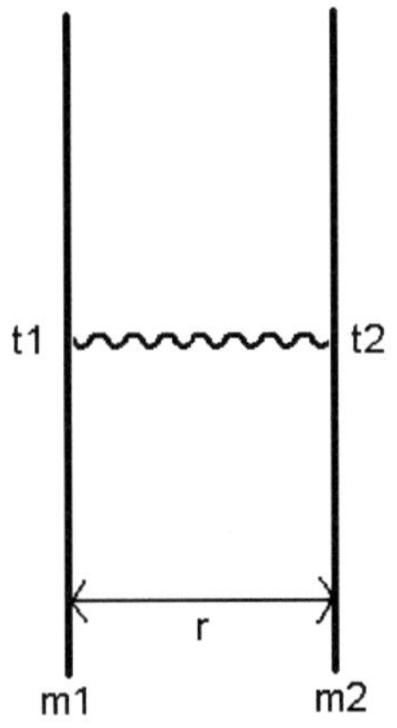

Figure 2: Worldlines of two static masses m_1 and m_2 which interact by exchanging virtual gravitons.

Changing variables to (t_1+t_2), (t_1-t_2) we find that the integral in (t_1+t_2) cancels the factor $1/T$. By integrating $\exp[-iE(t_1-t_2)]$ one obtains $\delta(E)$: this selects the static limit, *i.e.*, the exchanged gravitons have $E\cong 0$ (see below for a numerical estimate in a typical case; note that in equations (4.1) and (4.2) we use natural units $h/2\pi=c=1$). Finally we have

$$U(r) = -Gm_1m_2 \int d^3p \frac{e^{i\mathbf{pr}}}{p^2} = -\frac{2Gm_1m_2}{\pi r} \int_0^\infty dp' \frac{\sin p'}{p'} \quad \textbf{(4.2)}$$

with $p' = |\mathbf{p}|r$. (A similar reasoning also applies to quantum electrodynamics.) The last integral gives $\pi/2$ and the main contribution to the integration comes from the momentum region $p' < \pi$, *i.e.*, $pr < \pi$. This means that in the classical interaction of two masses at distance *r*, most exchanged virtual gravitons have momentum $p < \hbar/r$ (restoring $\hbar$), or wavelength $\lambda > r$. Such a large wavelength, larger than the propagation distance, means that the wavefunctions of these virtual particles are very different from plane waves. The momentum exchange between real and virtual particles is in this case more complicated than a scattering process. (And in fact a simple scattering could never be equivalent to an attractive force.) We shall give a more detailed discussion of this point in another paper.

In order to evaluate the typical magnitude order of the energies and momenta involved, consider the example of an object in free fall near the Earth. To stay close to the static case, we restrict our attention to the initial phase of the fall, with small velocity – say $v < 1$ m/s. The estimate above for the momentum of exchanged gravitons gives

$$p \le \frac{\hbar}{r}$$

With $r \sim 10^7$ m (Earth radius) this implies $p \le 10^{-41}$ kg m/s. The energy-momentum relation gives

$$\lambda f = \frac{hf}{\frac{h}{\lambda}} = \frac{E}{p} = \frac{1}{2}v$$

where the last equality is a consequence of the fact that the *E*/*p* ratio of the gravitons must match the *E*/*p* ratio of the targets. (With the limitations mentioned above, due to the condition $\lambda > r$.) But *v* is of the order of 1 m/s and this in turn implies that $f \sim 10^{-7}$ Hz. We see that the exchanged gravitons have indeed a very small frequency, in addition to the fact that the *E*/*p* ratio is very small itself (of order 1, while it is equal to *c* in real gravitons).

In the virtual process we are considering, involving the zero-modes, the frequency is determined by the pumping frequency and is much larger (up to 10^6-10^7 Hz for superconductors). The *E*/*p* ratio is approximately the same and the graviton wavelength is much smaller.

5. STIMULATED EMISSION

5.1. *A* and *B* Coefficients

We now have the necessary elements to estimate the process of stimulated emission. We know the *A* and *B* Einstein coefficients for the transition $\psi^+ \leftrightarrow \psi^-$. The dependence of the stimulated emission coefficient *B* on the frequency (equation (2.6)) is remarkable: in atomic transitions, the *B* coefficient typically decreases at larger frequency. It is also remarkable that *B* does not depend on the coupling *G*: the physical reason is that in the matrix element of the virtual mass dipole, m^2 is inversely proportional to *G*, for fixed *E* and *r*. (For instance, if *G* was smaller, then virtual mass dipoles would need to have larger mass to give the same energy shift, for fixed size *r*.) On the other hand, the dependence of *B* on *r* is similar to that in electrodynamics.

For dipole size $r \approx 10^{-9}$ m, we obtain $B \approx 10^{12}$ m³/Js² at frequency $\omega \approx 10^7$ Hz. This might be slightly over-estimated, because if we follow the analogy with electromagnetic atomic dipole emission, we find that the actual *B* coefficient observed in optical lasers does not correspond to an atomic matrix element $\langle \phi_1 | R | \phi_2 \rangle$ of the order of 10^{-10} m, as it should, but is smaller by at least one order of magnitude [10].

5.2. Pumping Rate

We know that the pumping is efficient, in the sense that the probability of the transitions $\psi^+ \rightarrow \psi^-$ under the action of the external coherent source H_Λ is large ($\frac{dP}{dt} \approx 10^{23}\,\text{s}^{-1}$, therefore the transition is almost instantaneous, in comparison with the other times involved). It remains to be seen how quickly the source can transfer the necessary energy. By considering the number of transitions of Cooper pairs in the intrinsic Josephson junctions of the superconductor for unit time and volume we found [8] that the maximum rate of transitions $\psi^+ \rightarrow \psi^-$ caused by the pumping interaction H_Λ per unit time and volume is

$$N = \frac{j}{2ez} \tag{5.1}$$

where *j* is the super-current density and *z* the layer thickness. With reasonable values for YBCO we have $N \approx 10^{34}\,\text{m}^{-3}\text{s}^{-1}$; then considering a super-current pulse of duration *T* and main frequency component $\omega \approx 1/T \approx 10^7$ Hz, in the pulse time $T \approx 10^{-7}$ s the total number of excited states produced by the pumping per unit volume is $NT \approx 10^{27}\,\text{m}^{-3}$. This is comparable with the density of excited atoms in active materials of solid state lasers.

Let us justify equation (5.1) in some more details. We have seen in [8] that the best way to produce large high-frequency variations in the vacuum-like energy density $\Lambda(t)$ of a superconductor is to send a current pulse through the superconductor. This works best in ceramic superconductors, where the pairs density has strong gradients across the layers. The passage of current in the c-direction can be described, in these materials, as a multiple Josephson tunnelling. We have done a simplified numerical simulation of this process by solving the Josephson equations of a few junctions connected in series. When a current pulse is sent to the junctions, they begin a synchronous oscillation with the frequency components present in the pulse. In this way, almost all the current is transmitted in the junctions as supercurrent; only about 0.1% of the current (at frequency 1 MHz) is transmitted as normal current, giving a voltage drop ΔV on the junction which matches the external frequency.

Now, it is well known that the power emitted by a Josephson junction is of the order of $I\Delta V$. This is easily shown to be the same, as magnitude order, as the number of pairs which tunnel through the junctions each second, multiplied by the Josephson energy gap hf. This leads to equation (5.1). As a eBook-keeping criterium, we can say that each tunnelling of a pair in the intrinsic junctions of the superconductor supplies the energy for the emission of a virtual graviton. This holds, however, only in the sense of the overall energetic balance. In fact, according to our quantum-gravity model, the gravitons are not emitted in a transition between different states of pairs in the superconductor (like in Fontana's model, involving a contact between different superconducting materials where the pairs are in *s* and *d* states – see Chapter 3). We suppose that virtual graviton emission occurs in all the bulk of a single superconductor of the YBCO type; the pairs oscillations in the intrinsic junctions cause excitations in the gravitational vacuum, then the decay of these excitations leads to emission of off-shell gravitons, which after possible multiplication by stimulated emission are absorbed in a target.

5.3. Population Inversion Condition

Even if the pumping rate is good, a crucial parameter is the density of available excited states ψ^- : this determines the occurrence of population inversion. We have seen that this density of states is indeed of $10^{27}\,\mathrm{m}^{-3}$ (single states have size of 10^{-9}m at most, but states smaller than 10^{-9}m can be disregarded because at small scale the B coefficient decreases quickly and this hampers stimulated emission). Therefore the magnitude orders appear to be favourable to the occurrence of population inversion.

It might seem that the population inversion condition $NT \approx 10^{27}\,\mathrm{m}^{-3}$ is more easily satisfied for current pulses of larger duration T, and therefore with lower frequency components. In practice, however, the current in the pulse will be smaller if the pulse lasts longer, as such pulses are typically obtained by quick discharge of some fixed total charge Q, so that $j = \frac{I}{A} = \frac{Q}{AT}$ and NT stays constant as T varies, with no gain for longer pulses.

5.4. Amplification Factor

Let us now see how the cascades of stimulated emission develop in a single pass through the superconductor. For simplicity, suppose first that the process starts only after the pumping has completed the population inversion. The mean free path $\delta = c\tau$ of a graviton in the average time τ before it generates another emission is given by the elementary relation

$$\sigma c\tau = \frac{1}{NT} \tag{5.2}$$

The fraction $\frac{1}{NT}$ is easily recognized as the volume containing, on the average, one single excited state. The cross-section σ for stimulated emission is related to the B coefficient by [10]

$$\sigma = \frac{1}{c}B\Delta E = \frac{1}{c}B\hbar\omega \tag{5.3}$$

Therefore, using relation (2.6), with $r \approx 10^{-9}\,\mathrm{m}$, we have

$$\tau = \frac{1}{B\hbar\omega NT} \approx \frac{1}{10^{-27}\,\omega^2 NT} \tag{5.4}$$

With the value of NT estimated above on the base of "our" superconducting parameters j and z (roughly independent from T and ω, as mentioned) we find numerically

$$\tau \approx \frac{1}{\omega^2} \qquad \textbf{(5.5)}$$

(There is a dimensional proportionality constant of order 1 in equation (5.5).) For $\omega \approx 10^7$ Hz this gives $\tau \approx 10^{-14}$ s, *i.e.*, $\delta = c\tau \approx 10^{-5}$ m. This short free mean path allows a strong amplification in the superconducting material. In a thickness of 0.5 mm, for instance, one would ideally have an amplification factor $2^{50} \approx 10^{15}$. Again, the energetic condition on the pumping becomes crucial to sustain the cascade. Suppose, for instance, that the super-current density j is available on a superconductor cross-section of 1 cm^2. The volume corresponding to one step of the cascade is of the order of 10^{-8} m^3; this implies that the number of excited states in this volume is 10^{19}. This can support the above amplification factor $2^{50} \approx 10^{15}$ (from an initial value of 10^4 gravitons or higher, see below), but it would not support a complete amplification in a thickness twice as large (superconductor thickness 1 mm, amplification factor $2^{100} \approx 10^{30}$).

The probability of spontaneous emission in a layer with thickness z (to start the cascades) can be computed recalling that the lifetime of an excited state for spontaneous emission is of the order of 1 s, *i.e.*, $dP/dt \approx 1$ s^{-1}, or $P \approx 10^{-12}$ in the time τ. The excited systems in the layer are of the order of 10^{19}, as seen, therefore 10^7 spontaneous decays are available, on the average, to start the cascades.

The figures above are very approximate and a detailed rate equation should be written to account for the emission, which in fact starts as soon as the pumping begins, and not only after its completion. It seems clear, however, that the amplification process is possible. On one side the process is favoured by high frequency, because the mean free path for stimulated emission is shorter at higher frequency (equation(5.5)). On the other side, although a quick pumping (short pulse, with high-frequency components) allows quicker population inversion, high frequency can also increase too much the probability of spontaneous emission ($A \propto \omega^4$!), so only a detailed balance can show which frequency gives the largest amplification.

6. CONCLUSIONS AND DISCUSSION

We have shown that virtual gravitational stimulated emission can occur in layered superconductors like YBCO pumped with high-frequency supercurrent pulses ($f \approx$ 1-10 MHz, $j \approx$ 100 A/cm^2). Laser action can be obtained in a single passage, like in a superfluorescent nitrogen laser, because the mean free path for stimulated emission is about 10^{-5} m and this leads to large amplification factors already in a thickness of few millimeters.

Our theoretical model has been inspired by Podkletnov's gas discharge experiment ([11]; see Chapter 1 and Chapter 9) and the figures mentioned above for the frequency and the current density actually refer to that set-up. The reported space coherence of the anomalous radiation beam strongly suggests a stimulated emission mechanism, while the observed energy-momentum ratio in the beam ($E/p \approx$ 1 m/s) suggests the concept of virtual radiation and has also been used in the calculations. A further strong experimental indication concerned the role of the superconductor, modelled here as a vacuum-energy-like source which thanks to its unique gravitational coupling is able to support the "pumping" process.

Though linked to experiment, our model has a wider scope and introduces several new theoretical concepts related to the gravitational vacuum and its virtual mass fluctuations (see below). On the other hand, it should be acknowledged that many details of the observed phenomenology (for instance, threshold values in voltage and frequency) can not yet be explained by our picture. Rate equations for the laser action appear presently out of reach.

According to our model, there is a connection between the emission of virtual gravitons and the tunnelling of the Cooper pairs in the intrinsic Josephson junctions of YBCO, but the connection is quite loose and mainly useful in keeping track of the energetic balance. The Cooper pairs do not directly cause the graviton emission. The essential requisite for that is the activation of gravitational excited states containing large virtual oscillating masses (the "oscillation" being a consequence of the quantum superposition of states localized at different points). The activation (pumping) of excited states can only occur through the interaction of the gravitational vacuum with matter described by a macroscopic wavefunction. This is because single particles, coupled to gravity in the standard way, have an exceedingly small probability of producing excited states (those with the virtual masses in anti-symmetric superposition).

It can be helpful to summarize here the technical hypotheses on which this active-vacuum model is based.

- We start from Einstein gravity, regarded in its quantum version as an effective theory for low energies. Unlike other approaches (see for instance the extended Heim theory in Chapter 11), we do not postulate any new fundamental physics. The gravitational zero-modes with virtual mass should be regarded as emergent particles, much like the Cooper pairs in the BCS theory. We do not investigate the (still unknown) fundamental mechanism which makes our spacetime flat and keeps it flat in spite of vacuum energy contributions of various kind (including the zero-modes themselves).

- We have used for some perturbative estimates a special version of the Einstein-Maxwell equations, and discussed the problem of the "spin-1 gravitons" apparently coming from their quantization (Appendix 2).

- We have computed the Feynman path integral for a class of static spherically-symmetric metrics with constant g_{00} component, confirming the result of our previous computation of the zero-modes. The result is intepreted as follows: if the average $\langle g_{rr}^2(0)\rangle$ computed on that subset of metrics vanishes, then there is a finite probability that a zero-mode is formed at the origin (and therefore at any point, due to Lorentz invariance).

- We have written the processes of virtual graviton generation, propagation and absorption in a target as a quantum sum over intermediate states. We have supposed that this sum can be factorized like in a succession of real events, because the intermediate states are sharply defined. It follows that the single probabilities can be computed to first order with the Fermi golden rule. An inherent disadvantage of this method is that it does not give any information on the temporal evolution of the process and on the propagation velocity of the pulse (compare Chapter **8**).

- We have supposed that the vacuum-energy-like term Λ, because of its non-linearity in the field g, can have non-zero matrix elements at first order.

APPENDIX 1: GRAVITATIONAL ZERO-MODES

We call "gravitational zero-mode" a field configuration for which the Einstein action $S_E = -\frac{1}{8\pi G}\int d^4x\sqrt{g}R$ is zero. Since the density $\sqrt{g}R$ is not positive-definite, this does not necessarily implies that $R = 0$, but only that the density is a function with null integral, *i.e.*, intuitively, there are regions of positive and negative scalar curvature. We are interested into the zero-modes because, in the Feynman path integral, field configurations with the same action tend to interfere constructively and so to give a contribution to the integral distinct from the usual classical contribution of the configurations near the stationary point of the action. An elementary example of this phenomenon is given in App. 3.3 (over-dampened oscillator). The zero-modes can only give a significant contribution to the path integral if they are not isolated configurations, but a whole full-dimensional subset of all the possible configurations.

The gravitational zero-mode condition has the form of an integral equation and is very difficult to tackle in its general form. We can only look for solutions with some special symmetry, or with some constant components. Consider the most general static spherically symmetric metric

$$d\tau^2 = B(r)dt^2 - A(r)dr^2 - r^2(d\theta^2 + \sin^2\theta d\phi^2)$$

where $A(r)$ and $B(r)$ are arbitrary smooth functions. We add the requirement that outside a certain radius $r_{\rm ext}$, $A(r)$ and $B(r)$ take the Schwarzschild form, namely

$$B(r) = \left(1 - \frac{2GM}{c^2 r}\right); \quad A(r) = \left(1 - \frac{2GM}{c^2 r}\right)^{-1} \quad \text{for } r \geq r_{ext} \qquad \textbf{(A1.1)}$$

This requirement serves two purposes: (1) It allows to give a physical meaning to these configurations, seen from the outside, as mass-energy fluctuations of strength M. For $r>r_{\rm ext}$ their scalar curvature is zero. (2) More technically, the York-Hawking-Gibbons boundary term of the action is known to be constant in this case [3]. Even with only the functions A and B to adjust, the condition $S_E=0$ is very difficult to satisfy. We do find a set of solutions, however, if we make the drastic simplification $g_{00}=B(r)=const$. The scalar curvature multiplied by the volume element becomes in this case

$$L = \sqrt{g}R = -8\pi\sqrt{|BA|}\left(\frac{rA'}{A^2} + 1 - \frac{1}{A}\right) \qquad \textbf{(A1.2)}$$

Apart from the Newton constant G, L is the lagrangian density of the Einstein-Hilbert action, computed for this particular metric. Let us fix arbitrarily a reference radius $r_{\rm ext}$, and introduce reduced coordinates $s=r/r_{\rm ext}$. Define an auxiliary function $\alpha=A^{-1}$. Regarding $L(s)$ as known, equation (A1.2) becomes an explicit first-order differential equation for α:

$$\alpha' = \frac{1}{s} - \frac{\alpha}{s} + \frac{L\sqrt{|\alpha|}}{8\pi s\sqrt{|B|}} \qquad \textbf{(A1.3)}$$

The boundary conditions (A1.1) are written

$$B(s \geq 1) = \left(1 - \frac{\tilde{M}}{s}\right); \quad A(s \geq 1) = \left(1 - \frac{\tilde{M}}{s}\right)^{-1} \qquad \textbf{(A1.4)}$$

where $\tilde{M}$ is a free parameter, the total mass in reduced units: $\tilde{M} = 2GM / c^2 r_{ext}$. In the following we shall take $\tilde{M} < 0$, in order to avoid singularities. For $r < r_{ext}$, we have $B = B(1) = 1 - \tilde{M}$.

It is interesting to note that putting $L=0$ in equation (A1.3) we can easily find an exact solution, ie a non-trivial static metric with $R=0$. Namely, if $1-\alpha>0$, then $\alpha=1-e^{\rm const}/s$, which does not satisfy the boundary condition; if $1-\alpha<0$, then $\alpha=1+e^{\rm const}/s$, implying $e^{cost} = -\tilde{M}$. The resulting g_{rr} component has the same form on the left and on the right of $s=1$, namely

$$g_{rr} = \left(1 + \frac{|\tilde{M}|}{s}\right)^{-1} \qquad \textbf{(A1.5)}$$

while g_{00} is constant and equal to $(1+|\tilde{M}|)$ for $s<1$, and is equal to $(1+|\tilde{M}|/s)$ for $s \geq 1$. Note that $g_{\rm rr}$ goes to zero at the origin.

For large $\tilde{M}$ and small L, the last term in equation (A1.3) is a small perturbation. Since α never diverges and α^{-1} does not appear in the equation, the perturbed solution is not very different from (A1.5). For values of $\tilde{M}$ of order 1 or smaller, the equation can be integrated numerically. If we choose a function $L(s)$ with null integral on the interval (0, 1), we obtain a metric which is a zero-mode of the action but not of the lagrangian. One can take, for instance, $L(s)=L_0\sin(2\pi ns)$, with n integer.

In conclusion, we have found a family of metrics with null scalar curvature, depending on a continuous parameter $\tilde{M}$. Furthermore, we have built a set of metrics close to the latter, by solving equation (A1.3) with L arbitrary but having null integral. These metrics do not have zero scalar curvature, but still have null action. They make up a full-dimensional subset of the functional space (see proof in [3]).

APPENDIX 2: WAVE SOLUTIONS OF THE EINSTEIN-MAXWELL EQUATIONS

In Section 2 we have written the linearized Einstein equations in the so-called Einstein-Maxwell form. In this approximation the relevant components of the gravitational field reduce to the two vectors $\mathbf{E}_g$ and $\mathbf{B}_g$, which satisfy equations very similar to Maxwell equations; the main difference is the minus sign in front of the sources ρ and **j**. The force on slow test particles in a static field is also very similar to the Lorenz force $\mathbf{F}=q\mathbf{E}+q\mathbf{v}\wedge\mathbf{B}$; depending on the conventions, there can be a factor 2 or 4 in the magnetic term and an overall negative sign. In general, the "gravitational gyromagnetic ratio" is different from the electromagnetic ratio. When the fields depend on time the force is given by an expression in which the gravitomagnetic term is the same, while the gravitoelectric term is replaced by a term of the form $(\nabla\phi_g+2\partial_t\mathbf{A}_g)$, which in general is not equal to the $\mathbf{E}_g$ field (see [12]; if ϕ_g can be disregarded, then $2\partial_t\mathbf{A}_g=4\mathbf{E}_g$).

In the following, however, we shall not be interested into the equations of motion for test particles. While the most frequent applications of the Einstein-Maxwell equations concern the motion of test particles in static fields, we shall mainly be interested into their wave solutions and into the coupling of the wave solutions to multi-polar sources.

In the wave equations we can not disregard, of course, the time derivatives of the fields. This leads to some complications. The problem has been recently discussed by several authors, including Pascual-Sanchez and Harris [13], but it is already clear in the textbook by Wald [14]. Wald derives first the field equations and the Lorenz-like force equation in the hypothesis that the time derivatives of the spatial components h_{ij} of the metric are negligibly small. (The correspondence between metric components and Einstein-Maxwell potentials is $\phi_g=h_{00}$, $\mathbf{A}_g=\{h_{0i}\}$.) Then, relaxing this hypothesis, he applies a residual gauge transformation to obtain the "radiation gauge" $\bar{h}=0$, $\bar{h}_{0i}=0$ in a region without sources (and also $\bar{h}_{00}=0$, if there are no sources anywhere in space). This gauge transformation shows that the field appearing in the Einstein-Maxwell equations are physically equivalent in vacuum to the spatial components h_{ij} of the gravitational waves in the transverse traceless gauge. Nevertheless, the fact that this correspondence is only valid in regions without sources prevent us from using the Einstein-Maxwell equations to study the generation of waves.

A related problem concerns the spin of the solutions. Even admitting that the Einstein-Maxwell equations have valid wave solutions in vacuum, the resulting vector waves seem to carry spin 1. Chiao, who makes use of the Einstein-Maxwell equations to describe gravitational waves, realizes the problem and states that the equations must be used together with an additional physical condition on the sources [15].

In order to describe the matter-radiation interaction (which is our concern here) Peng [16] has developed "generalized" Einstein-Maxwell equations in which the fields $\mathbf{E}_g$ and $\mathbf{B}_g$ have additional tensor indices. This formalism is shown to be completely equivalent to the metric formalism in the transverse traceless gauge. Ciubotariu [17] employed Peng's generalized equations to investigate the absorption of gravitational waves in matter obeying "gravitational Ohm law". These equations, however, are more complicated than the Maxwell equations, and the standard quantization procedure of the electromagnetic waves in a finite volume can not be directly applied to them.

We shall then choose another technique, which we could call the "Adler gauge". Adler [18] considers the Einstein-Maxwell equations in a system of "isotropic" coordinates, where the diagonal components of the metric are all equal, namely $h_{00}=h_{ii}$ ($i=1, 2, 3$). Then, using only the linearization hypothesis, he proves the wave equations

$$\Omega\phi = -4\pi G\rho$$
$$\Omega\mathbf{h} = 16\pi G\rho\mathbf{v}, \quad \mathbf{h} = \left(h_{01}, h_{02}, h_{03}\right)$$

from which the Einstein-Maxwell equations follow. One obtains in this way a complete analogy with electromagnetism; the price to pay is that the transformation to isotropic coordinates is a local rotation at any point (this diagonalizes h_{ij}), followed by local different dilatations in the three directions. For time-dependent fields, this transformation is also time-dependent. The great liberty of coordinate transformations in General Relativity allows this, but it is also clear that any considerations about the spin of the fields lose their meaning when the coordinates at each point of space are subjected to a different, time-dependent rotation. We thus solve the paradox of the graviton spin 1 *versus* spin 2, and we avoid the consequences of the well-known theorem [19, 20], according to which any spin-1 field in a Lorenz-invariant theory transmits a repulsive force between particles with "charges" of the same sign. The usual physical meaning of the field components is lost in this gauge, but this does not affect our quantum computations of probabilities or matrix elements.

APPENDIX 3: GENERAL PATH-INTEGRAL TREATMENT OF ZERO-MODES

A.3.1. Introduction: Propagators and Normalized Averages

We are going here to illustrate the concept of zero-modes of the action in quantum mechanics, in order to better justify our basic assumptions in [3], which led to the identification of the gravitational zero-modes (App. 1).

We start from the principle that the dynamics of any system is described by the Feynman path integral. In its simplest one-dimensional formulation we can write the probability amplitude for a particle to go from the initial position q_i at time t_i to the final position q_f at time t_f is given by the integral over all possible classical paths having those extrema, weighed with the phase factor $e^{\frac{i}{\hbar}S[q]}$, where $S[q]$ is the action of the path $q(t)$:

$$\left\langle q_f, t_f \left| e^{iH(t_2-t_1)} \right| q_i, t_i \right\rangle = K\left(q_f, t_f; q_i, t_i\right) = \int d\left[q\right] e^{\frac{i}{\hbar}S[q]} = \int d\left[q\right] e^{\frac{i}{\hbar}\int dt L(q(t))} \quad \textbf{(A3.1)}$$

The operator $e^{iH(t_2-t_1)}$ in the matrix element is the temporal evolution operator built with the system's Hamiltonian. [*Errata*: t_2 and t_1 must be replaced resp. by t_i and t_f.] L is the Lagrangian, for instance $L(q,\dot{q}) = \frac{1}{2}m\dot{q}^2 - U(q)$ in the simplest cases. The function K is called the "propagator". The symbol $\int d\left[q\right]$ denotes the integration over all the functions $q(t)$ and the conditions $q(t_i) = q_i$, $q(t_f) = q_f$ are implicitly assumed in the integration in this case.

It is straightforward to discretize this one-dimensional system and re-write the path integral as the product of ordinary integrals. In the end one obtains

$$\int d\left[q\right] e^{\frac{i}{\hbar}S[q]} = \lim_{N\to\infty}\left(\frac{m}{2\pi i\delta}\right)^{N/2} \int \prod_{k=1}^{N-1} dq_k \exp\left\{\frac{i\delta}{\hbar}\sum_{j=0}^{N-1}\frac{1}{2}m\left(\frac{q_{j+1}-q_j}{\delta}\right)^2 - U\left(q_j\right)\right\} \quad \textbf{(A3.2)}$$

The integration occurs only over the "internal" variables $q_1, \ldots, q_{N-1}$, while $q_0 = q_i$ and $q_N = q_f$. In the continuum limit $N \to \infty$ one takes advantage of the relation $N\delta = T$, where $T = t_f - t_i$.

In the following we shall use for comparison the propagator of the harmonic oscillator. This can be computed exactly, because the integrals are Gaussian, and turns out to be, with $L = \dot{q}^2 - \omega^2 q^2$,

$$K\left(q_f,t_f;q_i,t_i\right)=\sqrt{\frac{\omega}{2\pi i\hbar\sin(\omega T)}}\exp\left\{\frac{i\omega}{2\hbar\sin(\omega T)}\left[\left(q_f+q_i\right)^2\cos(\omega T)-2q_iq_f\right]\right\} \quad \textbf{(A3.3)}$$

Note that in certain special cases the propagator diverges, for instance when $\omega T=\pi$, *i.e.*, when $T=\frac{1}{2}T_0$, where $T_0=\frac{2\pi}{\omega}$ is the natural oscillation period of the classical oscillator. This divergence can be interpreted by observing that in that case there exist infinite classical paths connecting the two end-points q_i and q_f; these paths have of course different values of the total energy and different velocities at the initial and final points.

In order to avoid this kind of divergences, let us consider in the following not the simple propagator, but normalized averages (Feynman-Hibbs "transition elements") like

$$\left\langle f[q]\right\rangle=\frac{\int d[q]e^{\frac{i}{\hbar}S[q]}f[q]}{\int d[q]e^{\frac{i}{\hbar}S[q]}}$$

In the following, for instance, we shall compute the average of the square of an internal middle coordinate $\left\langle q_m{}^2\right\rangle$. In quantum field theory one often computes in this way the correlation functions of the fields:

$$\left\langle\phi(x)\phi(y)\right\rangle=\frac{\int d[\phi]e^{\frac{i}{\hbar}S[\phi]}\phi(x)\phi(y)}{\int d[\phi]e^{\frac{i}{\hbar}S[\phi]}}$$

where x and y are 4D spacetime coordinates.

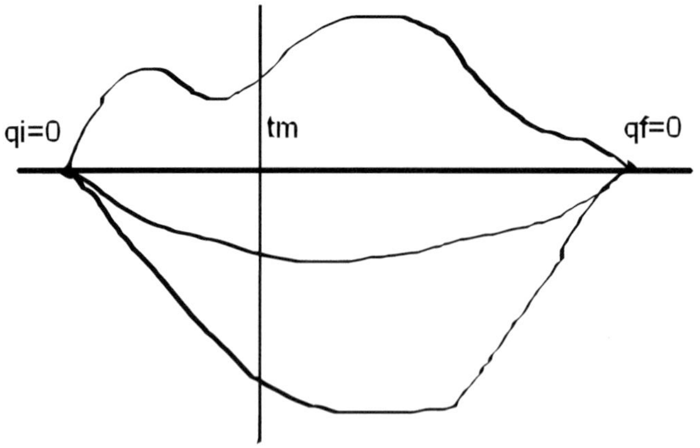

Figure 3: When $t_f-t_i=\frac{T_0}{2}$, there are infinite classical paths and both the propagator and $\left\langle q_m{}^2\right\rangle$ diverge.

A.3.2. Example: Calculation of the Normalized Average $\langle q_m^2\rangle$ for the Harmonic Oscillator

Before writing the path integral of a "singular" system with relevant zero-modes, let us see an example of a computation of an average value for the harmonic oscillator, both in the case of finite and infinite result.

Consider an intermediate time t_m $(t_i < t_m < t_f)$, call q_m the value of the coordinate at the time t_m and compute $\langle q_m{}^2 \rangle$, with the conditions $q_i = q_f = 0$ (Fig. **3**). Let us break the path integral in two parts, over the intervals (t_i, t_m) and (t_m, t_f). We use the propagator formula (A3.3) to obtain

$$\int d[q] e^{\frac{i}{\hbar}S[q]} q_m{}^2 = \int dq_m q_m^2 K(0, t_i; q_m, t_m) K(q_m, t_m; 0, t_f)$$
$$\int d[q] e^{\frac{i}{\hbar}S[q]} = K(0, t_i; 0, t_f)$$

For simplicity we then suppose that t_m is exactly centered between t_i and t_f, *i.e.*, $t_m - t_i = t_f - t_m = \tau = T/2$ (but it is easy to generalize the result). We obtain

$$\frac{\int d[q] e^{\frac{i}{\hbar}S[q]} q_m{}^2}{\int d[q] e^{\frac{i}{\hbar}S[q]}} = \sqrt{\frac{\omega \cos(\omega\tau)}{\pi i \hbar \sin(\omega\tau)}} \int dq_m q_m^2 \exp\left\{ \frac{i\omega}{\hbar \sin(\omega\tau)} q_m^2 \cos(\omega\tau) \right\} \tag{A3.4}$$

It is straightforward to check that the result is of order $\hbar/\omega$, as one would expect from quantum fluctuations in the ground state of the harmonic oscillator using the relations $E_0 = \frac{1}{2}\hbar\omega = \frac{1}{2}\left\langle \frac{1}{2} M\omega^2 q^2 \right\rangle$; note that in the formulas above we always have unit mass $M = 1$. However, for $\omega\tau = \frac{\pi}{2}$ one has a divergence, as expected, because in that case we are in the situation with infinite classical trajectories between the initial and final position.

To compute the Gaussian integrals we used the formulas

$$\int dx \exp\left(-\frac{1}{2}\alpha x^2 + i\beta x \right) = \sqrt{\frac{2\pi}{\alpha}} \exp\left(-\frac{\beta^2}{2\alpha} \right)$$
$$\int dx x^2 \exp\left(-\frac{1}{2}\alpha x^2 \right) = \sqrt{\frac{2\pi}{\alpha^3}}$$

A.3.3. A System with Relevant Zero-Modes: The "Totally Dampened" Oscillator

After this long introduction, necessary to fix the formalism and see some results for an "ordinary" system like the harmonic oscillator, we turn to a system for which the classical dynamics is trivial, while the zero-modes dominate the quantum dynamics. Since we want to prove this explicitly through the path integral, we must choose a simple system, whose path integral can be computed exactly. We analysed several cases and in the end we took the following "totally dampened" harmonic oscillator with Lagrangian linear in $\dot{q}$:

$$L = \frac{1}{2} M \mu \dot{q} - \omega^2 q^2 + q f(t)$$

We have added a coupling to an external applied force $f(t)$ in order to understand better the classical solutions, but in the path integral we shall set $f(t) = 0$. The classical equation of motion and its solution are

$$\frac{d}{dt}\frac{\partial L}{\partial \dot{q}} - \frac{\partial L}{\partial q} = 0 \Rightarrow q(t) = \frac{1}{2\omega^2} f(t)$$

The oscillator just follows the external force, and in the absence of an external force the only solution is $q(t)=0$. The $\dot{q}$ term in the Lagrangian has no effect, being a total time derivative.

The canonical quantization of this system is awkward, because the conjugate momentum $\partial L/\partial\dot{q}$ is trivially constant; the system is "constrained", the energy is not conserved, *etc.* But let us write the path integral and apply the recipe for calculating the physical average $\langle q_m^2\rangle$, with certain initial and final conditions $q_i(t_i)$ and $q_f(t_f)$ which for the moment we leave unspecified:

$$\int d[q]e^{\frac{i}{\hbar}S[q]}q_m{}^2 = \\ =\int\prod_{k=1}^{N-1}dq_k q_m^2 \exp\left\{\frac{iM\mu\delta}{2\hbar}\left[\frac{q_1-q_i}{\delta}-\omega^2q_1^2+\frac{q_2-q_1}{\delta}-\omega^2q_2^2+...+\frac{q_f-q_{N-1}}{\delta}-\omega^2q_f^2\right]\right\} \quad \textbf{(A3.5)}$$

The only terms linear in q_k which survive in the square bracket are $-q_i/\delta$ and q_f/δ. The exponential is completely factorized and when we divide all by $\int d[q]e^{\frac{i}{\hbar}S[q]}$, the only term left (independently from the initial and final conditions, or even, if we want, integrating over all initial and final conditions) is

$$\langle q_m^2\rangle=\frac{\int dq_m q_m^2\exp\left\{\frac{iM\mu\delta}{2\hbar}\left(-\omega^2q_m^2\right)\right\}}{\int dq_m\exp\left\{\frac{iM\mu\delta}{2\hbar}\left(-\omega^2q_m^2\right)\right\}}$$

Note that in equation (A3.5) we did not multiply by a pre-factor of the form $\left(\frac{m}{2\pi i\delta}\right)^{N/2}$ like in (A3.2), because that factor is obtained computing the matrix element (A3.1) through the system's Hamiltonian and then integrating on the momenta. In our case this is not possible, as seen, therefore we discretize directly the path integral without the pre-factor. Since we normalize the averages dividing by $\int d[q]e^{\frac{i}{\hbar}S[q]}$, the pre-factor is irrelevant. In other words, our aim is not to compute the propagator, but only normalized averages.

The Gaussian integrals are easily performed and the result is

$$\langle q_m^2\rangle=\frac{\hbar}{i\omega^2M\mu\delta}$$

In the continuum limit this is divergent. This result is not an anomaly of the continuum limit of our system, because other quantities have finite averages (see below); it shows that the system "runs away" to any large value of intermediate q, whatever the initial conditions (compare the harmonic oscillator in the special case of an interval (t_f-t_i) equal to the natural oscillation period).

For comparison, one can evaluate the average of $e^{-kq_m^2}$, which still gives a Gaussian integral; one easily obtains a finite result, which vanishes in the continuum limit. If we compute the average of a quantity like $q_m^2/(1+q_m^2)$, which tends to a finite value for large q_m, we obtain a finite and non-zero result in the continuum limit; to see this, one can exploit the known integral

$$\int dx\frac{x^2\exp\left(-\frac{i}{2}ax^2\right)}{x^2+1}=\pi e^{\frac{ia}{2}}\left\{-1+\text{erf}\left[\left(\frac{1}{2}+\frac{i}{2}\right)\sqrt{a}\right]\right\}+\frac{1-i}{\sqrt{a}}\sqrt{\pi}$$

and consider that the "Error function" erf(x) is finite for $a \to 0$, and tends to -1. In our case $a = const. \cdot \delta$, and $\delta \to 0$ in the continuum limit. The normalization factor at the denominator behaves like $\frac{1}{\sqrt{\delta}}$, therefore for $\delta \to 0$ the term $\frac{1-i}{\sqrt{a}}\sqrt{\pi}$ gives a finite contribution and the rest goes to zero.

In conclusion, the classical equations give the unique solution $q(t) = 0$. The quantum averages computed with the path integral tell us that for any initial and final conditions each intermediate coordinate q_m tends to grow without limit. This is clearly not just "quantum noise" (compare the corresponding average for the harmonic oscillator, equations (A3.4)). Quantum-mechanically, the system runs away to large q, far from the classical solution. What is the reason of this discrepancy? We believe this behaviour to be related to the existence of zero-modes of the action.

A.3.4. Zero-Modes of the Totally-Dampened Oscillator

Among the possible configurations of a totally-dampened oscillator, there exists a set of functions $q(t)$ which have action exactly zero or equal to an arbitrary constant. (This is an infinite-dimensional set, with non-zero measure as a subset of all possible configurations; not like, for instance, a line in 2D or a surface in 3D.) Such configurations do not interfere destructively in the path integral, but add in phase and give a relevant contribution.

Usually the main contributions to a path integral come, according to the principle of the stationary phase, from the classical paths, which make the action stationary, and from a narrow set of almost-classical paths (quantum fluctuations). The reason is that for usual physical systems the Euclidean Lagrangian is positively-defined and the zero-modes are trivial (see below). But this is true neither for our totally-dampened toy system nor for the gravitational action. Let us set equal to zero the action of the totally-dampened oscillator (then, to generalize, we can set the action equal to any function with zero integral on the given interval; this gives an infinite-dimensional class of associated zero-modes). For simplicity, we re-scale ω in such a way that $\mu M/2=1$ and obtain

$$\dot{q} - \omega^2 q^2 = 0$$

The solution is

$$q(t) = -\frac{1}{\omega^2 t + c},$$

with c real arbitrary constant. Depending on the value of c, this equation has bounded or unbounded solutions on a given interval, with possible singularities if the value $t = -c/\omega^2$ belongs to the interval. The only value which $q(t)$ can never assume is just $q = 0$. However, setting the action equal to a constant instead of zero (which still gives non-interfering paths) one can obtain also solutions which are zero at a given instant; for instance, the solution of $\dot{q} - \omega^2 q^2 = 1$ is

$$q(t) = \frac{1}{\omega}\tan(\omega t + c).$$

In the more general case

$$\dot{q} - \omega^2 q^2 = k$$

one finds

$$x = \tan\left(\sqrt{k}\omega t + c\right)$$

where we have changed variable to $x = \omega q / \sqrt{k}$ and taken k positive. This solution diverges in any fixed interval, provided k is large enough (compare with the solutions for the zero-modes of the harmonic oscillator!).

A.3.5. The Zero-Modes of the Harmonic Oscillator and of the e.m. Field are Trivial

We should check that the action of the harmonic oscillator does not admit zero-modes that cause a run-away of the system. Actually, we know in advance from the Euclidean treatment that such modes, if present, are physically irrelevant, because in the imaginary-time path integral we obtain expressions with real exponentials, like

$$\int d[q] \exp\left\{-\frac{1}{\hbar}\int dt \left(\frac{1}{2}\dot{q}^2 + \frac{1}{2}\omega^2 q^2\right)\right\}$$

These integrals receive relevant contributions only from the region $q \approx 0$. (The transition to imaginary time is impossible in our totally-dampened toy system and in quantum gravity.)

Going back to the real-time Lagrangian, the zero-mode equation for the harmonic oscillator is

$$L = \dot{q}^2 - \omega^2 q^2 = k \quad \textbf{(A3.6)}$$

The simplest case is k=0. The solution is

$$q(t) = c e^{\pm \omega t} \quad \textbf{(A3.7)}$$

with c a real constant. We see that in a finite time the variation of the exponential is bounded. The time interval used for the averages in the path integral (equation (A3.4)) is arbitrarily small, therefore only singular zero-modes which diverge at finite time give a run-away of the system. Solutions like (A3.7) are only a component of the usual quantum noise.

The general solution of (A3.6) is, in implicit form,

$$\ln\left(\sqrt{x^2+1} + x\right) = \omega(\pm t + c)$$

where we have changed to a variable $x = \omega q / \sqrt{k}$ and taken k positive (otherwise no solutions could exist which start from q=0 at time t=0). This solution starts from the origin and is also bounded in any finite time interval. Its behaviour for large time is $x(t) \approx e^{\pm \omega t + c}$.

Analogous situations are found in quantum field theory. While in gravitation the Euclidean action does not have a definite sign, and large static zero modes exist, in most other theories the Euclidean action is positive-definite and the zero-modes only represent small and short-lived quantum fluctuations. In electromagnetism, for instance, the zero-mode equation is

$$\varepsilon_0 \mathbf{E}^2 - \frac{\mathbf{B}^2}{\mu_0} = 0$$

Let us consider the static solutions. The independent variable is the four-potential $A_\mu = (V, \mathbf{A}) = (A_0, \mathbf{A})$, from which in the static case the fields $\mathbf{E} = -\nabla V$, $\mathbf{B} = \nabla \times \mathbf{A}$ are obtained. It is possible to fix the gauge as $\mathrm{A}_0 = 0$ and implement it in the functional integral [21]. It follows the equation

$$(\nabla \times \mathbf{A})^2 = 0$$

which has only the trivial solution $\mathbf{B} = 0$.

APPENDIX 4: GRAVITATIONAL PI FOR SPHERICALLY-SYMMETRIC METRICS

We shall now consider the path integral of Einstein gravitation, in order to evaluate the functional average of certain metric components and confirm the results of the zero-modes equations. The zero-modes equations tell us that relevant run-away configurations of vacuum exist, in which the metric is locally very different from its classical value $g_{\mu\nu}(x) = \eta_{\mu\nu}$.

Let us choose a spherical coordinate system. We integrate only over a sector of the functional space, namely over the spherically-symmetric metric configurations. If we obtain quadratic averages different from the classical null value, this allows us to reach our conclusion: there is a finite probability for the zero-mode to occur.

For these metrics the Einstein-Hilbert action is written (App. 1, [3])

$$S_E = -\frac{1}{8\pi G}\int d^4x\sqrt{g(x)}R(x) = \frac{4\pi}{G}\int dt \int_0^\infty dr \sqrt{|BA|}\left(\frac{rA'}{A^2} + 1 - \frac{1}{A}\right)$$

where $A = g_{rr}$ and $B = -g_{tt}$ are functions of *r*. Define a radius r_{ext}, the "external radius" of our configurations, on which we impose the boundary conditions

$$B(s \geq 1) = 1 - \frac{\tilde{M}}{s}, \quad A(s \geq 1) = \left(1 - \frac{\tilde{M}}{s}\right)^{-1}$$

being $s = r / r_{ext}$ the reduced coordinate and $\tilde{M} = 2GM / c^2 r_{ext}$ the reduced mass. Therefore we integrate over configurations which outside the radius r_{ext} appear like Schwarzschild metrics with mass *M*. In order to avoid singularities, we suppose $M < 0$. We can re-write the action as

$$S_E = \frac{4\pi}{G}\int dt \int_0^{r_{ext}} dr \sqrt{|BA|}\left(\frac{rA'}{A^2} + 1 - \frac{1}{A}\right)$$

because the scalar curvature of the Schwarzschild metric is zero. We can also add the Gibbs-Hawking-York boundary term, which in this case takes the form $S_{HGY} = -M\int dt$. For a fixed time interval, we can regard the integral $\int dt$ as a constant.

Supposing *B* constant ($B = 1 - \tilde{M}$), the path integral over these field modes is written

$$\int d[A] \exp\left\{\frac{i}{\hbar}(S_E + S_{HGY})\right\}$$
$$= \int d[A] \exp\left\{\frac{i}{\hbar}\frac{4\pi\sqrt{|B|}}{G}\int dt \int_0^1 ds \sqrt{|A|}\left(\frac{sA'}{A^2} + 1 - \frac{1}{A}\right)\right\} \exp\left\{-\frac{i}{\hbar}M\int dt\right\}$$

The second exponential can be disregarded in the functional averages, because it cancels with the normalization factor in the denominator. In the first exponential, let us define a constant factor

$\alpha = \frac{1}{\hbar}\frac{4\pi\sqrt{|B|}}{G}\int dt$ and discretize the integral in ds. We divide the integration interval [0, 1] in $(N+1)$ small intervals of length δ and replace the integral with a sum, where the derivative is written as a finite variation. We obtain

$$\int d[A] e^{\frac{i}{\hbar}S_E} = \int \prod_{j=0}^{N+1} dA_i \exp\left\{ i\alpha\delta \sum_{j=0}^{N} \sqrt{|A_j|} \left(\frac{j\delta\left(A_{j+1}-A_j\right)}{\delta A_j^2} + 1 - \frac{1}{A_j} \right) \right\}$$

The presence of the square root and of the fractions with A_j makes the integrals very complicated. Let us change variables. Suppose $A > 0$, which is physically a widely justified assumption (and remember we are looking for a sufficient condition, *i.e.,* we want to show that there exist a set of gravitational configurations for which the functional average of a quadratic quantity is different from the classical value). Define $\gamma = 1/\sqrt{A}$. This gives the new path integral

$$\int_0^{+\infty} \prod_{j=0}^{N+1} d\gamma_j \frac{2}{\gamma_j^3} \exp\left\{ -i\alpha\delta \sum_{j=0}^{N} \left(2j\delta \frac{\left(\gamma_{j+1}-\gamma_j\right)}{\delta} - \frac{1}{\gamma_j} + \gamma_j \right) \right\}$$

(Note that $\left(\gamma_j + \gamma_{j+1}\right) \approx 2\gamma_j$ in the continuum limit.) We want to use this to compute the average $\left\langle \gamma_m^2 \right\rangle$, where m is a fixed intermediate index. This is the average of the squared field γ^2 at the point $s = m\delta$, therefore in the continuum limit it gives the average of γ^2 at the origin. We know that the system has zero-modes for which $A \to 0$ at the origin, and therefore $\gamma \to \infty$. Therefore we would like to show that $\left\langle \gamma_m^2 \right\rangle \to \infty$ for $\delta \to 0$.

Consider the sum in the exponential. Writing explicitly some of its terms one can regroup terms with the same index coming from γ_{j+1} and γ_j, and finally one obtains an exponential of the form

$$\exp\left\{ i\alpha\delta\left(-\gamma_0 + \frac{1}{\gamma_0} + \gamma_1 + \frac{1}{\gamma_1} + \gamma_2 + \frac{1}{\gamma_2} + \ldots + \gamma_N + \frac{1}{\gamma_N} + 2N\gamma_{N+1} \right) \right\}$$

When we compute a normalized average, all the exponentials and their integrals are factorized and simplified, except the one with γ_m:

$$\left\langle \gamma_m^2 \right\rangle = \frac{I_N}{I_D} = \frac{\int_0^{+\infty} d\gamma_m \frac{2}{\gamma_m^3} \gamma_m^2 \exp\left\{ i\alpha\delta\left(\gamma_m + \frac{1}{\gamma_m} \right) \right\}}{\int_0^{+\infty} d\gamma_m \frac{2}{\gamma_m^3} \exp\left\{ i\alpha\delta\left(\gamma_m + \frac{1}{\gamma_m} \right) \right\}}$$

Remember that we are looking for a sufficient condition for the divergence of $\left\langle \gamma_m^2 \right\rangle$; we can impose a lower cut-off ω on the variable γ_m. This also makes sense physically, as it implies an upper cut-off on the metric, and thus non-singular configurations.

Consider first the integral at the numerator and rename the variable as $x = \gamma_m$. We obtain

$$I_N = \int_\omega^{+\infty} dx \frac{2}{x} e^{i\alpha\delta\left(x + \frac{1}{x}\right)}$$

For $\delta \to 0$ the integral diverges logarithmically at infinity. On the other hand, the denominator stays finite thanks to the x^{-3} behaviour. This implies that the g_{rr} component of the metric tends to zero at the origin and this is in agreement with the prediction of the formation of a zero-mode at the origin made by solving the classical equation of the zero-modes.

REFERENCES

Note: "arXiv" papers are available from Cornell's electronic preprint depository http://arxiv.org

[1] Halpern L, Laurent B. On the gravitational radiation of microscopic systems. Il Nuovo Cimento 1964; 33: 728-751.
[2] Fontana G. A possibility of emission of high frequency gravitational radiation from junctions between d-wave and s-wave superconductors. arXiv 1998: cond-mat/9812070
[3] Modanese G. The vacuum state of quantum gravity contains large virtual masses. Class Quantum Grav 2007; 24: 1899-1909.
[4] Carroll SM. The Cosmological Constant. Living Rev Rel 2001; 4:1 .
[5] Hamber HW. Quantum Gravitation. The Feynman Path Integral Approach. Berlin: Springer 2009. Lauscher O, Reuter M. Ultraviolet Fixed Point and Generalized Flow Equation of Quantum Gravity. Phys Rev 2002; D65: 025013. Reuter M, Saueressig F. Nonlocal Quantum Gravity and the Size of the Universe. Fortsch Phys 2004; 52: 650-654.
[6] Eastham DA. Atomic physics of lasers. Taylor and Francis 1986
[7] Modanese G. Local contribution of a quantum condensate to the vacuum energy density. Mod Phys Lett A 2003; 18: 683-690.
[8] Modanese G, Junker T. Conditions for stimulated emission in anomalous gravity-superconductors interactions. In: Classical and Quantum Gravity Research, Eds. Christiansen MN, Rasmussen TK. Nova Science Publishers 2008; pp. 245-269.
[9] Modanese G. Potential energy in quantum gravity. Nucl Phys B 1995; 434: 697-708.
[10] Eichler J, Eichler HJ. Laser: Bauformen, Strahlfuhrung, Anwendungen. Berlin: Springer 2002.
[11] Podkletnov E, Modanese G. Investigation of high voltage discharges in low pressure gases through large ceramic superconducting electrodes. J Low Temp Phys 2003; 132: 239-259.
[12] Bini D, Cherubini C, Chicone C, Mashhoon B. Gravitational induction. arXiv 2008: 0803.0390
[13] Pascual-Sanchez JF. The harmonic gauge condition in the gravitomagnetic equations. arXiv 2000: gr-qc/0010075. Harris EG. Am J Phys 1991; 59: 421.
[14] Wald RM. General Relativity. The University of Chicago Press 1984.
[15] Chiao RY, Mintera SJ, Kirk Wegter-McNelly K. Do mirrors for gravitational waves exist? Phys E 2010; 42: 234-255.
[16] Peng H, Gravitoelectric and gravitomagnetic waves and the T gauge. Preprint UAHDP-661 (1986). Presented at GR11, Stockholm, 1986.
[17] Ciubotariu CD. Absorption of gravitational waves. Phys Lett A 1991; 158: 27-30.
[18] Adler RJ, Silbergleit AS. A general treatment of orbiting gyroscope precession. Int J Theor Phys 2000; 39; 1291-1316. Adler RJ, Chen P. Gravitomagnetism in quantum mechanics. arXiv 2009: 0912.2814
[19] Low, FE. Classical Field Theory: Electromagnetism and Gravitation. New York: J. Wiley 1997.
[20] Deser, S. How special relativity determines the signs of the nonrelativistic Coulomb and Newtonian forces. arXiv 2004: gr-qc/0411026.
[21] Caracciolo G, Curci G, Menotti P. The propagator in the A_0=0 gauge. Phys Lett B 1982; 113: 311-314.

CHAPTER 6

Tajmar Experiments on "Condensing the Vacuum"

R.A. Lewis

Penn State University, Physics Dept. (retired), 1308 Andover Drive, Boalsburg PA 16827, USA

Abstract: Experiments by Tajmar *et al.* and others, where various devices were placed in the vicinity of superconductors to measure force fields resembling gravitoelectromagnetism, are reviewed. In some of these experiments, an acceleration field, 21 orders of magnitude stronger than that predicted by the GEM theory, was observed during angular acceleration of a niobium ring. The accelerometers picked up the field only when the temperature was below the 9.5 *K* critical temperature. Ring gyroscopes, sensitive to rotation fields comparable to gravitomagnetism, showed effects with a variety of cold (< 20 *K*) materials, including niobium, aluminum, Teflon and helium. The effects have a parity asymmetry, showing up only with clockwise rotation in Austria and counter-clockwise rotation in New Zealand. The rotation field measured with niobium in Austria is comparable to the Sagnac field due to rotation of the earth. Neither the acceleration nor the rotation fields are explainable with current theories of quantum mechanics or general relativity. Future experiments in Austria involve a cryostat which can be oriented at various angles relative to the earth's rotation axis.

Keywords: Gravitational forces, theories of gravitation, superconductors, high-Tc superconductors, low-Tc superconductors, laser gyroscope, general relativity, vacuum energy density, gravitomagnetism, gravity-like fields.

1. INTRODUCTION

General relativity and quantum mechanics have fundamental conflicts, both mathematically and physically. The source of a gravitational disturbance in general relativity must be described by a tensor representing locally conserved momentum and energy. The uncertainty principle does not accommodate a definition of the motion of a particle at a point in spacetime. The conflict is particularly problematic for a superconductor, which involves exchange forces among electrons separated by macroscopic distances. The search for the basic physics governing both microscopic and macroscopic systems must be experimental.

The Tajmar *et al.* experiments [1-5], described in this paper, utilize various detectors arranged near superconductors subjected to rotation and temperature variations. Two types of detectors are used; accelerometers, essentially cantilevered springs, measure acceleration fields. Ring gyroscopes measure rotation fields, which affect the phases of light beams traversing closed paths. The fields bear a superficial resemblance to electric and magnetic fields created by moving charged particles.

The results of the experiments are difficult to summarize. In some experiments the rotation field is in the same sense as the rotation of the superconductor. In other experiments the opposite polarity of the rotation field is observed. Changes in the mounting and shielding of the detectors result in changes in measured fields, as well as restrictions on rotation rates. To avoid biasing the discussion in favor of one of various theories (vacuum fields, higher dimensions), a chronological style is used to present the results.

To present a rational comparison among measurements, a modification to the GEM (gravitoelectromagnetism) approximation to general relativity is adopted [6]. The GEM equations give well-defined relations between mass currents and associated fields. The fields calculated are multiplied by an empirical enhancement factor, which is adjusted both in magnitude and sign to match individual measurements. Neither GEM nor general relativity has the ingredients to make definitive predictions of such enhancement factors. Nevertheless the term gravitomagnetism is used for the rotation field, and gravitoelectric field is used for the acceleration field.

***Address correspondence to R.A. Lewis:** Penn State University, Physics Dept. (retired), 1308 Andover Drive, Boalsburg PA 16827, USA; E-mail: r3l@psu.edu

Giovanni Modanese and Glen A. Robertson (Eds)

1.1. Condensing the Vacuum

The title of this chapter, "Tajmar experiments on condensing the vacuum", refers to a common theme among various theories which could possibly explain the effects:

- Local variations in the vacuum energy density [7, 8];
- 10^{-35} kg vacuum droplets [9] ;
- Three flavors of vacuum graviphotons [10].

String theory [11] relates the fluid properties of a superconductor with the quark gluon plasma (QGP) formed by melting the vacuum at the Relativistic Heavy Ion Collider RHIC. Since formation of the QGP involves high temperature, the phase transition is described as melting the vacuum. A superconductor is created at low temperatures, so that condensing the vacuum is appropriate terminology.

1.2. Theoretical and Experimental Conflicts

Various conflicts, both theoretical and experimental, were encountered in our analysis. Following the chronological approach, no attempt is made to resolve the conflicts.

The GEM [6] (gravito-electro-magnetism) equations, patterned after the Maxwell equations, predict gravitational induction by time-varying mass currents. The sign of the gravitoelectric field enhances the changes in mass current, in violation of Lenz's law. There are divergent views in the literature about the existence and sign of gravitational induction (besides its magnitude, which in no case is predicted to be so large). For instance, Costa [12] claims that Mashhoon [13] errs in including the gravitational induction term in the GEM equations.

Measurements of the gravitomagnetic field of a rotating superconductor were made independently in experiments by Graham *et al.* [14] at the Canterbury University in Christchurch New Zealand, which has a latitude of -43.5°. The Tajmar experiments [1-5] were made at Seibersdorf Austria, latitude +47.5°. Tajmar *et al.* and Graham *et al.* disagree on the interpretation of the data from the New Zealand experiments. Graham *et al.* say that the measurements are all within one standard deviation of a null result. Tajmar *et al.* say that the data show an eleven standard deviation signal. The dispute involves fluctuations observed during the 30 seconds when the superconductor was rotating counterclockwise.

Another dispute involves the relation between mass current and field. Tajmar *et al.* [3] claims that the data from the Graham *et al.* [14] experiment implies a stronger connection between gravitomagnetic field and mass current. The GEM equations say that the enhancement factor for the data point of the experiment by Graham *et al.* even including the fluctuation, is smaller than the enhancement factor for the data in the experiments by Tajmar *et al.* In summary, three disputes need to be settled by further experiments:

- Costa *vs.* Mashhhoon;
- Graham *vs.* Tajmar interpretation of UG2 data;
- Tajmar theory *vs.* GEM.

2. ACCELERATION FIELDS OF A ROTATING SUPERCONDUCTOR

The motivation for further experiments with rotating superconductors involves a discrepancy encountered in testing the niobium superconductor gyroscopes for Gravity Probe B [15]. The Cooper pair mass in niobium as determined from the London moment disagrees with theory. Tajmar and de Matos speculate [16] that the discrepancy is attributable to a thirty order of magnitude enhancement of the gravitomagnetic field in a rotating superconductor.

2.1. Superconducting Niobium Ring Experiment

An experiment by Tajmar *et al.* [1] was designed to measure an acceleration field induced by an angularly accelerated superconducting niobium ring. Accelerometers, essentially cantilevered springs, were used to measure force fields.

The predictions from the GEM [6] approximation provide a convenient framework for comparing measurements, since these equations relate force fields ($\vec{E}_g$, $\vec{B}_g$) to mass currents ($\vec{J}$)

$$\nabla\times\vec{B}_g = (-)\frac{4\pi G}{c^2}\vec{J} + \frac{1}{c^2}\frac{\partial\vec{E}_g}{\partial t}\,; \tag{1}$$

$$\nabla\times\vec{E}_g = -\frac{\partial\vec{B}_g}{\partial t}. \tag{2}$$

The (-) minus sign in equation (1) emphasizes sign differences in the relation between fields and sources for tensor and vector fields. The force between two masses is attractive for a field resembling gravitation, while the force between two like charges is repulsive in a vector field with two signs of charges. For comparison with data, the signs of GEM predictions are enclosed in parentheses. The terms gravitoelectric and gravitomagnetic are used for convenience.

In addition to analytic approximations, the gravitomagnetic field due to the rotating mass is calculated using numerical integration of the Biot Savart law,

$$\vec{B}_g = (-)\frac{G}{c^2}\int\frac{\vec{J}\times(\vec{x}-\vec{x}\,')}{|\vec{x}-\vec{x}\,'|^3}d^3x'\,f_{Enhanc}. \tag{3}$$

The enhancement factor f_{Enhanc} is adjusted in sign and magnitude for each measurement.

The azimuthal component of the gravitoelectric field $\vec{E}_g$ due to angular acceleration $\dot{\omega}$ of a ring of mass M is approximated for points near the rotation axis, using expressions from Jackson [17] for a circular current loop of radius a,

$$E_{g\varphi} \cong (+)\frac{\pi G}{c^2}M\dot{\omega}\frac{a^2R}{\left(a^2+R^2+z^2\right)^{3/2}}f_{Enhanc}. \tag{4}$$

The enhancement factor is computed for each measurement. Numerical integration of the Biot Savart law is used, for better accuracy than the approximations in equation (4). The mass density of 8.57 g/cm^3 is used for niobium.

Fig. **1** shows the geometry of the niobium and accelerometer arrangement, for experiments reported in 2006. Results are obtained from fits to dozens of data points.

Data are given for different directions of rotation and temperatures.

As expected, the signals from the in-ring accelerometer reverse sign when the direction of rotation is reversed. The signals are not observed when the niobium is above the 9.4 *K* critical temperature. In addition, signals up to 2.5×10^{-3} m/sec^2 are observed when Pb or Nb rings are warmed up (quenched) during rotation.

As noted in Tajmar *et al.* [1], the signs of the peaks observed during quenches are opposite to the signs obtained during angular acceleration. When the Nb is quenched while rotating at 550 *rad/sec*, the acceleration field should be positive; to be consistent with the sign of signals observed when the angular acceleration is negative.

Data from the 2006 experiments are summarized in Table **1**, in a form for comparison with later experiments.

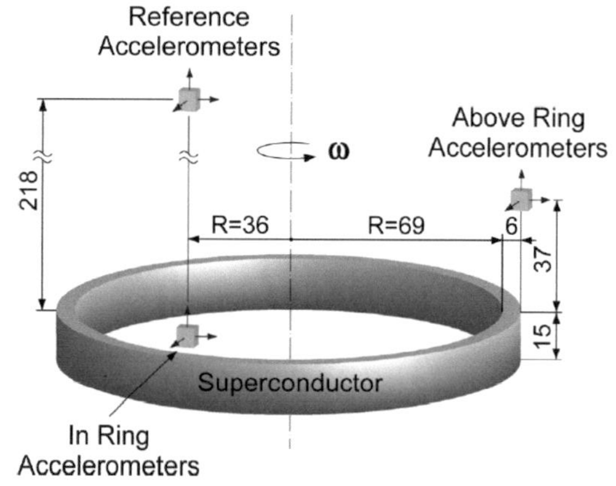

Figure 1: Rotating Nb ring and accelerometer positions (mm) [1].

Table 1: Accelerometer data, 2006 [1].

	$\dot{\omega}$	**GEM** ($E_{g\phi}$)	**Measured**	**Enhancement**
	rad/s^2	*m/s*^{2}	*m/s*^{2}	
In-ring	1000	$4.47\text{x}10^{-25}$	$(-9.45\pm0.27)\text{ x }10^{-4}$	$-21\cdot10^{20}$
Above ring	1000	$3.73\text{ x }10^{-25}$	$(-8.54\pm0.24)\text{ x }10^{-4}$	$-23\cdot10^{20}$
Reference	1000	$1.22\text{ x }10^{-26}$	$\pm2\text{ x }10^{-4}$	NA
In-ring 0.1 s quench	5500	$-2.45\text{x}10^{-24}$	$-2.71\text{ x }10^{-3}$	$11\text{ x }10^{20}$

The measured values represent fits to data, 111 data points in-ring and 83 points above ring. The $\pm 2\text{x}10^{-4}$ m/s^2 quoted for the reference accelerometer represents limitations of the sensitivity of the experiment.

The measured accelerations are a factor of 2.2 x 10^{21} larger than expected, assuming angular acceleration of 0.35 *kg* of mass.

3. PHYSICS OF THE LASER GYROSCOPE

The measurement by Graham *et al.* [14], using a UG2 laser ring at the Canterbury University in Christchurch New Zealand, is depicted in Fig. **2**. The results from this experiment are given in units peculiar to the apparatus, requiring conversion to compare with the experimental data from Tajmar *et al.* [2, 3].

The physics of the ring laser gyroscope [18] involves the phase shift of light due to a gravitomagnetic vector potential $\vec{h} = c\left(g_{01}, g_{02}, g_{03}\right)$ or a rotation field $\vec{h} = \vec{\omega}\times\vec{r}$. The phase change of light going around a closed path is related to the average gravitomagnetic field ($\vec{B}_g = \nabla\times\vec{h}$),

$$\Delta\varphi = \frac{4\pi}{\lambda c}\oint\vec{h}\cdot d\vec{s} = \frac{4\pi}{\lambda c}\int_{surface}\vec{B}_g\cdot\vec{n}dA. \tag{5}$$

Results from the Graham *et al.* [14] experiments are represented as shifts of the Sagnac frequency. The shift in the Sagnac frequency is defined as the ratio of the phase shift to the transit time of light around the periphery of the ring. The relation between Sagnac frequency and gravitomagnetic field is

$$\Delta f_S = \frac{4A\langle B_g \rangle}{\lambda P}, \tag{6}$$

where A = 21 x 39.7 m^2, P = 2 · (21+39.7) meters, λ = 632.8 *nm* (helium neon laser). At the minus 43.5° latitude of Christchurch, New Zealand, the vertical component of the earth's rotation field is

$$\omega_E = \frac{2\pi \text{ radians}}{1 \text{ day}} \sin\left(-43.5^o\right) = -5.01x10^{-5}\, rad/s\,; \tag{7}$$

corresponding to a Sagnac frequency of 2177 *Hz*. Shifts in frequency with the rotating lead superconductor are less than ±1 *mHz*, or $B_g = \pm 2.3 \cdot x10^{-11} rad/s$.

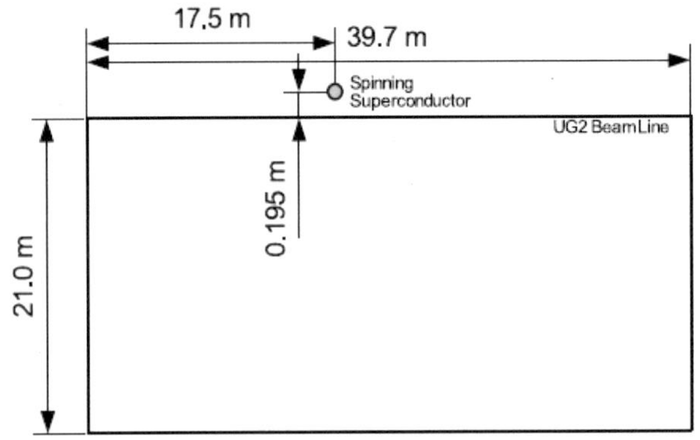

Figure 2: Graham *et al.* setup of the frame dragging experiment with the UG-2 laser (top view) [14]. (Figure copied with permission from Elsevier; licence number 2554280967940).

The ring gyros used by Tajmar *et al.* [2, 3] in 2008 are KVH DSP-3000 FOGs (Fiber Optics Gyroscopes) with dimensions of 88.9 *mm* x 58.42 *mm* x 33.02 *mm*. The Optolink SRS-1000 gyro [5] used in 2010 has 1000 meters of fiber optics wound into a spiral 150 *mm* in diameter. Tajmar reports all results in terms of B_g, *i.e.,* the rotation field, rather than raw data from the gyros.

4. LASER GYRO MEASUREMENTS OF GRAVITOMAGNETIC FIELD

In 2007, Tajmar *et al.* [2] tested the hypothesis that the anomalous signals are due to gravitomagnetic fields using laser gyroscopes, arranged as shown in Fig. **3**. Two gyros, reference and above-ring, are located at a radial distance of 5.4 *cm*, and axial distances of 21.6 *cm* (reference) and 4.3 *cm* (above-ring) measured from the top of the superconductor.

After addressing a vibration problem in the earlier version, the following accelerometer data were obtained in a new series in 2007.

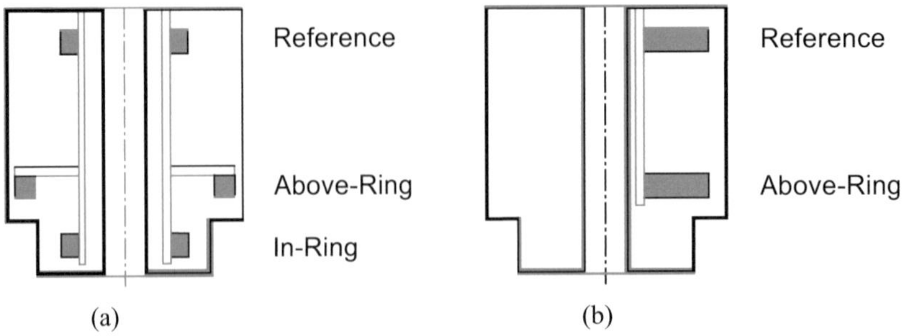

Figure 3: Tajmar *et al.* (**a**) accelerometer and (**b**) gyro experiments (The crosses are the SC ring) [2]. (Figure copied with permission from AIP Office of Rights and Permissions).

Table 2: Accelerometer data [2].

	$\dot{\omega}$	$E_{g\phi}$, **GEM**	**Measured**	**Enhancement**
	rad/s²	*m/sec²*	*m/sec²*	
In-ring	1000	$4.47\text{x}10^{-25}$	$(-1.41\pm0.27)\text{x}10^{-4}$	$-3.2\text{x}10^{20}$
Above ring	1000	$3.73\text{x}10^{-25}$	$(-5.94\pm0.98)\text{x}10^{-5}$	$-1.6\text{x}10^{20}$

The enhancements are a factor of 7 to 14 lower than the acceleration fields reported in 2006.

The gravitomagnetic field due to the rotating mass is calculated using numerical integration of the Biot Savart law, equation (3). As a cross-check, the axial component of the field near the axis is approximated using techniques from Jackson [17],

$$B_{gz}\left(R,z\right)=\left(-\right)\frac{2\pi GM\omega a^2}{c^2\left(a^2+z^2\right)^{3/2}}\left[1-\frac{R^2}{4}\frac{\left(12z^2-3a^2\right)}{\left(a^2+z^2\right)^2}\right]f_{Enhanc}\,. \tag{8}$$

In Table **3**, the signal at -70 *rad/s* quoted by UG2 is -0.45 ± 0.45 *mHz* frequency shift, corresponding to a gravitomagnetic field of $-1.0\pm1.0\text{x}10^{-11}$ *rad/s*. The value of B_g measured by Tajmar *et al.* [2] is deduced from a graph, since the value was not quoted in the text.

Table 3: Gyro measurements [2, 14].

		ω	B_g, **GEM**	B_g, **Measured**	**Enhancement**
		rad/s	*rad/s*	*rad/s*	
Tajmar	0.36 *kg* Nb,	600	$-4.86\text{x}10^{-24}$	$3\text{x}10^{-4}$	$-0.62\text{x}10^{20}$
Canterbury	2.78 kg Pb,	-70 (ccw)	$6.6\text{x}10^{-30}$	$-1.0\pm1.0\text{x}10^{-11}$	$-1.5\pm1.5\text{x}10^{18}$
Canterbury		100 (cw)	$-9.4\text{x}10^{-30}$	$0\pm0.55\text{x}10^{-11}$	$0\pm0.6\text{x}10^{18}$

Tajmar describes clockwise rotation as positive angular velocity, whereas the hands on an analog clock move opposite to the direction of increasing azimuth angle. The non-standard convention used in navigation is irrelevant, assuming that the same convention is applied to the interpretation of the accelerometer and gyro signals.

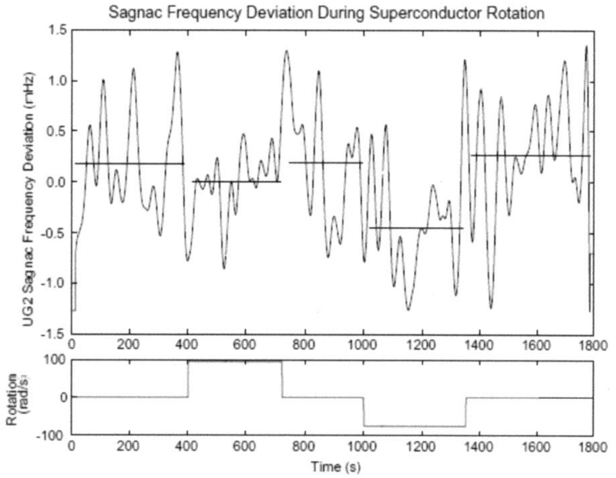

Figure 4: Graham *et al.* UG2 signals during rotations from -70 *rad/s* (ccw) to 100 *rad/s* (cw) [14]. (Figure copied with permission from Elsevier; licence number 2554280967940).

Data from the Graham *et al.* [14] UG2 experiments are shown in a graph in Fig. **4**, with results from a fit described in the text.

The physics detected by the accelerometers and ring gyroscopes is something other than simple enhancement of the GEM approximation to tensor gravitation. The GEM model [6] is consistent with $\nabla \times \vec{E}_g = -\partial \vec{B}_g / \partial t$ only if the enhancement factors are the same for E_g and B_g fields. Several mechanisms are required, to explain details such as the ratios of acceleration and gravitomagnetic fields, and the differing signs of the polarities.

5. EXPERIMENTS AFTER 2008

A more elaborate array of laser gyroscopes was used for further Tajmar *et al.* [3] experiments in 2008. Four gyros are mounted in three positions, illustrated in Fig. **5**. Two are mounted 45.8 *mm* above the rotating ring in opposite orientations to investigate offset problems. Two more gyros are 92.8 *mm* and 220.8 *mm* above the top surface of the ring. All sensors are mounted at a radial distance of 53.75 *mm*.

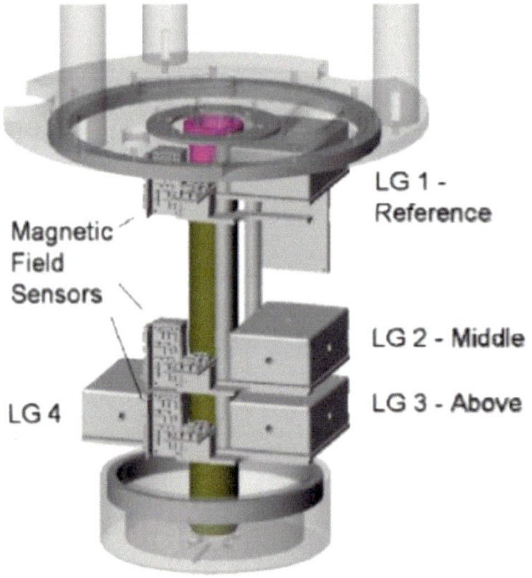

Figure 5: Geometry of gyros for 2008 experiments [3]. (Figure copied with permission from AIP Office of Rights and Permissions).

Two new features were observed with the arrangement. A signal is observed for aluminum, niobium and YBCO, but only for clockwise rotation. The second new feature was that the signals persisted for temperatures above the critical temperatures for the materials.

A possible physical mechanism for the parity asymmetry is that the signals are affected by the rotation of the earth. The Sagnac effect affects both the superconducting ring and the gyros. At the 47.5° latitude of Seibersdorf Austria the vertical component of the rotation field is 5.36×10^{-5} *rad/s*. The rotation field is comparable to the observed signals, which are summarized in Table **4**.

A hint (one standard deviation) of a parity-violating signal from a lead disc rotating counterclockwise is seen by Graham *et al.* [14] in the Canterbury, New Zealand data. The change in parity between northern (Tajmar *et al.*) and southern (Graham *et al.*) hemisphere measurements lends credence to an effect of the rotation of the earth. The factor $\vec{B}_g \cdot \hat{n}$ in equation (5) is of opposite signs north and south of the equator.

Tajmar *et al.* [3] claim that the magnitude of the Graham *et al.* [14] UG2 signal is larger than the signals in Austria. However the model used to compare the two measurements is not explained in detail. Another scorecard table, using the GEM [6] model, is constructed to compare measurements in Table **4**.

Table 4: Gyro measurements [3, 14].

	Material	ω	Calc., GEM	Measured	Enhancement
		rad/s	s^{-1}	s^{-1}	
LG3-LG4	Nb	400 (cw)	-3.24x10^{-24}	1.30x10^{-5}	-3.99x10^{18}
LG3-LG4	Nb	-400 (ccw)	3.24x10^{-24}	-1±1x10^{-6}	-0.3±0.3x10^{18}
LG2	Nb	400 (cw)	-1.26x10^{-24}	1.20x10^{-5}	-9.52x10^{18}
LG1	Nb	400 (cw)	-2.15x10^{-25}	1.90x10^{-5}	-8.84x10^{19}
UG2	Pb	-70 (ccw)	6.58x10^{-30}	-1.0±1.0x10^{-11}	-1.5±1.5x10^{18}

The notations cw and ccw refer to signals observed with clockwise or counterclockwise rotation. Significant signals are observed by Tajmar *et al.* [3] only for clockwise rotation.

The enhancement measured with the above ring gyros LG3-LG4 is an order of magnitude smaller than reported by Tajmar *et al.* [3]. The enhancement factor is larger for LG1 and LG2, since the measured signal is practically constant, rather than decreasing monotonically with distance. The enhancement for the Graham *et al.* [14], Canterbury, New Zealand UG2 signal is smaller than for the Tajmar *et al.* Seibersdorf, Austria signals, by factors of two to fifty.

Tajmar *et al.* [10] claim that the Graham *et al.* [14] Canterbury, New Zealand UG2 signal is even larger than the Seibersdorf, Austria signals. Tajmar *et al.* multiply the Graham *et al.* UG2 signals by a factor of 1.7x10^6 to estimate gravitomagnetic fields near the rotating lead sample. The mathematical model for this extrapolation is not specified by the authors.

6. EFFECTS OF HELIUM GAS FLOW

In 2009, Tajmar *et al.* [4] set up a more elaborate array of laser gyroscopes to investigate possible sources for the parity-violating signals as shown in Fig. **6**.

Setup Fig. **6a** is similar to Fig. **4** above. Anomalous signals were again observed, even using a Teflon ring. The authors remark that the signals (typically 2x10^{-5} *rad/s*) are as large as one third of the earth's rotation (5.36x10^{-5} *rad/s* in Austria).

In setup Fig. **6b**, an isolation pot between the spinning ring and the sensor chamber minimizes the helium gas contact with the sensor chamber. Signals comparable to those observed in setup A were observed with aluminum and stainless steel rings. However a wrong-sign signal was observed with ccw rotation, as well as the signal with cw rotation. The effect was reduced by an order of magnitude when a solid disk of aluminum was used instead of a ring.

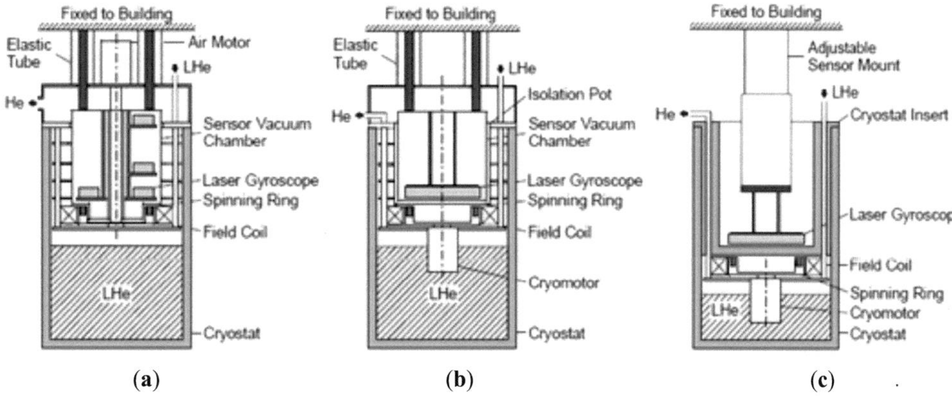

Figure 6: Setup (**a**), (**b**) and (**c**) [4]. (Figure copied with permission from IOP Publisihing Ltd.).

In setup Fig. **6c**, a cryostat insert was used to completely isolate the spinning ring from the gyro and to minimize helium gas flow along the side walls. The Al ring effect was reduced by nearly two orders of magnitude, demonstrating that the origin of the anomalous signals is related to the helium flow. A small signal was observed when the ring was replaced by a cup with fins. Results of measurements with niobium are summarized in Table **5**.

Unlike the parity asymmetric data reported in 2008 [3], with setup A the niobium ring gives comparable signals with both cw and ccw rotation. Parity asymmetric signals are reported in Tajmar *et al.* [4] with YBCO and Teflon rings.

Table 5: 2009 gyro data with niobium [4].

	ω	Calc. B_g, GEM	Measured	Enhancement
	rad/s	*rad/s*	*rad/s*	
A	400 (cw)	$-3.24x10^{-24}$	$2.28\pm0.16x10^{-5}$	$-7.0x10^{18}$
A	-400 (ccw)	$3.24x10^{-24}$	$-1.92\pm0.2x10^{-5}$	$-5.9x10^{18}$
C	100 (cw)	$-1.3x10^{-24}$	$-12\pm5x10^{-8}$	$9.2x10^{16}$
C	-100 (ccw)	$1.3x10^{-24}$	$12\pm8x10^{-8}$	$9.2x10^{16}$

The maximum angular acceleration of 33 *rad/s*^{2} was not adequate to test previous accelerometer results, with up to 1500 *rad/s*^{2}.

Table 6: Accelerometer data [4].

	$\dot{\omega}$	$E_{g\phi}$, GEM	Measured	Enhancement
	rad/s^2	*m/sec*^{2}	*m/sec*^{2}	
Above ring	33	$1.22x10^{-26}$	$0\pm3x10^{-4}$	$0\pm2.5x10^{22}$

Quote from Tajmar *et al.* [4]:

> "Compared to classical frame-dragging spin-coupling predictions, our signals are up to 18 orders of magnitude larger. This suggests that the observed phenomenon is new and without explanation so far".

Disclaimer from reference [4]:

> We also mounted 10 accelerometers on top of the gyroscope in tangential, radial and vertical direction at various radial locations in order to characterize the acceleration environment in a facility vibration-free environment (our dominant error source in the first accelerometer measurements [1-3]). We can put a limit on acceleration fields in all three directions to ±30 μg for all types of rings used (Nb, Al and YBCO) at 4 K and 77 K for YBCO with ring angular top speeds of 100 rad/s and accelerations of 33 rad/s2. Also magnetic fields generated by the field coil of 100 μT and 1 mT during the test runs did not show a change in the acceleration signals. Due to the much smaller angular acceleration compared to our earlier air motor measurements (≅1500 rad/s^2), we were not able to confirm or refute our previous accelerometer measurements [1-3] without the vibration error source. However, our results put strong limits on all proposed theories suggesting frame-dragging or vertical gravitational fields generated from spinning superconductors.

7. EFFECTS OF ROTATING LIQUID HELIUM

In 2010, Tajmar *et al.* [5] indicate:

> "In order to further investigate the real origin behind our signals, we decided to design a new experimental setup which allows rotating liquid and superfluid helium in a controlled manner and to continue mapping the anomalous gyro effect outside of the facility at different locations to study its expansion."

In Fig. **7a**, gyroscopes are located at the center, bottom and side positions near rotating liquid helium. Polarity of signal in center gyro is opposite to angular velocity, in agreement with GEM [6].

The pot inside can be filled with up to 30 liters of liquid helium. Its outer and inner radii are 211 *mm* and 93.5 *mm*, and the outer and inner heights are 271 *mm* and 177 *mm* respectively. Fig. **7b** shows typical signals (blue, black symbols) and the angular velocity (red). Results from the measurements are summarized in Table **6**.

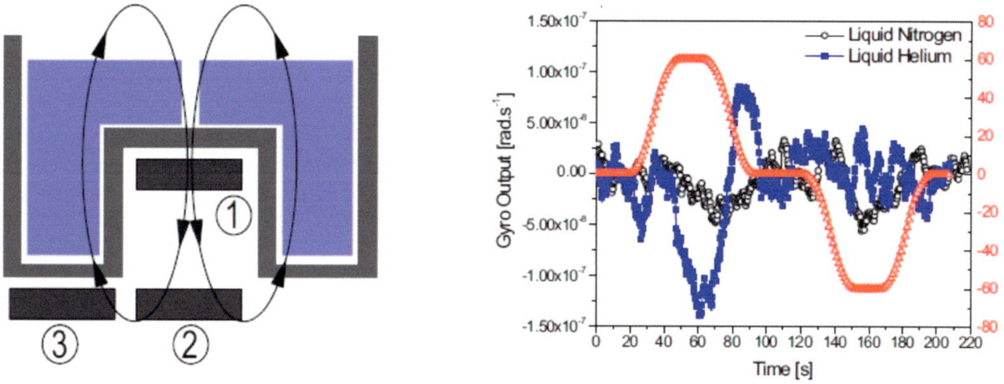

(**a**) (**b**)

Figure 7: Tajmar *et al.* (**a**) Gyroscopes and (**b**) Typical data [5]. (Figure copied with permission from AIP Office of Rights and Permissions).

Table 7: Gyro measurements [5].

	ω	**Clockwise**	**Counterclockwise**
	rad/s	s^{-1}	s^{-1}
Center	60	$-10.2\pm2.4\times10^{-8}$	$-1.2\pm1.8\times10^{-8}$
Bottom	60	$-0.6\pm1.2\times10^{-8}$	$-1.8\pm1.8\times10^{-8}$
Side	60	$6.0\pm1.8\times10^{-8}$	$-3.0\pm1.8\times10^{-8}$
Superfluid, Center	10	$1.9\pm3.4\times10^{-8}$	$0.1\pm2.0\times10^{-8}$

The signal in the side gyro was smaller than and opposite in sign to the signal in the center gyro (see Table **7**). According to Tajmar *et al.* [5], this would be expected from a dipolar field distribution. The Biot Savart numerical integration does not show a field reversal at the location of side gyro, assuming that the entire 3.7 *kg* of liquid helium rotates. Perhaps the signal arises from the thin niobium lining the helium container.

Table 8: Helium data [5].

	ω	**Calc.** B_g**, GEM**	**Measured**	**Enhancement**
	rad/s		*rad/s*	
Center	60 (cw)	-4.2×10^{-24}	$-10.2\pm2.4\times10^{-8}$	2.4×10^{16}
Center	-60 (ccw)	4.2×10^{-24}	$-1.2\pm1.8\times10^{-8}$	$-0.28\pm0.42\times10^{16}$

Rotating liquid helium produces anomalous results similar to those found in earlier experiments (Table **8**).

8. EFFECT OF ORIENTATION RELATIVE TO EARTH AXIS

Both the magnitude of the signals in the 2007 (Tajmar *et al.* [2]) data and the parity asymmetries in Austria and New Zealand suggest an interference with the Sagnac field associated with the rotation of the earth. A future modification of the Tajmar *et al.* [5] apparatus, shown in Fig. **8**, includes a mechanism for orienting the setup.

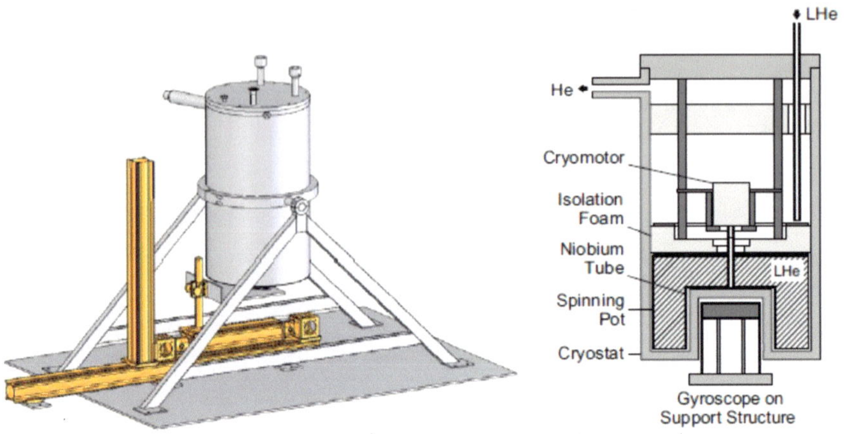

Figure 8: Future measurements include orienting apparatus relative to earth's rotation axis [5]. (Figure copied with permission from AIP office of Rights and Permissions).

9. MODELS FOR COMPARING MEASUREMENTS

The model used for comparing measurements in this write-up is that superconductive properties enhance the gravitomagnetic field normally associated with the warmer bulk material. Wherein, the Tajmar *et al.* [1-5] model indicates that

- The superconductor produces an enhanced gravitomagnetic field from surface currents which parallel electromagnetic currents, and
- The gravitomagnetic field inside the niobium equals the rotation field 2ω multiplied by the ratio of the mass of Cooper pairs to the mass of the niobium.

Either model produces the same ratios for gravitoelectric and gravitomagnetic fields, for a fixed ring composition and geometry. Tajmar *et al.* [2] state that the E_g and B_g results are consistent with the deMatos [9] model within a factor of two. Tables **2-5** show that the enhancement of B_g is consistently smaller than that for E_g, by factors of two to five.

Comparison between Canterbury, New Zealand and Seibersdorf, Austria is more problematic. Wherein,

- Table **2** says that the Graham *et al.* [14] UG2 measurements are a factor of 40 lower than the Tajmar *et al.* [2] data, and
- Table **5** says the Graham *et al.* UG2 data are a factor of 2.5 lower than Tajmar *et al.* [4].

Quote from Tajmar *et al.* [3]:

> "The coupling factor [B_g/ω] computed for the [Graham *et al.* [14] UG2 data] counter-clockwise rotation is $3.32 \pm 0.28 \times 10^{-7}$, which is nearly one order of magnitude above our measurements in the close vicinity of the superconductor."

Quote from Graham *et al.* [14]:

> "There have been recent claims [Tajmar *et al.* [3] data] of measured effects similar to frame dragging, of strength about two orders of magnitude lower than predicted by Tajmar and deMatos [16]. These effects are speculated to be non-parity-conserving and perhaps to have a non-dipole spatial distribution at least along the rotation axis. Our experimental results do not have the sensitivity to either confirm or refute these recent claims."

Three analyses come up with three different conclusions about the relation between the Graham *et al.* [14] UG2 and Tajmar *et al.* [2, 3] gravitomagnetic field measurements. Tables **2** and **5** show that the Graham *et al.* UG2 measurements indicate a significantly smaller field than the Tajmar *et al.* results. Graham *et al.* and Tajmar *et al.* have dissenting opinions.

10. TAJMAR EXPERIMENT SCORECARD

The enhancement factors from various measurements are summarized in Table **9**, using the following notations:

- cw, ccw – clockwise (cw) or counterclockwise (ccw) rotation of the superconducting ring.
- quench – the SC ring was quenched, *i.e.,* heated above the critical temperature, while rotating.
- 2009A – the above-ring gyro as depicted in Fig. **6a** [4].
- 2009C – cryostat insert blocked helium gas from reaching a more sensitive gyro, centered on the rotation axis [4].
- Positive enhancement factors imply polarity predicted by GEM [6].

Five distinct patterns are discernible. For brevity denote GEM+ as measurements with sign (chirality) in agreement with GEM [6], and GEM- as opposite sign:

- Acceleration fields from a quench have GEM+ polarities, and are associated with superconductivity.
- Acceleration fields from angular acceleration have GEM-, and are associated with superconductivity.
- Rotation fields with large enhancement factors have GEM- chirality, and are associated with non-superconducting materials such as warm (15k) Nb, aluminum, Teflon.
- Rotation fields with small enhancement factors have GEM+ chirality.
- Instrumental effects, which do not change sign with cw and ccw rotation.

The GEM- signals decrease with improvements in the apparatus. Like electric and magnetic fields in a plasma, the GEM- fields could be partially shielded by structures in the apparatus.

Fields which appear to violate parity symmetry could be mixtures of GEM- fields and instrumental effects. A genuine signal from the rotating material is masked by instrumental effects giving the appearance of parity violation. Tajmar *et al.* [3] claim that the Graham *et al.* [14] Canterbury, New Zealand UG2 measurements support a parity violation associated with the rotation of the earth. Table **8** says that the Graham *et al.* UG2 measurements are within one standard deviation of a null result.

Table 9: Summary of result comparisons

Measurement	**Material**	$\dot{\omega}$		**Enhancement factor**	**Location**	**Year**	**References**
		rad/s²					
Tajmar *et al.*							
In-ring accel.	0.36 kg, Nb	1000		$-21x10^{20}$	Austria	2006	[1]
Above ring	0.36 kg, Nb	1000		$-23x10^{20}$	Austria	2006	[1]
In-ring, 0.1 sec quench	0.36 kg, Nb	5500		$+11x10^{20}$	Austria		[1]
In-ring, 0.1 sec quench	0.36 kg, Nb	-5500		$+11x10^{20}$	Austria		[1]
In-ring accel.	0.36 kg, Nb	1000		$-3.2x10^{20}$	Austria	2007	[2]
Above ring	0.36 kg, Nb	1000		$-1.6x10^{20}$	Austria	2007	[2]
Above ring	0.36 kg, Nb	33		$0 \pm 2.5x10^{22}$	Austria	2009	[4]
		ω					
		rad/s					
Tajmar *et al.*; gyro	0.36 kg, Nb	600		$-0.62x10^{20}$	Austria	2007	[2]
Graham *et al.*; UG2	2.78 kg, Pb	-70	(ccw)	$-1.5 \pm 1.5x10^{18}$	Canterbury	2008	[14]
Graham *et al.*; UG2	2.78 kg, Pb	100	(cw)	$0 \pm 0.6x10^{18}$	Canterbury	2008	[14]
Tajmar gyro (Nb)							
	0.36 kg, Nb	400	(cw)	$-3.99x10^{18}$	Austria	2008	[3]
	0.36 kg, Nb	-400	(ccw)	$-0.3 \pm 0.3x10^{18}$	Austria	2008	[3]
	0.36 kg, Nb	400	(cw)	$-7.0x10^{18}$	Austria	2009A	[4]
	0.36 kg, Nb	-400	(ccw)	$-5.9x10^{18}$	Austria	2009A	[4]
	0.36 kg, Nb	100	cw	$+9.2 \pm 3.8x10^{16}$	Austria	2009C	[4]
	0.36 kg, Nb	-100	ccw	$+9.2 \pm 6.1x10^{16}$	Austria	2009C	[4]
He, 3.7 kg							
	3.7 kg, He	-60	cw	$+2.4 \pm 0.6x10^{16}$	Austria	2010	[5]
	3.7 kg, He	60	ccw	$+0.28 \pm 0.42x10^{16}$	Austria	2010	[5]

11. CONCLUSIONS AND DISCUSSION

The largest acceleration fields were observed when Nb or Pb rings were quenched while rotating. Unlike measurements with fixed temperature during rotation, the direction of the acceleration field is consistent with the GEM [6] prediction.

Acceleration fields accompanying angular acceleration of a niobium ring have signs opposite to those expected from the GEM [6] prediction. In the 2006 (Tajmar *et al.* [1]) data, the accelerometers detect a signal 21 orders of magnitude larger than the GEM predictions, only with the temperature below the critical temperature. Smaller acceleration fields were reported in 2007 (Tajmar *et al.* [2]); attributed to alleviating a vibration issue. Attempts to measure acceleration fields in later experiments were inconclusive, as the modifications did not allow sufficient angular acceleration.

The measurements of a rotation field are more difficult to characterize. For example, the fields were reduced an order of magnitude when helium gas flow was inhibited. One possibility is that the helium gas participated in the generation of the rotation field. The hypothesis that the helium gas affected the ring gyros is hard to accept, since the field was detectable only when samples were rotated clockwise.

Using the GEM [6] model to compare ring gyro measurements leads to the conclusion that the Graham *et al.* [14], New Zealand signals are a factor of 2.5 to 40 too small to be consistent with the Tajmar *et al.* [2, 3], Austrian signals. The authors, Graham *et al.* and Tajmar *et al.* disagree on the interpretation of the signals.

The parity asymmetries as well as the magnitude of gyro signals suggest an interaction involving the rotation of the earth. The interaction is not clearly associated with superfluids, since gyro signals were observed with materials such as aluminum, Teflon, and helium, which are not superfluids at the temperatures sampled.

Neither the acceleration fields nor the rotation fields are explained with current theories involving quantum mechanics and general relativity. The physics detected by the accelerometers and ring gyroscopes is something more elaborate than enhancement of the GEM [6] approximation to tensor gravitation. Possible physics models involving vacuum fields are described in references [7-10].

12. APPENDIX A: GENERAL RELATIVITY *VS.* GRAVITOELECTROMAGNETISM

The GEM equations are used in this document only as a phenomenological model, to relate measurements taken with varying geometrical arrangements. The relations between the GEM equations and general relativity are described in this appendix.

12.1. Historical Relation Between GR and GEM

Historically the GEM equations [19] pre-date general relativity by several decades. Heaviside condensed the Maxwell theory to the four equations of electromagnetism still in use today. Similar equations for gravitation were gleaned from Faraday and Maxwell. In 1893 gravity was a modified vector field, with gravitoelectric and gravitomagnetic fields comparable to electric and magnetic fields in the Maxwell equations.

Nordstrom and Einstein developed scalar and tensor versions of gravitation [20] around 1918. The tensor field won out in subsequent observations, including astronomical and laboratory tests. Remnants of the scalar and vector theories can be found in aspects of the tensor theory.

12.2. Structure of Gravitational Forces

The Einstein equation involves the metric tensor $g_{\mu\nu}$. In the weak field approximation, the metric tensor can be written as $g_{\mu\nu} = \eta_{\mu\nu} + h_{\mu\nu}$ where $\eta_{\mu\nu} = diag(1,-1,-1,-1)$ is the Minkowski metric, and $h_{\mu\nu}$ is a metric perturbation with elements small compared to unity. Following Bertschinger [21], write the components of the metric as follows.

$$g_{\mu\nu} = \begin{pmatrix} 1 & 0 & 0 & 0 \\ 0 & -1 & 0 & 0 \\ 0 & 0 & -1 & 0 \\ 0 & 0 & 0 & -1 \end{pmatrix} + \begin{pmatrix} \Phi & h_x & h_y & h_z \\ h_x & s_{11} + \frac{1}{3}\Psi & s_{12} & s_{13} \\ h_y & s_{21} & s_{22} + \frac{1}{3}\Psi & s_{23} \\ h_z & s_{31} & s_{32} & s_{33} + \frac{1}{3}\Psi \end{pmatrix} \quad \textbf{(9)}$$

Subsets of the metric can be considered as scalar, vector and tensor fields, useful under certain conditions (weak fields, quasi-static sources).

Scalar fields:

$$\Phi = h_{00} \quad \textbf{(10)}$$

$$\Psi = trace(h_{ij}) \ \textbf{(i,j = 1,2,3)} \quad \textbf{(11)}$$

Vector field:

$$\vec{h} = \left(h_x, h_y, h_z\right) \tag{12}$$

Tensor field:

$$s_{ij} = h_{ij} - \tfrac{1}{3}\Psi\delta_{ij} \quad \textbf{(i,j = 1,2,3)} \tag{13}$$

A test mass in a gravitational field follows a geodesic path $x^{\mu}(s) = (ct, x^1, x^2, x^3)$, where s is any parameter (such as time) which increases monotonically with distance along the path,

$$\frac{d^2x^{\mu}}{ds^2} + \Gamma^{\mu}_{\nu\lambda}\frac{dx^{\lambda}}{ds}\frac{dx^{\nu}}{ds} = \mathbf{0} \tag{14}$$

The connection field Γ is defined as

$$\Gamma^{\sigma}_{\mu\nu} = \tfrac{1}{2}g^{\sigma\lambda}\left(\frac{\partial g_{\nu\lambda}}{\partial x^{\mu}} + \frac{\partial g_{\mu\lambda}}{\partial x^{\nu}} - \frac{\partial g_{\mu\nu}}{\partial x^{\lambda}}\right) \tag{15}$$

Choosing parameter s as laboratory time,

$$\frac{dx^{\sigma}}{ds} = \left(c, v_x, v_y, v_z\right) = \mathbf{(c, v^1, v^2, v^3)} \tag{16}$$

For a stationary test mass only the dx^0/ds component of the 4-velocity is non-zero. Factors of c enter through the relation $dx^0 = cdt$. To avoid confusion with the summation convention, express the component of force in the z direction. The force (acceleration) in the x^3 direction is then given by the Γ^3_{00} connection, using Weinberg notation, and ignoring terms of second order in $h_{\mu\nu}$,

$$a^3\left(static\right) = c^2\Gamma^3_{00} = \tfrac{1}{2}g^{33}\left(\frac{\partial g_{03}}{\partial x^0} + \frac{\partial g_{03}}{\partial x^0} - \frac{\partial g_{00}}{\partial x^3}\right)c^2 = -c\frac{\partial h_z}{\partial t} + c^2\frac{\partial \Phi}{\partial z} \tag{17}$$

The time derivative of the vector potential appears to correspond to gravitational induction. Costa claims that the gravitational induction vanishes upon further analysis. This claim is disputed.

For a test mass moving with a non-relativistic velocity $\vec{v}$, the first order terms in the z-component of acceleration are as follows.

$$a^3\left(velocity\right) = -cv^j\Gamma^3_{j0} = -\tfrac{1}{2}g^{33}v^j\left(\frac{\partial g_{03}}{\partial x^j} + \frac{\partial g_{j3}}{\partial x^0} - \frac{\partial g_{j0}}{\partial x^3}\right)c \tag{18}$$

$$= \tfrac{1}{2}cv^j\left[\frac{\partial h_3}{\partial x^j} - \frac{\partial h_j}{\partial x^3}\right] + \tfrac{1}{2}v^j\frac{\partial s_{j3}}{\partial t} + \tfrac{1}{6}v^3\frac{\partial \Psi}{\partial t}.$$

The first term in equation (18) can be related to the Lorentz force expressed in terms of the gravitomagnetic field $\vec{B}_g = c\nabla\times\vec{h}$,

$$\vec{F} = m\vec{v}\times\vec{B}_g = mc\vec{v}\times\left(\nabla\times\vec{h}\right) \tag{19}$$

In component form, equation (19) can be written as

$$F^i = mc\varepsilon_{ijk} v^j \varepsilon_{kmn} \partial_m h_n = mcv^j \left[\delta_{in}\delta_{jm} - \delta_{im}\delta_{jn} \right] \partial_m h_n = mcv^j \left[\frac{\partial h_i}{\partial x^j} - \frac{\partial h_j}{\partial x^i} \right] \tag{20}$$

Except for a factor of ½ , equation (20) is the same as terms in equation (18) .

The second term in equation (18) involves time derivatives of the scalar ψ in the Bertschinger notation of equation 3.

12.3. Structure of Gravitational Sources

The Einstein equation describes the relation between a gravitational field and sources,

$$R_{\mu\nu} - (1/2) g_{\mu\nu} R = -(8\pi G / c^2) T_{\mu\nu} - \Lambda g_{\mu\nu} \tag{21}$$

In the weak field approximation, the equation can be written (Weinberg equation10.1.10)

$$\Box h_{\mu\nu} = -16\pi G S_{\mu\nu} = -16\pi G \left[T_{\mu\nu} - \tfrac{1}{2} \eta_{\mu\nu} T^{\lambda}{}_{\lambda} \right] \tag{22}$$

with solution

$$h_{\mu\nu}(\vec{x}, t) = 4G \int d^3x' \frac{S_{\mu\nu}\left(\vec{x}', t - |\vec{x} - \vec{x}'|\right)}{|\vec{x} - \vec{x}'|} \tag{23}$$

where

$$S_{\mu\nu} = T_{\mu\nu} - \tfrac{1}{2} \eta_{\mu\nu} T^{\lambda}{}_{\lambda} \tag{24}$$

Techniques for evaluating integrals such as equation (23) have been developed for electrodynamics [17].

To clarify terminology, consider a source consisting of two point masses and a spring (Fig. **9**).

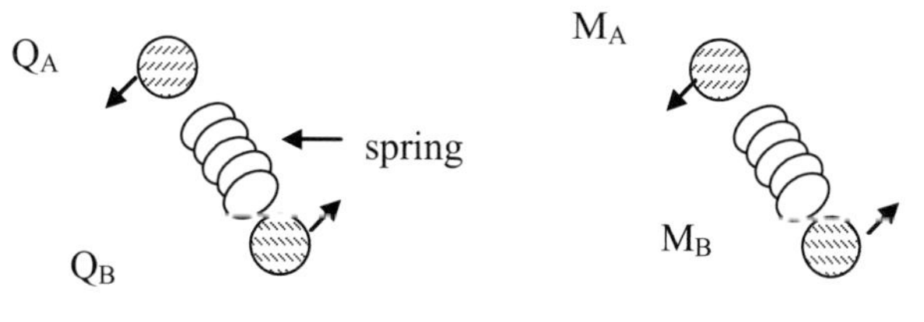

Figure 9: Oscillating, rotating electromagnetic and gravitational sources.

The stress tensor at a point in space and time is a linear combination involving the two masses and the spring. The stress tensor for a particle, from Low [22] 6.6.17, is

$$T_P^{\mu\nu}(m) = m u^\mu u^\nu \sqrt{1 - v^2}\, \delta^3\left(\vec{x} - \vec{x}_P(x_0)\right) \tag{25}$$

where the four-velocity from Jackson [17] page 532 is

$$u^{\mu} = (\gamma c, \vec{v}\gamma) \qquad \textbf{(26)}$$

The stress tensor for an elastic medium consists of two parts [23], corresponding to mass currents and internal potential energies. The concepts are explained in more detail by Williams [24]. Both Salt and Williams cite DeWitt [25] for a more elegant presentation.

Describing an elastic solid involves a strain field $\vec{u} = (u, v, w)$ and a coordinate system $\vec{x} = (x, y, z)$. An atom at some location $\vec{x}$ experiences a force towards an equilibrium point $\vec{x} - \vec{u}$. However if the atom moves to its equilibrium location, it alters the equilibrium of neighboring atoms. The potential energy in a small region of the material can be expressed in terms of strain field gradients,

$$U = \tfrac{1}{2} E \frac{\partial u_i}{\partial x^j} \frac{\partial u_i}{\partial x^j} \quad \textbf{(sum i,j = 1,2,3)} \qquad \textbf{(27)}$$

This expression ignores the distinctions among various definitions of the elastic moduli E, such as Young's modulus, shear modulus and bulk modulus, which would require a rank 4 tensor.

Most solids have elastic moduli around 1 Mbar (10^{12} dynes/cm^2, or 10^{11} newtons/m^2), comparable to the degeneracy pressure [26] for a material with a density of 1 free electron per cubic Angstrom.

For clarity, at the expense of accuracy, the stress tensor to be considered is expressed without terms of second order in velocity. Such terms would be needed to check energy conservation. Even though the stress energy terms appear to be comparable to v^2 terms, their derivatives are crucial. Since F=ma, a force has a contribution comparable to an accelerating particle. The mass density ρ refers both to the continuous mass distribution in the spring, and to point particles, such as $\rho = M_A \delta^3 (\vec{x} - \vec{x}_A(t))$ for mass A.

$$T_{\mu\nu}(\vec{x}, t) = \begin{pmatrix} \tfrac{1}{2}\rho c^2 + \tfrac{1}{2} E \frac{\partial u_i}{\partial x^j} \frac{\partial u_i}{\partial x^j} & \rho c v_x & \rho c v_y & \rho c v_z \\ \rho c v_x & E \frac{\partial u}{\partial x} & \tfrac{1}{2} E \left(\frac{\partial u}{\partial y} + \frac{\partial v}{\partial x} \right) & \tfrac{1}{2} E \left(\frac{\partial u}{\partial z} + \frac{\partial w}{\partial x} \right) \\ \rho c v_y & \tfrac{1}{2} E \left(\frac{\partial u}{\partial y} + \frac{\partial v}{\partial x} \right) & E \frac{\partial v}{\partial y} & \tfrac{1}{2} E \left(\frac{\partial v}{\partial z} + \frac{\partial w}{\partial y} \right) \\ \rho c v_z & \tfrac{1}{2} E \left(\frac{\partial u}{\partial z} + \frac{\partial w}{\partial x} \right) & \tfrac{1}{2} E \left(\frac{\partial v}{\partial z} + \frac{\partial w}{\partial y} \right) & E \frac{\partial w}{\partial z} \end{pmatrix}. \qquad \textbf{(28)}$$

Retaining only the mass sources is not legitimate, due to the constraints of locally conserved energy and momentum. Since all other terms in the Einstein equation equation (21) have zero divergence, the stress tensor must be divergenceless,

$$\frac{\partial T_{\mu\nu}}{\partial x^{\mu}} = 0 \qquad \textbf{(29)}$$

For example consider the ν=3 row of $T_{\mu\nu}$,

$$\frac{\partial T_{03}}{\partial x^0} - \frac{\partial T_{j3}}{\partial x^j} = \rho \frac{\partial v_z}{\partial t} - \tfrac{1}{2} E \left[\frac{\partial^2 u}{\partial z \partial x} + \frac{\partial^2 w}{\partial x^2} + \frac{\partial^2 v}{\partial z \partial y} + \frac{\partial^2 w}{\partial y^2} + 2 \frac{\partial^2 w}{\partial z^2} \right] = \mathbf{0} \qquad \textbf{(30)}$$

Equation (30) resembles Kittel [27] equation 4.25, ignoring the distinctions among the elastic moduli c_{11}, c_{12} and c_{44} for a cubic crystal.

Constraints need to be applied, for example to ensure that a force $\vec{F}_{sA}$ exerted by the spring on mass A is accompanied by an equal and opposite force exerted on mass B.

12.4. Calculation of Fields from Sources

The vector potentials $\vec{A}$ and $\vec{h}$ seem to be expressible in terms of the currents, for points far from the source, and neglecting retardation effects , using tricks from Jackson page 185 to approximate the integrals. For the oscillating charges,

$$\vec{A}(\vec{x}) = \frac{\mu_0}{4\pi}\int \frac{\vec{J}_Q(\vec{x}\,')}{|\vec{x}-\vec{x}\,'|}d^3x' \;\cong\; \frac{\mu_0}{4\pi|\vec{x}|^3}\vec{m}\times\vec{x} \;\textbf{(wrong!)} \qquad \textbf{(31)}$$

where $\vec{m}$ is the magnetic moment of the current distribution,

$$\vec{m} = \tfrac{1}{2}\int \vec{x}\,'\times\vec{J}_Q(x')d^3x' \;=\; \tfrac{1}{2}(\vec{x}_A\times Q_A\vec{v}_A + \vec{x}_B\times Q_B\vec{v}_B) \qquad \textbf{(32)}$$

The relation between vector potential and current in equation (31) is valid only for steady currents. For time-varying currents, Jackson page 242 describes the vector potential in terms of the transverse current. The complication occurs because the simple relation in equation (31) between vector potential and current requires zero divergence $\nabla\cdot\vec{J}=0$. Time-varying currents are usually accompanied by charge density variations $\partial\rho/\partial t \neq 0$.

Similarly for the oscillating masses,

$$\vec{h}(\vec{x}) = \frac{G}{c^2}\int \frac{\vec{J}_M(\vec{x}\,')}{|\vec{x}-\vec{x}\,'|}d^3x' \;\cong\; \frac{G}{c^2|\vec{x}|^3}\vec{L}\times\vec{x} \;\textbf{(wrong!)} \qquad \textbf{(33)}$$

where $\vec{L}$ is the angular momentum of the mass currents,

$$\vec{L} = \vec{x}_A\times M_A\vec{v}_A + \vec{x}_B\times M_B\vec{v}_B \qquad \textbf{(34)}$$

Fixing up the relations for the time-varying electromagnetic fields is straightforward, thanks to a century of development of the theory. In particular the forces can be attributed to electrically neutral devices, which do not contribute to the electromagnetic fields. Fixing up the relations for time-varying gravitoelectromagnetic fields is more complicated, because the forces causing time-varying sources must be taken into account.

In one reference [28] varying angular momentum in an isolated source induces mass currents in a tubular detector. Newton is rolling over in his grave! Besides ignoring external torques, the analysis also ignores contributions of internal forces in the source.

Weinberg considers post-Newtonian effects in a collection of objects held together by gravitation. On page 220, equation 9.1.68 reads as follows:

$$\Gamma^i_{00} = \frac{\partial}{\partial x^i}\left(2\phi^2+\psi\right) + \frac{\partial \varsigma_i}{\partial t} \qquad \textbf{(35)}$$

The fields ϕ and ψ are scalar fields, similar to eqns. (10) and (11) above. The vector potential $\vec{\varsigma}$ is similar to $\vec{h}$ in equation (12) above.

Examine equation 9.1.69 from Weinberg, involving the gravitomagnetic field:

$$\Gamma^{i}_{0j} = \frac{1}{2}\left(\frac{\partial \varsigma_i}{\partial x^j} - \frac{\partial \varsigma_j}{\partial x^i}\right) - \delta_{ij}\frac{\partial \phi}{\partial t} \quad \textbf{(36)}$$

This is the same as equation (18) above.

Fields involving derivatives of the vector potential $\vec{h}$ appear in the Weinberg example. In a sense induction fields exist. However relations between gravitoelectric and gravitomagnetic fields are more complicated than their counterparts in electrodynamics, because of derivatives of scalar potentials.

12.5. Tidal Gravitational Forces on Extended Objects

Time-varying sources produce effects on distant detectors through the tensor field s_{ij} of equation 1. However the forces cannot be computed using the geodesic equation 14, since the relevant Christoffel elements are zero. Although Weinberg describes the response of a Weber bar, another reference [29] is needed to understand the physics. Finn uses the notation h_{ij} instead of s_{ij} for the tensor field.

Two masses in an extended object are initially separated by a distance L. When the gravitational wave arrives at time t=0 two effects occur:

$$\delta\left(L^2\right) = L'^2 - L^2 = h^{TT}_{jk}\left(x^j_B - x^j_A\right)\left(x^k_B - x^k_A\right) \quad \textbf{(37)}$$

Equation (37) represents a hair-splitting definition of the distance between two points in an extended object. The trick is to switch to Fermi normal coordinates, choosing some point such as the center of mass as the origin of coordinates.

The acceleration of a point at $\vec{x} = \vec{x}_{cm} + \vec{X}$ in the object can be expressed as follows:

$$\frac{d^2 X_i}{dt^2} = \frac{\partial^2 g_{ij}}{\partial t^2} X^j + O\left(v^k/c\right) \quad \textbf{(38)}$$

Equation (38) is the tidal force for a field with time-varying spatial components of the metric. Weinberg treats the Weber bar as an extended harmonic oscillator, with the tidal force as the driving force.

Costa states that all gravitational forces are tidal forces. Possible interpretation: all observers are in an accelerated reference frame.

13. CONCLUSIONS

Costa states that there is no gravitational analogue to Faraday's law of induction, the $\partial \vec{B}_g/\partial t$ term in GEM equations. Bertschinger equation 39 includes a $\partial \vec{B}/\partial t$ term in the expression for the curl of the gravitoelectric field, but claims that the gravitational displacement current $\partial \vec{E}/\partial t$ is absent in the expression for the curl of the gravitomagnetic field.

Weinberg equation 9.1.68 implies a force on a stationary test mass due to a time-varying vector potential. Costa might argue that the induction field appears because of the coordinate system (gauge) chosen to express the fields. Bertschinger can argue that in a different gauge the displacement current vanishes.

The physics of the problem lies in the derivatives of the scalar terms in equations (17) and (18). These terms represent non-local energy densities, from many-body interactions. The basic physics, the Pauli exclusion principle [26], is quantum mechanical. No experimental evidence exists to assess whether disembodied stress forces contribute to time-varying gravitation. Since F=ma, disembodied stress forces are as important as accelerating mass in general relativity.

DeWitt [25] uses the terminology "quantization of the gravitational field". It is stated at the outset that there is no experimental motivation for the investigation of this problem whatever.

In the Tajmar experiments, the forces involved in allowing electrons to follow different paths than background ions would have to be taken into account. However describing a Cooper pair condensate with local forces is ill-defined. It is possible for some novel mechanism [30] to enhance gravitational coupling of the condensate, taking advantage of a superconductor as a non-local source.

Until some such theory is developed, the GEM equations provide simple, well-defined relations among mass currents, gravitoelectric and gravitomagnetic fields, ignoring stress forces.

REFERENCES

Note: "arXiv" papers are available from Cornell's electronic preprint depository http://arxiv.org

[1] Tajmar M, Plesescu F, Marhold K, de Matos CJ. Experimental detection of the gravitomagnetic London Moment. arXiv2006: gr-qc/0603033.

[2] Tajmar M, Plesescu F, Seifert B, Marhold K. Measurement of gravitomagnetic and acceleration fields around rotating superconductors. In : Proc. Of Space Technology and Applications International Forum (STAIF) 2007. Am Inst Phys 2007; CP880: 1071 .

[3] Tajmar M, Plesescu F, Seifert B, Schnitzer R and Vasiljevich I. Investigation of frame-dragging-like signals from spinning superconductors using laser gyroscopes. In: Proc. Of Space Technology and Applications International Forum (STAIF) 2008. Am Inst Phys 2008; CP969: 1080.

[4] Tajmar M, Plesescu F, and Seifert B. Anomalous fiber optic gyroscope signals observed above rotating rings at low temperature. J Phys Conf Ser 2009; 150: 032101.

[5] Tajmar M, Plesescu F. Fiber-optic-gyroscope measurements close to rotating liquid helium. In: Proc. Of Space, Propulsion & Energy Sciences International Forum (SPESIF) 2010. Am Inst Phys 2010; CP1208: 220.

[6] Wald RM. General Relativity. The University of Chicago Press 1984.

[7] Modanese G, Junker T. Conditions for stimulated emission in anomalous gravity-superconductors interactions. In: Christiansen MN, Rasmussen TK Eds. Classical and Quantum Gravity Research. Nova Science Publishers 2008; pp. 245-269.

[8] Wu N. Gravitational shielding effects in gauge theory of gravity. arXiv 2003: hep-th/0307225. Wu N. Symmetry breaking and anomalous gravitational effects. In: Modanese G, Robertson GA, Eds. Gravity-Superconductors Interaction: Theory and Experiment. Bentham 2012; Chapter 14.

[9] de Matos CJ, Beck C. Possible measurable effects of dark energy in rotating superconductors. arXiv 2007: 0707.1797.

[10] Hauser J, Droescher W. Emerging physics for novel field propulsion science. In: Proc. of Space, Propulsion and Energy Sciences International Forum (SPESIF) 2010. Am Inst Phys 2010; CP1208: 168. Hauser J, Droescher W. Emerging Physics for Gravity-Like Fields. In: Modanese G, Robertson GA, Eds. Gravity-Superconductors Interaction: Theory and Experiment. Bentham 2012; Chapter 11.

[11] Chui G. A first: String theory predicts an experimental result. Symmetry Breaking, 2009. Available from http://www.bnl.gov/rhic/stringTheory.asp

[12] Costa LFO and Herdeiro CAR. Gravitoelectromagnetic analogy based on tidal tensors. Phys Rev 2008; D 78: 024021.

[13] Mashhoon B. Gravitoelectromagnetism. arXiv 2000: gr-qc/0011014.

[14] Graham RD, Hurst RB, Thirkettle RJ, Rowe CH and Butler, PH. Experiment to detect frame dragging in a lead superconductor. Physica C 2008; 468: 383.

[15] Tate J, Felch SB, Cabrera B. Determination of the Cooper-pair mass in niobium. Phys Rev 1990; B 42: 7885.

[16] Tajmar M, de Matos C. Gravitomagnetic fields in rotating superconductors to solve Tate's Cooper pair mass anomaly. In: Proc. Of Space Technology and Applications International Forum (STAIF) 2006. Am Inst Phys 2006; CP813: 1415.

[17] Jackson JD. Classical Electrodynamics. John Wiley & Sons 1999.
[18] Stedman GE. Ring-laser tests of fundamental physics and geophysics. Rep Prog Phys 1997; 60: 615.
[19] Heaviside O. A gravitational and electromagnetic analogy. The Electrician 1893; 31: 81. Available from http://www.wbabin.net/historical/heaviside.pdf
[20] Weinberg S. Gravitation and cosmology. Wiley 1972.
[21] Bertschinger E. Gravitation in the weak-field limit. MIT Physics, 2000. Available from http://web.mit.edu/edbert/GR/gr6.pdf
[22] Low FE. Classical field theory, electromagnetism and gravitation. Wiley 1997.
[23] Salt D. Action principles for the elastic solid and the perfect fluid in general relativity. J Phys A 1971; 4: 501.
[24] Williams DN. The elastic energy-momentum tensor in special relativity. Annals Phys 1989; 196: 345-360.
[25] DeWitt BS. The Quantization of Geometry. In: Witten L. ed. Gravitation: an introduction to current research. John Wiley and Sons 1962; pp. 266-381.
[26] Chandrasekhar S. The maximum mass of ideal white dwarfs. Astrophys J 1931, 74: 81.
[27] Kittel C. Introduction to solid state physics. Wiley 1961.
[28] Bini D, Cherubini C, Chicone C, Mashhoon B. Gravitational induction. arXiv 2008: 0803.03920.
[29] Finn LS. Gravitational radiation sources and signatures. arXiv 1999: gr-qc/9903107.
[30] Fontana G. High Temperature Superconductors as Quantum Sources of Gravitational Waves: the HTSC GASER. In: Modanese G, Robertson GA, Eds. Gravity-Superconductors Interaction: Theory and Experiment. Bentham 2012; Chapter 3.

CHAPTER 7

The Chiao Gravity-Superconductor EM Transducer: An Overview

R.A. Lewis*

Penn State University, Physics Dept. (retired); 1308 Andover Drive, Boalsburg PA 16827, USA

Abstract: The technique used by Raymond Chiao for producing a new type of radiation from a superconductor is reviewed. In this technique, superconductors are used both for the source and detector of the radiation, on the hypothesis that the radiation effects are enhanced in superfluid materials. The coherent conversion in superconductors between electromagnetic and gravitational-like radiation requires modifications to standard model theories. The requirements for such modifications are discussed, assuming that they exist. For example, the gravitational-like disturbance needs to be expressed in terms of a vector field disturbance in a macroscopic number of particles, and where the coupling must be typically 20 orders of magnitude stronger than Newtonian gravity. Analogies with particle physics suggest that superconductors also probe vacuum fields comparable to the QCD (quantum chromodynamics) vacuum or higher dimensions.

Keywords: Gravitational forces, theories of gravitation, superconductors, high-Tc superconductors, type-II superconductors, general relativity, vector radiation, scalar radiation, gravitational waves, quadrupole antenna, microwaves, gravitomagnetism, gravity-like fields.

1. INTRODUCTION

In this paper, the technique used by Raymond Chiao for producing a new type of radiation from a superconductor is reviewed. It is noted that, results from the measurement with YBaCuO superconductor imply that the coupling of the new type of radiation is at most 5 orders of magnitude weaker than electromagnetic coupling. No measurements have been made with other superconductors, such as lead, which have better long-range coherence properties than YBaCuO. Such measurements are needed to confirm or deny the existence of the Chiao effect.

The Chiao hypothesis is that an electromagnetic wave can set up mass currents in a superconductor which coherently produce gravitational radiation. In experiments, normal materials, that are transparent to gravitational radiation, can be used as shields around a receiving superconductor to preclude transmission of electromagnetic radiation.

Chiao argues that a vector field associated with a gravitational wave in a superconductor material coherently interacts with an electromagnetic wave. The coupling of the gravitational currents is so strong that only a thin film of superconductor would suffice to serve as a reflector of the radiation. An argument for assuming tensor rather than vector radiation is that the detector could be used to search for Standard Model astronomical sources of gravitational radiation. Chiao ignores the possibility of other forms of gravitation-like radiation.

The Chiao experiment challenges the notion that general relativity describes gravitational-like interactions in a superconductor. The textbook treatment of general relativity uses ad hoc methods to construct sources which conserve energy and momentum locally. The treatment is mathematically inconsistent, since material stresses involve non-local exchange forces. In a semi-classical model, the stress tensor is replaced [1] by a quantum expectation value $\langle\psi|\hat{T}_{\mu\nu}|\psi\rangle$. This model encounters mathematical problems associated with gauge invariance and collapse of the wave function. Vacuum fields are invoked to construct a consistent theory. However quantum gravity [2] (string theory, loop quantum gravity) is not developed sufficiently to predict what type of radiation might emerge from a superconductor.

***Address correspondence to R.A. Lewis:** Penn State University, Physics Dept. (retired); 1308 Andover Drive, Boalsburg PA 16827., USA; E-mail: r3l@psu.edu

Giovanni Modanese and Glen A. Robertson (Eds)

1.1. Background

The basic apparatus used by Chiao [3] involves a source excited by microwaves, and a detector shielded from electromagnetic radiation. The basic "shining through the wall" arrangement used in searches for other types of radiation, such as axions [4] is illustrated in Fig. **1**.

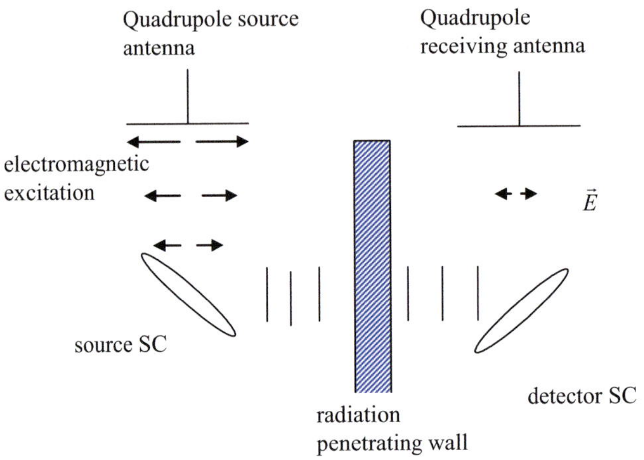

Figure 1: The superconducting source produces radiation sensed by a detector shielded by a wall.

In Fig. **1**, both the source and detector are YBaCuO discs, 1 inch in diameter and ¼ inch thick. A T-shaped antenna is used to excite electromagnetic vibrations in a quadrupole pattern in the source. Another antenna is used to search for electromagnetic signals produced in the detector.

2. QUADRUPOLE SOURCE, RECEIVER ANTENNAE

Chiao [3] assumes the new type of radiation produced by the source in Fig. **1** to be tensor, with properties similar to gravitational radiation. Such radiation requires an experiment involving a quadrupole excitation pattern. In order to generate and receive a quadrupole field pattern, Chiao constructed two T-shaped antennae. Measurements of the radiative coupling between the antennae were made, with the antennae placed directly facing each other at a fixed distance, as shown in Fig. **2**.

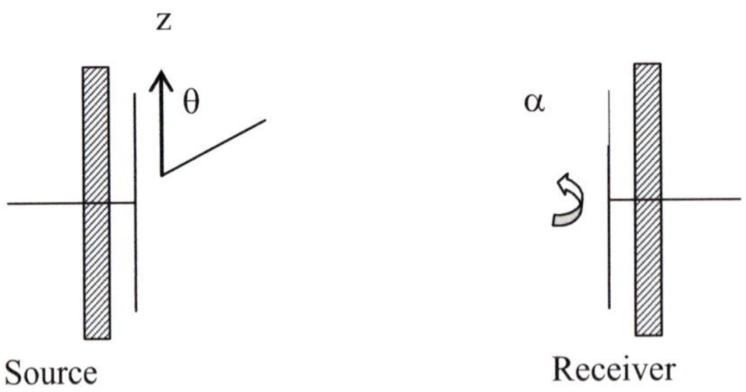

Figure 2: Source, receiver antennae test for quadrupole radiation.

Further, one inch lengths of wire, matching the wavelength of 12 *GHz* microwaves, were soldered to the center conductor of an SMA coaxial cable. The joint extends a quarter wavelength beyond a ground plane. Resonance properties of the antenna were measured with an HP 8720 network analyzer. The quality factor of the resonance was Q=10.

In this experiment, the source resembles a collinear back-to-back arrangement of dipoles, oriented in the *z*-direction. This collinear quadrupole approximation is used to describe a qualitative model of the radiation pattern. The amplitude pattern from a collinear quadrupole varies as the $\sin(2\theta)$, where θ is angle from the *z*-axis. The amplitude vanishes both on the z-axis and in the *x-y* plane. The electric field points in opposite directions at finite distances above and below the *x-y* plane. This behavior gives the desired quadrupole excitation in a superconductor, with current oscillations 180° out of phase in opposite halves of the disc.

Chiao performed tests by rotating the receiver antenna around the line joining the antennae, labeled angle α in Fig. **1**. Extinction occurred for α =45°, while maximum transmission was observed for α =90° (and presumably for α =0° also).

Most likely Chiao meant to claim that the pattern imitated helicity $h=\pm 2$ radiation, which is required for gravitational waves to satisfy the equivalence principle [5]. For helicity 2 radiation, the polarization rotates by an angle 2α when the coordinates are rotated by an angle of α about the direction of propagation.

Chiao does not present any quantitative results from the tests of the antennae. However the simple collinear quadrupole model indicates that the source antenna has the desired property, of inducing a quadrupole pattern of currents in a superconductor. The helicity argument is unimportant, since a quadrupole source can produce scalar, vector or tensor radiation.

3. MICROWAVE TRANSMISSION

Chiao generated the 12 *GHz* microwave beam, starting with a 6 *GHz* "brick" oscillator (Frequency West model MS-54M-09), with an output of 13 *dBm* (20 *mW*) at 6 *GHz*. This 6 *GHz* signal was amplified, and then doubled in a second harmonic mixer (MITEQ model MX2V080160), in order to produce a 12 *GHz* microwave beam with a power level of 7 *dBm* (5 *mW*).

The 12 *GHz* beam was transmitted to the source antenna, using SMA coaxial cable.

4. MICROWAVE RECEIVER

Signals from the receiver superconductor were picked up by another antenna. The antenna signal was amplified using a low-noise preamp (Astrotel model PMJ-LNB KU). This preamp is used for receiving 12 *GHz* microwave satellite communications.

The noise figure (NF) of the amplifier is 0.6 *dB*. This NF determined the overall sensitivity of the measurement. The front-end LNB (Low-Noise Block) assembly, besides having a low-noise preamp, also contained inside it an internal mixer that down-converted the amplified 12 *GHz* signal into a standard 1 *GHz* intermediate frequency (IF) band.

The IF signal was fed into a commercial satellite signal level meter (Channel Master model 10051FD), which both served as the DC power supply for the LNB assembly by supplying a DC voltage back through the center conductor of a F-style IF coax cable into the LNB assembly, and also provided amplification of the IF signal. Its output was then fed into a spectrum analyzer (Hewlett-Packard model 8559A).

The HP 8559A spectrum analyzer has a frequency range of 0.01 to 21 *GHz*. The noise level is -98 *dBm/kHz* for 9.0 to 15 *GHz*, and -111 *dBm/kHz* for 0.01 to 3 *GHz*. Resolution bandwidths from 1 *kHz* to 3 *MHz* can be selected.

5. SIGNAL SENSITIVITY CALIBRATIONS

Chiao calibrated the sensitivity of the source-receiver system in a separate experiment, in which the antennae were replaced by a 70 *dB* attenuator. The resulting spectrum was a signal of -77 *dBm* (2×10^{-11} *watts*), which was 25 *dB* above the noise floor.

In effect, the noise floor of -102 *dBm* ($6.3x10^{-14}$ *W*) represents an upper limit to the signal which could be observed. However spectra were shown only for the configuration in which the Astrotel preamp, LNB assembly, and signal level meter were used. As a cross-check, properties of various components are listed in Table **1**.

Table 1: Power levels, gain, attenuation.

Component	Power Level	Gain/Attenuation	Noise Floor
MS-54M-09	13 dBm (20 mW)		
MITEQ		-13 dB	
Amplifier		X	
Beam power	7 dBm (5 mW)		
MITEQ to HP 8559A	-77 dBm	-70 dB attenuator	-102 dBm
Cable loss		-14 dB	
Astrotel		56 dB	
Channel Master		X	
Channel Master to HP 8559A	-38 dBm (leakage)		-55 dBm
Channel Master to HP 8559	-55 dBm		-55 dBm

The power level measured with the MITEQ connected directly to the HP 8559 through a 70 *dB* attenuator should have been 7-70 = -63 *dBm*, instead of -77 *dBm* measured. The 14 *dB* discrepancy is listed as "cable loss" in the table. Chiao does not comment on the discrepancy.

The only effect of the Astrotel and Channel Master was to raise the noise floor by 57 *dB*. The Astrotel accounts for 56 *dB* amplification. No data could be found for the Channel Master. Presumably the signal was also amplified by 57 *dB*. In effect, Chiao uses the noise floor as a calibration for the signal sensitivity.

6. UPPER LIMIT ON SIGNAL

For the Chiao experiment, an upper limit on a signal from the detector is illustrated by the spectrum observed in the HP 8559, as shown in Fig. **3**. It was noted that the Faraday cages surrounding the source and detector were compromised, allowing a small leakage signal to mark the 1.3 *GHz* frequency of the expected signal. No features were observed at 1.3 *GHz*, with the shielding restored. Hence the electromagnetic wave emerging from the detector had a power less than $6.3x10^{-14}$ *watts*.

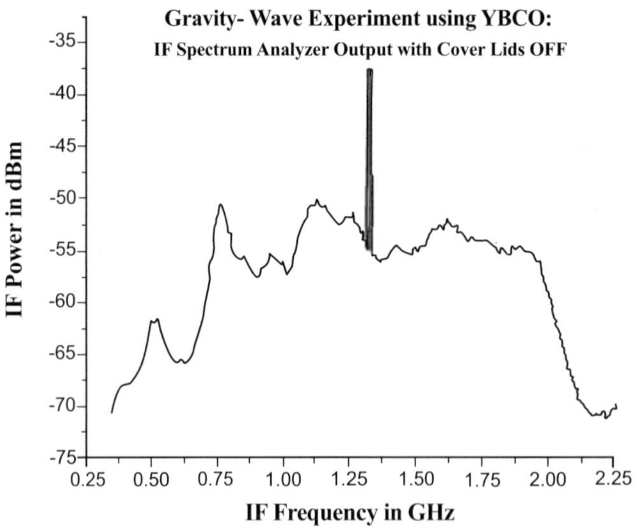

Figure 3: Spectrum observed from detector [3].

However, the input signal impinging on the source was -7 *dBm* ($2x10^{-4}$ *watts*), such that, the product of efficiencies for conversion between em and GW radiation is

$$\eta^2 = 6.3*10^{-14} / 2*10^{-4} = \left(1.8 \cdot 10^{-5}\right)^2 . \quad \textbf{(1)}$$

7. SCALAR, VECTOR, TENSOR RADIATION FROM A QUADRUPOLE SOURCE

The Chiao effect requires that a superconductor produces some type of excitation field in the surrounding space. Field equations are required to express relations between source currents and fields sensed by external detectors. Consistent theories have been worked out for three types of classical field:

- Scalar (one number at each point in spacetime),
- Vector (three numbers),
- Tensor (nine numbers).

The polarization of forces accompanying the various field patterns is calculated, to assess the sensitivity of the Chiao detector arrangement for new physics.

A source is specified in terms of a distribution $\rho(\vec{x},t)$ representing mass density, vacuum energy fluctuations, or other parameters depending on the theory. Fig. **4a** represents the quadrupole pattern of currents set up by microwaves impinging on the Chiao superconductor source. To simplify the mathematics, the Chiao superconductor source is modeled as two point masses, oscillating in a linear quadrupole pattern, as depicted in Fig. **4b**.

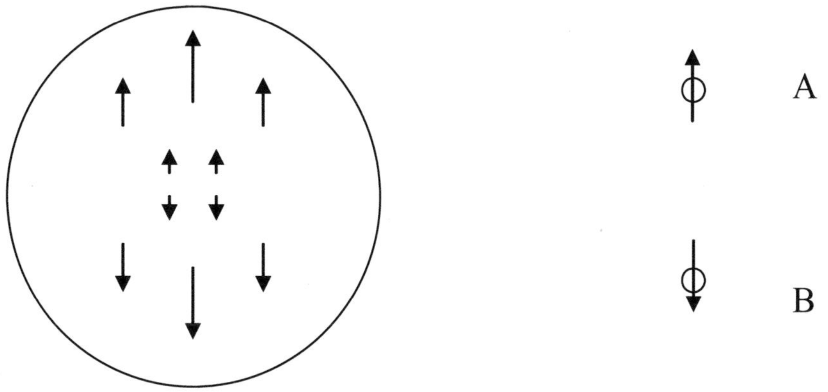

Figure 4: Current patterns in Chiao superconductor, (A) actual currents, (B) point mass approximation.

The motion of the two masses is

$$\vec{x}_A = \left(0, 0, L + se^{i\omega t}\right); \quad \textbf{(2)}$$

$$\vec{x}_B = \left(0, 0, -L - se^{i\omega t}\right), \quad \textbf{(3)}$$

where the amplitude $s<<L$ is small compared to the separation. The source distribution can be expressed as follows:

$$\rho(\vec{x},t) = m\delta^3\left(\vec{x} - \vec{x}_A(t)\right) + m\delta^3\left(\vec{x} - \vec{x}_B(t)\right). \quad \textbf{(4)}$$

7.1. Scalar Radiation

The field equation for a massless scalar field can be written as follows,

$$\Phi = {4\pi G_S \rho}/{c^2} \ . \tag{5}$$

The scalar field produced by a time-varying source is given by the retarded potential,

$$\Phi(\vec{x}_o, t_o) = \frac{4G_S}{c^4} \int d^3x \frac{\rho\left(\vec{x}, t_o - {|\vec{x}_o - \vec{x}|}/{c}\right)}{|\vec{x}_o - \vec{x}|} . \tag{6}$$

The radiation pattern for the linear quadrupole, assuming a scalar field, is given by

$$\Phi_2(\vec{x}_o, t_o) = -\frac{16G_S}{Rc^4} \cos^2(\theta) mLs\omega^2 e^{i\omega t - ikR} . \tag{7}$$

Forces produced by the field can be expressed in terms of the gradient,

$$\vec{E}_\Phi = c^2 \nabla \Phi_2(\vec{x}_o, t_o) = ik\vec{n}\frac{16G_S}{Rc^2}(\vec{n}\cdot\hat{z})^2 mLs\omega^2 e^{i\omega t - ikR} , \tag{8}$$

where $\hat{n}$ is a unit vector pointing to an observation point.

The polarization of the force field is

$$\vec{\varepsilon}_\Phi = (\vec{n}\cdot\hat{z})^2 \vec{n} \tag{9}$$

in the direction of propagation. The polarization is invariant under rotations about the direction of propagation. Ergo the scalar field has zero helicity (see Weinberg [3], page 257).

The phase of the scalar field radial gradient varies by 2π over a distance scale of 2.5 cm. Hence the scalar force is capable of inducing a dipole and quadrupole pattern of currents in the detector SC.

7.2. Vector Radiation

The technique for evaluating retarded potentials and fields for vector radiation from the setup in Fig. **1** is based on the derivation of Lienard Wiechert potentials in Jackson [6] and Low [7]. The time-varying part of the vector potential is

$$A^\mu_{rad} = \frac{G_V mLs}{4\pi cR} k_z k_0 (1,0,0,1) e^{i\omega t - ikR} , \tag{10}$$

where $k_0 = \omega/c$ and $k_z = \vec{n}\cdot\hat{z}\,\omega/c$.

Forces produced by the vector field can be expressed in terms of the combination

$$\vec{E}_V = \nabla A^0 - \frac{\partial \vec{A}}{\partial t} = \frac{G_V m}{4\pi Rc^3}\omega^3 sL(\vec{n}\cdot\hat{z})(\vec{n}\times(\vec{n}\times\hat{z}))e^{i\omega t - ikR} . \tag{11}$$

The polarization of the force field is

$$\vec{\varepsilon}_V = (\vec{n}\cdot\hat{z})\vec{n}\times(\vec{n}\times\hat{z}). \tag{12}$$

The helicity is a mixture of h=+1 and h=-1. The vector radiation is linearly polarized, which is a mixture of left and right circular polarizations.

7.3. Tensor Radiation

The Finn [8] recipe for calculating gravitational radiation involves the quadrupole moment tensor of the source. The procedure spelled out by Finn uses the Transverse Traceless gauge.

$$h_{ij}^{TT}(\vec{x}_o,t_o) = \frac{2G}{rc^5}\frac{d^2}{dt_o^2}Q_{ij}^{TT}(t_o - r); \tag{13}$$

$$Q_{ij}^{TT} = P_{ik}(\vec{x}_o)Q_{kl}P_{lj}(\vec{x}_o) - \tfrac{1}{2}P_{ij}(\vec{x}_o)Q_{lm}P_{lm}(\vec{x}_o); \tag{14}$$

$$Q_{ij}(t) = \int d^3x\rho(\vec{x},t)(x_ix_j - \tfrac{1}{3}x^kx_k\delta_{ij}); \tag{15}$$

$$P_{jk}(\vec{x}_o) = \delta_{jk} - n_jn_k. \tag{16}$$

In equation (16), $\vec{n}$ is a unit vector in the direction of the observation point.

Tensor radiation can be described in terms of a tensor field h_{ij}, where i and j denote three spatial indices. Following the derivation of the transverse-traceless gauge by Finn, the tensor field can be written as

$$h_{ij}^{TT} = \frac{2G_T}{Rc^5}\tfrac{1}{3}mLs\omega^2 e^{i\omega t - ikR}\varepsilon_{ij}, \tag{17}$$

where the polarization tensor is

$$\varepsilon_{ij} = \begin{pmatrix} n_x^2n_z^2 - n_y^2 & n_xn_y(n_z^2+1) & -n_xn_z(1-n_z^2) \\ n_xn_y(n_z^2+1) & n_y^2n_z^2 - n_x^2 & -n_yn_z(1-n_z^2) \\ -n_xn_z(1-n_z^2) & -n_yn_z(1-n_z^2) & (1-n_z^2)^2 \end{pmatrix}. \tag{18}$$

The tensor field has helicity $h = \underline{+}2$.

7.4. Application to Chiao Detector

Both the scalar and vector forces have phase variations over the one inch diameter of the detector SC, so that dipole and quadrupole current patterns can be induced. Since the receiving antenna in Fig. **1** has quadrupole symmetry, it would be sensitive only to the quadrupole component of the current pattern.

Tensor radiation from the source SC has the quadrupole symmetry to induce quadrupole current in the receiving SC.

The coupling constants G_S, G_V and G_T have the same dimensions as the Newtonian gravitational constant. Comparable expressions for the near fields of scalar, vector and tensor interactions are described by Nieto [9]. Coupling to mass is used for convenience, although the coupling could be to other aspects of matter such as lepton number, hypercharge, or local variations in the vacuum [10].

In the Fontana theory [11], wave-function coherence among N Cooper pairs amplifies the probability of a gravitational disturbance by a factor of N^2. Fontana equation 12 describes two possible polarization patterns, for radiation generated in the z-direction by non-linear effects after excitation of a superconductor,

$$h^{\mu\nu} = A(\vec{x},t)\begin{pmatrix} 1 & 0 & 0 & 0 \\ 0 & -1 & 0 & 0 \\ 0 & 0 & -1 & 0 \\ 0 & 0 & 0 & -1 \end{pmatrix} + B(\vec{x},t)\begin{pmatrix} 1 & 0 & 0 & 1 \\ 0 & 0 & 0 & 0 \\ 0 & 0 & 0 & 0 \\ 1 & 0 & 0 & 1 \end{pmatrix} \qquad \textbf{(19a)}$$

In the Bertschinger [12] classification scheme, the first term is a combination of two scalars, $\Phi = g^{00} - 1$ and $\Psi = trace\left(h^{ij}\right)$. The second term represents a vector disturbance, since h^{03} is the z-component of the gravitomagnetic vector potential.

8. GRAVITATIONAL INTERACTION WITH MATERIAL STRESSES

The relation between the gravitomagnetic field and mass current is controversial [13]. Some references dispute whether a time-varying gravitomagnetic field can be produced from a time-varying mass current.

$$\nabla \times \vec{B}_g = (-)\frac{4\pi G}{c^2}\vec{J}_m + \frac{1}{c^2}\frac{\partial \vec{E}_g}{\partial t} \qquad \textbf{(19b)}$$

and whether the time-varying magnetic field implies gravitational induction,

$$\nabla \times \vec{E}_g = -\frac{\partial \vec{B}_g}{\partial t}. \qquad \textbf{(19)}$$

The gravitoelectromagnetic (GEM) approximation lacks a prescription for including material stresses, which are crucial for general relativity.

The mass current can be expressed in terms of fluid motion,

$$\vec{J}_m = \rho\vec{v}. \qquad \textbf{(20)}$$

The time derivative of the mass current is accompanied by material stress. For a perfect fluid, the stress can be expressed as a pressure gradient,

$$\dot{\vec{J}}_m = \frac{\partial\left(\rho\vec{v}\right)}{\partial t} = -\nabla p. \qquad \textbf{(21)}$$

The stress tensor, which is the source for gravitation in general relativity, must conserve energy and momentum locally,

$$T^{\mu\nu}{}_{,\nu} = 0. \qquad \textbf{(22)}$$

Material stresses must be included in the stress tensor to make it divergenceless.

Standard textbooks [14] do not discuss the contribution of material stresses to gravitation in detail. Such stresses involve non-local exchange forces, which are outside the validity of general relativity. For example, the basic mechanism in hydrodynamics, atom-atom scattering, relies on the Pauli Exclusion Principle to prevent electron wave functions from overlapping. The uncertainty principle does not allow an unambiguous meaning to the stress tensor at a point. A semi-classical model encounters mathematical problems described by Ford [1].

Known sources of gravitational waves such as binary pulsars [15] do not rely on quantum mechanical forces. More esoteric sources in the universe might produce forms of radiation not restricted by local conservation of energy and momentum.

9. CHIAO THEORY AND PROPOSED EXPERIMENTS

The main concept guiding the Chiao experiment is described in a document [16] composed before the experiments were performed. The Hamiltonian describing the interactions among Cooper pairs, electromagnetic and gravitomagnetic fields was

$$H = \frac{1}{2m_{2eff}}\left(\vec{p} - e_2\vec{A}(t) - m_2\vec{h}(t)\right)^2. \tag{23}$$

The gravitomagnetic field is a vector field, rather than the tensor field of gravitation. The Hamiltonian includes a term $e_2\vec{A}(t)m_2\vec{h}(t)/m_{2eff}$ in which the gravitomagnetic and magnetic fields are coupled. Chiao interprets this term to mean that electromagnetic radiation can be converted into gravitoelectromagnetic radiation in the presence of a superconductor medium. The concept is similar to Vector Dominance [17] in particle physics, wherein a photon converts into a ρ or ω (vector) meson in a nuclear environment.

Chiao runs into trouble in computing the matrix element for the em to GW conversion, ascribing helicity 2 to electromagnetic radiation. Helicity 2 electromagnetic radiation is unknown physics. The interaction integral

$$\Delta E_{\text{int}}^{(1)} = \left(\frac{m_2}{m_{2eff}}\right)\left\langle\psi\left|e_2\vec{A}(t)\cdot\vec{h}(t)\right|\psi\right\rangle \tag{24}$$

vanishes unless the fields have the same helicity. A more logical approach is to consider *h(t)* as a vector field.

In the document describing experimental results [3], Chiao replaces the gravitomagnetic field with a tensor field.

$$H = \frac{1}{2m}\left(p_i - eA_i - \tfrac{1}{2}\sigma_{ab}\Omega_i^{ab}\right)^2 + V, \tag{25}$$

where i is a spatial index, a,b are spatial tetrad indices, σ_{ab} is a two-by-two matrix-valued tensor representing the spin. The tensor Ω_i^{ab} represents the components of a spin connection, related to the Lorentz transformation of spin. The potential energy V includes the Coulomb interactions between electrons in the quantum fluid.

The concept developed by Chiao parallels problems encountered in the development of quantum chromodynamics. A special vacuum field, the Pomeron [18], was invented to allow a particle to move through nuclear material coherently while its mass was changing. The particle cannot exchange any quantum numbers with the nuclear material without losing coherence. Chiao attempts to allow a vector photon to convert coherently into a tensor graviton within a superconductor. The conversion requires exchange of quantum numbers (angular momentum) with the condensate, resulting loss of coherence.

In the language of particle physics, diffraction is basically particle scattering. A particle passing near an obstacle exchanges a Pomeron with the obstacle, resulting in deflection. Integrating over all possible exchanges produces the same result as the Huygens principle.

The superconducting condensate has an ingredient absent from particle physics, essentially a target consisting of 10^{23} coherent particles spread out over a macroscopic distance. Chiao addresses this aspect by describing gravitational radiation in terms of Fermi normal coordinates [19]. A gravitational wave produces a disturbance which can be expressed in terms of a vector field excitation of the superconducting medium.

The strain fields produced by a gravitational field in an extended object can be expressed in a way which appears to be forces on individual particles. Suppose two masses are initially separated by a distance *L*. A hair-splitting definition [8] of the distance between two points implies that this distance is affected by the gravitational wave,

$$\delta\left(L^2\right) = L'^2 - L^2 = h_{jk}^{TT}\left(x_B^j - x_A^j\right)\left(x_B^k - x_A^k\right). \tag{26}$$

By choosing the center of mass to be the origin, the implications of equation (27) can be expressed in terms of a force on a stationary test mass,

$$F_k = m\frac{\partial^2 h_{jk}}{\partial t^2}x^j . \tag{27}$$

Chiao describes the tidal force in terms of a velocity field,

$$N_a = \frac{1}{2}\int_0^{X_b}\frac{\partial h_{ab}}{\partial t}\,dX^b . \tag{28}$$

The canonical momentum of a Cooper pair is then expressed as

$$\vec{\Pi} = \vec{p} - e\vec{A} - m\vec{N} . \tag{29}$$

Equation (30) gives the appearance of two vector fields, rather than a vector and a tensor field, accompanying particle momentum.

In string theory [20], gravitons with polarizations that lie entirely within the brane are effective spin 2 objects while scalar or vector states result if the polarizations are partially or completely perpendicular to the brane. The possibility of a vector interaction between a superconductor and gravitation cannot be ignored.

The multitude of particles produced in heavy ion collisions comes closer to a parallel with radiation interacting with 10^{23} coherent particles in a superfluid. The Relativistic Heavy Ion Collider RHIC was constructed at Brookhaven Labs to probe the QCD (quantum chromodynamics) vacuum. The sigma scalar field was originally hypothesized [21] to explain relations between beta decay and strong nuclear forces. In subsequent developments the sigma scalar field had to be a global scalar vacuum field, accompanying the Higgs and QED (quantum electrodynamics) vacuum as another component of zero point energy.

In the early universe, quarks and gluons were free particles. In order to form particles such as mesons and nucleons, the quarks had to agree on certain phase angles. Like the Ginsburg Landau field in a superconductor, the sigma field was originally the order parameter in the quark condensate phase transition. As the universe expanded and cooled, the sigma field developed into a tachyon field [22]. In quantum mechanics a tachyon field cannot be localized, averting the possibility for violating causality. Some of the effects of the vacuum quark condensate are summarized:

- The sigma field has a vacuum energy density [23] of 57 MeV/fm^3, which contributes to nuclear binding by a Casimir mechanism.

- Excitations of the sigma field show up as effects on pions produced in particle collisions, described [24] as a pole in the scattering amplitude, $\sqrt{s_{Pole}} = 466 - i466\ MeV$.

- The high temperature phase of the quark-gluon plasma turns out [25] to be a quantum liquid.

Explaining the viscosity to entropy ratio for both high T_c superconductors and the quark-gluon plasma is cited as a successful prediction from string theory [26], specifically the Maldacena AdS/CFT correspondence relating thermodynamics of a 5 dimensional black hole in anti-deSitter (AdS) space and conformal field theory (CFT).

In summary, Chiao proposes to produce a coherent conversion between electromagnetic and gravitational radiation by expressing the gravitational effect as a vector disturbance in a superconductor. A classical model for the conversion is depicted as the linear quadrupole of Fig. **2**. With a suitable choice of geometry a quadrupole pattern of currents can generate or absorb scalar, vector or tensor radiation. A quantum mechanical description of the process has ingredients comparable to concepts in the development of quantum chromodynamics in high energy physics.

10. INTERPRETATION OF SIGNAL

The results of the Chiao experiment are interpreted, assuming a signal of $6.3\text{x}10^{-14}$ *watts*. The input power impinging on the source was -7 *dBm* ($2\text{x}10^{-4}$ *watts*). The geometric mean of input and signal, $\sqrt{2x10^{-4} \times 6.3x10^{-14}}$ = $3.5\text{x}10^{-9}$ *watts*, is treated as if it were a gravitational wave. This assignment is consistent with the argument that source and detector have equal efficiencies for converting between electromagnetic and gravitational radiation.

The source and detector antennae are approximated as linear quadrupoles, as shown in Fig. **5**.

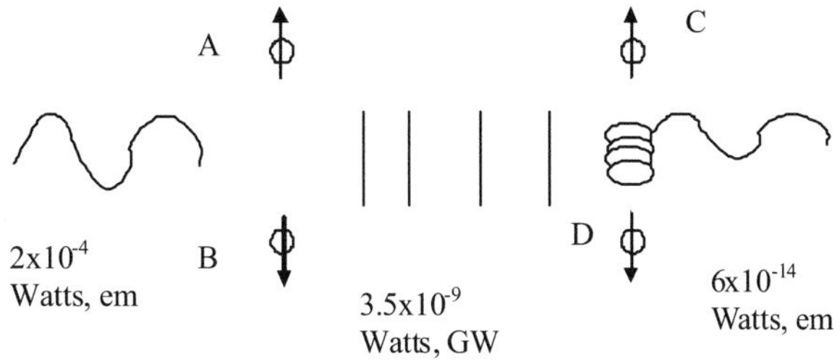

Figure 5: Quadrupole model of source, detector radiation.

The detector antenna is equipped with a spring, to connect with the Weinberg description of a Weber bar.

The four mass points oscillate in the z direction in a quadrupole pattern, given by

$$z_A = L + s_s e^{i\omega t}; \tag{30}$$

$$z_B = -L - s_s e^{i\omega t}; \tag{31}$$

$$z_C = L + s_d e^{i\omega t}; \tag{32}$$

$$z_D = -L - s_d e^{i\omega t}. \tag{33}$$

The electric charges are equal to one half of the charge of electrons within a depth of λ_L =170 nanometers (London penetration depth [27]) of an exposed surface,

$$N_e = \pi\left(0.5 inches\right)^2 \times \left(170\ nm\right) \times \left(10^{22} / cm^3\right) = 4.3\text{x}10^{17}\text{ e-}. \quad \textbf{(34)}$$

The density of 10^{22} *electrons/cm*^{3} is a compromise among various temperatures and compositions of YBaCuO [28]. The charge in each of the four mass points in Fig. **4** is

$$Q_e = 0.07\ Coulombs. \quad \textbf{(35)}$$

The corresponding (effective) mass M_e is to be computed, based on the 3.5x10^{-9} *watts* of detected gravitational radiation.

The power radiated by a quadrupole is related to the third derivative of a quadrupole moment tensor,

$$Q_{ij} = \int\left(x_i x_j - \tfrac{1}{3} x_k x^k \delta_{ij}\right)\rho(x) d^3x = 2M_e\left(L^2 + 2Lse^{i\omega t}\right) diag\left(-\tfrac{1}{3}, -\tfrac{1}{3}, \tfrac{2}{3}\right) \quad \textbf{(36)}$$

for small amplitude $s<<L$ oscillations. The separation parameter is taken as L=1 *cm*, roughly the gyroradius of a bar one inch long.

The power radiated as gravitational radiation is cribbed from Finn [8] equation (31),

$$L_{GW} = \frac{G}{5c^5}\left\langle \frac{d^3Q_{ij}}{dt^3}\frac{d^3Q^{ij}}{dt^3}\right\rangle = \frac{2G}{15c^5}\left(M_e\omega^3 Ls\right)^2. \quad \textbf{(37)}$$

The power radiated as electromagnetic radiation is adapted from Jackson [6, (*section* 9.49)], substituting Q_e for M_e in equation (36),

$$L_{em} = \frac{1}{40}\frac{1}{4\pi\varepsilon_0}\left\langle \frac{d^3Q_{ij}}{dt^3}\frac{d^3Q^{ij}}{dt^3}\right\rangle; \quad \textbf{(38)}$$

$$L_{em} = \frac{Q_e^2 L^2 s^2 \omega^6}{240\pi\varepsilon_0 c^5}. \quad \textbf{(39)}$$

The amplitudes of oscillation are calculated in terms of power radiated,

$$s = \sqrt{\frac{240\pi\varepsilon_0 c^5 L_{em}}{Q_e^2 L^2 \omega^6}}; \quad \textbf{(40)}$$

$$s_s = 6.1x10^{-15}\ meters\left(source\right); \quad \textbf{(41)}$$

$$s_d = 1.08x10^{-19}\ meters\ \left(\det ector\right). \quad \textbf{(42)}$$

According to Chiao [3] (Fig. **3**), the source and detector superconductors were placed a distance of twelve inches apart. The cross section for receiving GW radiation cannot exceed the geometrical area, which is comparable to the $\pi\ \lambda^2$ unitarity limit for a resonant receiver. Hence the GW power emitted from the source must have been

$$L_{GW}(source) = L_{GW}(measured) \times \frac{4\pi\left(12\ inches\right)^2}{\pi\left(0.5\ inches\right)^2} = 8.1x10^{-6}\, watts. \quad \textbf{(43)}$$

In order to match the power emitted, the mass associated with the oscillating electrons must be

$$M_e = \sqrt{\frac{15c^5 L_{GW}}{2G\omega^6 L^2 s_s^2}} = 5.7x10^7 \ kg \,. \tag{44}$$

To put this effective mass into perspective, calculate the mass per electron,

$$m_{eff} = {M_e}/{N_e} = 0.13 \ nanograms \,. \tag{45}$$

This mass is much larger than the 9.1x10^{-31} *kg* mass of an electron, but smaller than the Planck mass M_{PL} = 22 *micrograms*.

The amplitude $|h_{ij}|$ of the GW at *R*=30 *cm* can be computed from the energy flux,

$$S_{GW} = \frac{\omega^2 c^3}{8\pi G}|h_{ij}|^2 = \frac{L_{GW}(source)}{4\pi R^2} = 7.1x10^{-6} \ watts/m^2 \,. \tag{46}$$

Solving for the amplitude,

$$|h_{ij}| = \sqrt{\frac{8\pi G S_{GW}}{\omega^2 c^3}} = 2.8x10^{-31} \,. \tag{47}$$

This amplitude is nearly equal to the amplitude of relic cosmological gravitational radiation [29] at 12 *GHz*.

11. SUPERCONDUCTING SHEETS REFLECT GRAVITATIONAL WAVES

In subsequent documents Chiao proposes experiments [30] using thin type I superconducting film. Chiao argues that the force acting on a Cooper pair is

$$\vec{F}_{tot} = q\vec{E} + m\vec{E}_G \,. \tag{48}$$

Assuming that the electromagnetic and gravitoelectric forces are linearly related, equation (47) can be written as

$$\vec{F}_{tot} = \Xi q\vec{E} = q\vec{E}\left(1 - \frac{4\pi\varepsilon_0 G m_e^2}{e^2}\right). \tag{49}$$

The correction to the factor Ξ is the ratio of gravitational and electrostatic forces within a Cooper pair.

A gravitational wave with gravitoelectric field $\vec{E}_G$ incident on a superconductor produces an electric field

$$\vec{E} = \frac{1}{\Xi - 1}\frac{m}{q}\vec{E}_G = \frac{q}{m}\frac{1}{4\pi\varepsilon_0 G}\vec{E}_G \,. \tag{50}$$

From this it follows that the corrected gravitational inductance length scale is given by a microscopic quantity instead of the astronomically large length scales given by general relativity. Chiao concludes that laboratory scale superconducting mirrors for GR microwaves exist.

Quotes from MIT Technology Review [31]:

"That's a tantalizing quote but what are they hinting at? One experiment springs to mind that might be capable of seeing this effect. Gravity Probe B is a space-based experiment designed to spot the geodetic effect and frame dragging".

"The Gravity Probe B experiment has been throwing out anomalous results ever since it was launched. The weird effect seen on GpB is very similar in nature (size, direction) to that seen by Tajmar. The mechanism suggested by Droscher & Hauser has to be the main contender, as it was an a-priori prediction as commented on in the cover story of New Scientist Jan 26, 3 months before Tajmar's announcement on the ESA homepage. Other explanations are rather ugly a posteriori tack-ons to the Standard model".

The gyroscopes in Gravity Probe B are slightly misaligned, so that a gravitomagnetic field generated by one gyroscope affects the others [32]. The drift of the gyro's axis unit vector tip can be expressed as a function of the misalignment angle ψ (compare Fig. **6**) as

$$\Omega = \tfrac{1}{2} B_g \sin\psi \,, \tag{51}$$

where B_g is the frame-dragging-like field at the location from one gyro caused by the others.

The model overestimates drift rates by a factor of 25-30. The model was developed for the ring geometry in Tajmar experiments, whereas the superconductors in GPB consist of thin spherical shells.

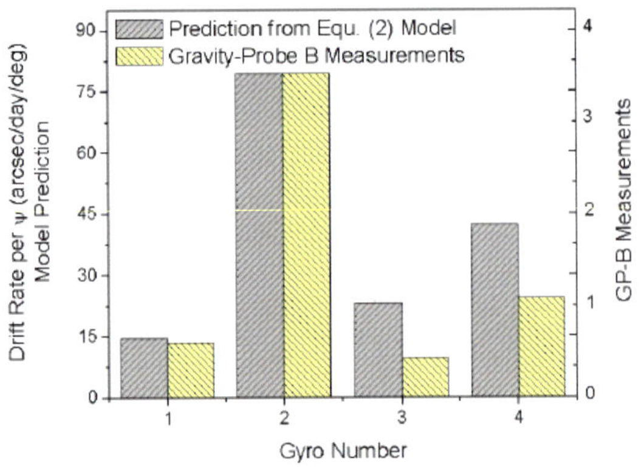

Figure 6: Drift rate of GpB gyroscopes due to B_g field from misalignments [32]. Reproduced with permission from AIP.

12. LIMITS ON GRAVITOELECTROMAGNETISM, ELECTROMAGNETISM CONVERSION

Chiao gives the impression that equation (49), relating gravitoelectric and electric fields, applies to external fields incident on a superconductor. However energy conservation imposes constraints on the conversion of one type of field to another.

The energy density in an electric field is given by

$$U_{em} = \tfrac{1}{2}\varepsilon_0 \vec{E}_{em}^2 \,. \tag{52}$$

The energy density in a gravitoelectric field is

$$U_G = \frac{1}{8\pi G}\vec{E}_G^2 \,. \tag{53}$$

The field energy densities are equal for field strengths

$$E_{em0} = \frac{E_{G0}}{\sqrt{4\pi\varepsilon_0 G}}. \quad \textbf{(54)}$$

Suppose that an external gravitoelectric field much larger than the gravitational attraction between electrons in a Cooper pair is imposed. According to equation (49) the electric field generated is

$$E_{em} = \frac{q}{m\sqrt{4\pi\varepsilon_0 G}} \frac{E_G}{\sqrt{4\pi\varepsilon_0 G}}. \quad \textbf{(55)}$$

Equation (54) is consistent with energy conservation only if the electron has an effective mass

$$m_{eff} = \frac{q}{\sqrt{4\pi\varepsilon_0 G}} = \sqrt{\alpha}M_{Pl} = 1.85\ \mu g. \quad \textbf{(56)}$$

where $\alpha = 1/137$ is the fine structure constant, and $M_{Pl} = 2.17\text{x}10^{-8}$ *kg* is the Planck mass. Alternatively the effective gravitational coupling *G* could be enhanced by 21 orders of magnitude.

Perhaps Chiao could argue that the electron has a much larger effective mass when interacting with a high frequency gravitational field. The theory implies shielding of a static gravitational field, which is inconsistent with observations. Therefore, it is concluded that the theory is incomplete.

13. CONCLUSIONS AND DISCUSSION

The amplitude of the GW radiation at the location of the Chiao detector could have been as large as $|h_{ij}|$ = 10^{-31}, competitive with relic cosmological gravitational waves [29] at 12 *GHz*.

Chiao mentions other properties of YBaCuO, which detract from its performance as an electromagnetic to gravitational wave transducer, such as dissipation by normal electrons. Chiao mentions the use of lead [30], for the transducer, but evidently did not obtain the resources to carry out the experiment. The peculiarities of YBaCuO may help to explain why Chiao did not observe a large signal, using only 77 *K* liquid nitrogen temperature.

The Chiao prediction that a thin film of superconductor acts as a gravitational wave reflector received attention at MIT. The connection with experiments is more convincing than the theory. The theory is incomplete, in the sense that the frequency dependence needs to be addressed to avoid conflicts with energy conservation and observations with static fields. Chiao does not spell out the hidden requirement, that the product $m_e \cdot G$ of gravitational coupling times electron mass must be enhanced by 21 orders of magnitude at high frequency for 100% conversion efficiency. The measured upper limit of $1.8*10^{-5}$ conversion efficiency implies that the product $m_e\ G$ is enhanced by at most 16 orders of magnitude assuming tensor radiation, or around 17 orders of magnitude assuming scalar or vector radiation.

The Chiao experiment using microwaves needs to be extended to come up with a definitive result on whether the em/GW conversions occur. For example the converters should be type I superconductors, to ensure coherence over the entire volume of the superconductor. This coherence is required to generate the desired quadrupole mass current pattern, matching the model depicted in Figs. **4** and **6**, and the assumptions of the Fontana [11] model.

REFERENCES

Note: "arXiv" papers are available from Cornell's electronic preprint depository http://arxiv.org

[1] Ford LH. Gravitational Radiation by Quantum Systems. Ann Phys 1982; 144: 238-248.

[2] Smolin L. The Trouble with Physics. Houghton Mifflin 2007.

[3] Chiao RY, Fitelson WJ and Speliotopoulos AD. Search for quantum transducers between electromagnetic and gravitational radiation: A measurement of an upper linit on the transducer conversion efficiency of yttrium barium copper oxide. arXiv 2003: gr-qc/0304026.
[4] Zavattini E, Zavattini G, Ruoso G, Polacco E, Milotti E, Karuza EM, Gastaldi U, Di Domenico G, Della Valle F, Cimino R, Carusotto S, Cantatore G and Bregant M. Experimental observation of optical rotation generated in vacuum by a magnetic field. Phys Rev Lett 2006; 96: 110406.
[5] Weinberg S. Gravitation and cosmology, principles and applications of the general theory of relativity. Wiley 1972.
[6] Jackson JD. Classical Electrodynamics. John Wiley & Sons 1998.
[7] Low FE. Classical field theory, electromagnetism and gravitation. Wiley 1997.
[8] Finn, LS. Gravitational radiation sources and signatures. arXiv 1999: gr-qc/9903107.
[9] Nieto MM, Macrae KL, Goldman T and Hughes RJ. Phenomenological aspects of new gravitational forces, III. Slowly rotating astronomical bodies. Phys Rev 1987; D 36: 3694.
[10] Modanese G. Local contribution of a quantum condensate to the vacuum energy density. Mod Phys Lett A 2003; 18: 683-690.
[11] Fontana G. High Temperature Superconductors as Quantum Sources of Gravitational Waves: the HTSC GASER. In: Modanese G, Robertson GA, Eds. Gravity-Superconductors Interaction: Theory and Experiment. Bentham 2012; Chapter 3.
[12] Bertschinger E. Gravitation in the weak-field limit. MIT Physics, 2000. Available from http://web.mit.edu/edbert/GR/gr6.pdf
[13] Costa LFO and Herdeiro CAR. Gravitoelectromagnetic analogy based on tidal tensors. Phys Rev 2008; D 78: 024021.
[14] Misner CW, Thorne KS, Wheeler JA. Gravitation. W.H. Freeman 1973. Landau LD, Lifschitz EM. The classical theory of fields. Butterworth Heinman 1996.
[15] Weisberg JM and Taylor JH. The relativistic binary pulsar B1913+16. arXiv 2002: astro-ph/0211217.
[16] Chiao RY. Superconductors as quantum transducers and antennas for gravitational and electromagnetic radiation. arXiv 2002: gr-qc/0204012.
[17] Diebold R and Poirier JA. Vector dominance model comparison of π^+ photoproduction with ρ^0 production by pions. Phys Rev Lett 1968; 20: 1532.
[18] Levin E. An introduction to Pomerons. arXiv 1998: hep-ph/9808486.
[19] Licht AL. Gravitational waves on conductors. arXiv 2004: gr-qc/0402109.
[20] Atwood D, Bar-Shalom S and Soni A. Dijet production at hadron colliders in theories with large extra dimensions. Phys Rev D 2000; 62: 056008.
[21] Gell-Mann M, Levy M. The axial current in beta decay. Nuovo Cimento 1960; 16: 705.
[22] Ripka G. Quarks bound by chiral fields: the quark structure of the vacuum and of light mesons and baryons. Oxford Clarendon Press 1997.
[23] Weise W. Nuclear aspects of chiral symmetry. Nucl Phys 1993; A553: 59c.
[24] Patkos A, Szep Z and Szepfalusy P, Second sheet σ -pole and the threshold enhancement of the spectral function in the scalar-isoscalar meson sector. Phys Rev D 2002; 66: 116004.
[25] Harris J. Evidence for a quark-gluon plasma at RHIC. Available from http://www.ccsem.infn.it/issp2006/docs/harris1.pdf
[26] Chui G. A first: String theory predicts an experimental result. Symmetry Breaking, 2009. Available from http://www.bnl.gov/rhic/stringTheory.asp
[27] Tinkham M. Introduction to superconductivity. McGraw-Hill 1996.
[28] Poole CP, Farach HA and Creswick RJ. Superconductivity. Academic Press 1995.
[29] Grishchuk LP, Discovering relic gravitational waves in Cosmic Microwave Background radiation. arXiv 2007: gr-qc/07073319.
[30] Chiao RY, Minter SJ and Wegter-McNelly K. Laboratory-scale superconducting mirrors for gravitational microwaves. arXiv 2009: gr-qc/09033280.
[31] MIT Technology Review. The Physics arXiv Blog, March 2009. Available from http://www.technologyreview.com/blog/arxiv/23198/
[32] Tajmar M, Plesescu F, Seifert B, Schnitzer R and Vasiljevich I. Investigation of frame-dragging-like signals from spinning superconductors using laser gyroscopes. In: Proc. Of Space Technology and Applications International Forum (STAIF) 2008. Am Inst Phys 2008; CP969: 1080.

CHAPTER 8

Study of Light Interaction with Gravity Impulses and Measurements of the Speed of Gravity Impulses

Evgeny Podkletnov[1,*] and Giovanni Modanese[2,*]

[1]Tampere University, Korkeakoulunkatu 1, FI-33720 Tampere, Finland and [2]Free University of Bolzano, Faculty of Science and Technology, P.zza Università 5, 39100 Bolzano, Italy

Abstract: An attempt has been made in this work to study the scattering of laser light by the gravity-like impulse produced in an impulse gravity generator (IGG) and also an experiment has been conducted in order to determine the propagation speed of the gravity impulse. The light attenuation was found to last between 34 and 48 ns and to increase with voltage, up to a maximum of 7% at 2000 kV. The propagation time of the pulse over a distance of 1211 m was measured recording the response of two identical piezoelectric sensors connected to two synchronized rubidium atomic clocks. The delay was 63±1 ns, corresponding to a propagation speed of 64*c*. The theoretical analysis of these results is not simple and requires a quantum picture. Different targets (ballistic pendulums, photons, piezoelectric sensors) appear to be affected by the IGG beam in different ways, possibly reacting to components of the beam which propagate with different velocities. Accordingly, the superluminal correlation between the two sensors does not necessarily imply superluminal information transmission. Using the microscopic model for the emission given in Chapter 5, we also have estimated the cross-sectional density of virtual gravitons in the beam and we have shown that their propagation velocity can not be fixed by the emission process. The predicted rate of graviton-photon scattering is consistent with the observed laser attenuation.

Keywords: Theories of gravitation, superconductors, high-Tc superconductors, type-II superconductors, superluminal quantum correlations, x-shaped waves, graviton-photon scattering, virtual gravitons, piezoelectric sensors, rubidium atomic clocks, gravity-like fields.

1. INTRODUCTION

This paper represents the continuation of our 2003 work on a pulsed force beam generator based on high-voltage gas discharges through an YBCO electrode [1]. The device was called "Impulse Gravity Generator" because the beam was found to act upon targets of different mass and composition with a force proportional only to their mass, resulting in target velocities of the order of 1 m/s, with weak dependence on the discharge voltage. We also reported in [1] preliminary results on the interaction of the IGG beam with a laser beam. When the angle formed by the two beams was small (≈0.1°) and the interaction region long (57 m) a clear attenuation of the light intensity was observed (7-10%) for a time of the order of 10^{-7} s or less.

The results of the interaction with light have been confirmed by further measurements (Sections 1 and 2). The laser light attenuation was found to last between 34 and 48 ns and to exhibit a steady dependence on the discharge voltage. Moreover, we have been able to measure the propagation time of the pulse over a long distance (1211 m), recording the responses of two identical piezoelectric sensors connected to two synchronized rubidium atomic clocks. Repeated measurements consistently yielded a delay of 63±1 ns, corresponding to a propagation speed of (64±1)*c*. This delay was found to be independent from all the electric parameters of the discharge. The design of the experiment is simple and robust, as it obviously allows to exclude any spurious electromagnetic correlations of the detectors. All measurements were made shortly after the publication of our previous results.

The theoretical interpretation of these results is more recent and has turned out to be quite hard. Phenomena of superluminal propagation have been previously observed with microwaves, and our effect could be the

***Address correspondence to Eveny Podkletnov:** Tampere University, Korkeakoulunkatu 1, FI-33720 Tampere, Finland; E-mail: epodkletnov@hotmail.com **and Giovanni Modanese:** Free University of Bolzano, Faculty of Science and Technology, P.zza Università 5, 39100 Bolzano, Italy; Tel. 0471-017003; E-mail: giovanni.modanese@unibz.it

Giovanni Modanese and Glen A. Robertson (Eds)

analogue on a larger scale. To our knowledge, however, there are no solutions of the classical gravitational equations corresponding to the observed field configuration. In a quantum picture, the force beam consists of virtual particles with a finite lifetime, defined by the Heisenberg time-energy uncertainty principle. The main result of our analysis (Section 5) is that the beam is dispersive and targets of different nature absorb virtual particles with different propagation velocities. The corresponding computations are straightforward when the targets are ballistic pendulums or photons, but the case of the piezoelectric sensors (the only where the propagation velocity has been measured) is much more complex. An idealized representation of the sensors with overdamped harmonic oscillators gives inconsistent results and is clearly not adequate (Section 5.2). Using the microscopic model for the emission given in Chapter 5 [2-5], we also have estimated the cross-sectional density of virtual gravitons in the beam and we have shown that their propagation velocity can not be fixed by the emission process (Section 4). For completeness, in Section 3 we briefly recall the results on superluminal propagation of microwaves and the debate on (the absence of) causality violation for this kind of "pseudo-signals". Section 6 comprises our conclusions and hints of possible alternative interpretations. Note that for convenience of exposition the sections of this paper do not follow a strictly logical order; nevertheless, the logical connections should be clear from the context. For instance, the cross-sectional density of virtual gravitons is not needed for general considerations on the propagation velocity and is actually a consequence of a particular microscopic model; however, it is placed in Section 4 because it is necessary to assess the photon-graviton scattering in Section 5.1 and the absorption rate of an harmonic oscillator in Section 5.2.

2. EXPERIMENTAL

In order to study the interaction of a gravity impulse with light we used the experimental setup that is shown schematically in Fig. **1**. The gravity projection area has the diameter of 100 mm or 4 inches and the direction of propagation is marked with an arrow. Laser beams were directed at a certain small angle to the projection of the impulse in order to increase the interaction time as placing the laser at 90 degrees to the gravity beam produces effects below the detection level of the sensor. The interaction occurred along the distance of 57 meters. A ruby laser of 694 nm with the power of 930 mW was applied and the spot size on the sensor was about 4 mm in diameter. Also a blue laser with a wavelength of 473 nm and the power of 450 mW was used and the spot size diameter was equal to 3 mm. A high resolution optical sensor was used to measure possible variations of the laser light intensity. In order to achieve a good temporal resolution we used a Femtosecond Optically Gated fluorescence kinetic measurement system FOG 100 (best resolution 100 fs).

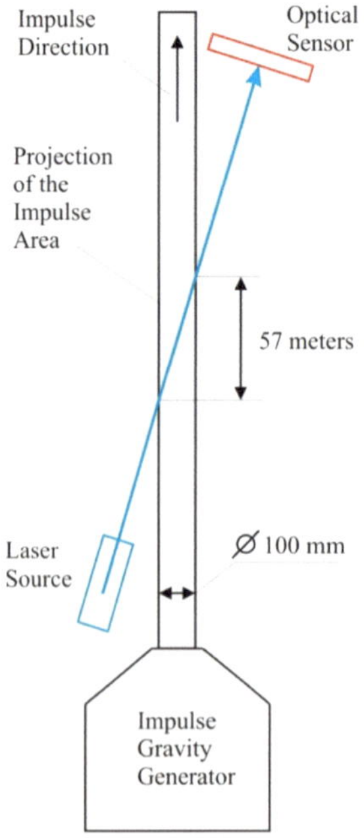

Figure 1: Experimental setup for the investigation of the interaction of the gravity impulse with light. Details are not to scale.

In order to study the propagation speed of the impulse two sensors were used. One of them was placed close to the gravity generator and another one at the distance of 1211 m away (measured with a laser meter). Piezoelectric thin film sensors based on molecular imprinted polymers operating at 450 MHz were used to register the impact of the impulse. The pressure shock of the impulse changed the frequency response and the moment of the change was registered by the matching transducer (an RF mixer using a 450 MHz carrier frequency). The output of each sensor was connected *via* a transducer to a portable rubidium atomic clock. The two atomic clocks used for the setup were synchronized before the measurements and all measurements were performed in a time period of 72 hours.

The main task with the sensors was to choose two units that gave equal delay of the signal when subjected to a shock wave. Only the time of the start of the frequency change was registered and thus the response time of the sensor was within nanoseconds range. After the discharge of the IGG the impulse activated the first gage and then a second gage; the delay was registered by the atomic clocks. To ensure that the delay is not affected by the piezoelectric gages the measurements have also been repeated after the sensors were interchanged. The geometry of the setup is shown in Fig. **2**. All measurements have been carried out using the same type of superconducting emitters with diameter of 100 mm and thickness of 12 mm. The measurements have been repeated with two similar emitters using discharge voltage from 750 kV to 2000 kV.

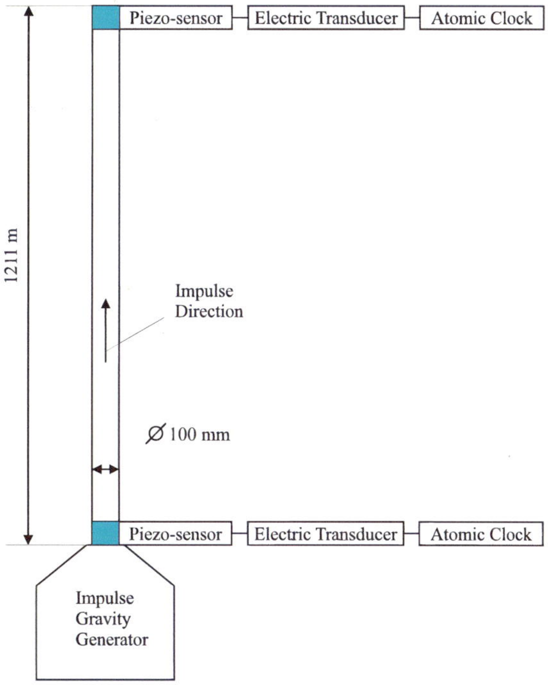

Figure 2: Experimental setup for the measurements of the speed of the gravity impulse. Details are not to scale.

3. RESULTS

In the first experiment it was noticed that during the discharge of the impulse generator the intensity of the signal from the optical sensor decreased from 2.8% to 7.5%, depending on the discharge voltage, and then returned quickly to the baseline. Our measuring system indicated that the duration time of this process varied from 34 to 48 ns. The dependence of the intensity of the laser beam measured by the optical sensor on the discharge voltage is listed in Fig. **3**. The duration of the pulse depends mainly on the voltage but also on the current; it also depends to some extent on the distance between the emitter and the target electrode. Practically the same results have been observed for both types of lasers and no difference within the error of the measuring system have been found. All measurements have been reproduced several times (usually 8-10) for each discharge voltage and both emitters, the deviation of the measured values was within 2-3%. The decrease of the light intensity was almost the same for both emitters, as shown in Fig. **3**.

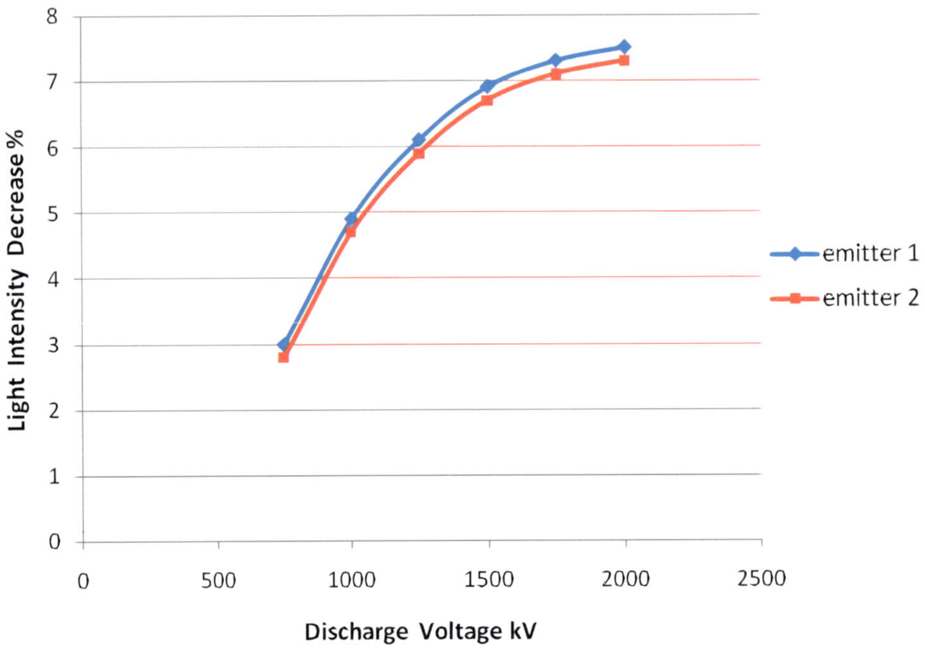

Figure 3: Dependence of the light intensity decrease on the IGG discharge voltage.

In the second experiment the measured value of the delay between the two sensors was 63±1 ns. This means that the gravity impulse propagated about 64±1 times faster than the speed of light. No delay difference was observed between the two emitters and the impulse speed did not depend on the value of the discharge voltage. Interchanging of the piezo-electric sensors also had negligible effect on the measurements. The magnetic field on the emitter, the voltage risetime, the pulse current all had negligible effects on the measurements. From multiple runs we know that the force of the beam depends on the voltage and its risetime, but the pulse speed is independent from all the electric parameters, the size of the emitter, the thickness of layers in the emitter and the distance between the electrodes.

4. POSSIBLY RELATED PHENOMENA REPORTED IN THE LITERATURE

Phenomena of electromagnetic wave propagation with superluminal group velocity have been observed in several laboratories in the last years and can be grouped in two categories: evanescent waves and Bessel beams of so-called "X-shaped waves" (see [7] for the most famous experiment and for general references). The peculiar features of these waves do not allow the transmission of true signals and information with superluminal speed, so the causality principle is not endangered by their existence. (See further discussion below). The phenomenon described in this paper differs from those cited above under several aspects. Let us first point out the following two differences: (a) The propagation distance is much larger, and still yet compatible with the uncertainty principle applied to the particles of the beam (Section 4.1). (b) There are no known solutions of the classical field equations corresponding, for shape and amplitude, to the observed beam. In fact, the theoretical model set out in Chapter 5 of this eBook [2] and adopted here describes the IGG beam as consisting of virtual gravitons generated in a stimulated emission process "pumped" by the superconductor.

Reference [7] demonstrated experimentally the superluminal propagation of localized microwaves over a distance of 1 m or more. A carrier signal of frequency 8.6 GHz was modulated with rectangular pulses. The waves were first sent by a horn antenna (launcher) on a circular slit with mean diameter d placed in the focal plane of a circular mirror. In this way, a so-called Bessel beam was generated by interference. The field of the beam can be considered as formed by the superposition of pairs of X-shaped plane waves. These move along the axis of the mirror with velocity approximately 5-7% larger than the light velocity, for a slit diameter d of 20 cm, or up to 25% for d=10 cm. A similar experiment was performed in the optical range [8], but a clear observation of superluminal propagation was impossible in that case. Superluminal effects for evanescent waves were demonstrated in tunnelling experiments in both the optical domain and microwaves range; these effects can be revealed, however, only over short distances, typically a few centimetres for microwaves (the most favourable case).

Several other papers [7, 9] discuss the issue of signal transmission. The question is, if superluminal propagation effects can be used to convey information at superluminal speed, and the answer is generally that they can't, though it also depends on what is exactly meant by signal. A typical argument is that waves with superluminal group velocity are always accompanied by a "precursor wave" propagating at light speed. Some authors speculated, however, that in certain cases the superluminal wave could overtake the precursor. Other authors argued that the violation of causality by true superluminal signals is only apparent and could be avoided through the so-called Feynman-Stueckelberg reinterpretation principle. We do not intend to enter in this debate. We just observe that the phenomenon of superluminal propagation reported in this paper is on one hand impressing, for its long range and large velocity; on the other hand, it can not be easily controlled, and it is presently impossible to say if the pulse can be modulated. The generation of the force pulse is a complex phenomenon, occurring in a single-shot gas discharge whose timing is only approximately triggered from the outside. Moreover, the discharge involves a superconductor, whose macroscopic wavefunction obeys an evolution equation with some characteristic time, possibly exceeding the propagation time of the pulse.

5. FIELD STRENGTH IN THE BEAM *VS.* MOMENTUM FLOW AND CROSS-SECTIONAL DENSITY OF VIRTUAL PARTICLES

In this Section we summarize the observed features of the IGG beam using the concepts of field strength and flux of virtual particles which are emitted from the superconductor and carry energy and momentum to the targets. Here and in the following sections the suffix "g" (E_g, p_g...) always means "graviton", while "t" is for "target".

The first crucial observation is that the velocity of ballistic pendulums exposed to the beam is independent from their mass and composition. (This will hold in principle up to a certain "maximum mass" - see below). Let us denote with p_t the momentum of a target pendulum of mass m after the action of the beam: $p_t=mv_t$, where v_t is the target velocity. If the beam acts for a time Δt, it exerts an average force $F=p_t/\Delta t$, with acceleration $g=F/m=v_t/\Delta t$. The average field strength in the beam in the time Δt is by definition equal to g. We want to relate this field strength to the number of virtual gravitons absorbed by the target. We outline here the main argument, leaving some details for the next subsections.

We know that the average energy of the virtual gravitons emitted by the superconductor is of the order of $E_g \approx 10^{-27}$ J (Section 4.1). Furthermore, the pendulum data support the hypothesis of an elastic absorption in the targets (Section 4.2) and give an energy-momentum ratio $E_t/p_t \approx 1$ m/s. This implies that the momentum of a single graviton is $p_g \approx 10^{-27}$ kg m/s. The number of gravitons absorbed by a target is $N=p_t/p_g$. From the relation $p_t=mv_t$ we see that, being v_t constant, N is proportional to the mass of the target. When $v_t=1$ m/s, each gram of mass in the target absorbs ca. 10^{24} gravitons. The same proportionality between the mass and the number of absorbed virtual gravitons is respected in the usual static interaction of two masses m_1 and m_2, described by the Newton force law (see [2], Section 4). In that case, the interpretation of the gravitostatic force as due to an exchange of virtual gravitons implies a gravitons flux proportional to m_1m_2, with unitary probability of emission and absorption. A similar rule applies, of course, to the electrostatic force, with a virtual photon flux proportional to the charges q_1q_2.

This absorption rate proportional to the target mass, and not to its cross-section, establishes a clear distinction between our "virtual beam" and beams of real particles. Nevertheless, the virtual gravitons of the IGG beam are generated in the superconducting emitter through elementary virtual processes which have a well-defined maximum density per surface element. Let us denote by ΔV the voltage on the emitter during the discharge and by I the total current crossing the emitter (this is almost entirely supercurrent, but also comprises a small percentage of normal current, less than 0.1% [3]). The maximum number of virtual particles with energy E_g emitted is given by $I\Delta V\Delta t/E_g$. The product $I\Delta V\Delta t$ has been estimated through a microscopic model of the superconductor based on its intrinsic Josephson junctions and is of the order of a few milli-Joules. (A direct measurement of the emitter voltage is very difficult, for several practical reasons [4, 5]). Let us suppose, in order to fix the ideas, that the maximum energy emitted from each square centimeter of the emitter is 10 mJ, corresponding to a maximum graviton flux $\Phi=10^{25}$ gravitons/cm^2. We shall call this quantity "cross-sectional density of virtual gravitons" in the beam. Being the beam strongly directional, this is also the cross-sectional density of gravitons which can hit the target.

This means, for instance, that cubic or spherical targets with linear size about 1 cm and mass density below 10 g/cm^3 will respect the relation $p_t=mv_t$, with constant v_t. Such targets can be regarded as "good test masses", such that the field strength they measure is independent from their mass. There exists however, in principle, a "maximum test mass" for a given beam. If the beam has a cross-sectional density Φ of 10^{25} virtual gravitons/cm^2, then cubic or spherical targets with linear size of 1 cm and mass density larger than 10 g/cm^3 will not receive enough momentum to reach the constant v_t measured with the test masses. The same will happen with targets having density 10 g/cm^3 but size larger than 1 cm, because the cross-section does not increase in proportion to the volume. All the figures above are just an indication and must be corrected for obvious geometrical factors. This "maximum test mass" effect could not yet be confirmed experimentally, since the pendulums employed were well below the limit. It is probably related, however, to the observed absence of absorption of the beam by walls and other large rigid obstacles.

5.1. Energy and Propagation Range of the Virtual Gravitons

In a quantum picture, the beam is composed of virtual particles with energies belonging to a wide band around 10^{-27} J. The particles are emitted in a process with a continuous frequency spectrum, pumped by a supercurrent pulse with main components of the order of 10 MHz. The time-energy uncertainty relation $\Delta E \Delta t \geq \hbar$, when applied to virtual particles of energy ΔE, is usually interpreted as giving the largest possible lifetime of the particles (see for instance [6]). With $\Delta E/\hbar \approx 10^7$ Hz, a propagation velocity $v=64c$ implies a propagation range of the order of 10^3 m. This is compatible with the observations and suggests that the present distance of 1211 m could perhaps be increased, but not by magnitude orders (at least for pulses of this velocity; our analysis shows that the beam is dispersive and its velocity differs for different E/p ratios). Note that the indetermination relation also applies to the electromagnetic X-shaped superluminal microwaves (Section 3). In that case the frequency is $\approx 10^4$ times larger, the velocity only slightly larger than c, and the range is of the order of 1 m. An improvement of their range through reduction of the frequency is clearly impossible, because electromagnetic waves with frequency, say, 10 MHz are impossible to focalize in Bessel beams. The real puzzle, actually, is how the IGG beam can be so focalized. We think the reason for this lies in its generation through a stimulated emission process [2].

5.2. Elastic Absorption in the Target and E/p Balance

The virtual particles of our beam carry, for the same energy, much more momentum than photons, gravitons or other "on-shell" particles with zero mass ($E/p=c$ in that case). This small E/p ratio emerged clearly from the experimental data already in 2003 [1]: the action of the beam on ballistic pendulums showed an energy-momentum ratio in the beam of the order of $E_g/p_g \approx 0.25$ m/s. This was deduced from the free particle relation

$$\frac{E_t}{p_t} = \frac{\frac{1}{2} m v_t^2}{m v_t} = \frac{1}{2} v_t \tag{5.1}$$

and from the observed pendulum velocities of the order of 0.5 m/s (depending on the voltage; a typical figure was an elongation of 0.1 m, in a pendulum of length 0.8 m, with a discharge voltage of 1500 kV). Observations also showed that no heat was dissipated in the pendulums, so the beam absorption could be regarded as an elastic process, and we have

$$\frac{E_t}{p_t} = \frac{E_g}{p_g} \tag{5.2}$$

This does not exclude the possible occurrence of dissipative processes after the absorption in other targets, like a dampened harmonic oscillator (Section 5). Note that if the momentum of the targets originated from ordinary radiation pressure, the associated energy would be of the order of $mv_t c \approx 10^6$ J, while we have seen that the superconducting emitter can produce no more than 10 mJ/cm^2.

There are still two points to explain: (a) What determines the magnitude order of v_t? Smaller values were still compatible with the energetic balance and would only imply a larger "maximum test mass". (b) What causes the slight dependence of v_t on the discharge voltage? For Point (a), we refer to Section 4 of [2], where we argued that $v_t \approx 1$ m/s is the velocity resulting from the elementary absorption of single virtual gravitons by single nucleons. Concerning Point (b), we believe that larger voltages give a shorter pulse risetime, and thus higher pumping frequency and higher graviton energy. This can be checked with a detailed circuit analysis [5].

5.3. Relation Between the Propagation Velocity and the Energy-Momentum Ratio of the Virtual Gravitons

Extensions of Special Relativity exist, compatible with the relativity principle, which provide a complete framework for the kinematics and dynamics of superluminal particles [9]. We only need here a few basic relations. The relation between energy, momentum and mass of a particle is valid also for $v>c$, and implies that the mass is imaginary. The mass of the particles composing the IGG beam is imaginary, because being their energy-momentum ratio E/p of the order of 1 m/s, one has

$$m^2c^4 = E^2 - p^2c^2 < 0 \quad \textbf{(5.3)}$$

Let us define a variable $\alpha = p/E$, in order to allow for different values of the E/p ratio. We shall consider values of α such that $1/\alpha < c$, *i.e.,* the mass is imaginary and the radiation pressure is larger than light radiation pressure for the same energy. The relation between E, m and v also keeps valid for superluminal particles, namely

$$E = \frac{mc^2}{\sqrt{1-\beta^2}}, \qquad \beta = \frac{v}{c} \quad \textbf{(5.4)}$$

Re-writing (5.3) with $p=\alpha E$ and comparing with the square of (5.4), we obtain

$$1-\beta^2 = 1-\alpha^2c^2 \quad \Rightarrow \quad \alpha = \frac{v}{c^2} \quad \textbf{(5.5)}$$

With the observed value $v=64c$, we find in our case $E_g/p_g \approx 4.7 \cdot 10^6$ m/s. Now, an important task of any theoretical model is to explain the $v=64c$ data. Supposing that equation (5.5) is valid, there are two possibilities. A first possibility is that the energy-momentum ratio in the beam is fixed by the emission process. Here we discuss this possibility using our microscopic model for the emission [2], and reach a negative answer: the emission process can not fix α. In the next section we shall discuss the second possibility, namely that the propagation velocity is fixed "a posteriori", through equation (5.2), by the ratio E_t/p_t of the energy and momentum absorbed in the target.

According to our microscopic model, the virtual gravitons are generated in an elementary process in which gravitational zero-modes (pairs of virtual masses) decay from an excited antisymmetric state ψ^- to the symmetric ground state ψ^+. The energy gap between the two states depends on the virtual masses and has a continuous spectrum. The excitation of the states ψ^+ occurs in the interaction of the gravitational vacuum with the quantum condensate of the pairs in a superconductor crossed by a supercurrent pulse. Since the pulse has a broad frequency band, the excited zero-modes have a corresponding energy spread. The momentum of the emitted gravitons is balanced by the recoil of the zero-modes. It can be easily shown that even if the energy was sharply fixed by some resonance in the emitter or in the detector, the energy-momentum ratio would not be fixed. The conservation equations give

$$\begin{cases} Mv_r^2 + E_g = \Delta E \\ 2Mv_r - p_g = 0 \end{cases} \quad (5.6)$$

where ΔE is the energy gap, v_r is the recoil velocity of the zero-mode and $2M\approx 10^{-13}$ kg is the zero-mode mass. After replacing $p_g=\alpha E_g$, the system (5.6) leads to the equation

$$\frac{1}{M\alpha^2}E_g^2 + E_g - \Delta E = 0\,, \quad (5.7)$$

which has a positive solution $E_g\approx\Delta E$ independently from α. Furthermore, the recoil velocity v_r turns out to be always non-relativistic. This means that the recoil of the zero-modes can always ensure conservation of momentum, independently from the value of the energy-momentum ratio. Therefore the propagation velocity can not be fixed by the emission process. The situation is summarized in Fig. **4**.

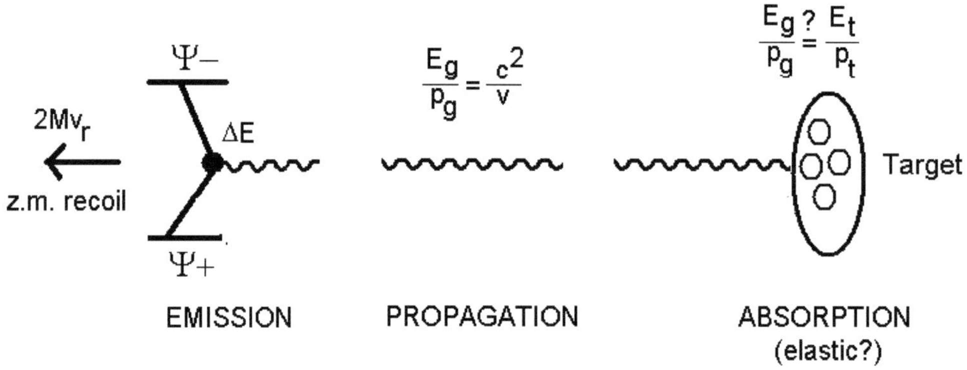

Figure 4: What sets the velocity of the virtual particles of the IGG beam at the observed value of 64*c*? Is it the emission or the absorption process? Our model for the emission allows an arbitrary energy-momentum ratio E_g/p_g, even if the gap ΔE or the graviton energy E_g are sharply fixed. Therefore the velocity must be fixed "a posteriori" by the absorption process. This process is elastic when the target is a pendulum or a laser beam, but inelastic in the case of a piezoelectric sensor.

6. RELATION BETWEEN THE E_G/P_G RATIO AND THE TARGET

We have seen that the emission and propagation of the IGG beam should be described as a virtual quantum process, ending with elastic absorption in the targets. It follows that the absorption process can determine the energy-momentum ratio of the beam particles, and consequently their propagation velocity. This would be inconceivable for a beam of real particles. The possible targets studied experimentally and theoretically are three.

a) Ballistic pendulums. In this case $E_g/p_g=E_t/p_t=v_t/2$, where v_t is the velocity of the pendulum bob after exposure to the beam. Experimentally one observes $v_t\approx$ 1 m/s and equation (5.5) gives a very large propagation velocity. A direct measurement of the propagation velocity was not done in this case, and is probably impossible, because of the long response time of the pendulums.

b) The photons of a laser beam. Theoretically, the photons can absorb elastically the virtual gravitons only if the latter have $E_g/p_g=c$ and thus $v=c$. Neither direct measurements of the propagation velocity were made in this case, nor simultaneous observations of the effect of the IGG beam on the laser and the pendulums. The latter situation is conceptually very interesting, because it seems to require, in order to produce an effect on both targets, the presence in the beam of two kinds of virtual gravitons with different velocities and energy-momentum ratios.

c) The piezoelectric sensor. This is the only case in which the propagation velocity (and thus indirectly the E/p ratio) have been measured, and turn out to be very well defined. The sensor is based on a thin PVDF film. The ratio between the absorbed energy and momentum is determined by the specific transfer function of the transducer and by the internal dissipation rate. In the following, we shall analyse for illustration purposes the idealized case of a detector made by an harmonic oscillator with proper frequency of the order of the reciprocal pulse risetime.

Case (a) has already been treated. Cases (b) and (c) are considered in the following Subsections.

6.1. Interaction with a Laser Beam

In this Subsection we give a theoretical interpretation of the observed effect of the IGG beam on a laser beam. First we discuss which kind of interaction may occur, then we estimate the strength of this interaction.

The effect of the IGG on a laser beam can be understood neither within a classical model (gravitational field plus geometric optics), nor within a semi-classical model (gravitational field plus photons). The predicted effect would in both cases be far too small. We have discussed this in [1], where we gave an estimate of the classical curvature associated with the IGG beam; the field strength and its spatial derivatives were deduced in that estimate from the observed effect of the beam on ballistic pendulums. Also considered in [1] were possible indirect effects of the IGG on the laser beam, like those of air turbulence. We concluded that air turbulence could not account for the observed quick variations in the laser intensity. Actually, being the laser insensitive to any electromagnetic interference, the laser detection of the IGG beam is one of the safest possible demonstrations.

Therefore we make the hypothesis that the observed drop in laser light intensity is due to a quantum scattering between the virtual gravitons of the IGG beam and the photons. Repeated collisions expel a part of the photons from the beam. We suppose that in each collision the virtual graviton is completely absorbed by the photon; the process is elastic and both energy and momentum are conserved. This total absorption condition is necessary for a process involving a virtual particle. Only under this condition can the probability of the process be taken equal to one, like in the interaction of two masses through virtual gravitons [2] (Section 4). The graviton/photon scattering cross section for real particles is known to be exceedingly small [10].

Conservation of the E/p ratio in the absorption implies in turn that the gravitons involved in this process (unlike those collected by the piezoelectric sensors) must have standard ratio $E/p=c$ and propagate with light velocity. From the laser attenuation data, we know that the interaction of the IGG beam with the laser beam lasts on the average about 40 ns; this means that the IGG beam is in fact a bunch extended on the average for a length of 12 m. The two lasers employed have a power of the order of 1 W and the photon energy is of the order of 10^{-19} J. Therefore in the interaction time $\Delta t \approx 40$ ns the photons involved are $N_\gamma = \Delta t \cdot P / E_\gamma \approx 10^{11}$. The density of virtual gravitons in the IGG beam is of the order of 10^{25} per square centimetre (Section 4). It is not clear which effective cross-section should be taken in order to compute the interaction of the two beams. The cross-section of the laser beam is of the order of 0.1 cm^2, but the IGG beam, at the long graviton wavelengths involved in the photon interaction ($E_g \approx 10^{-27}$, $v=c \Rightarrow \lambda_g \approx 10^3$ m) could suffer considerable diffraction. (For comparison, the wavelength of the virtual gravitons absorbed by the ballistic pendulums is of the order of 10^{-7} m, and the beam appears to be well collimated in that case).

Disregarding diffraction losses for the moment, take a cross-section of 0.1 cm^2 and so a total number of virtual gravitons of 10^{24}. This gives a ratio of 10^{13} gravitons per photon. The momentum of a graviton, however, is about 10^7 times smaller than the photon momentum (because $E/p=c$ for both, but the photon energy is 10^7 times larger). This means that a photon must absorb several gravitons, before its momentum changes so much that is misses the optical detector. A simple vector model, shown in Fig. **5**, allows to compute the maximum deviation of a photon in a single graviton absorption. The angle ϕ between the beams corresponds to a deviation of ca. 10 cm in 57 m, *i.e.*, $\phi \approx 0.0018$ rad. The resulting deviation θ of the photon is $\theta \approx 10^{-7}\phi$.

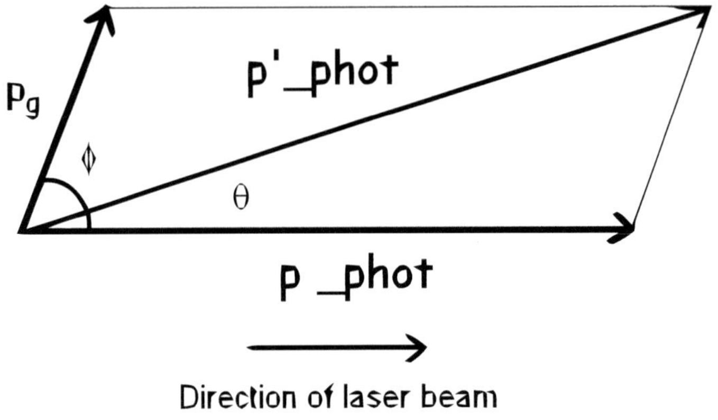

Figure 5: Deviation of an optical photon after absorption of a virtual graviton in the experimental configuration of Fig. **1**. The vectors are not to scale; in reality $p_\gamma \approx 10^7 p_g$, the angle ϕ between $\mathbf{p}_g$ and $\mathbf{p}_\gamma$ is 0.0018 rad, the deviation angle is $\theta \approx 10^{-7}\phi$.

Therefore if the effects of subsequent absorptions are summed, the ratio of ca. 10^{13} gravitons for each photon would be more than enough to ensure the expulsion of all photons from the laser beam (the minimum angular deviation for the expulsion, given by the ratio between detector diameter and distance, is about 20 times smaller than ϕ). But since the observed attenuation in the laser intensity at the detector is at most 7%, some other factors must be at play, which reduce the efficiency of the process. Such factors could be

- Diffraction of the virtual gravitons beam, with ensuing drastic reduction of the density of gravitons per cm^2.

- A smaller emission rate of gravitons with ratio $E/p=c$, in comparison to those with ratio $E/p \approx 1$ m/s detected by the ballistic pendulums. We recall that in the microscopic model the wavelength of the gravitons affects the spontaneous emission coefficient A.

6.2. Energy-momentum balance in the action of an impulsive force on a harmonic oscillator

Consider a damped harmonic oscillator with mass m, proper frequency $\omega = \sqrt{k/m}$ and dampening coefficient $\gamma > 1$, initially at rest, subjected to an external impulsive force $F(t)$. Let us suppose that there is a single pulse, with short risetime τ (comparable to the oscillator's period T), followed by a slower return to zero. It is quite intuitive that the rise of the pulse will cause the mass to oscillate, passing to it a momentum p_t and an energy E_t, which will then be dissipated by dampening. We are interested into the ratio E_t/p_t and into its dependence on the other variables. The dynamics of such a system is well known in general form, but usually this ratio is not of particular interest. We would like to find a manageable expression tailored to our specific case.

The differential equation of the system is

$$\ddot{x} + 2\gamma\omega\dot{x} + \omega^2 x = \frac{F(t)}{m} \qquad \textbf{(5.1)}$$

One can easily write the general solution as a Fourier transform and plot numerical solutions which give an idea of the dynamical evolution. In order to compute E_t in general form, we write the integral

$$E_t = \int_0^{t_f} F(t)\dot{x}(t)dt \qquad \textbf{(5.2)}$$

where t_f is the instant when $F(t)$ has returned to zero. Integrating by parts and replacing $x(t)$ through the equation of motion (5.1), we find

$$E_t = \left[x(t)F(t)\right]_0^{t_f} - \int_0^{t_f} x(t)\dot{F}(t)dt = -\int_0^{t_f}\left[\frac{F(t)}{m\omega^2} - \frac{\ddot{x}(t)}{\omega^2} - \frac{2\gamma\dot{x}}{\omega}\right]\dot{F}(t)dt \qquad \textbf{(5.3)}$$

After a further integration by parts, and noting that all finite terms evaluated at t=0 or t=t_f are zero, we finally obtain

$$\begin{aligned} E_t &= -\frac{1}{m\omega^2}\int_0^{t_f} F(t)\dot{F}(t)dt - \frac{1}{\omega^2}\int_0^{t_f}\dot{x}(t)\ddot{F}(t)dt - \frac{2\gamma}{\omega}\int_0^{t_f} x(t)\ddot{F}(t)dt \\ &= -\frac{1}{2m\omega^2}\left[\left(F(t)\right)^2\right]_0^{t_f} - \frac{1}{\omega^2}\int_0^{t_f}\dot{x}(t)\ddot{F}(t)dt - \frac{2\gamma}{\omega}\int_0^{t_f} x(t)\ddot{F}(t)dt \\ &\approx -\frac{1}{\omega^2}\ddot{F}\int_0^{t_f}\dot{x}(t)dt - \frac{2\gamma}{\omega}\ddot{F}\int_0^{t_f} x(t)dt \end{aligned} \qquad \textbf{(5.4)}$$

In the last step we have supposed that the second derivative of $F(t)$ is approximately constant, which is true, for instance, in the simple case of a parabolic increase followed by linear decrease (see Fig. **6**). Note that the integral $\int_0^{t_f}\dot{x}(t)dt = x\left(t_f\right) - x\left(0\right)$ is null in this approximation, if the oscillator is overdamped and $t_f >> \tau$.

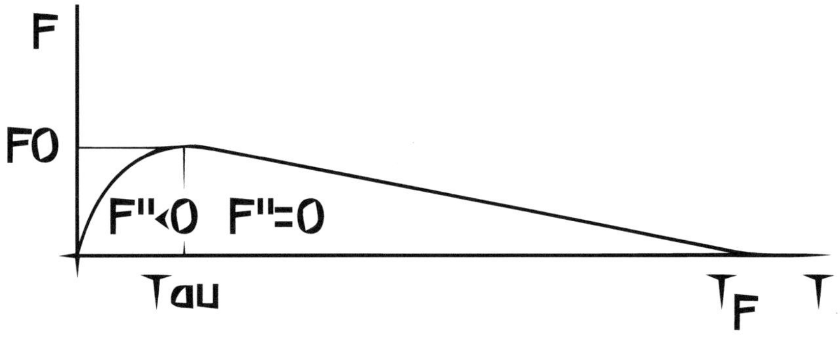

Figure 6: Simple example of a force pulse with risetime τ, total duration t_f and piecewise constant value of $\ddot{F}(t)$.

In the computation of the momentum transferred from the beam to the mass m we need to take into account that a part of this momentum is passed to the rigid body holding the spring. Suppose that the momentum p of the beam has positive direction (see Fig. **7**)

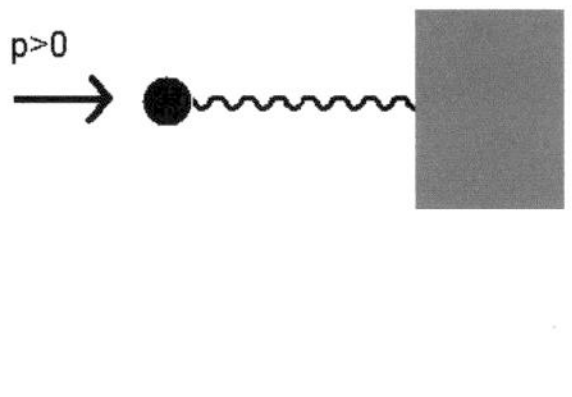

Figure 7: Momentum transferred from the beam to a target harmonic oscillator. Part of the momentum is passed to the rigid body supporting the oscillator (see equation 6.5).

In a certain time interval dt, the momentum dp_t lost by the beam is equal to the variation dp of the momentum of the target mass m plus the momentum $-F_{spring}dt$ transferred to the rigid body by reaction force. We have

$$\frac{dp}{dt} = m\ddot{x} + kx \tag{5.5}$$

and from the equation of motion

$$\frac{dp_t}{dt} = F(t) - m\gamma\omega\dot{x} \tag{5.6}$$

The total transferred momentum is therefore

$$p_t = \int_0^{t_f} \left(F(t) - m\gamma\omega\dot{x} \right) dt \tag{5.7}$$

The integral of $m\gamma\omega\dot{x}$, however, gives a contribution $m\gamma\omega\left(x(t_f) - x(0)\right)$ and this vanishes for an overdamped oscillation if $t_f >> T$.

In conclusion, the ratio between energy and momentum absorbed by the oscillator is

$$\frac{E_t}{p_t} = \frac{-\frac{2\gamma}{\omega}\ddot{F}\int_0^{t_f} x(t)dt}{\int_0^{t_f} F(t)dt} \tag{5.8}$$

This includes the simplifications that $\ddot{F}(t)$ is constant and the pulse is overdamped.

A numerical estimate of this expression is not trivial. Denote by F_0 the maximum value of the force pulse and by x_0 the maximum displacement of the mass in the overdamped oscillation. Since γ is of magnitude order 1 and $\ddot{F} \approx -F_0\omega^2$ (because $\omega=2\pi/T$ and $T\approx\tau$), one has $E_t \approx F_0 x_0$. It is known that $x_0 \approx F_0/(m\omega^2)$ (from the general solution for the motion of an overdamped oscillator), whence $E_t \approx F_0^2/(m\omega^2)$. The denominator in equation (5.8) can be estimated as $p_t \approx F_0 t_f$, so the ratio is

$$\frac{E_t}{p_t} \approx \frac{F_0}{m\omega^2 t_f} \tag{5.9}$$

Here ω and t_f are well known experimentally: $\omega/2\pi \approx 500$ MHz, $t_f \approx 40$ ns. The mass m and force F_0, on the contrary, can be only guessed. Suppose first that the piezoelectric film has a surface of the order of 0.1 cm^2 and a thickness of 20 μm (see Table **1**). With reasonable assumptions on the density, we find $m \approx 10^{-9}$ kg. With $E_g/p_g \approx 4.7 \cdot 10^6$ m/s and with the cross-sectional gravitons density given in Section 4 we find that the detector receives a momentum $p_t \approx 10^{-17}$ m·kg/s, whence $F_0 \approx p_t/t_f \approx 10^{-9}$ N. From (5.9) we then obtain a target ratio $E_t/p_t \approx 10^{-9}$ m/s, which is much smaller than the supposed graviton ratio E_g/p_g. Possible adjustments of m and F_0 do not alter this conclusion.

6.3. Electrostatic Energy of the Piezo Effects of the Shock Wave in Air

Is it possible to match the graviton ratio E_g/p_g with the target ratio E_t/p_t by considering inelastic energy absorption in the piezoelectric film in the form of electrostatic energy? Given the short duration of the pulse, we are not really in an electrostatic situation. Since, however, the energy absorption from the beam and the energy-momentum balance also occur in that short time interval, the electrostatic energy $W=(1/2)C\Delta V^2$ of the film gives at least a useful approximation. In order to estimate the film capacitance C we need to know its thickness. This is usually related to its resonance frequency (first harmonic of the stationary ultrasonic wave in the film). A 500 MHz frequency corresponds to a thickness of less than 1 μm.

Taking a surface of the order of 0.1 cm^2, it follows that the film capacitance is at least 10^{-9} F. If the film voltage ΔV was of the order of 0.1-1 V, the resulting energy ($W \approx 10^{-10}$ J) would give a ratio $E_t/p_t \approx 10^7$ m/s, compatible with the ratio E_g/p_g and the propagation velocity 64c. An output voltage of 0.1-1 V is quite common in piezoelectric ultrasonic detectors [13]. In our case, however, such a voltage is incompatible with the other data. Namely, there are two possibilities:

1. The force on the sensors is directly caused by the virtual gravitons: $F_0 \approx 10^{-9}$ N (Section 6.2). It is then straightforward to check, using the piezoelectric constant of the film and its Young modulus (see Table **1**), that the voltage output is smaller than 1 V, by several magnitude orders. Consequently, the energy E_t is too small.

2. The force on the sensors is due to the pressure shock in air caused by the IGG beam. Since the air molecules behave as free targets, in order to be coherent with our model and with the pendulum data we must suppose that the air molecules absorb virtual gravitons with ratio $E_g/p_g \approx 1$ m/s and huge propagation velocity $v \approx c^2$. The observed delay of 63 ns between the two detectors would then be explained as due to some spurious factor, like for instance a systematic difference in air pressure or temperature around the sensors, affecting the propagation of the shock wave. Or, alternatively, attenuation of the shock wave with distance could cause a small systematic delay in its action on the far sensor. Such systematic factors are independent from the electrical parameters of the discharge and from the features of the emitter. This could explain why the observed delay is independent from those parameters and features.

Table 1: Typical properties of PVDF piezo film [11, 12].

Parameter	**Value**	**Units**
Thickness	9 -110	μm
Piezo strain constant d_{33}	-33	10^{-12} V/m or C/N
Young's modulus	2-4	10^9 N/m^2
Relative permittivity	12-13	
Mass density	1.78	10^3 kg/m^3
Speed of sound	2.2	10^3 m/s

7. CONCLUDING REMARKS

The laser observations are compatible with a quantum picture of the IGG beam as composed of virtual particles with definite energy and momentum (Section 6.1). The explanation of the propagation velocity measured with the piezoelectric detectors, however, is more complex. Our analysis has considered several possibilities, and none appears to be entirely convincing at present.

The ratio between energy and momentum absorbed in the sensors can not be measured directly. It is therefore hard to test experimentally the relation (5.5) between the propagation speed and the E/p ratio of the virtual particles in the beam. As a first example of theoretical modelling of the sensor, we have schematized it with an overdamped harmonic oscillator (Section 6.2). This calculation illustrates well the many factors at play, but the result is that for an harmonic oscillator the target ratio E_t/p_t is totally incompatible with the beam ratio E_g/p_g. Considering the inelastic absorption of electrostatic energy improves the situation (Section 6.3), but still the various data are not compatible. Considering a possible shock wave created by the IGG beam in air suggests that the propagation velocity would be even larger than 64c and the observed delay could be due to spurious causes. These preliminary results open the way to further possible alternatives, namely:

- The peculiar features of the piezoelectric sensor and the transfer function between the sensitive element and the transducer may cause the sensor to behave very differently from an overdamped oscillator. Note that the observed propagation velocity is consistent with equation (5.5) only if the energy dissipation in the transducer is much larger than in a harmonic oscillator. Furthermore, the transfer function must give a well-defined E_t/p_t ratio, if it has to match the E_g/p_g ratio and the velocity with the observed precision of 1.5%.

- The relation (5.5) might not hold true, the particles of the beam could be completely different and possess some intrinsic property which set their propagation velocity. We do not see, however, any reasonable alternatives to the extended Special Relativity, from which equation (5.5) is derived.

- Finally, it is possible that our microscopic model for the emission process is not correct, and that the E_g/p_g ratio is indeed fixed at the emission. In this case, if the response of the detector has a peak at a certain frequency, the propagation velocity would be fixed, independently from the absorption ratio E_t/p_t.

The use of different kinds of detectors in future trials will be crucial to select the right alternative.

REFERENCES

[1] Podkletnov E, Modanese G. Investigation of high voltage discharges in low pressure gases through large ceramic superconducting electrodes. J Low Temp Phys 2003; 132: 239-259.

[2] Modanese G. Quantum Gravity Evaluation of Stimulated Graviton Emission in Superconductors. In: Modanese G, Robertson GA, Eds. Gravity-Superconductors Interaction: Theory and Experiment. Bentham 2012; Chapter 5.

[3] Modanese G, Junker T. Conditions for stimulated emission in anomalous gravity-superconductors interactions. In: Christiansen MN, Rasmussen TK, Eds. Classical and Quantum Gravity ResearChapter Nova Science 2008: 245-269.

[4] Junker T. Test of experimental techniques for replication of the impulse gravity generator experiment. In: Modanese G, Robertson GA, Eds. Gravity-Superconductors Interaction: Theory and Experiment. Bentham 2012; Chapter 9.

[5] Modanese G. Evolution of the IGG concept at IGF from 2004 to 2007. arXiv 2012: 1102:5447. Available from http://arxiv.org

[6] Zee A. Quantum Field Theory in a Nutshell. Princeton University 2003.

[7] Mugnai D, Ranfagni A, Ruggeri R. Observation of Superluminal Behaviors in Wave Propagation. Phys Rev Lett 2000; 84: 4830-4833.

[8] Saari P, Reivelt K. Evidence of X-Shaped Propagation-Invariant Localized Light Waves. Phys Rev Lett 1997; 79: 4135. Durnin J, Miceli JJ, Eberly JH. Diffraction-free beams. Phys Rev Lett 1987; 58: 1499.

[9] Recami E, Fontana F, Garavaglia R. Special relativity and superluminal motions: a discussion of some recent experiments. Int J Mod Phys A 2000; 15: 2793-2812. Recami E, Zamboni-Rached M, Nobrega KZ, Dartora CA, Hernandez-Figueroa HE. On the localized superluminal solutions to the Maxwell equations. arXiv:0709.2191. Available from http://arxiv.org

[10] Skobelev VV. Graviton-photon interaction. Russ Phys J 1976; 18: 62-65.

[11] Piezo Film Sensors Technical Manual. Measurements Specialties Inc. 1999. Available from www.msiusa.com

[12] Rathod VT, Mahapatra DR, Jain A, Gayathri A. Characterization of a large-area PVDF thin film for electro-mechanical and ultrasonic sensing applications. Sensors and Actuators A 2010; 163: 164–171.

[13] Olympus Transducer C310 Data Sheet. Panametrics NDT 2007. Available from www.olympusndt.com

CHAPTER 9

Setup for the Impulse Gravity Generator Experiment Replication

Timo Junker*

Göde Stiftung – Goede Foundation, Am Heerbach 5, 63857 Waldaschaff, Germany

Abstract: Over the years, several reports of experiments have been published claiming evidence of an unusual interaction between superconductors and gravitational force fields. One of the best-known examples is the work of E. Podkletnov *et al.* the "Impulse Gravity Generator" (IGG) experiment. Due to its potential importance in physics, a setup for replication of the IGG experiment has been constructed. The following article describes the experimental setup used to verify or contradict the claims made. The replication follows the requirements specified by E. Podkletnov *et al.* as closely as possible. Additional modifications, where made, enhanced the experimental possibilities. Superconducting emitters with two or more layers are fabricated, cooled down to 35K and charged with different high-voltage generators to realize different characteristic discharge curves. Two discharge chambers are employed to realize various superconductor gas transitions, as well as superconductor metal transitions. Various sensors are used to measure disturbances and vibrations. As a result, energy from a possible anomalous radiation can be proved to nine orders of magnitude smaller than the smallest "gravity impulse" energy reported by E. Podkletnov *et al.*

Keywords: Superconductors, high-Tc superconductors, superconducting emitters, type-II superconductors, layered superconductors, melt-textured YBCO, high-voltage generators, high-voltage gas discharges, laser triangulation sensor, gravity-like fields.

1. INTRODUCTION

In 1992 E. Podkletnov and R. Nieminen published the first experimental results claiming that an unusual interaction between superconductor and gravitational force fields could exist [1]. In 1997 Podkletnov published an extended version of his experiment with spinning superconductors [2]. The validity of this project remained untested until G. Hathaway, in the scope of his replication found no measurable evidence of a gravity-like force [3].

In 2001 E. Podkletnov, with G. Modanese, published a new model of the superconductor experiment. Here he studied high-voltage discharges over a two-layered superconductor and discovered an "anomalous radiation" which resulted in a "short repulsive force on small movable objects along the propagation axis" [4]. An extension of this work was published by E. Podkletnov and G. Modanese in 2003 [5].

The results of both experiments remain controversial to this day, as the Impulse Gravity Generator experiment has yet to be independently confirmed (to the author's knowledge) [4, 5].

The following article describes a scalable setup, closely based on, but modified from, the published specifications of the experimental setups. A replication setup of the type specified herein should ultimately help to verify or contradict E. Podkletnov's claims.

2. ANALYSIS OF EXPERIMENTAL SETUPS BY E. PODKLETNOV *et al.*

The history of the development of the Impulse Gravity Generator Experiment is shown in the reports [4, 5]. In the first two experimental setups, the high voltage discharge was created using the Van De Graaf generator. The third setup, using a Marx generator, was a completely new design. In all the setups presented, Podkletnov *et al.* report that an anomalous radiation appears during the experiment. Common to

***Address correspondence to Timo Junker:** Göde Stiftung – Goede Foundation, Am Heerbach 5, 63857 Waldaschaff, Germany; Tel: +49-6095-999-198; E-mail: timo_junker@gravitation.org

Giovanni Modanese and Glen A. Robertson (Eds)

all three setups is a high-voltage discharge from a superconductor over a spark gap. The setups differ in superconductor fabrication and generators used.

Not all of the reported descriptions are consistent – many questions remain open. For example there exists no statement or graph describing the characteristic curve of the high-voltage discharge pulse. Because of the missing information it is unclear whether the pulse is a function of a critical dampening or oscillating discharge. Both possibilities make diverse demands on the equipment and materials used. Therefore an assessment must be made based on published electrical parameters to clarify which of the discharges is fundamentally possible.

3. HIGH VOLTAGE DISCHARGE CIRCUIT

3.1. Van de Graaf Setup

In the first two experimental setups reported by Podkletnov *et al.* a Van de Graaf generator is used to create high-voltage current. The superconductor is made by a Plasma-Spray-Process placed on a sphere. The sphere with superconductor is charged to 500kV relative to the counter electrode and the wall of the vacuum chamber. The exact design of the discharge is unclear since the capacitance of the Van de Graaf Generator (the sphere with the applied superconductor) is continually charged. The moment current flows, the voltage of Van de Graaf Generator breaks down. This means that the capacity needs to be well isolated against premature discharge and drain currents. It is reported, however, that the discharge chamber is only evacuated to a rough vacuum by means of a rotary vacuum pump [4]. In practice, rotary vacuum pumps yield a pressure of 10^{-2} to 10^{-3}mbar. This is the pressure range in which a gas discharge takes place, which can be described by Paschen's law. Depending on pressure and distance to the ground potential/counter electrode, a minimum arc voltage can result, in which a potential from a few hundred to a thousand Volt is sufficient to ignite a discharge. For this reason the capacitance of the electrostatic generator can never be charged to a higher potential. The Paschen minimum for Helium is at a $(pd)_{min}$ -product of 53.2 bar µm [6] (p. 168). The minimum breakdown voltage amounts to 155V. Converted to a spark gap of 250mm the Paschen minimum reaches a pressure of ~$2*10^{-1}$mbar, with a gap of 2000 mm, $2.7*10^{-2}$mbar. This correlates to the ultimate pressure normally attainable with rotary pumps.

At very low values of the pd product (less than $(pd)_{min}$), there are not enough gas molecules available for a collision ionization to form an avalanche breakdown. Theoretically, an extremely high breakdown voltage would be achieved at these low values of a pd product. This is not the case, however, since the high field intensity leads to field emission on the electrode surfaces. In rough vacuum range, indicating high pd product value, the breakdown voltage rises to a few thousand volts. Since here, however, the cooling of the superconductor as described in the report is hardly reasonable (no insulating vacuum), it is assumed for the sake of further evaluation that the experiment was performed in a high vacuum. This enables the superconductor sphere to be charged to 500kV by the Van de Graaf generator. The discharge is supplied by the capacitance of the superconductor sphere.

Since no statement is made about the size of the discharge chamber, a minimum and maximum capacitance estimation was considered. For the assessment of the minimum capacitance *C*, a single sphere is observed with concentric counter electrodes (radius→∞) far away [7] (p. 28 continuing).

$$C = 4\pi\varepsilon_r\varepsilon_0 r \qquad \textbf{(1)}$$

$\varepsilon_0 = 8.854187817*10^{-12}$F/m

ε_r =1.0 (vacuum)

The spheres used by E. Podkletnov *et al.* have a diameter 2*r of 25cm – 50cm with an estimated capacitance C_{25cm} of ~14pF and C_{50cm} of ~28pF. For the maximum capacity *C'*, a concentric spherical capacitor is presumed [7] (p. 28 continuing).

$$C = 4\pi\varepsilon_r\varepsilon_0 \frac{Rr}{R-r} \tag{2}$$

The diameter $2R$ of the outer sphere (the chamber) is assumed to be at 2.5m. The electric field E_r between the spherical shells is proposed to be at ~ 25kV/cm [6].

$$E_r = \frac{U}{r}\frac{R}{R-r} \tag{3}$$

It follows that the capacitance of a spherical capacitor in concentric constellation is C'_{25cm}~16pF and C'_{50cm}~35pF. The real capacitance will lie between both extremes.

In E. Podkletnov's experimental setup the distance between the two spheres is given as 0.25m to 2m. In the case of a breakdown, the current path length is estimated to be ~2m to 5m. The inductance of a conductor can be assumed to be at 1µH/m (estimated from [8] p.122, [9] p.100). That means that the circuit inductance (Fig. **1**) is 2µH - 5µH.

Since no damping element is indicated for either the generator setup or the discharge circuit, it has to be concluded that the total energy stored in the generator is transferred to the discharge circuit.

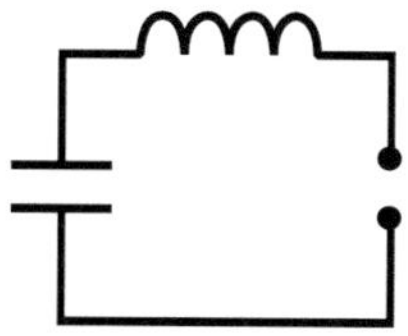

Figure 1: Discharge circuit; the resistance of the spark gap during discharge is neglected.

$$\frac{1}{2}LI^2 = \frac{1}{2}CU^2 \tag{4}$$

$$I_{\max} = U\sqrt{\frac{C}{L}} \tag{5}$$

I_{max} (5), with the above mentioned capacitance and inductance, is estimated at ~ 0.9 to 1.3kA (depending on the inductance of the setup).

The frequency f of an oscillating discharge is calculated as:

$$f = \frac{1}{2\pi\sqrt{LC}} \tag{6}$$

The reachable frequencies of an undamped discharge with the capacitance of C=35pF and a circuit inductance of 5µH is ~12MHz.

A critical circuit dampening is necessary for a damped discharge [8] (p. 310):

$$R_D = 2\sqrt{\frac{L}{C}} \tag{7}$$

This correlates to the damping resistance in the discharge circuit of the Van de Graaf generator in the range from 1.1kΩ to 1.5kΩ (depending on the capacitance and the current path length).

3.2. Marx Generator Setup

The construction of a fused silica vacuum chamber and use of a Marx generator represent a newer form of the Van de Graaf setup. According to the report, E. Podkletnov used a Marx generator with C = 1.25nF [4, 5]. The maximum peak current is I_{max} = 10kA at a U = 2MV total charging voltage. No information was given for exact construction or model size, nor about load resistors or an impedance matching of the circuit discharge. This makes it difficult to determine what discharge modes were used.

Due to the construction of the Marx generator, even with a gas discharge at Paschen minimum, it is possible to transfer a lightning impulse at 2MV to the superconductor. This is because the superconductor serves as emitter in a spark gap and not as the impact capacitance as it does in the Van de Graaf generator.

Generally Marx generators are produced to generate damping discharges with a standard characteristic curve. Oscillating discharges would be counter to the functional principle of conventional Marx generators (reverse voltage!). Specially constructed Marx generators, depending on their adaption to the discharge circuit, are able to generate oscillating and damping discharges.

The report for the Marx generator setup does not mention circuit inductance or information about damping resistors. However, the inductance can be estimated from (4).

$$L = C\frac{U^2}{I^2} \tag{8}$$

The overall circuit inductance is ~ 50μH.

For the estimation of the inductance of the circuit discharge, the entire current path can be assumed at 1μH/m as in 2.1. The current path length of a discharge circuit (at 2MV) with the compact Marx generator is estimated at 7m (Fig. **2**); in contrast, using a conventional Marx generator, estimated at 13m. Therefore, the inductance of the discharge circuit is between 7 and 13μH. When the inductance of the discharge circuit is added to the inductance of the conventional Marx generator (inductance of the capacitors and the current path), the sum of 50μH is a realistic overall inductance.

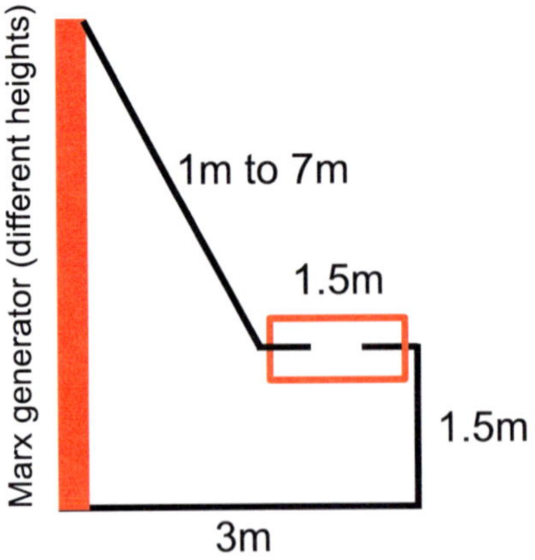

Figure 2: Current path of marx setup. The left red column represents the marx generator. Standard marx generators are ~ 8 to 9 m high by 20 stages. The yellow rectangle stands for the vacuum chamber.

The frequency of an oscillating undamped discharge (6) is 0.6MHz. The resistance needed for the critical damping circuit according to (7) is in the region of 400 Ω. It cannot be assumed that the equivalent serial resistance of the capacitors and the 20 spark gaps (including those of the superconductor) reach this value in order to critically damp the discharge circuit. The Marx generator used for this case must be designed for this load. In the event of an oscillating discharge, the amount of possible frequency transmissions was estimated with consideration to different models in order to use a comparable generator.

4. DETECTORS USED BY E. PODKLETNOV *et al.* [4]

For the detection of the gravity impulse E. Podkletnov *et al.* used a pendulum mounted in a vacuum chamber. The pendulum had replaceable spheres having a diameter of 10 to 20mm and of different mass and materials. He used a ruler 2mm below the spheres to measure the pendulum amplitude. The pendulum spheres were connected to a cotton string with a length of 80cm in most cases. The pendulum, shielded by a brick wall, was situated 6m away from the discharge location. A second pendulum system was used 150m away.

5. BI-LAYERED SUPERCONDUCTORS USED BY E. PODKLETNOV *et al.* [4]

The main features of the method used by Podkletnov *et al.* is described in his reports [4, 5]. He used the melt texturing growth method and the sintering method for his bi-layer superconductor. The raw material is produced from a solid state reaction. It is then melt textured, grinded and sieved. The material for the superconducting layer is made of $YBa_2Cu_3O_{7-y}$ with small amounts of Ce and Ag. In the normal conducting layer, Yttrium is substituted by rare earth elements Re ($Y_{1-x}Re_xBa_2Cu_3O_{7-y}$). Both of the different layers in his 100mm diameter discs get textured together using the OCMTG (oxygen controlled melt texture growth) procedure to become the final emitter. For the OCMTG procedure he applied crystal seeds of Sm123 which were placed 15mm apart on the top side of the two layered disk. For the larger bi-layered discs (120mm diameter) he is using the sintering method.

Summary of requirements for a replication:

- Surface potential of the superconductor of >500kV.
- Peak current of 10e4A (at 2MV; superconductor diameter 100mm).
- Use of He Gas at low pressures.
- Cooling of the superconductor at 40K (use of liquid He).
- Trapped magnetic field in the superconductor.
- Magnetic field of 0.9T during discharge (not implemented in this replication yet).
- Different discharge gap distances.
- Pendulum in vacuum as sensor.
- Production of double-layer superconductor through OCMG-texturing or sintering.

6. SETUP FOR INDEPENDENT REPLICATION OF THE EXPERIMENT

6.1. Overview of the Replication Setup

A replication setup has been independently built based upon Podkletnov's *et al.* published specifications. The aim of the replication setup was to follow the defaults closely to allow reproduction of the experiments in question. Where made, modifications to the originally specified setups are intended to make the setup as flexible as possible and to enhance the experimental possibilities.

The experimental setup (Fig. **3**) is built on the ground floor of the building. The high voltage part of the setup is built in an EMC chamber; the detectors, set up in a wind shield, and the data acquisition (DAQ) are outside of the EMC chamber. The discharge path (superconductor - counter electrode) and the detectors are aligned in the beam axis of the gravity impulse. Several control positions are included, the control unit (CU) outside the chamber being the main one. Next to it, a EMC window is implemented into the EMC chamber to observe the processes inside from a safe place. A more detailed description is outlined in the following paragraphs.

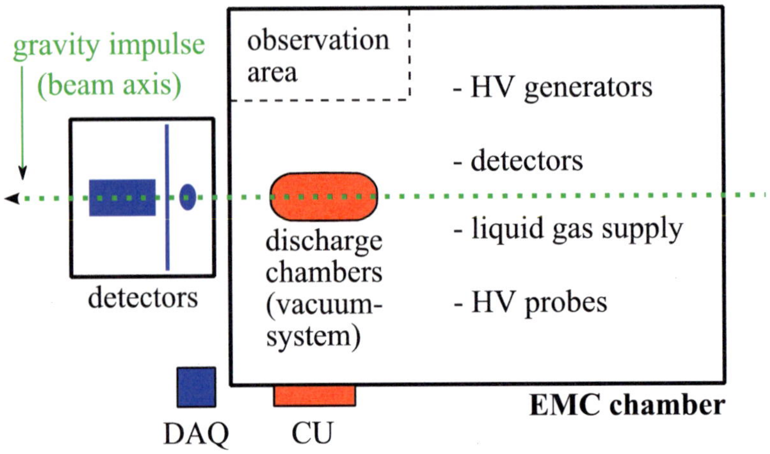

Figure 3: Draft of main parts.

7. DISCHARGE DEVICES

To produce high voltage discharge, at present two generators are being used which allow diverse forms of characteristic curves (Fig. **4**). Both generators can intentionally discharge over an ignition spark gap. In addition, the start of the discharge can be measured with an optical fiber used as a trigger signal for data acquisition. With both generators many variations of discharge are possible such as lightning impulse discharge, square-wave discharge and oscillating discharge.

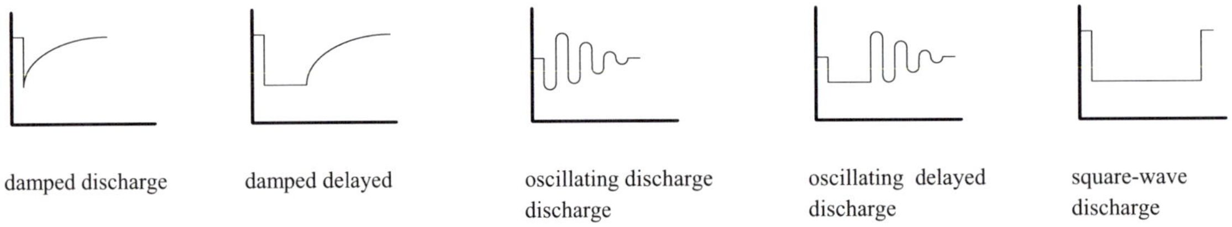

Figure 4: Voltage or current (ordinate) over time (abscissa) drafts of diverse characteristic curves.

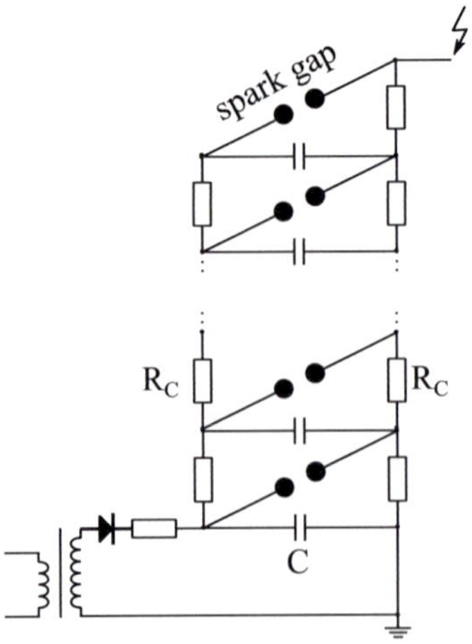

Figure 5: Marx generator (impulse generator).

7.1. Marx Generator

The current Marx generator (Fig. **5**) consists of 10 stages. The overall capacitance is 15nF, the maximum impulse voltage 800kV. The Marx generator consists of stacked capacitors that are charged in parallel over charging resistors R_C. The arrangement of the spark gaps enables the individual stages to be discharged in series so that the individual stage voltage is added. The total charging voltage can be adjusted over the distance of the spark gaps. A battery-powered pulse is led to the first spark gap for ignition in order to start the discharge on purpose. The ignition spark gap is controlled outside the EMC test chamber over an optical fiber. A Marx generator that can easily be modified was chosen. The spark gaps, charging resistors and the charging generator remain and enable the capacitors to be simply replaced. There are numerous companies worldwide that deliver special capacitors for specific requirements.

7.2. Pulse-Forming Network

Square-wave discharges are produced with a pulse-forming network (PFN; see Fig. **6**). The PFN consists of 10 low inductive capacitors that are able to be charged in parallel and "successively" discharged through adjustable inductances. The layout of the inductances enables a single impulse voltage discharge to be stretched to a width of 4μs in order to generate a square-wave discharge. The maximum charging voltage is 70kV, the overall capacitance 1.3nF. The discharge gap is ignited over a triggered spark gap or a Hydrogen-Thyratron.

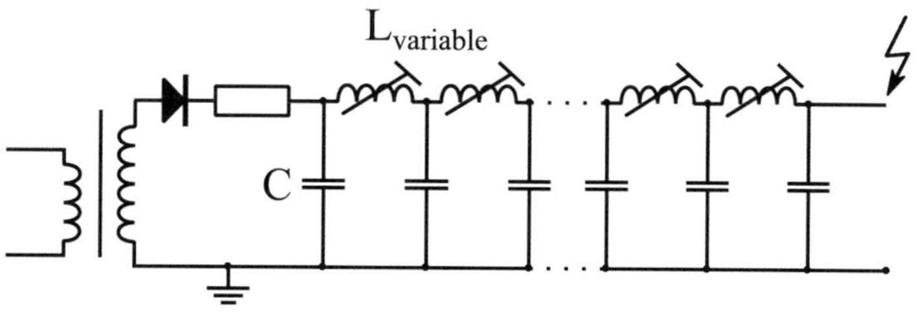

Figure 6: Pulse-forming network.

8. DISCHARGE CIRCUITS

The design of both discharge chambers enables different types of discharges through the superconductor. One option is to arrange the superconductor in a closed discharge circuit (Fig. **7a**). Another one includes a spark gap. As long as the discharge over the spark gap has not taken place, the superconductor can be held in a potential state. (Fig. **7b** and **c**). Variation c) complies with Podkletnov's *et al.* chosen setup. Since it is unclear what role the superconductor-gas-junction has on the formation of the described radiation, all three possibilities were considered.

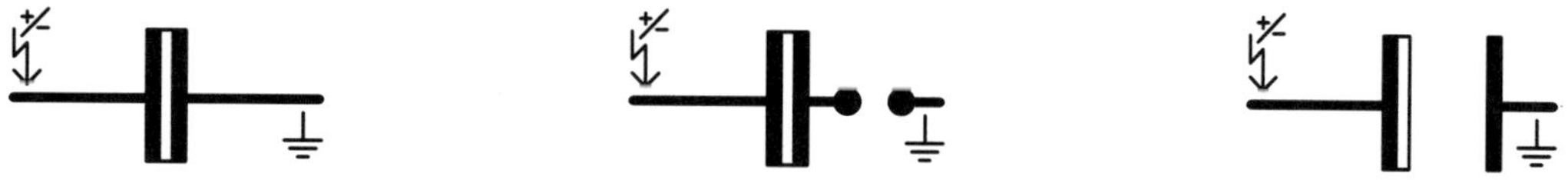

a) Closed discharge circuit

b) Open discharge circuit (SC on high-voltage potential before breakdown)

c) Discharge through gas medium (SC on high-voltage before breakdown)

Figure 7: Discharge circuits. In the 3 possible variations a), b) and c) permanent magnets/solenoids for pinning the superconductors are not shown. Yellow = superconductor.

9. HIGH VOLTAGE MEASUREMENT TECHNIQUE

9.1. Voltage Measuring

The voltage measuring of the lightning pulse and the square-wave discharge is carried out with an ohmic voltage divider (Fig. **8**). The voltage divider is designed for 1.2MV and consists of a high-voltage resistor

R_{HV} with 10kΩ constructed from a low-inductance wound resistor. The design of a low-inductance low-voltage resistor R_{LV} allows the maximum possible output voltage (1000V) to be transferred through the coaxial cable to the oscilloscope. Depending on the voltage levels, different resistance values R_{LV} are installed in a shielded casing. On the side of the oscilloscope, the coaxial cable is terminated with an attenuator. The chosen attenuator matches the maximum input-voltage of the oscilloscope. Measurements have shown that the signal-to-noise ratio improves markedly as a result.

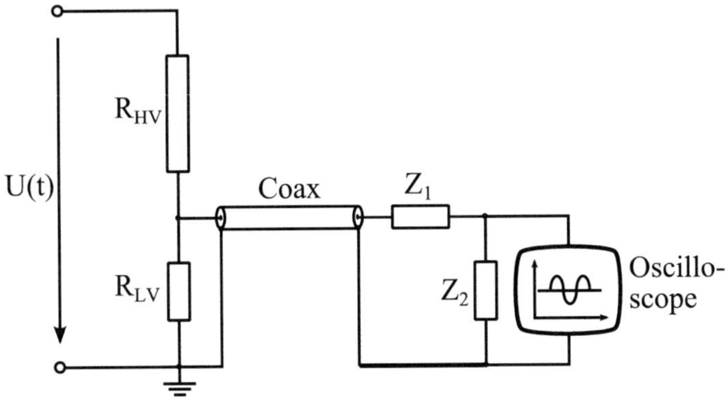

Figure 8: Voltage divider.

9.2. Current Measuring

For the measurement of fast changing currents, a current-viewing resistor and a Pearson current monitor are used (Fig. **9**). The Pearson current monitor is electrostatically shielded, the maximum peak current is 10kA with a rise time of 5ns. The probe is fit concentric with the conductor and connected to the oscilloscope through a shielded coaxial cable.

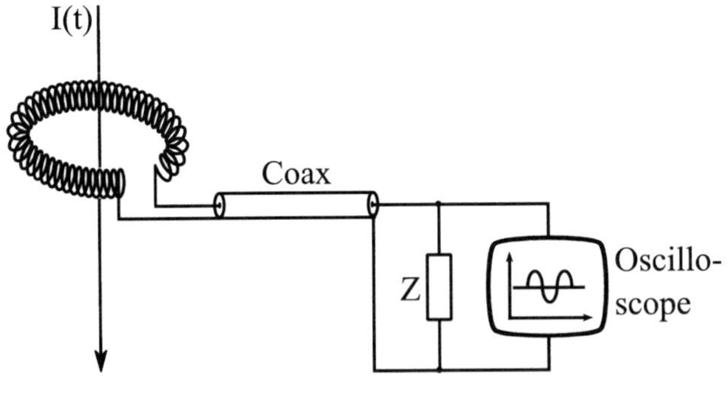

Figure 9: Pearson probe.

The double coaxial design of the current-viewing resistor (Fig. **10**) enables it to reduce noise and electromagnetic interference. The probe is fit concentric with the conductor and connected to the oscilloscope through a shielded coaxial cable. The measuring resistance value R_m is 5mΩ and designed for a peak current of 20kA with a usable rise time of 9ns.

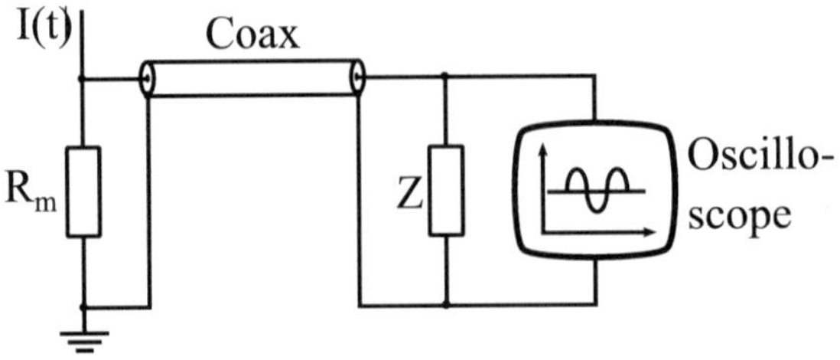

Figure 10: Current-viewing resistor.

The coaxial cables used are of the same type and length so that cable delay time is irrelevant.

9.3. Trigger Signal

The Marx generator is triggered by an ignition spark at the spark gap. After that, the 10 spark gaps ignite consecutively. The flash occurring at the ignition of the spark gap is monitored by an optical fiber, and detected in a photodiode outside of the EMC chamber. The photodiode produces a voltage signal which is amplified and used as a trigger signal for data acquisition.

10. DISCHARGE CHAMBERS

Two different discharge chambers were built for the experiments (Fig. **11**). The first chamber contains the superconductor which is immersed in liquid nitrogen under atmospheric pressure. In the simplest case, the liquid gas container is made of polystyrene. The current leads are made of stainless steel, either as sheets (low-inductive) or as pipes. The sheet/pipe is optimized in thickness, taking into account the skin effect (effective current-carrying plate thickness), to reduce the heat input on the liquid gas and the superconductor.

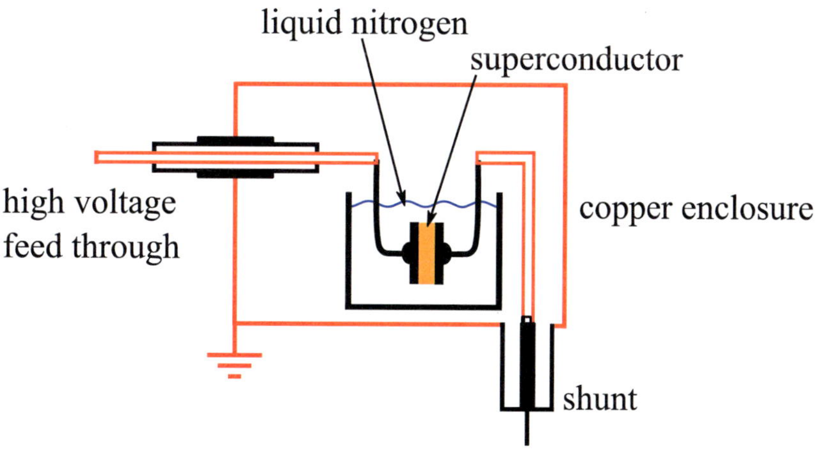

Figure 11: Bath discharge chamber.

The second chamber is implemented as a vacuum chamber. The vacuum chamber has a modular structure made of Duran glass. The use of glass as chamber material enables the observation of the discharge over the whole discharge path length and serves as a simple isolation between the high-voltage potential and earth. Because of the modular design, it is possible to create different chamber sizes quite fast (in diameter and length). The counter-electrode is movable. This enables the spark gap to be set at a distance between closed contact and 50 centimeters (basic setup). Attention was paid to low resistance electrical contacts as well as flexibility in the supply of gas and liquid-gas into the chamber.

Currently both of the chambers are optimized on a superconductor diameter of 50mm and extensively tested for their functionality. The use of various chambers facilitates high-voltage discharges in gases with different “densities” (Fig. **12**).

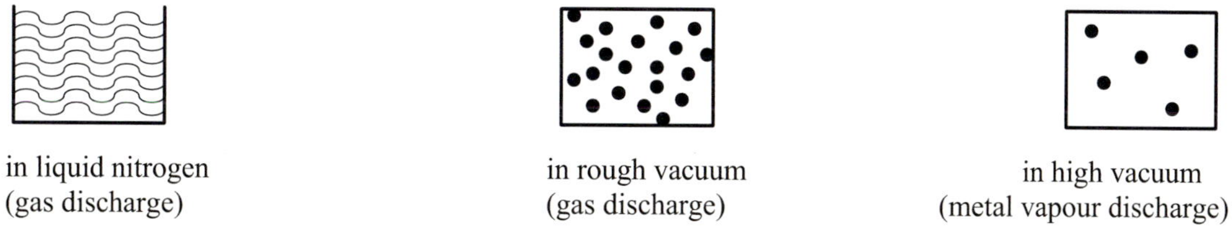

Figure 12: Discharges in gases with different “densities”.

11. VACUUM SYSTEM

The vacuum system (Fig. **13**) consists of discharge chamber, pumping system and associated measurement device. In combination with the vacuum chamber, discharges in different gases (*e.g.,* He, Ar, N_2, H_2, O_2, air) and pressures are possible – from overpressure (1bar) to high vacuum pressure (10^{-8} mbar). The gas supply can be used either on the high-voltage side or on the earth side.

In addition to gas transfer vacuum pumps, sorption pumps are used to produce different gas pressures and atmospheres in the discharge chamber. A cryopump is used to absorb water vapour and condensable gases in order to avoid (or minimize) deposition on the superconductor. All the pumps are designed for continuous operation, even during the high-voltage discharges, by having suitable EMI shielding.

Different vacuum gauges (Table **1**) are applied to measure the gas pressure in the chamber according to the pressure range. Because of its low measurement uncertainty, the capacitive vacuum is used in the range of 1 and 10^{-3} mbar (gas discharge, pd-product). All vacuum gauges are working continuously even during the high-voltage discharge. For each of these gauges including the current and instrument leads, EMI shieldings were developed. The shieldings avert serious defects of the gauges which by nature are sensitive to strong EMI. The shieldings have been tested for their functionality but with no further specification.

A quadrupole mass spectrometer serves to evaluate the composition of the gas inside the chamber and to control the "water content".

Table 1: Survey of the used vacuum gauges.

Principles of Measurement	**Used in the Range**	**Measurement Uncertainty**
Relative pressure vacuum gauges according to Bourdon (independent of type of gas)	10 to 1000 mbar	1% (full scale)
Capacitance vacuum gauge (independent of type of gas)	$1.3*10^{-2}$ to 1.3 mbar	0.9% to 0.2% of means. value (at 1°C temp. drift)
Hot cathode ionization vacuum gauge according to Bayard-Alpert (dependent of type of gas)	10^{-8} to 10^{-2} mbar	15% of meas. Value
Thermal conductivity vacuum gauge according to Pirani (dependent of type of gas)	10^{-2} to 10^{2}	not specified

After every reconstruction of the vacuum chamber (*e.g.,* change of the superconductor), the vacuum chamber and all vacuum components are baked out under high vacuum for several days. Afterwards they are cleaned by plasma discharge. As a result, the virtual leak rate decreases notably due to desorption of absorbed gases on the inside of the vacuum chamber. This makes it possible to maintain the pressure over a longer period of time (several minutes) without the use of vacuum pumps.

Pressure control (Fig. **14**) is necessary in order to attain the desired pressures. For this purpose, the upstream-pressure control (mass flow-controlling, quasi-stationary) and the downstream-pressure control (pumping speed) were tested. The quasi-stationary method has proved to be most reliable when working with high-voltage. In this case an active-pressure control is unnecessary and no additional effort is needed to shield the controller. To achieve quasi-stationary pressure control, first a high vacuum (bake out and plasma cleaning) is generated in the chamber. Then the chamber is slowly filled by a defined orifice plate with the required gas. The rise in pressure is regulated relative to the size of the orifice (dependent on pumping speed). The discharge can be started when the desired pressure is reached. If the fill rate is adapted to the repeat time of the Marx generator and data acquisition, many consecutive discharges are possible over a wide pressure range.

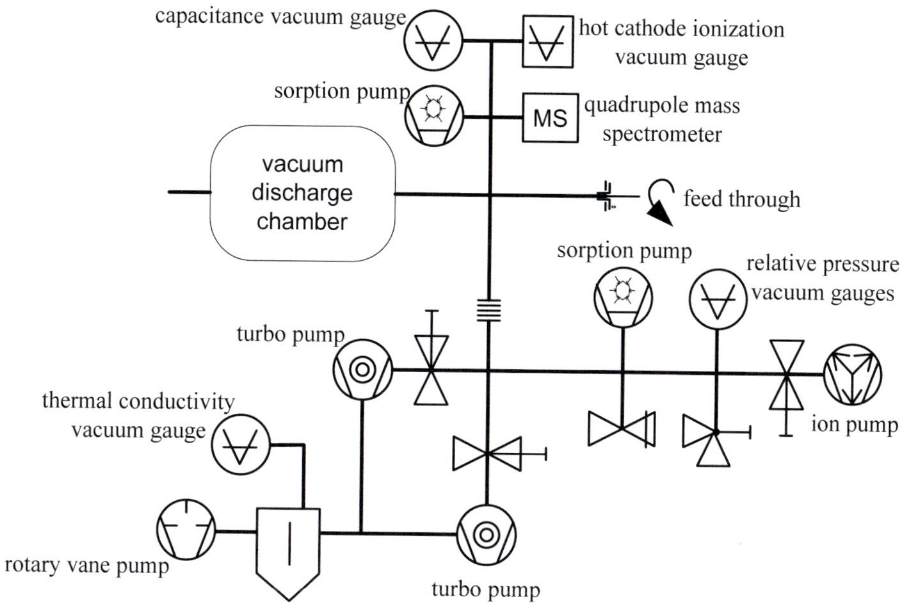

Figure 13: Vacuum system.

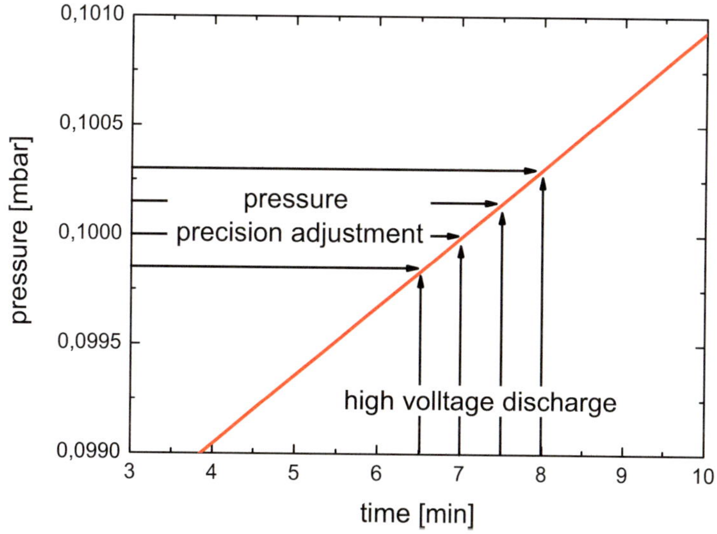

Figure 14: Quasi-stationary pressure control.

12. CRYO-SYSTEM

The cryogenic system cools the superconductor as well as the sorption pumps and the radiation shields under high-voltage. Depending on the discharge chamber, different cooling systems for the superconductor are necessary. The easiest option is the "bath discharge chamber" in which the superconductor is submerged in liquid nitrogen. The temperature of the superconductor complies with the boiling temperature of the liquid nitrogen; as long as the storage vessel is full, cooling is certain. In contrast, the cooling of the superconductor in the vacuum discharge chamber is more complex. The superconductor is installed on a mounting support on the heat exchanger of a cryostat. To maintain the cooling of the superconductor, the heat exchanger has to be permanently supplied with liquid gas.

There are two variations for the supply of liquid helium or nitrogen from a Dewar vessel:

a) If the Dewar vessel is at ground potential, then the liquid gas supply must be electrically insulated. The feed line is made of a nylon hose when liquid nitrogen is used. If on the other hand liquid Helium is applied, all transfer lines from the Dewar to the heat exchanger are made of double-walled transfer lines with an isolation vacuum. To separate the high-voltage potential from the Dewar vessel, a double-walled electrically insulating adapter is implemented. This adaptor is made of Krylit, an epoxy fibreglass composite that is tested helium leak-proof and has a good high-voltage insulation characteristic.

b) If the Dewar vessel has the same potential as the superconductor, the entire transfer line can be made electrically conductive. In this case the Dewar vessel is situated on an isolated base. But this configuration changes the capacity of the discharge circuit and with it, the characteristic of the high-voltage discharge.

The lowest reachable temperature of the superconductor with liquid nitrogen is 85K (with a permanent magnet for flux pinning), with Helium 35 K (Fig. **15**). The thermal load from the surrounding (vacuum chamber made of glass) and the load of the thermal conductivity loss from the cryostat junctions (from the heat exchanger to the cold head, to the superconductor) are already apparent. By use of a special-built radiation shielding (a PTFE coil in the discharge chamber flushed with liquid nitrogen), it is possible to cool the superconductor's temperature down to 80 K. With the inclusion of the radiation shield, the chamber cannot be used for every discharge configuration (sparkover to the radiation shield amongst special pressure/discharge ranges). This radiation shield is only used when necessary.

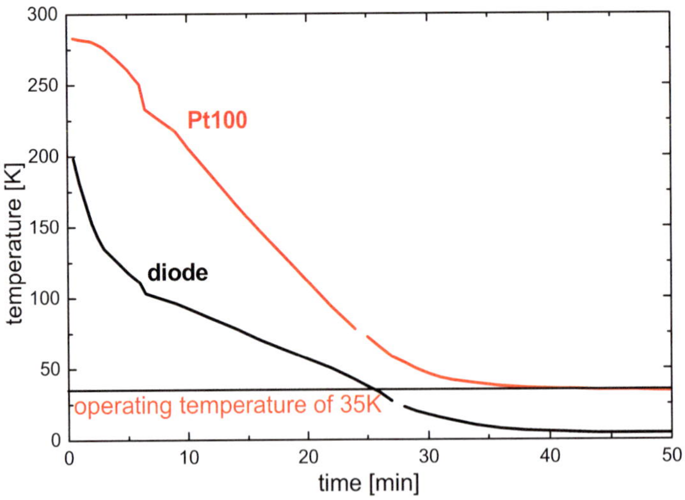

Figure 15: Achievable temperature of the superconductor (permanent magnet in the mounting support) with liquid helium without a radiation shield.

There have also been tests with Dewar bath cryostats (Fig. **16**). The superconductor is mounted to a wall of a vessel filled with liquid gas.

The advantage in using a liquid gas supplied heat exchanger cryostat is that it is possible to adjust the temperature of the superconductor through additional heating and by regulation of the liquid gas flow rate.

12.1. Superconductor Mounting Support

The mounting support connects the superconductor with the heat exchanger of the cryostat. It contains a permanent magnet to enable flux pinning to the superconductor. For this purpose there are different mounting supports, depending on the cooling system and the heat exchanger. The superconductor itself is coated with silver or gold and soldered to the support with indium [10].

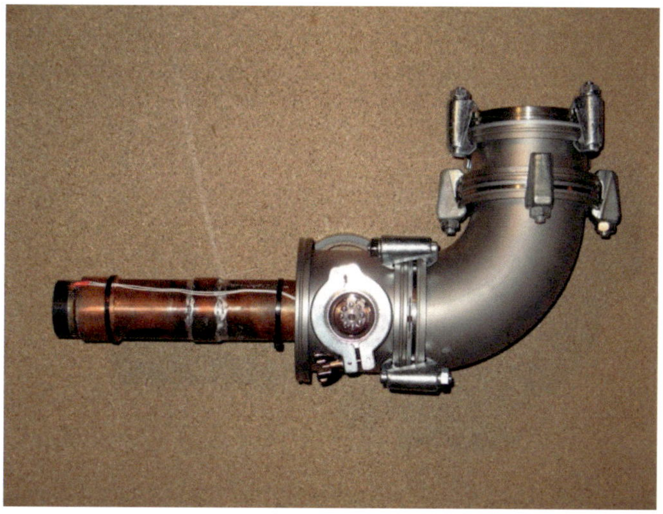

Figure 16: Bath cryostat with superconductor.

Figure 17: Mounting support with superconductor.

To improve the thermal conductivity the permanent magnet is embedded in the holder. By using gallium it is easily reversible.

The mounting support developed for this experiment (Fig. **17**) enables the pinning of the superconductor with different magnetic field configurations (Fig. **18**):

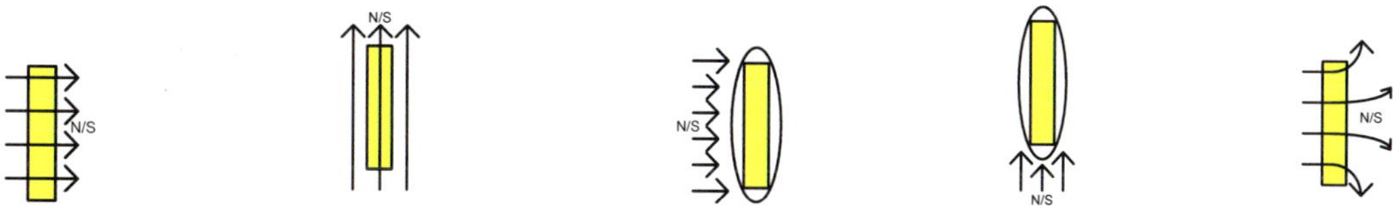

axial magnetic field pinning; JJ-alignment

diametrical magnetic field pinning; JJ-alignment

zero field in HTSC with axial magnetic field (shielding currents)

zero field in HTSC with radial magnetic field (shielding currents)

inhomogeneous field axial, HTSC on NdFeB pinning, JJ-alignment

Figure 18: Different pinning of the superconductor.

12.2. Structure of the Continuous Flow Cryostat

A standard cryostat was rebuilt for the vacuum discharge chamber. The superconductor/magnet- mounting support is installed on the heat exchanger. Indium foil or gallium (removable casting compound) was implemented to improve the thermal conductivity. Additional conductors were set annularly around the cryostat to improve the current supply from the high-voltage connection to the superconductor. The conductors are made of stainless steel to minimize the heat transfer from room temperature to the cooled superconductor.

A diode is installed on the heat exchanger to measure the temperature. On the superconductor itself, there is a PT100 which serves to define the transition temperature (T_c). The Pt100 has to be calibrated to determine the transition temperature of the superconductor mounted on the fully configured cryostat. The cryostat system is set on a measuring vacuum chamber so that the superconductor rises in a coil system. The coil system consists of the detector coil which is surrounded by the stimulating coil. The superconductor is cooled or warmed slowly whereby the change of the diamagnetic state of the superconductor (Tc) is measured through a change of the induction on the detector coil (Fig. **19**). The measured resistance of the PT100 serves as the indicator of the transition temperature.

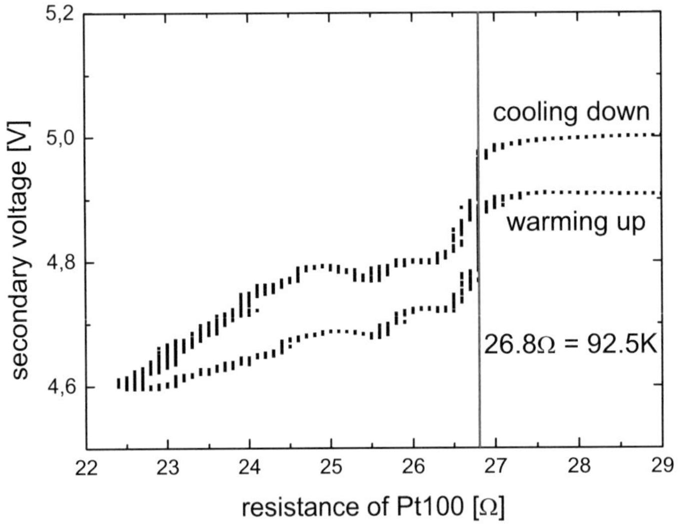

Figure 19: Measured transition temperature of the YBCO emitter.

13. EMI SHIELDING

The complete replication setup, consisting of the high-voltage generators, high-voltage measurement equipment, as well as the discharge chambers and vacuum system, is built up in an EMC chamber. This metal chamber is incorporated in the lab. It protects the detectors and the data acquisition system against high-frequency electromagnetic interference which occurs during high-voltage discharges. The EMC chamber allows excellent shielding of electromagnetic frequencies up to 40GHz. It is used in combination with power line filters to reduce the interference of other electronic equipment to a minimum. The shielding effectiveness is ~80dB at 10kHz, ~ 100dB and more in the range of 100kHz to several GHz. All the feed lines (measurement leads, power supplies) entering the chamber are surrounded by corrugated pipes that shield against EMI. The EMI shielding also includes the measurement devices and the vacuum system.

In addition, an observation and control position is created in the chamber. This position is protected against direct lightning by means of a metal grid enclosure. Another control position is located outside of the chamber. A window makes it possible to observe the processes inside the EMC chamber.

Great importance is given to the grounding inside the EMC-chamber. To create a defined earth potential, the ground connections to all the high-voltage generators, the voltage divider and the vacuum system are combined with a large surfaced copper foil. The connection of this foil to the EMC chamber is of good conductivity. As a result, the interference currents, which are able to occur between the earth points of the devices, are expected to be diminished or prevented.

14. DETECTORS

The detectors are installed outside the EMC chamber on a vibration-damped table in a wind shield. Next to highly sensitive pendulums, the collection of detectors contains pressure sensors, E/M-field sensors and temperature sensors. The detectors can possibly measure interactions with the anomalous radiation and can validate the signals against disturbances (acoustic, vibration, E/M-interferences).

14.1. Pendulum

For E. Potkletnov *et al.* superconductor emitter with a thickness of 4mm and a discharge of 500kV, an energy of the anomalous radiation was measured at 1.8*10-4 J (18.5 g mass, 80 cm pendulum length, 40mm deviation) [4]. The pendulums used in this replication, are constructed to achieve a very high detection limit.

The pendulum consists of a thin-walled glass ball (7.8g) with a 50mm diameter suspended from a 80cm long tungsten filament with a diameter of 10 μm. The pendulum itself is in a vacuum tight lidded glass tube. The pendulum object is interchangeable (different materials and masses) and the thin-walled glass ball can be filled with different materials. The glass ball is located in the beam path of the energy pulse. Pendulum movements are detected with laser triangulation sensors (500nm resolution at 2kS). The measuring resolution is determined by the resolution of the sensor and the measuring fault due to room disturbances.

The measuring fault/disturbance depends mainly on the following factors:

- Building movements caused by wind (tilting of the foundation plate).
- Building vibrations due to persons and machines in the building.
- Vibration caused by opening and closing of the EMC chamber door.
- Sound (lightning discharge → vibration of the EMC chamber → resonance with the pendulum string).
- Rotation of the pendulum sphere is interpreted as an amplitude change.

The biggest influence on the pendulum movement is the opening and closing of the heavy EMC chamber door (30μm pendulum vibration); the biggest acoustical influence is the lightning discharge (few μm) which appears as second harmonics on the fundamental oscillation of the pendulum. If measurements have to be taken immediately after preparations, the resolution of the pendulum is 30μm (peak to peak). After 10 to 20 minutes pendulum movement subsides down to 3-7 μm (peak to peak). This is valid for the movement of the pendulum in air under standard pressure. Further error sizes are not considered for the first assessment.

The deviation of the pendulum induces a height change of the pendulum sphere. Because the triangulation sensor is installed at a stationary height (sphere centred in neutral position), an error results in the measurement of the deflection. Since this error is effect-diminishing (apparently lower amplitude), it is not considered in this estimation.

The detection limit of the energy of the anomalous radiation (Table **2**) is calculated as follows:

$$W_p = mgh \quad \textbf{(9)}$$

$$\sin\frac{\alpha}{2} \approx \frac{\alpha}{2} \approx \frac{h}{s};\ with\ \alpha = \frac{s}{l}$$

$$W_p = mg\frac{s^2}{2l} \quad \textbf{(10)}$$

(Pendulum mass, $m = 7.80*10^{-03}$kg; pendulum length, $l = 0.8$m; gravity acceleration, $g = 9.81\text{m/s}^2$.)

For the detection of small energies the pendulum swing is very small. Hence the small-angle approximation can be used for simplification.

The following chart shows the measurement resolution and the detection limit of the anomalous radiation energy of the present detector version:

Table 2: Detection limit of the anomalous radiation.

Waiting period after closing the EMC door	**Resolution pendulum swing**	**Energy resolution**
[min]	[m]	[J]
0	$3*10^{-05}$	$4.3*10^{-11}$
10	$8*10^{-06}$	$3.1*10^{-12}$
20	$4*10^{-06}$	$7.7*10^{-13}$
30	$2*10^{-06}$	$1.9*10^{-13}$
Beam energy of E. Podkletnov's experiment at 500kV		$1.8*10^{-04}$

Even under disadvantageous conditions, an energy of the anomalous radiation can be proved that is nine orders of magnitude smaller than the smallest "gravity impulse" energy (500kv, 4 mm emitter) that E. Podklenov *et al.* reported [4].

There are two main pendulums in the beam axis for the detection of the anomalous radiation. One of the main pendulums is in the direction of the anomalous radiation impulse. The second main pendulum is opposite, on the other side of the discharge chamber. Because this pendulum is close to the high voltage discharge, it is set up in a specially constructed EMC chamber that can be evacuated. The power and sensor lines for the laser triangulation sensor are also shielded. The detection limit of energy is lower by a factor of 100 mainly because of the absent vibration damping and acoustic absorption.

14.2. Further Sensors

Apart from the pendulums, there are more sensors in a noise-control box in the beam axis (Fig. **20**). In addition to the acceleration sensors (2,3,5,6,8), a pressure sensor (7) and an ultrasonic senor (4) are used. The sensors that are arranged in the beam axis accomplish two functions. One purpose is to examine whether they can be stimulated by a possible "gravity impulse". The other is simply to measure disturbances and vibrations. Due to the way the sensors were constructed, it is uncertain if they can be stimulated by a short impulse in their resonant frequency. The acceleration sensors are of ceramic shear mode design; the pressure sensor works with a flexible membrane. The latter is installed into the innermost chamber made of lead and should serve to measure the air movement caused by any anomalous radiation. Because a high sampling rate was chosen, the data acquisition is also designed to acquire the resonant frequency of the sensors. On the detector board a seismic sensor (ceramic shear acc.) is installed, and at the head of the pendulum chamber an acceleration sensor (ceramic shear acc.). The seismic sensor (3) serves to detect the vibration of the detector board (vertical alignment). The acceleration sensor (2), at the upper end of the pendulum chamber, detects a movement of the pendulum suspension and therefore an apparent amplitude of the pendulum mass (horizontal alignment, parallel to beam axis). Movements transverse to the pulse axis are measured by the tri-axial acceleration sensor (6). A microphone outside the sound insulation box records the acoustic signals of the discharge.

To improve the coupling during high frequency, the acceleration sensors are attached to the base by stud mounting (lead box, pendulum chamber flange and breadboard). Temperature sensors are used to assist the assessment of the long-term behaviour of the installed detector sensors. An E/M field coil is installed into the beam axis in order to detect electromagnetic disturbances outside of the EMC chamber caused by the high-voltage discharge.

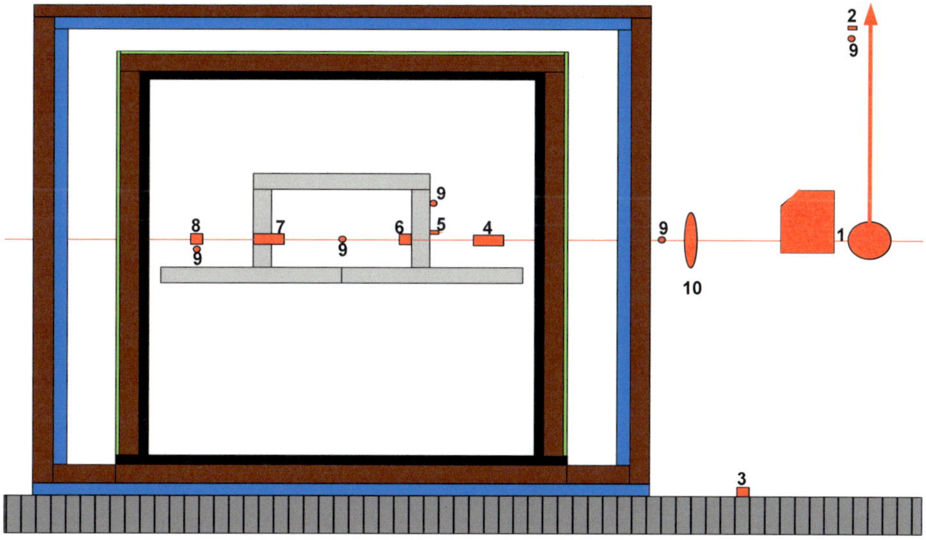

Figure 20: Detector system. Shows the configuration of the sensors with one noise-control box fitting into the other. The vacuum chamber that contains the pendulum (1) is not shown.

14.3. Acoustic Shielding, Vibration Isolation of the Sensors

The detectors (pendulum and detector box) are mounted on a breadboard which is damped against vibration. The detector system is outside the EMC chamber. Elastomeric isolators are attached between the table chassis and the breadboard. The table chassis is loaded with 200kg of panels. The board is located directly on the building foundation and is nowhere in contact with the false floor or the EMC chamber.

To obtain a high detection limit it is necessary to hold the suspension point of the pendulum as stable as possible. In preliminary experiments, an effective attenuation of vibrations and acoustic disturbances had to be discovered. The results had shown that a semi-active pneumatic vibration damping gives no advantage over a simple elastomeric isolation. To the contrary, the resonance frequency of the pneumatic vibration is in the range of the natural frequency of the pendulum (0.56Hz), whereas the natural frequency of the elastomeric damper is about 10Hz.

The advantages and disadvantages of the board suspension had also been analysed. Based on the measurement results, the variant having the table chassis standing on the building foundation was preferred to that of suspending the breadboard from the ceiling.

The aim was to achieve an optimal acoustic damping (damping factor is still not specified). The acoustic disturbances caused by the high-voltage discharge can be clearly seen in the measurement signals of the pendulum. Due to the run time the disturbances can be easily distinguished from an anomalous pulse.

The complete detector system including the pendulum chamber is housed in a heavy wind shield box. This box shields air movements in the lab and slightly damps acoustic disturbances. The wind shield is located on the false floor and is at no point in contact with the EMC chamber or the detector table. A 40% reduction of the pendulum swing caused by disturbances is reached using this wind shield.

14.4. Alignment of the Beam Axis

The alignment of the beam axis to the detectors is achieved through a system of levelling lasers. A horizontal line laser projects the horizontal level of the beam axis past the EMC chamber door. The vertical level (direction) of the beam axis (and the detectors) is oriented to two vertical line lasers facing each other, one in and one outside of the chamber. The line lasers are adjusted by a closable aperture in the EMC chamber.

This orientation is only necessary when there are building modifications or a replacement of the discharge chamber. Because of the distance between the superconductor discharge gap and the detectors, an alignment less than 2 degrees is necessary to obtain a congruence between the beam axis column and the detectors that is higher than 50%.

15. DATA ACQUISITION

DAQ-modules and digital storage oscilloscopes are used for the acquisition of the sensor signals (Fig. **21**). These are controlled by LabVIEW. The analog sensors are read out in groups according to the needed sampling rate. The digital sensors (laser triangulation sensors) are recorded through serial interfaces. Temperature sensors are read out in accordance to their settling time with a low sampling rate. Acceleration and pressure sensors are acquired with a sampling rate five times that of their natural resonance frequency. The high-voltage measuring devices and the ultrasonic sensor are acquired with a digital storage oscilloscope. The data is sent to the measurement program after the acquisition is finished.

The synchronization within a sensor group occurs on the measuring module itself. These measurement modules are clocked simultaneously by the PXI-Chassis bus. Because the time stamp of the serial signals is dependent on the operating system and the storage oscilloscopes are single devices that are not synchronized by the PXI-bus, a synchronization of all measurement devices is necessary. Therefore the trigger signal is recorded on a channel of every measurement card/measurement system to synchronize all the measuring groups. To prevent earth loops, every sensor signal is acquired differentially.

Because of the high sampling rate, the measurement program saves the data in the computer's memory. After the acquisition is finished, the data get stored on the hard disk. During this transaction, the measuring program processes the synchronization of the time axis of each measuring channel.

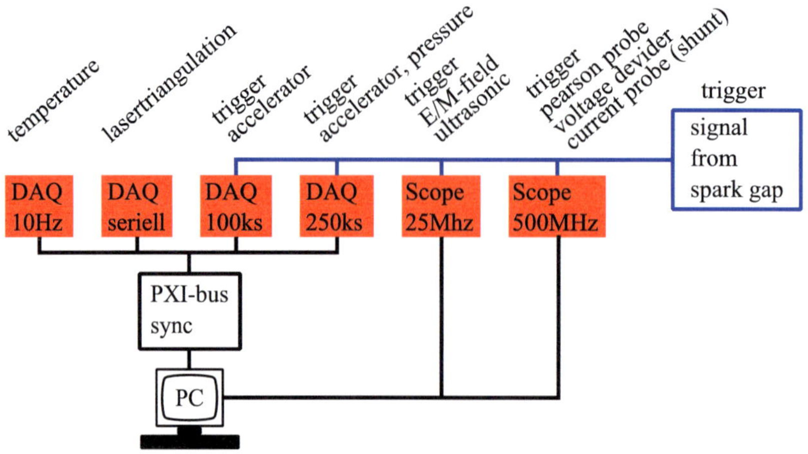

Figure 21: DAQ.

15.1. Superconductor Fabrication Methods

The double-layer superconductors made for this replication setup (Fig. **22**) follow the mentioned method where possible [4]. To achieve superconductor-disks with a higher density, the particle size distribution of the sintered emitters are changed. A high-temperature furnace is used for texturing and sintering; an annealing oven was converted for the oxygen treatment. The texturing curve has been optimized to achieve a single domain superconductor size of 40mm using the top seeded melt texture growth. Because E. Podkletnov *et al.* applied several seed crystals for a disk 100mm in diameter, an undisturbed domain growth of 15-20mm is sufficient. The superconductors were examined using levitation measuring and field mapping. Their quality easily matches that of the superconductors mentioned in the literature. The edge of the superconducting disk was lathed to a radius of 4mm and the surface was final polished with a 2mm diamond suspension.

Figure 22: Sintered, bi-layered superconducting disc ($YBa_2Cu_3O_{7-y}$ over $Y_{1-x}Re_xBa_2Cu_3O_{7-y}$) not machined.

16. TEST OF THE REPLICATION SETUP

To verify the proper function of the replication setup, high-voltage discharges were produced by a superconductor and a metal disc (as dummy) during the construction phase. These tests were not conducted to verify or contradict the claims by E. Podkletnov *et al.* A single-layer single-domain superconductor was used to test the interaction between high-voltage current, the vacuum system and the cryogenic system. These tests also served as an evaluation of error sizes and disturbing influences.

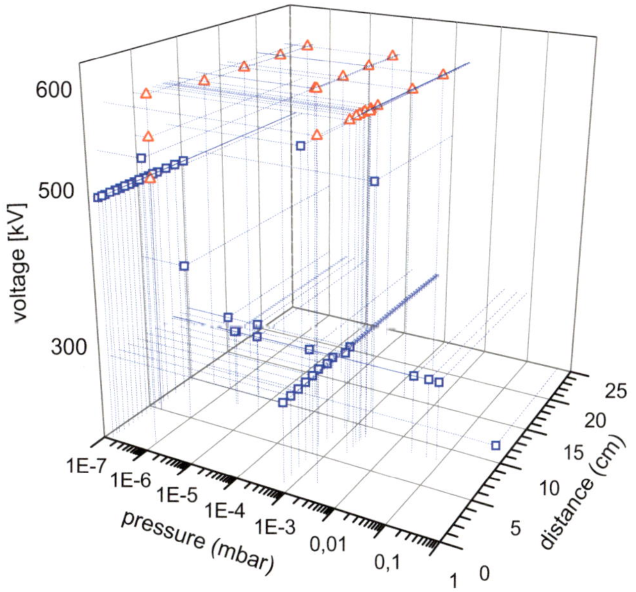

Figrue 23: Shows partial results of the preliminary experiments. Discharges up to 600kV with different pressure ranges (gas discharges and high-vacuum discharges) with a distance up to 20cm between the superconductor emitter and the counter electrode are shown. The blue dots are the projections perpendicular to the axis. The red triangles represent discharges with the superconductor cooled with liquid helium, the blue squares the superconductor cooled with nitrogen.

For this examination a single layer $YBa_2Cu_3O_{7-y}$ superconductor was used. The superconductor was melt-textured in the standard top-seeded melt texture growth process. It is a single domain superconductor with 50mm diameter. The edge was lathed to a radius of 4mm, coated with silver on the back side and soldered to the mounting support using indium. The mounting support contains a permanent magnet (NdFeB) with axial magnetization. A Pt100 was glued to the superconductor and its resistance at the transition temperature was examined before the cryostat was mounted to the vacuum discharge chamber. After several hours of backing out the chamber, a short duration of plasma cleaning of the inner surfaces of the chamber was executed. The superconductor was cooled down after the sorption pumps reached working temperature (liquid nitrogen). The chamber was flushed with helium. The desired pressure was applied using air that was reduced in water content through a cold trap. To determine the discharge characteristics high-voltage discharges were made in a pattern with different pressure, distance and voltage potential (Fig. **23**).

The experiments proved that the system is working properly. All the devices, sensors and detectors work, even under high-voltage discharges.

The results of the replication of the Impulse Gravity Generator experiment, using bi-layered emitters will be published soon.

REFERENCES

[1] Podkletnov E, Nieminen R. A possibility of gravitational force shielding by bulk $YBa_2Cu_3O_{7-x}$ superconductor. Physica C 1992; 203: 441-444.

[2] Podkletnov E. Weak gravitation shielding properties of composite bulk $YBa_2Cu_3O_{7-x}$ superconductor below 70 K under e.m. field. arXiv 1997: cond-mat/9701074. Available from http://arxiv.org

[3] Hathaway G, Cleveland B, Bao Y. Gravity modification experiment using a rotating superconducting disk and radio frequency fields. Physica C 2003; 385: 488–500.

[4] Podkletnov E, Modanese G. Impulse Gravity Generator Based on Charged YBa2Cu3O7-y Superconductor with Composite Crystal Structure. arXiv 2001: physics/0108005. Available from http://arxiv.org

[5] Podkletnov E, Modanese G. Investigation of high voltage discharges in low pressure gases through large ceramic superconducting electrodes. J Low Temp Phys 2003; 132: 239-259.

[6] Küchler A. Hochspannungstechnik. Berlin Heidelberg: Springer-Verlag 2005.

[7] Hasenpusch O. Hochspannungstechnik. München: Franzis-Verlag GmbH 1991.

[8] Küchler A. Hochspannungstechnik. Düsseldorf: VDI Verlag GmbH 1996.

[9] Schwab AJ. Elektromagnetische Verträglichkeit. Berlin: Springer-Verlag 1996.

[10] Ekin J. B5.1 Superconductor contacts. In: Handbook of Superconducting Materials. IOP Publishing 2003; pp. 857-887.

CHAPTER 10

Experimental Considerations in Superconductor Gravity Experiments

George Hathaway[1,*] and Harald Reiss[2]

[1]*Hathaway Consulting Services, 1080 19th Sideroad, King City, Ontario, Canada L7B 1K5 and* [2]*Department of Physics, University of Wuerzburg, Am Hubland, D-97074 Wuerzburg, Germany*

Abstract: Reports of experimental modifications of gravity over rotating superconductors or of a weight increase or decrease during cool-down of superconductors have prompted many researchers to consider designing and performing their own experiments. However, many of these reports have not dealt in sufficient depth with the considerable difficulties attendant on this type of experiment. In general, there is a large class of phenomena that can mask the sought-after effect and produce spurious or artefactual results which are actually due to prosaic and mundane albeit subtle phenomena. Some of the proposed experiments deal with superconductor-mediated detection of gravitational waves. These will not be dealt with specifically in this paper. Rather, we concentrate on those experiments designed to detect mass anomalies and gravity field anomalies. However, many of the issues raised herein are germane to both types of experiments. We will describe some general types of experiments concentrating on instrumentation and interpretation of results rather than the theory leading to the experiments. We examine specific and subtle thermal phenomena occurring during the transition of superconductors near T_{crit}. We then enumerate the multitude of experimental pitfalls that lay before the researcher. Examples provided clearly indicate that minimum standards of experimental precision, accounting for boundary conditions, an understanding of thermal exchange between cryogen and superconductor, error analysis and thorough reporting of the experiment are necessary to distinguish a true anomaly from prosaic explanations and artefacts.

Keywords: Gravitational forces, superconductors, high-Tc superconductors, low-Tc superconductors, type-II superconductors, superconductors cool-down, weight anomalies, gravity field anomalies, thermal exchanges in superconductors, error analysis, artefacts in gravitational experiments, gravity-like fields.

1. INTRODUCTION

This paper presents a brief analysis of experiments designed to confirm or reject claimed gravitational anomalies. We will not discuss conditions under which true anomalies possibly could arise nor the theoretical underpinning of such experiments but highlight detection methods and suggest some standards for their detection should they exist. We mainly focus on experiments involving liquid cryogen-cooled high-temperature superconductors (HTSC).

These experiments can be divided roughly into two groups: 1) experiments to determine mass anomalies assuming constant gravity ([1, 2]), and 2) experiments investigating gravity anomalies assuming constant mass [3, 4]. Weight anomalies are claimed to be either due to an intrinsic *materials* property (group 1) or that HTSC materials induced local variations in the *gravity field* (group 2). Experiments designed to produce or detect gravitational waves involving HTSC will not be discussed. Other papers in this eBook address gravitational waves. Here we focus primarily on force detection.

2. MASS ANOMALIES

Group 1 experiments can be subdivided into two classes: measurement of total weight of the complete experimental set-up; an indirect method that requires very large corrections to the data and compared to other approaches is more or less useless (*e.g.,* Fig. **1**) or direct detection of apparent weight of just the sample (plus suspension & fixtures-*e.g.,* Fig. **2**).

***Address correspondence to George Hathaway:** Hathaway Consulting Services, 1080 19th Sideroad, King City, Ontario, Canada L7B 1K5; Tel: 905-795-9382; Email: ghathaway@ieee.org

Giovanni Modanese and Glen A. Robertson (Eds)

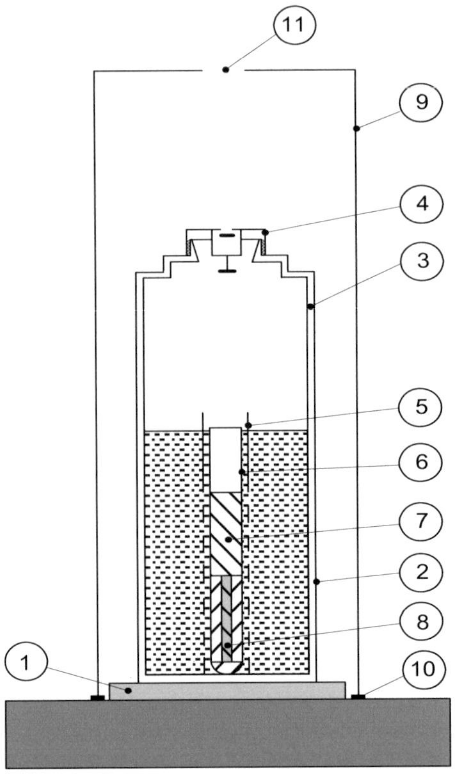

Figure 1: Set-up for measurement of total apparatus weight (Example of a low cost but completely unqualified set-up for gravity experiments); (1) Balance, (2) Stainless steel dewar, (3) Dewar insulation, (4) Plastic cap, (5) Cylindrical cavity, (6) Sample holder, (7) Glass fibre insulation, (8) Sample, (9) Hood, (10) Rubber seal, (11) Vent.

2.1. Detection of Sample Mass from Apparent Total Weight of Experimental Set-Up

Instead of directly measuring apparent weight of the sample, measurements of total weight of sample plus cryostat and coolant all positioned on an appropriate, mechanically stable balance could be performed (Fig **1**). As an advantage, no buoyancy or free convection or surface roughness or wetting phenomena have to be taken into account. However, substantial quantities of the coolant are evaporated when the sample is dipped into the liquid whilst suspended on a filament. Contrary to similar methods employing liquid cryogens where uncertainties in the height of the liquid result in very small buoyancy effects experienced by the thin filament, here the reduction of the liquid level results in a strong and fluctuating decrease of total weight. The liquid level cannot accurately be controlled and restored to the original height at $t = t_0$. A continuous, compensating supply of liquid to the cryostat would seriously disturb weight measurement. Liquid level height determines local temperatures on the inner wall of the cryostat and thus local heat flow through the cryostat's super-insulation Thus the evaporation rate of the coolant is not constant, masking the weight anomaly.

To avoid explosions, such measurements have to be taken in open atmosphere. This means condensation of water vapour and ice formation, this time not on a thin filament but on parts of the cryostat close to the neck, with much larger dimensions than a thin filament and correspondingly more condensate. In summary, from a thermal design standpoint, measuring of total weight of an experimental set-up is not recommended and data achieved in this way would likely be of little value.

2.2. Direct Detection of Sample Mass

There are four experimental criteria that provide the preferred means of direct detection of sample mass at cryogenic temperatures: (a) the sample is cooled solely by radiative exchange with a cold external heat sink, (b) the weight measurement is performed in vacuum, (c) sample weight data are taken using a magnetic suspension balance, (d) sample temperature is continuously monitored. If these conditions are met, there is neither buoyancy in a fluid or gas nor sample wetting by a cooling liquid. Radiative cooling can be realised by allowing the sample to radiate away its heat to cryogenically-cooled surroundings through the vacuum. However in this case, cool-down can be very slow.

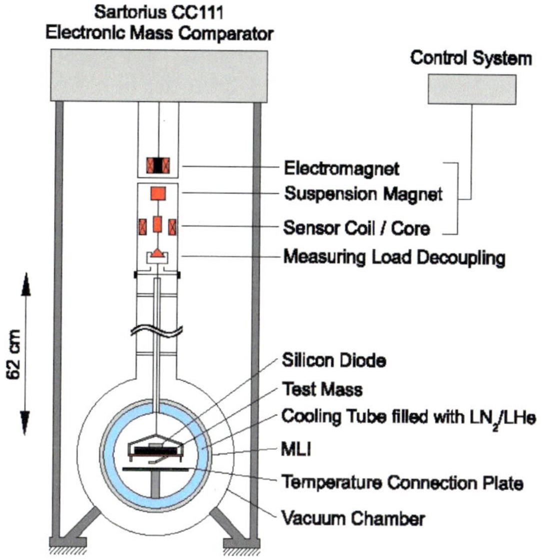

Figure 2: A superior experimental set-up for *Direct* Detectionof Sample Weight after Tajmar. Reprinted by Permission of *Meas. Sci. Tech.*, 21 (2010) 015111.

If radiative cooling through a vacuum is insufficient and cooling must occur by direct physical contact between the sample and a "cold head", weight and usually temperature data can be taken only at discrete times during non-contact, at which time the sample is warming up, hopefully at a well-defined rate. Note that a weight anomaly might occur just during contact, when the sample is coldest, and thus be missed. This is a particular problem when T_{crit} is close to the cryogen boiling point as with liquid nitrogen. This cooling method not only requires an interruption of the thermal contact between sample and cold head during data acquisition, but during contact also induces temperature gradients over the surface and through the thickness. Temperature gradients can result in only partial phase changes of the HTSC material which could substantially reduce the amplitude of possibly existing weight anomalies.

If (a) and (b) cannot be met in their entirety, cool-down can be realized by introducing a heat transfer gas, *e.g.*, helium, into the evacuated space surrounding the sample, allowing the sample to cool down more rapidly than by radiative transfer. As a consequence, buoyancy reduces sample weight. Further, if there is a small amount of water vapour in addition to the transfer gas, condensation of water vapour onto the sample and support filament will lead to a steady increase of sample weight, and the flow of water vapour to the cold sample can in turn induce convection. Application of high precision suspension balances then would be useless, and conventional balances would have to be used. A possible solution is to quickly pump out the transfer gas immediately after cool-down. Pumping of large volumes of helium is time consuming and can allow the sample to warm up during pump down.

These problems can be mitigated somewhat by clever design of the weighing chamber. To enable weighing in vacuum, any heat transfer gas must be pumped out quickly, necessitating the use of a small volume and minimum distances between sample and surrounding cold sink. Accurate temperature measurement at reasonable cost require physical contact between a thermocouple or cryogenic diode and the outside world by means of conductive wires which conduct unwanted heat to the sample. Using an intermittent physical conductive contact permits the sample to be cooled and weighed between direct temperature measurement. This also allows the sample to be suspended by the thinnest, least thermally-conductive suspension fibre suitable for the job.

Tajmar [5] has designed and implemented such an improved scheme with which he was able to cool by a combination of radiation and conduction (liquid helium transfer gas) solid samples to near liquid helium temperatures, accurately measured, and perform sensitive weight measurements in vacuum whilst the balance was self-calibrating. The sample was placed on a small weighing platform and a temperature-measuring diode was thermally attached to the sample. The platform was suspended by a single suspension fibre to a magnetic suspension system all enclosed in a vacuum chamber (see Fig.**2**). The suspension system allowed the whole platform to be raised and lowered a few millimetres at pre-determined times, during which balance calibration took place. The wires from the diode were attached to connector pins protruding from the bottom of the sample platform which, then the platform was in the lowered position, contacted a fixed base plate with appropriately-placed electrical contacts leading to the temperature readout circuit. Surrounding but not touching the weighing platform was a tightly-wound copper tube through which liquid cryogen was circulated in order the cool the sample. Helium gas was leaked into the small insulated weighing chamber to cool the sample by conduction and then quickly pumped out to allow accurate weighing in vacuum.

Theoretically the only means of warming the sample in a pure vacuum (never actually reached in practice) is conduction down the suspension fibre and radiation from hotter regions of the chamber walls.

While the evacuated magnetic suspension balance itself, as a sensitive scientific apparatus no doubt provides an excellent potential for weight measurements, precise understanding of thermal boundary conditions for gravity experiments during cool-down or warm-up periods could become a problem in successfully competing with the high weighing accuracy provided by the balance. An example numerical analysis is presented in Section 2.3 in which, among other experiments, the warm-up period of Tajmar's experiment [5] was modelled by a numerical solution of Fourier's differential equation for the temperature field of samples and their surroundings. While [5] claims radiation as the sole heat transfer in this experiment, indications of thermal leakages other than radiation were found during the analysis (Section 2.3.4). We further draw the attention of experimenters to possible shortcomings of this or any other experimental set-up that are related to small Biot numbers which can have significant influence on the results. This potential conflict arises from a possible mismatch between relaxation times for the phonon and electron responses to thermal disturbances resulting during warm-up or cool-down periods. This effect will be addressed in Section 3.

The Tajmar set-up [5] involved an expensive magnetic suspension balance and considerable machine fabrication. For most experimenters this option may not be available so direct contact between liquid cryogen and sample may be the only alternative. Though this certainly reduces experimental effort and cost, the results may be of little use however. The setup shown in Fig. **1** is a striking example.

In these cases, additional problems arise if the sample is cooled by dipping in a boiling, cryogenic liquid and the liquid is then allowed to boil off, exposing the sample to cold vapours. Immediately after dipping of the sample at time t_0, into the liquid, strong boiling (due to the heat capacity of the sample) causes the sample which is usually hanging freely on a fibre to swing violently in the liquid for some minutes. During this period, sample weight anomalies can become completely obscured. After thermal equilibrium is achieved at time $t_1 >> t_0$ the data curves become smooth (Fig. **3**). The strong weight fluctuations immediately after dipping have little effect on the experimental results provided the height of the liquid is kept at approximately constant level and the HTSC material at $t = t_1$ has not already experienced a phase change. The time scale during which a phase change occurs can be analysed numerically if heat transfer at the liquid/solid interface and thermal diffusivity of the sample material are known (compare Section 2.3.3). While boiling and convective heat transfer involving various materials have been extensively studied in the literature, little information is available about heat transfer at the liquid/sample interface for HTSC materials. In these situations, a fit to experimental data would be necessary.

Heat transfer not only determines the temperature excursions of the sample but also the rate of free convection or two-phase (*i.e.*, gas bubble/liquid) flow in liquid layers close to the sample surface. Weight *vs.* time data taken from typical HTSC dipping experiments show the super-position of a variety of contributions including rest mass of the sample, buoyancy, convection in the liquid, ejection of bubbles, temperature gradients and de-wetting. While buoyancy can be compensated to a greater or lesser extent, it can impose a significant error source if a sample holder housing the superconductor samples is used for the set-up and its material experiences phase transitions, with a corresponding uncontrollable change of volume of the sample holder. Usually de-wetting results in temperature gradients in the sample which cause a phase change to occur in slices rather than in the entire sample at once. No firm conclusions concerning existence or non-existence of a gravity anomaly can be taken from this type of data.

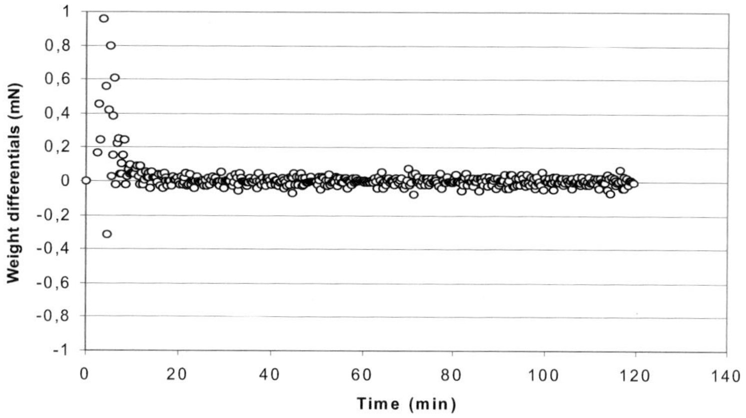

Figure 3: Weight Differentials (Apparent Weight Differences $AW(t_{N+1}) - AW(t_N)$) Between Single Data Points Taken at Discrete Time Steps $t_N < t_{N+1}$ Using a Pile of 10 YBaCuO Pellets in a Sample Holder that at t = 0 is Dipped into LN_2. Reprinted by Permission of *Physics Essays Publication*, from *Physics Essays*, 16, 236-253 (2003).

In the case of the sample being completely immersed in liquid for the duration of the experiment, "dummy" weights can be used as standards [1] that under no circumstances should undergo changes of their rest mass. Identical surface quality and heat transfer properties of the HTSC samples and dummy samples can be obtained by housing them in a sample holder made from a material such as PTFE (Teflon), whose thickness and material is selected to shift the phase change of the HTSC samples to times greater than t_1, but again there may be phase changes of the PTFE material which would uncontrollably alter buoyancy. It may also be possible to chemically treat the HTSC samples to be non-HTSC, eg by addition of a tiny quantity of Zn to the precursor powders, but otherwise identical.

2.3. Numerical Analysis of Warm-up and Cool-down Experiments

A numerical FE (finite element) analysis has been performed to obtain sample temperature evolution, T(x,y,z,t), under strongly differing heat transfer conditions in cool-down and warm-up experiments. In terms of thermal resistances to a liquid coolant (at 77 K) or to a radiative environment (at RT), temperature excursion of the HTSC pellets is analyzed with respect to four conditions, namely the following items (see also Fig. **4**):

1) **Cool-down**, with only liquid boiling/liquid convection at the sample/coolant interface; heat transfer is modeled with temperature dependent (and thus time-dependent) boiling and convection heat transfer coefficients at the interface to the coolant (resistance R1); there is no further thermal resistance between coolant and sample (Section 2.3.1),

2) **Warm-up**, with no liquid phase present but a convection resistance (R21) at the interface to ambient and, in series, a conductive resistance (R22, of a sample holder) between interface and sample (a weak radiation contribution is included in R22) (Section 2.3.2),

3) **Cool-down**, with conductive resistance between pellets and sample holder walls (R_{31}) and, in series, a liquid boiling/convection resistance (R_{32}) at the interface to the coolant (Section 2.3.3); the sample holder is the same as in item 2,

4) **Warm-up**, solely radiative heat transfer between sample and radiative environment, with solely a thermal radiation resistance (R4 only - no further resistance between environment and sample), with the sample exposed under vacuum. This part of the diagram schematically describes the inner heat transfer rates in the suspension balance experiment reported by [5] (Section 2.3.4).

The total thermal resistance increases from items (1) to (4).

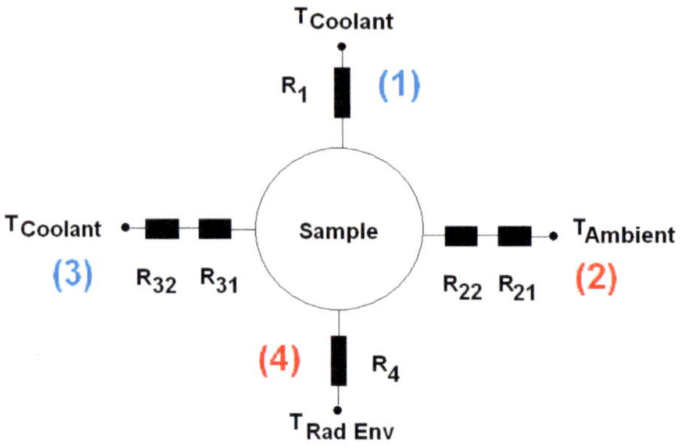

Figure 4: Thermal resistances (schematic) indicating the conditions under which finite element calculations for items (1) to (4) were performed. Indices of the individual resistances, R_{kl}, are assigned according to direction of heat flow (hot to cold).

2.3.1. Cool-down Experiment (Simulation of the Meissner-Ochsenfeld effect)

Previously [6], we reported on the time dependence of the Meissner-Ochsenfeld effect. Numerical analysis shows that it takes a superconductor pellet after start of cool-down a time interval, Δt, in the order of tens of seconds, depending on the anisotropy of its diffusivity, to get a sufficient number of volume elements, N_{SC}, superconducting (Fig. **5**); otherwise there is no stable levitation position where total levitation force balances gravity. In this analysis the size of each of the N volume elements was 200 x 200 x 20 μm^3 for simulation of a polycrystalline YBCO material (because of cylindrical symmetry and under mapped meshing, the number of finite elements is different from N).

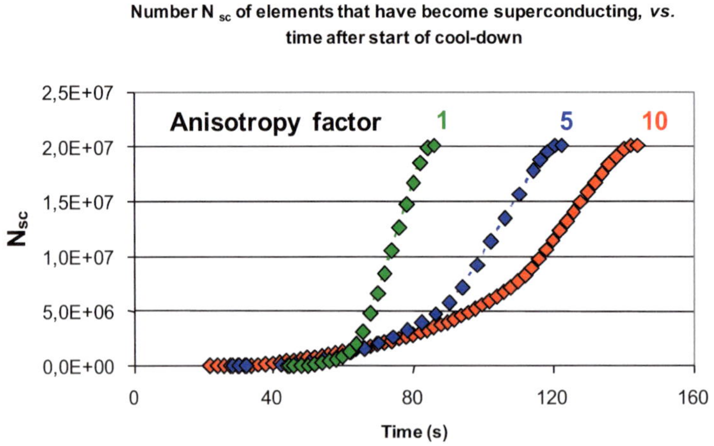

Figure 5: Number $N_{sc}(t)$ of elements for 3 different anisotropy ratios of thermal conductivity (see text) that have become superconducting, *vs.* time after start of cool-down. The curves are calculated from numerical (finite element) results for the temperature profiles T(x,y,z,t) during the cool-down period of a YBCO pellet. The figure shows part of the data reported in *Cryogenics* 49 (2009) 433-448.

The sample is directly wetted by LN_2, with no thermal resistance other than from boiling and convection between coolant and sample surface. Results of the simulations compared well with levitation experiments using a YBCO pellet of exactly the size as simulated in the finite element calculations (the calculated time for obtaining a stable levitation position was indeed confirmed in the experiment). This in turn confirmed that simulation of the external heat transfer (violent boiling followed by liquid convection during field cooling) was adequate.

2.3.2. Warm-up Experiment using small YBCO Pellets in a Sample Holder

This the reverse of the cool-down (dipping) experiment reported in [1] (the corresponding numerical analysis of the cool-down period is described in the next Subsection). In the present case, warm-up was in ambient atmosphere. Comparison of the considerably longer time needed for the sample to arrive at room temperature was roughly confirmed by the calculations although extrapolations were made, since extending the simulation to very long time intervals consumes much computer time. External heat transfer was by convection in natural atmosphere and by radiation to the warm environment using heat transfer coefficients from standard texts.

Fig. **8** (red full and open diamonds and squares) shows derivatives dT/dt of pellet temperature at positions on top and bottom of the pile of 18 pellets, respectively.

2.3.3. Cool-down Experiment using small YBCO Pellets in a Sample Holder

In this experiment, samples (pellets) were housed in a cylindrical sample holder. We will not comment on tentative conclusions from the unexplained observation of a weight increase of the pellets but simply analyse the thermal aspects of the experimental setup as an example where a large thermal resistance represents the interface between the coolant and samples.

Since [5] designed a replication experiment, it is worthwhile to emphasize the difference between the experiments reported in [1, 5]: Tajmar measured weight dependence on temperature of a single pellet under vacuum, in zero applied magnetic field and during *warm-up*, with no buoyancy effects to be compensated. Temperature distribution in pellet and cavity is almost perfectly homogeneous, due to the weak external heat transfer (although it was not solely radiative, see Section 2.3.4). Transition of the whole pellet volume, from superconducting to normal state, is *sharp in time* (overall warm-up rate is rather quick). The dipping experiment [1], on the other hand, used a sample holder that had an internal cavity housing up to 18 small YBCO or BSCCO pellet partly in a local external magnetic field, with the sample holder dipped into liquid nitrogen, *i.e.,* the measurement was done during *cool-down* and with buoyancy in the liquid that had to be compensated. Heat transfer from the inner wall of the sample holder to the pellets is much stronger than under radiation heat transfer, though the strong heat transfer fluctuations (violent boiling, followed later by convection) at the outer sample holder surface is damped by the thermal resistance of the PTFE wall. This means cool-down proceeds fast, and there will be strong, non-vanishing radial temperature gradients in the pellets. Transition from normal to superconducting state in the pellet volume accordingly will be *gradual in time.*

Overall design of sample holder and an internal cavity is shown in Fig. **4** of [1] with a detailed description not repeated here. Cool-down of the sample holder is described by heat transfer coefficients, α, for boiling and convection under liquid nitrogen. We used the same α-values reported in [6], where the heat transfer coefficients, although obtained from the literature from measurements on a metallic surface, were confirmed by levitation experiments. The thermal resistance and heat capacity of the walls of the sample holder serve to shift a possibly-existing weight change during phase transition to a time after which the initial violent movements of liquid and sample holder under boiling will have ceased.

Temperature evolution during cool-down and radial temperature gradients are illustrated as a contour diagram in Fig. **6**. Transition of the HRSC pellets (red) from normal to superconducting state has not started until well after 60 s, and is not accomplished until after t = 720 s after start of cool-down (see Fig. **7**).

Fig. **8** (blue full and open diamonds and squares) shows derivatives dT/dt of pellet temperature at positions on top and bottom of the pile of 18 pellets, respectively. Also shown is the dT/dt derivative of the purely radiative heating condition of item 4 (black circles). The derivatives dT/dt of the cool-down experiment (item 3) are orders of magnitude larger than the values calculated for the suspension balance experiment (radiative heating, item 4).

Radial temperature differences in the pellets in the cool-down (item 3) experiment shown in Fig. **9** are quite large due to conductive thermal resistance introduced by the wall of the sample holder and also a few thin layers of glass fibre paper between sample holder and pellets.

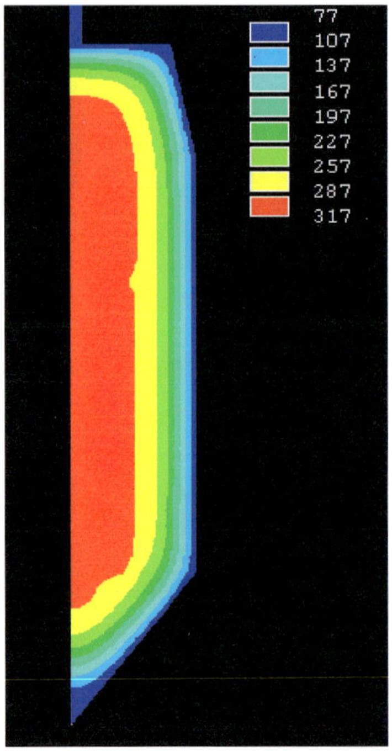

Figure 6: Temperature distribution in the sample holder used in the cool-down (dipping) experiment reported in [1] calculated at t = 60 s after start of cool-down under boiling and convection in LN_2. For details of geometry of sample holder and superconducting pellets housed in an internal cavity see [1].

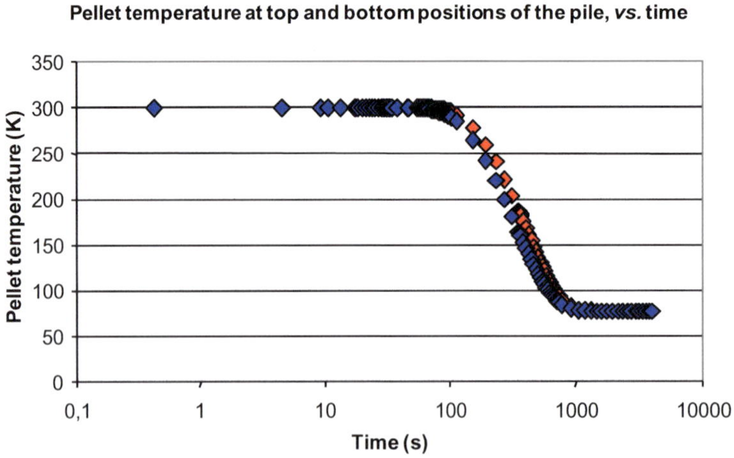

Figure 7: Pellet temperature evolution calculated at central pellet positions for the uppermost (red diamonds) and lowermost pellets (blue diamonds) of the pile consisting of in total 18 YBCO pellets in the experiment reported in [1] *vs.* time after start of cool-down.

We also compared calculated temperature evolution in the case where the pellets were replaced by ceramic powder in the cavity with the temperature measurement reported in Fig. **8** of [1]. There, T = 77 K is reached in the cavity after about t = 300 s. This is roughly confirmed by the numerical simulation if we assume an uncertainty between 0.1 to 1 K of the temperature measurement using a Pt-100 thermometer. The same heat transfer coefficients for cooling with LN_2 have been applied as before thus confirming the validity of the set of α-values prepared in [6] from the literature.

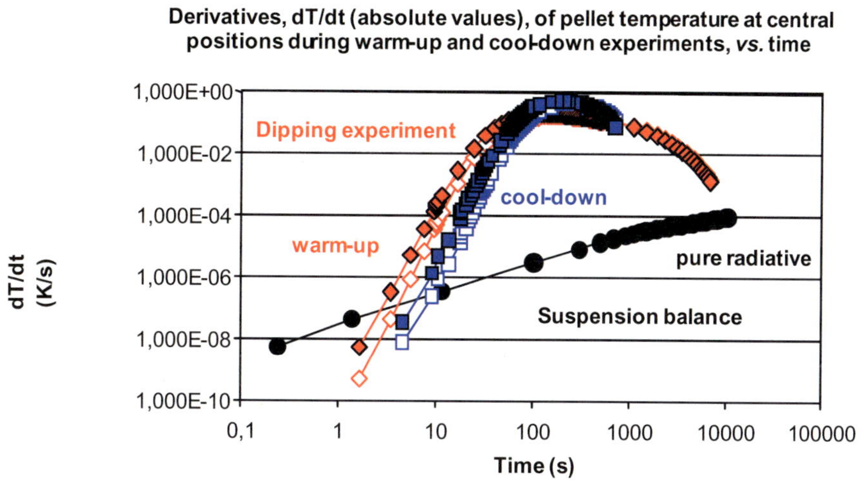

Figure 8: Derivatives, dT/dt (absolute values), of pellet temperature at central positions of pellet during warm-up and cool-down periods in suspension balance (black circles, pure radiative heat transfer) and dipping experiments (full and open diamonds and squares, for the dT/dt at top and bottom positions of the pile, respectively), *vs.* time.

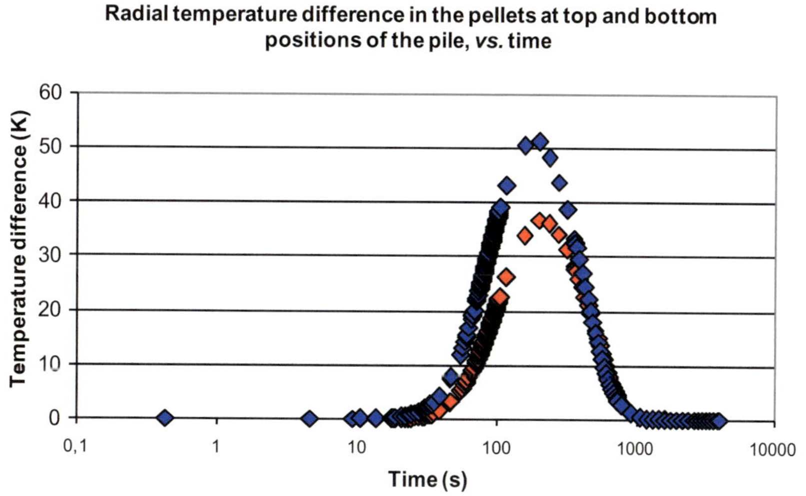

Figure 9: Radial temperature differences for pellet temperature evolution calculated at central pellet positions for the uppermost (red diamonds) and lowermost pellets (blue diamonds) of the pile consisting of in total 18 YBCO pellets, *vs.* time after start of cool-down.

2.3.4. Warm-up Experiment using a YBCO Pellet in a Suspension Balance

As above, we will not comment on the interpretation of the results of the experiment reported by Tajmar [5]. Tajmar's innovative setup is used simply as an example, in this case for the high radiative resistance between YBCO pellets and a radiative environment the temperature of which increases slowly during warm-up from cryogenic to room temperature.

This experiment, essentially an attempt to replicate the "dipping" experiment described in the previous subsection (item 3), applies pure radiation heat transfer to a superconducting pellet. The authors [5] correctly claim that weak external heat transfer (like radiation exchange) in relation to much stronger (conductive) internal heat transfer, reflected in a small Biot number, is responsible for uniform temperature evolution in the sample, which means all volume elements of the sample pellet will arrive at critical temperature at almost the same time after start of warm-up.

While application of Equations. (2) to (5) in [5] was made to quantify this expectation, these equations are valid under static conditions only. But it was correct when the authors more intuitively assumed that sample temperature evolution was uniform. The phase transition in the whole sample volume accordingly is completed instantaneously which means a very large decay rate of electron pairs to quasi-particles will be required to satisfy the phonon aspect of evolution of the disturbance (this aspect will be addressed in Section 3). But from an initial viewpoint, experiments [1, 5] cannot be declared identical, or item (4) a replication of item (3), simply from thermal boundary and internal heat transfer considerations. Contrary to a statement made in [2] (a first attempt made by the same authors to replicate the dipping experiment), the dipping (cool-down) experiment reported in [1] was *not* performed in gaseous but solely under *liquid* nitrogen, and most of the error analysis in [1] had to account for exactly this problem.

In the present analysis, we simulate the neighbourhood of the pellet by two concentric cavities with closed ends. Tajmar [5] reports length and inner radius of the cavity is 92 and 25 mm, respectively. To provide an upper limit of the derivative, dT_{Cav}/dt, of cavity temperature, T_{Cav}, we assume that the cavity has a wall thickness of 2 mm; increased thickness would increase thermal mass of the cavity und thus reduce dT_{Cav}/dt.

For the simulation, we use a thermal emissivity, ε_{ML} = of 0.0315 at cryogenic temperatures, of 30 metallized Mylar foils. For this ε_{ML} see [7] and references cited therein. The simulated sample is a $YBa_2Cu_3O_{7-\delta}$ (YBCO) pellet of 12.5 mm radius and 3 mm thickness. During warm-up, the YBCO pellet is freely hanging inside the cavity. In the experiment, the pellet is connected to the suspension balance by a single thin V2A-wire of 0.125 mm radius and of 62 cm length. In the simulation, the wire near the pellet surface is split into 3 more wires (to avoid meshing problems that would arise near a centrally located point of contact), each with the same radius as before, and the 3 wires are attached to a sample holder that consists of a 0.5 mm thick plate of radius 12.5 mm (same as pellet radius, an assumption with no effect on the results). Pellet and sample holder are positioned 3 mm above a connecting plate of 15 mm radius that besides radiative heat exchange is thermally coupled to the cavity wall by a conducting solid support column of 1 mm radius. Cavity, sample holder, connecting plate and column are all made of Cu. It should be emphasized that (but small, if any) deviations of the simulated experimental setup from the real laboratory setup do not have any significant influence on the simulation results.

Thermal conductivity, λ, and specific heat, c_p, of YBCO and Cu used in the simulations are the same as reported in [6]. Anisotropy of thermal conductivity of YBCO is accounted for again by the ratio $\lambda_{ab}/\lambda_c = 10$ in the ab-plane and c-axis direction. The emissivity, ε_{SC}, of the YBCO pellet and the ε_{Cu} of copper are about 0.9 and 0.1, respectively. The ε_{Cu} is taken from [8] for cryogenic temperatures, but little information is available for the emissivity of YBCO, especially in the SC state. We use the value $\varepsilon_{SC} = 0.9$ which is typical for ceramics (YBCO in the normal conducting state) in the interval between cryogenic to low temperatures, [9]. All emissivities, ε, used in this paper are hemispherical total values.

2.3.4.1. Radiative Energy Balance (Analytical Expressions)

Radiative heat transfer through the multi-foil super-insulation, in cylindrical geometry, is calculated in two steps: The first step involves the relation for radiative transfer between two concentric, infinitely extended cylinders with surfaces A_1 and A_2 and with no foils in-between. Here the outer surface, A_2, constitutes the environment outside the cavity (a vacuum chamber made of V2A stainless steel fixed into the suspension balance). The inner surface, A_1, is the cavity wall. The V2A steel wall of the vacuum chamber is at T_2 = 300 K (constant), while the intervening cavity wall is at transient temperature $T_{Cav} = T_1 = T_1(t)$ during warm-up. With ε_{V2A} indicating the emissivity of the vacuum chamber and σ the Stefan-Boltzmann constant, we have the well-known relation for heat flow:

$Q_{Rad,0}(t) = -\sigma A_1 (T_{Cav}^4 - T_2^4)/[1/\varepsilon_{Cu} + (A_1/A_2)(1/\varepsilon_{V2A} - 1)]$, **(1a)**

see *e.g.*, [10]. All surfaces are assumed as grey diffuse radiators, and all $Q_{Rad}(t)$ in Eqs. (1) to (3) are given in W. The minus sign in Equation (1a) accounts for a positive radiative heat flow from the environment A_2 to surface A_1 during warm-up of the sample.

Second, we use the "shielding efficiency", ζ, of evacuated multi-layers [11] to reduce $Q_{Rad,0}(t)$ from Equation (1a) to

$Q_{Rad}(t) = \zeta\, Q_{Rad,0}(t)$ **(1b)**

with

$\zeta = \{1 + (\eta_{Cav}/\eta_{ML})\, r_1 \sum 1/[r_1 + (n/(N+1))(r_2 - r_1)]\}^{-1}$ **(1c)**

in cylindrical geometry. The dimensions of the cavity require application of Equation (1a-c), and by no means would the result be correct if one would erroneously apply corresponding expressions for plane-parallel walls.

In Equation (1c) we use N = 30 the number of foils, and the summation is taken over $1 \le n \le N$. In this equation, η_{Cav} and η_{MF} denote reduced (dimensionless) emissivity, $\eta_{Cav} = \varepsilon_{Cu}/(2-\varepsilon_{Cu})$ and $\eta_{ML} = \varepsilon_{ML}/(2-\varepsilon_{ML})$, respectively, and r_1, r_2 the radii of surfaces A_1 and A_2 (r_1 = 25, r_2 = 250 mm). For simplicity, we have assumed $\varepsilon_{Cu} = \varepsilon_{V2A}$, at cryogenic temperatures.

The residual error contained in Equation (1b,c), *i.e.,* neglecting the end surfaces, will be compensated by the finite element model described in the next Subsection. Division of $Q_{Rad}(t)$ by A_1 and the temperature difference

$\Delta T(t) = T_2 - T_{Cav}(t)$ **(2a)**

between ambient and cavity yields a transient heat transfer coefficient

$\alpha_{Rad}(t) = Q_{Rad}(t)/[\,A_1\, \Delta T(t)]$ **(2b)**

for radiative exchange between ambient (the vacuum chamber) and the outer wall of the cavity. As for many evacuated super-insulations, the α_{Rad} are very small, below 2 W/m^2. Under transient conditions, heat flow $Q_{Rad}(t)$ through the super-insulation onto the cavity wall splits into

a) Radiation (emitted or reflected, because all $\varepsilon < 1$) that is distributed internally, *i.e.*, from the cavity inner wall to sample, sample holder, connecting plate and support column,

b) Thermal conduction that distributes heat also in the cavity wall according to its heat capacity.

In the subsequent energy balances, and in all numerical simulations reported in the next Subsection, temperature dependent quantities (density, specific heat, thermal conductivity as in [6]) have been applied.

Assume tentatively for the moment, and as a first order of magnitude estimate, that there will be *no* radiation exchange inside the cavity. This yields an upper limit for the rate of change of (medium) cavity temperature

$dT_{Cav}/dt \le Q_{Rad}(t)/(V_{Cav}\, \rho_{Cav}\, c_{Cav})$ [K/s] **(3)**

with V_{Cav} the volume and ρ_{Cav} and c_{Cav} density and specific heat of the cavity walls at cryogenic temperatures, respectively.

Fig. **10** (solid diamonds) shows dT_{Cav}/dt calculated from Equation (3) from which we can roughly estimate the period of time, $\Delta t'$, needed for the cavity with *no* radiation exchange inside, to arrive at T = 160 K (the value indicated in Fig. **4** of [5] after about 3 hrs into the warm-up period),

$$\Delta t'[(T_{Cav}(t)] = [T_{Cav}(t{=}3\ hrs)\_T_{Cav}(t{=}0)]/\ dT_{Cav}/dt \qquad \textbf{(4)}$$

Since the rates dT_{Cav}/dt decrease with increasing time, we will take dT_{Cav}/dt at 120 K (about 5.6 10^{-4} K/s) from Fig. **10** as an average value. This yields $\Delta t'[(T_{Cav}(t)]$ of about 41 hrs instead of the 3 hrs indicated in Fig. **4** of [5], which strongly suggests that radiation was not the *sole* heat transfer mechanism when the experiment was performed. Obviously, pure radiative heat transfer would be too weak to confirm the temperature increase to the claimed 160 K.

For a second, again order of magnitude estimate with the same analytical model, assume that the same thermal power, $Q_{Rad}(t)$ from Equation (1a-c), is delivered to the *whole* assembly (cavity plus the inner components like sample, sample holder *etc.*). We will take dT_{all}/dt at 85 K (about 7.2 10^{-4} K/s), again from Fig. **10** (solid squares), to estimate $\Delta t'[(T_{all}(t)]$. This would take all components about 6 hrs to reach the critical temperature, T_{crit} = 92 K, of standard YBCO, still definitely longer than the 3 hrs indicated in Fig. **4** of [5]. After this period of time, the cavity temperature necessarily is at 92 K (the evacuated multi-foils act like a thermal mirror), not at the claimed 160 K. This is also a natural consequence of the isotropic temperature distribution occurring in a cavity furnished with evacuated thermal super-insulation, under static or as is the case here, under quasi-static conditions. Already these comparatively simple but obvious energy balances deliver results for the expected temperature evolution that are in clear contrast to what is reported in [5].

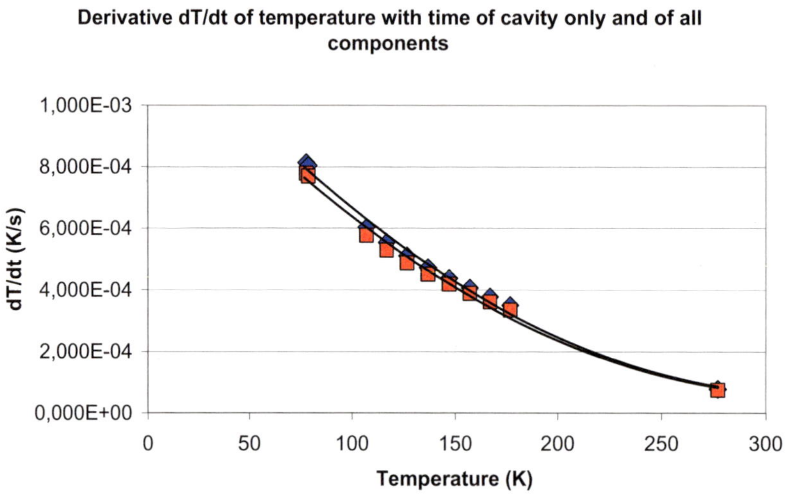

Figure 10: Derivatives w.r.t. time, dT_{Cav}/dt and dT_{all}/dt, of temperature of cavity (red squares) and temperature of the whole assembly (blue diamonds that for cavity plus pellet, sample holder *etc.* almost coincide) used in the warm-up (suspension balance) experiment reported in [5] *vs.* cavity temperature (all data are mean values).

It makes no difference for these analytical order of magnitude-estimates whether a YBCO sample or a Cu-pellet is introduced into the cavity, and all these expectations are confirmed in the numerical modelling described in the next Subsection.

2.3.4.2. Numerical Modelling

A standard finite element (FE) program and mapped meshing was applied to YBCO-pellet, sample holder and connecting plate, while free meshing was allowed for the cavity including its end faces. The four wires connecting the sample holder to the balance, and the column to support the connecting plate, were modelled with 1D-thermal elements. For all other components, 3D-solid elements were used. Radiation heat transfer inside the cavity was modelled with 2D-surface elements using radiation exchange factors, the different emissivity of all components, and a radiation matrix element that incorporates all these data and the corresponding view factors in-between.

The calculated temperature excursions successfully reproduced not only experimental surface temperature excursions, but the results are also in very good agreement with levitation experiments performed by one of the authors (H. R.). Both observations confirm the heat transfer simulations made in the FE simulations of the four experiments (items 1-4). In particular, the FE analysis provides a look into the *interior* of the sample and the thermal relaxation time.

As in the previous Subsection, neither cavity nor sample temperature claimed in Fig. **4** of [5] can be confirmed by the results of the numerical simulations at a time 3 hrs after start of the warm-up period if we assume solely radiative heat flow from the vacuum chamber to the cavity. Instead, an effective, external heat transfer coefficient not confined to radiation exchange had to be defined to speed up temperature increase so as to conform to the data measured by Tajmar. Calculations started at T = 77 K and were uniform for cavity, sample and sample holder, connecting plate, column and wires. The end of the long wire that connects the sample holder to the suspension balance was kept at 300 K constant.

The effective heat transfer coefficient describes *external* heat transfer, *i.e.,* from ambient, that is,. the vacuum chamber, to the cavity only, while *internal* heat transfer, *within* the cavity, is *solely* by radiation. A realistic external heat transfer coefficient may result from a variety of heat leakages like solid/solid contacts through the super-insulation, or from non-insulated parts of the assembly, or from residual gas pressure that was not sufficiently removed before and during the warm-up experiment (note the strong fluctuations of gas pressure seen in Fig. **4** of [5]). Various constant values of the effective heat transfer coefficient were then assumed in the numerical calculations (α_{eff} = 0.1, 0.25 and 1 W/(m^{2}K)). For comparison, a hypothetical free air convection around a horizontal cylinder was modelled using the relations reported in [12]. Free convection α-values for ambient air, under this geometry, amount to about 10 W/(m^{2}K), which, as can be seen, is certainly too large. Another option was to assign α_{eff} = 0.1 W/(m^{2}K) to regions near the ends of the cavity and 0.25 W/(m^{2}K) to its central parts (indicated by "0.1/0.25" in the following figures).

Fig. **11a** and **b** shows numerically calculated cavity and pellet temperature *vs.* time for various values of the effective heat transfer coefficient. Comparison with the *cavity* temperature reported in Fig. **4** of [5] shows that after t = 3 hrs the curve calculated with α_{eff} = 0.25 W/(m^{2}K) provides the best approximation to the reported data, while comparison of calculated *pellet* temperature with the value (about 95 K) reported from the measurements is best when using α_{eff} = 0.1 W/(m^{2}K). It is again obvious, now from Fig. **11a** and **b**, that *solely* radiative heat flow to the cavity explains neither pellet nor cavity temperature evolutions reported in Fig. **4** of [5].

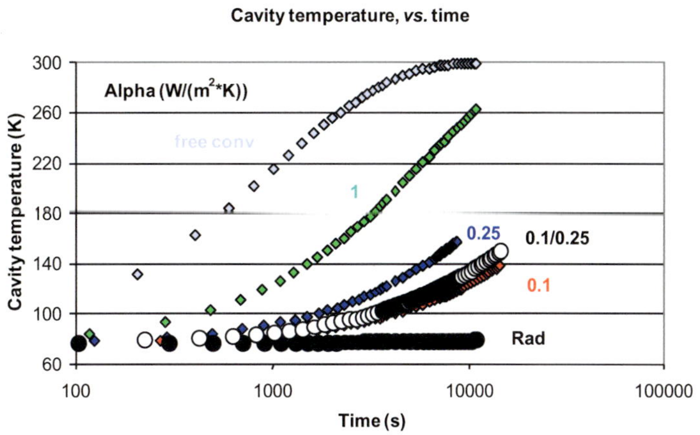

Figure 11a: Calculated cavity temperature evolution *vs.* time, for the warm-up (suspension balance) experiment. Data come from the finite element model with different values of the effective external heat transfer coefficient, α, describing heat exchange either solely by radiation ('Rad') between vacuum chamber and cavity through multilayer super-insulation, of using effective α-values (0.1 to 1 W/(m^{2}K)). Also shown for comparison is free convection of ambient air surrounding a horizontal cylinder. The α-value incorporates possibly existing heat leakages and is adjusted to reproduce the temperature evaluation reported in [5]. The '0.1/0.25' curve refers to α-values of 0.1 and 0.25 W/(m^{2}K) applied at the end and central sections of the cavity, respectively.

Fig. **12** shows temperature distribution on the upper surface of the pellet obtained for purely radiative heat flow soon after the start of warm-up. At cryogenic temperature, radiative heat transfer is weak as in the experiment in [5] but even temperature disturbances caused by the 3 thin wires attached to the sample holder can be made visible on the pellet surface (compare the temperature scales indicated in the figure). As was to be expected, the temperature distribution on the pellet is very homogeneous, within few 10^{-3} K, for all heat transfer options tested for simulation of this experiment and confirms the expectation (temperature homogeneity) in [5].

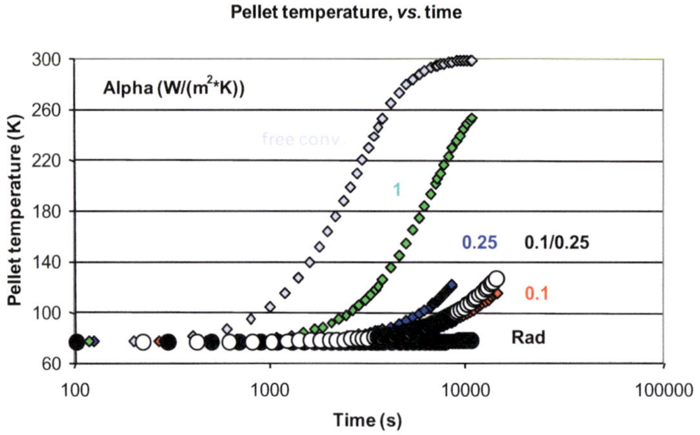

Figure 11b: Calculated pellet temperature evolution *vs.* time in the warm-up (suspension balance) experiment reported in [5]; results are given for different values of the external heat transfer coefficient, α (same as in Fig. **11a**). For details see text.

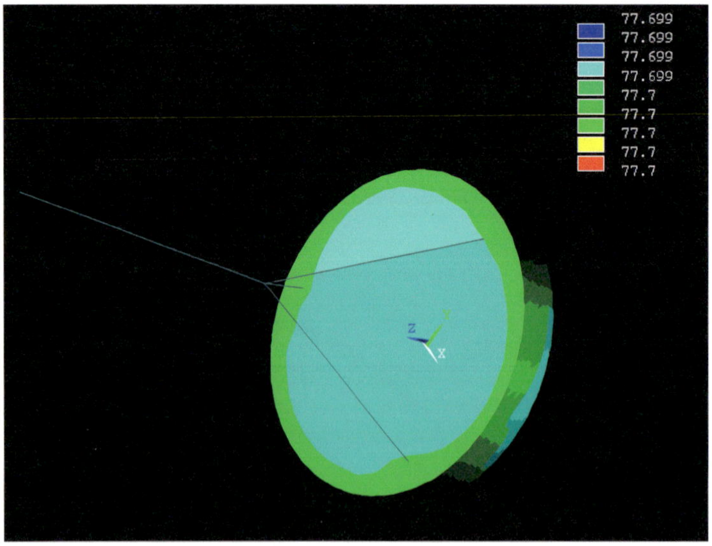

Figure 12: Temperature distribution on the upper surface of pellet soon after start of the warm-up period in the suspension balance experiment reported in [5] calculated for solely radiative external heat transfer. The thin lines indicate the 1D-elements used for the 4 thin wires that attach the sample holder to the balance. The xyz-triad located at the origin of the pellet is given for orientation (the co-ordinate z points in vertical direction).

Pellet and cavity surface temperature distribution is given in Fig. **13a** and **b** for the pellet at a time immediately before the phase transition at 92 K. At the phase transition, there are almost no radial temperature gradients in the pellet.

With the same volume elements as before (width and length of 200 µm but 50 µm thickness), we have about N = 7.3 10^5 elements in the pellet volume of 25 mm diameter and 3 mm thickness. Counting the number $N_{SC}(t)$ of elements with temperatures $T < T_{crit}$ then yields the curves given in Fig. **14**. As is to be expected from the very homogenous temperature distributions in the pellet, the transition from superconducting to normal conducting state is very sharp for this condition ("warm-up 4"). The curve falls within a Δt of 4 s from the first to all elements becoming normal conducting.

Note in Fig. **14** that the temperature interval, ΔT, between start temperature in the experiments and phase transition is between 77 K and T_{Crit} in the warm-up experiments (ΔT about 15 K) while it is between RT and T_{Crit} during cool-down (ΔT almost 210 K), respectively. Together with the thermal resistances in the pellets and those at the interface to the coolant or to radiative environment, this explains the location (in time) of the phase transition (saturation of the curves at maximum $N_{SC}(t)$ during cool-down, or at $N_{SC}(t) = 0$ during warm-up, respectively). The shift observed from curve (1) to curve (3) results from the additional conductive resistance in (3), and the corresponding shift from (2) to (4) is due to a very large radiative resistance in the suspension balance experiment (4).

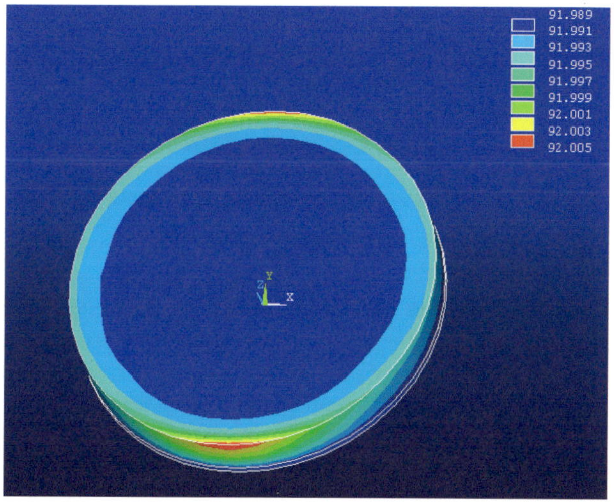

Figure 13a: Temperature distribution on upper surface of pellet calculated for the external heat transfer coefficients, $\alpha = 0.1$ and 0.25 W/(m^{2}K) applied to the end and central sections of the cavity, respectively, at t = 7120 s after start of warm-up phase in the suspension balance experiment reported in [5]. Data are given in the immediate neighbourhood of the superconductor/normal phase transition. The xyz-triad located at the pellet origin is given for orientation.

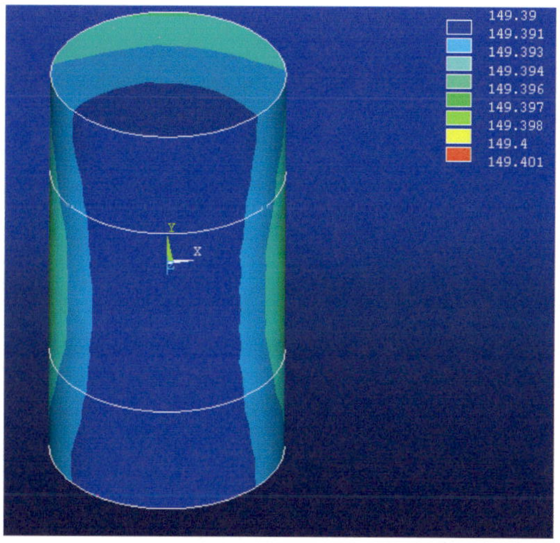

Figure 13b: Cavity surface temperature distribution (with closed ends) calculated for the external heat transfer coefficients, $\alpha = 0.1$ and 0.25 W/(m^{2}K) applied to the end and central sections of the cavity, respectively, at t = 4 hrs after start of warm-up phase in the suspension balance experiment reported in [5]. The xyz-triad located at the pellet origin is given for orientation.

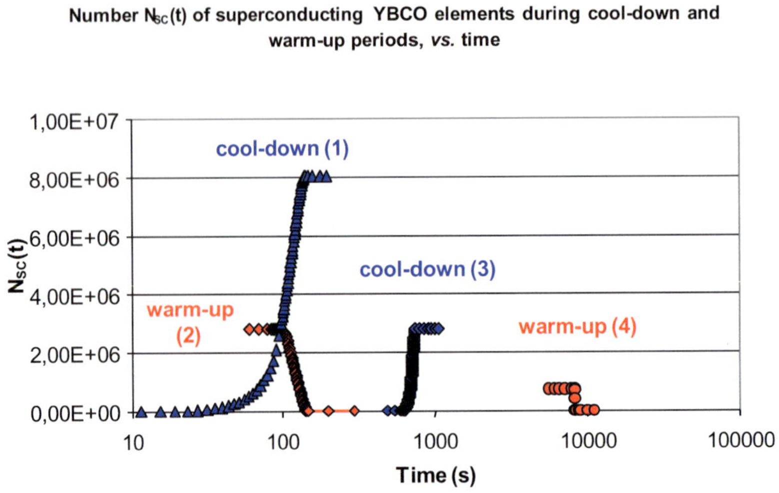

Figure 14: Number $N_{SC}(t)$ of superconducting (SC) of a total of N volume elements (width and length each 200 μm, thickness 50 μm) in the cool-down and warm-up experiments (items 1 to 4) for the finite element simulation of temperature excursion of polycrystalline YBCO pellet material *vs.* time, to identify the location (in time) of the phase transition after start of cool-down. See Fig. **4** for identification of the corresponding thermal resistances.

2.3.4.3. Summary of Section 2.3.4.

We have confirmed the expectation made in [5] that sample temperature evolution would be highly uniform during the warm-up period of item (4). It appears that either radiation was not the sole heat transfer mechanism in the suspension balance experiment, or else radiative heat transfer was rather non-uniform outside and inside the cavity. This situation might be improved if measures are taken against obviously existing thermal leakage. On the other hand, measures like these would hardly change the dT(t)/dt of the pellet, and the principal question is whether it is really an advantage in a gravity experiment of group 1 of the above to obtain a *sharp* superconductor/normal conductor phase transition of the whole sample (see next Section).

3. LIFETIMES OF THERMAL EXCITATIONS

An aspect that may become important for any of the experiments mentioned in the previous sections covers a possible mismatch between relaxation times of

a) Propagation of a thermal disturbance by conduction in a superconductor sample during cool-down or warm-up experiments, and

b) The response of the electron system of the superconductor to the same disturbance.

3.1. Two Competing Relaxation Times

Thermal diffusivity determines a characteristic time, τ_{Ph}, for propagation of thermal waves in solids after a thermal excitation. At very low temperature the superconducting electronic system is decoupled from propagation of thermal waves but reflects its own dynamic response to this or other specific excitations by corresponding relaxation times, τ_{El}. Both relaxation times, τ_{Ph} and τ_{El}, to the same disturbance (in the present case, a temperature increase or decrease) are not necessarily identical. And it should be mentioned that in superconductors there is also a relaxation time with respect to diffusion of magnetic flux density the comparison of which with thermal relaxation time in superconductor and matrix material has lead to development of multi-filament composite wires.

The situation can be compared with current injection experiments (see for example [13, 14]): There, the decay of electronic excitations is studied after an artificial increase of the number of quasi-particles. Here, we have an increase of this number from decay of electron pairs if sample temperature, T, increases (but T < T_{Crit}), with the total number of particles conserved. Also, current injection experiments deal with thin films, here we investigate solid samples (YBCO pellets).

During levitation experiments, traditionally a decision between "levitation or not" is based entirely upon observation of *bulk temperature*, T(t), below critical temperature, T_{Crit}. We instead in [6] made this decision on the basis of the *internal* temperature excursions, T(x,y,z,t) which generate a non-zero levitation force from each of the N volume elements only if the corresponding element temperature is below T_{Crit} and the vector sum of all contributions to levitation finally overcomes the sample weight. The question then is whether during cool-down the re-arrangement of the corresponding wave function (anti-symmetric for electron pairs), as a response of the electron system of the superconductor to the thermal disturbance, really keeps step with the temperature excursion of the lattice (the phonon aspect) to re-establish dynamic equilibrium between generation of electron pairs and their decay to quasi-particles.

For a provisional comparison of τ_{Ph} and τ_{El}, consider a thin YBCO film of d = 0.1 μm thickness onto which a very short thermal pulse (a "Dirac pulse" from a laser) is delivered. The small thickness, d, is assumed just to compare with thin superconducting metallic films in current injection experiments. Assuming the thermal diffusivity, a, of the YBCO film is constant (independent of temperature) and considering only the dominating (n = 1) term in a series expansion of T(t) (see *e.g.*, [15]), the time constant $\tau_{Ph} = d^2/(a\,\pi^2)$ of the film amounts to about 0.3 ns at T = 90 K. This is definitely smaller than τ_{El}, the lifetime of disturbed states which is on the order of 0.1 to tens of μs [13, 14] observed in current injection experiments using Al or Sn films of same thickness.

Measurement of quasi-particle recombination time (lifetime of the disturbed electron system) in superconducting tin after current injection from a current pulse or from laser pulse excitation was performed in [14] with current i_P = 10.1 mA and t_p = 1 μs the length of the current pulse. There are about 6.3×10^{10} unit charges injected into a volume of at least 3×10^{-14} m^3, which means an injected particle density of about 2.1×10^{24} $1/m^3$ (see [14], Fig. **3**). On the other hand, after a thermal disturbance in YBCO, we have a probability for exciting quasi-particles from pairs proportional to exp[-ΔE(T)/kT]. Here we use the equilibrium expression. Assuming that a HTSC sample was heated locally from 90.5 to 91.999 K and using the electron density at temperature close to absolute zero of about 6×10^{27} $1/m^3$ of which about (½) (1/10) is available for pair formation, the volume in which current was injected in [14] and a standard (BSC) temperature dependence of ΔE(T) [16], this yields the number of additionally excited electron states in the same volume of about 4.7×10^{26} $1/m^3$, or about two orders of magnitude larger than in the current injection experiment, as a lower limit.

Therefore, a naive estimate of the lifetime $\tau_{El} = \tau_{El,therm\ excit}$ could be made by $\tau_{El,therm\ excit} = M\ \tau_{El,Current\ inject}$, with $\tau_{El,Current\ inject}$ on the order of 0.1 to 1 μs and $M \geq 10^2$. This would result in a τ_{El} of at least 10 to 100 μs for the thermal disturbance, much larger than τ_{Ph}, if the temperature is not too close to T_{Crit}. Note that the current injection experiments were performed in [14] at temperature below 2 K, safely below T_{Crit} of Sn of 3.72 K and that the quasi-particle diffusion length in the present experiments is much smaller than the sample thickness which means there will only be little escape of phonons.

To improve understanding of experiments focussed on possible weight anomalies in superconductors during phase transition, the generation rate of electron pairs as bound states in superconductors and their decay to quasi-particles as a dynamic equilibrium process or under thermal excitation and recombination thereof, have to be analysed. Like a variety of other excitations in superconductors (absorption of micro-waves, steady-state or pulsed tunnel diode experiments, absorption of incident phonons by the lattice), decay of thermal excitations of the electronic system can be described as relaxation of strongly populated non-equilibrium states.

Dynamic (experimental) equilibrium generation or decay rates, $G_{exp}(t)$, of electron pairs (EP) initiated by propagation of a thermal disturbance resulting entirely from phonon interactions were estimated from

$dN_{SC}(t)/dt$ using $N_{SC}(t)$ from Fig. **14** and the equilibrium density, ρ_{EP}, of this material for the 4 experimental conditions (items 1-4). Results are shown in Fig. **15**.

Note that the rates shown in Fig. **15** are given per unit of sample volume. If taking instead the differentials dV_{SC} of the volume elements that have become superconducting or normal conducting, respectively, the rates of items (1) to (4) of course would be identical. But it takes the superconductor a finite time, Δt, to induce screening super-currents over the *whole* sample surface (which again reflects the time interval to re-arrange the wave function of the whole electron system) so that the lifetime of the excited states in the sample (4) must be short enough to satisfy the rates $G_{exp}(t)$.

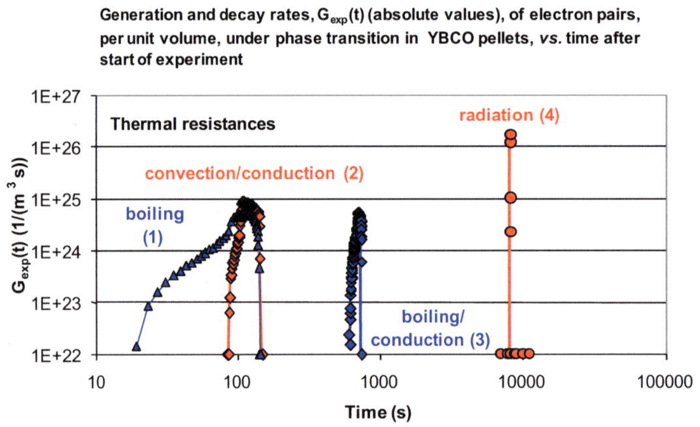

Figure 15: Generation and decay rates, per unit sample volume, *vs.* time after start of the experiment, with the location of the phase transition on the time axis resulting from the thermal resistances, items (1) to (4), and the corresponding temperature intervals, ΔT between T_{Crit} of the sample and corresponding environment (see text). The rates $G_{exp}(t)$ are calculated from the product $\rho_{EP}\, dN_{SC}(t)/dt$ using the $N_{SC}(t)$ from Fig. **14**. Only rates exceeding 10^{22} $1/(m^3\ s)$ are plotted in this figure.

Up to this point, we have considered only temperature excursion, $T(x,y,z,t)$, *i.e.,* the *phonon* aspect of propagation of a thermal disturbance and its consequences. That is, $N_{SC}(t)$, its derivative $dN_{SC}(t)/dt$, and the equilibrium density of pairs - a temperature dependent quantity. Calculation of the theoretical decay rates, $G_{NE}(t)$, the response of the *electron* system to the disturbance, and the decay of the *result* of this disturbance (eg during warm-up: decay of electron pairs to an increased number of quasi-particles) now follows.

3.2. Non-Equilibrium Decay Rates

Non-equilibrium rates, $G_{NE}(t)$, are modelled as sequential generation or decay events with lifetimes estimated either from the uncertainty principle or by analogy to an aspect of the nucleon-nucleon, pion-mediated Yukawa interaction. In a later publication we will report on a diffusion model approach where we imagine that the concentration of bound electron states finally arrives at equilibrium when it “diffuses” through the sample volume. This is also related to diffusion of magnetic flux density in superconductors. The number of excitations, N_{Exc}, is proportional to $\exp[-\Delta E/kT]$, with ΔE the width of the energy gap. Though an equilibrium quantity, it can be taken as an approximate measure of the “disturbance” of the wave function that has to be re-established to its dynamical equilibrium shape. Lifetime of an excitation then is understood as the (relaxation) time, τ_{El}, that the disturbed anti-symmetric wave function, determined by N_{Exc} among other dependencies, will require to be re-arranged to properly describe the return to dynamical equilibrium while respecting the Pauli exclusion principle. Lifetime, τ_{El}, thus determines average rates, $G_{NE}(t) \sim N_{Exc}/\tau_{El}$, to which the experimental $G_{exp}(t)$ finally are compared.

Comparing the experimentally-confirmed simulations to the theoretical modelling (Fig. **16**), a discrepancy appears between both rates, $G_{exp}(t)$ and $G_{NE}(t)$, if temperature is very close to critical temperature (left section of

Fig. **16**) and, in particular if the pellet is warmed-up solely by radiation, the highest thermal resistance situation, which means a small Biot number, taking account the thermal conductivity of the YBCO pellet material.

The theoretical rates, $G_{NE}(t)$ indicated by diamonds, were obtained for constant temperature meaning also the rates are constant. For comparison, maximum values of the experimental $G_{exp}(t)$ are also plotted (taken from Fig. **15**), indicated by blue and red horizontal dashed lines which refer to boiling/convection/conduction and radiative resistance (items 1 and 4, the extreme cases among the above), respectively. Open and closed diamonds indicate initial and average decay rates of excitations, respectively, the average taken over a volume V_C the size of which is given by a screened Coulomb potential between electrons that couple to **S** = 0, **L** = 2 pairs, as is typical for unconventional superconductors. Generally, the problem is that because of numerical reasons a very small volume has to be taken for calculating averages of $G_{NE}(t)$, but it must be large enough to reasonably compare with the $G_{exp}(t)$. Provisionally, screening factors between 1 and 0.01 of the Coulomb potential for two single unit electron charges have been assumed in the calculations but even the 0.01 value may still be too large in comparison with a Thomas-Fermi potential. Here the problem is to identify a reasonable value of the Thomas-Fermi scattering length of HTSC from the literature.

If the energy gap, ΔE, was constant and independent of temperature, the logarithmic dependence of the decay rates shown in Fig. **16** would be linear in $\Delta E/kT$, as is the case in a typical Arrhenius plot. But ΔE depends on temperature so that a linear relationship between the decay rates and $\Delta E/kT$ is observed only in a log-log plot.

Apart from considerable uncertainty of the screening factor, Fig. **16** suggests the theoretical rates (full and open diamonds) decrease strongly when temperature ($T < T_{Crit}$) approaches T_{Crit} during warm-up periods (left section of Fig. **16**, see figure caption).

The results given in Fig. **16** are obtained at *constant* temperature. If temperature differs from T_{Crit} by *e.g.,* only 10^{-6} K, and if T is constant, the "driving force" (the differences in free enthalpy of the system) approaches zero, which means in the limit $T = T_{Crit}$ the system finally has returned to dynamic equilibrium, now at the phase transition, as a state with only statistical fluctuations of the populations of pairs and quasi-particles. Intuitively, one would say the system then is frozen: the more the temperature approaches T_{Crit}, the smaller the variation of both populations and the longer the lifetime of the (residual) disturbed state, as is predicted by Fig. **16**. However, $T = T(t)$ during cool-down or warm-up periods and the theoretical rates (G_{NE}, constant in Fig. **16**) will change with time. The interplay of two time constants, in the present case τ_{Ph} and τ_{El}, has to be taken as a function of time.

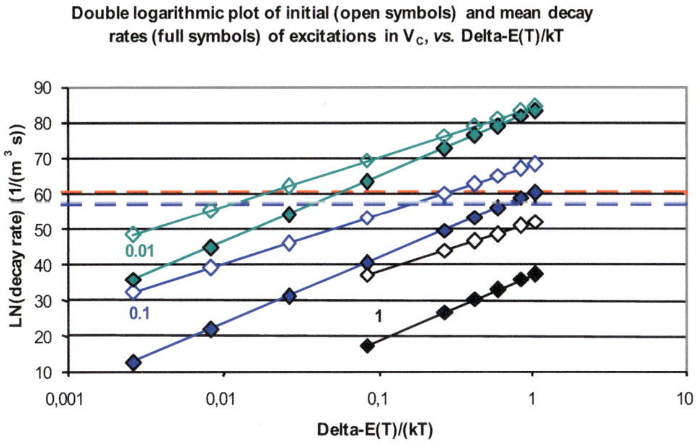

Figure 16: Comparison of experimental, $G_{exp}(t)$ (dashed lines on the ordinate), and theoretical decay rates, $G_{NE}(t)$ (diamonds), assuming different screening factors of the screened Coulomb potential between two electrons in a solid, with ΔE the energy gap of YBCO. Data are given for *constant* temperature, which means the calculated rates are constant at the corresponding temperature. Open and solid symbols indicate initial (immediately after the start of the disturbance) and mean (average taken over the whole time interval needed for return to equilibrium) decay rates, respectively. The curves are given between abscissa values from very close to $\Delta E(T)/kT_{Crit}$ (left section) to $\Delta E(T)/kT$ with T = 91.5 K (right section of the figure).

As a consequence of the difference between τ_{Ph} and τ_{El}, or the difference between both experimental and theoretical rates, we observe a "dead time" interval on the T(t)-scale, which may dissolve or smear out signals that otherwise would be clearly observable (levitation, break-down of super-currents, or the jump of the specific heat at the phase transition, to mention only a few). The same conclusion applies if it is expected that a correlation (if any) of gravity with superconductivity could be detected at the phase transition. In particular, if smear-out occurred just during the interval when temperature was measured in the suspension balance experiment [5], the signal (if any) simply would be lost. For an initial estimate of the dead time interval, we could take the lifetime estimated above from current injection experiments, but this needs more investigation using the above mentioned models to properly indicate its value.

The same situation arises if in any experiment the external heat transfer to the sample is weak in comparison to a much stronger internal heat conduction. In Fig. **15**, this case is reflected by the tiny half-width of the decay rate obtained under boundary conditions of item (4). Note that the peak value of this curve dominates in Fig. **15**. Pure radiation heat transfer means a small Biot number so that in a warm-up experiment, all volume elements *simultaneously* become normal conducting as soon as the temperature field, T(x,y,z,t), for all x,y,z within the sample volume, reaches T_{Crit}. This in turn indicates the theoretical decay rate per unit volume must be enormous to keep pace with the phonon excitations.

Initial simulation results (Fig. **16**) indicate that it is uncertain if the theoretical decay rates will be large enough very near to the phase transition. There is no doubt that all volume elements of the pellet used in, for instance, the suspension balance [5] finally arrived at normal conduction. But the real question is; after what length of time after start of warm-up did the electronic states during the experiment arrive at dynamic equilibrium? Or, reversing the question; is it really desirable to look for intermittently occurring *sharp* signals in this or similar experiments to detect correlations of gravity with phase transition? Or would it be more reasonable to look for differences between initial and final sample weight in gravity experiments (group 1 of the above)? More investigations are underway to address these questions, and details and results will be reported in another contribution to this volume (H. Reiss, Impact of Lifetime and Decay Rates of Thermally Excited States in Superconductors on a Gravity Experiment).

4. CORRELATION OF MASS ANOMALIES WITH INTRINSIC PROPERTIES OF SUPERCONDUCTORS

If superconductivity can be considered as a pathway to *mass* anomalies, a rigorous correlation, not just coincidence, between two *intrinsic* materials properties has to be demonstrated: a negative free enthalpy difference between normal and superconducting states, and a change of rest mass. If on the other hand a weight anomaly should have its origin in a gravity field anomaly, correlation between an *intrinsic* materials (*e.g.,* HTSC) property with an *extrinsic* variable (the gravity field) has to be verified.

The free enthalpy difference of type I or type II superconductor materials is maximum between the normal conducting and Meissner states. If mass anomalies resulting from being in the superconducting state really existed, they would be destroyed if the material leaves the Meissner state. If the existence of the Meissner state cannot be demonstrated in, for example, binary, ternary or quaternary diborides or borocarbides or other new classes of HTSC such as recently-reported zero-resistance in nano-sized Au-wires [17], none of them is a superconductor. By analogy, if it cannot be demonstrated that mass anomalies are rigorously coupled to the existence of the Meissner state, the observed anomalies cannot have their origin in superconductivity. Existence of a correlation between the two intrinsic variables has been investigated when samples were exposed to a magnetic field strong enough to destroy the Meissner state ([1] is apparently the only experiment that specifically was designed to determine such correlations).

5. GRAVITY ANOMALIES

In Podkletnov's experiment [3] the weight anomaly in a test sample, eg SiO_2 can be attributed only to a variation of an *extrinsic* variable (the gravity field). The SiO_2 test mass temperature or other intrinsic thermodynamic properties should not have had any (direct) impact on the experimental result. It was not clear in this experiment whether the total external field on the rotating superconductor resulting from both the levitation solenoid and the multi-phase RF coils was smaller or larger than the lower critical field of the rotating HTSC disk material itself. Both the Meissner effect and Shubnikov phases of superconductors allow levitation, with different levitation heights, however. The resulting field at the disk's surface might well have been above the lower critical magnetic field and the disk then would have experienced hysteresis losses during rotation which would likely have increased the disk's temperature. However temperature measurements were not reported and the only indication the disk was in a superconducting state was its apparent levitation above the levitation solenoid.

Proof of a gravity field anomaly in the Podkletnov experiment thus must concentrate solely on the elimination of possible existing disturbances of the weight measurement such as convection or buoyancy superimposed on gravity. In his published report, Podkletnov used a thin foil to shield the test mass from direct contact with up-streaming evaporated helium gas and gas convection arising from gas/solid friction at the surface of the rotating disk. But the foil is cooled by the up-streaming He vapour to temperatures well below room temperature. Condensation of water vapour and a corresponding down-streaming of air from the atmosphere onto the upper foil surface then is superimposed onto gravity, leading to complicated flow patterns in the vicinity of the freely hanging SiO_2 sample. Considerable convective effects would prevail even if the sample is shielded in a protective tube but filled with atmospheric air. As in Hathaway [4] a good thermal insulator should have been used as a shield instead of a simple thin, thermally transparent foil.

In order to reveal a possible correlation in the Podkletnov experiment, the origin of the claimed gravity anomaly has to be identified. Simple replacement of the HTSC disks by normal- or non-conductors would not modify the gravity field to any measurable extent. This replacement *a priori* would not be able to either confirm or reject a suspected correlation. By analogy, a correlation between the operation of two thermodynamic "machines" A and B in which A assumes two different states (like A_1: Meissner, A_2: non-conducting) cannot be confirmed or rejected if one of the two machines (A) is replaced by a third, C, that can assume only one state, in this case a non-conducting disk. The claims reported in Podkletnov [3] would thus reduce to simply a coincidence between levitation and rotation of a HTSC disk and a suspected gravity anomaly.

6. EXPERIMENTAL ISSUES IN THE SEARCH FOR MASS OR GRAVITY ANOMALIES: TRAPS AND PITFALLS

Experiments designed to search for mass anomalies generally involve the use of high–sensitivity balances operating either in atmosphere, or preferably, in vacuum. Such balances should ideally be isolated from disturbances arising from seismic influences in the vicinity and may even necessitate taking account of tidal or diurnal effects depending on their sensitivity. Electronic balances incorporating electrically-driven strain-detection elements must be shielded from electromagnetic fields arising from sources in the vicinity, either from the experiment itself or the power and control electronics nearby. This, of course, also applies to balances used in the search for gravity field anomalies. As the use of HTSC materials necessitates extremes of temperature in the neighbourhood of the balance, special care has to be taken in balances based on a moving arm or torsion element to ensure that the various structures are not affected either by thermally-induced length changes or by even minuscule amounts of vapour condensation.

The search for gravity field anomalies involves experiments designed to look for anomalous forces on the matter making up test masses. The experimenter must be aware that there are conventional explanations for most of the apparent anomalous forces seen in such experiments. Since most balances used for such experiments are operating at the limit of their sensitivity, many subtle influences come into play which are not always immediately obvious. In general, the two most powerful external effects are magnetostatic and electrostatic. As superconductors are forced above and below their transition temperature, trapped fields can be suddenly released, affecting magnetic materials in the vicinity. This is true for static magnetic fields operating on magnetic balance or container materials and for quasi-static fields inducing Lenz effects on conductors. The use of high voltages, *e.g.,* for powering levitation coils, can induce surface charges on non-conducting test masses which have been insulated both electrically and thermally from their surroundings.

If nearby support structures and containers are not completely free of polar molecules such as water films, there can be electrostatic forces between the films and test masses which may result in unusual balance readings which coincide with some activation of a portion of the experiment. Charged elements such as test masses or balance components can be attracted to or repelled from electric field gradients as in electrophoresis. Typically, however, forces proportional to v x B are usually negligible.

Samples which are non-uniform in density or composition result in spatial gradients of thermal properties which can be affected by subtle but conventional thermal influences. Thus precise descriptions of materials preparation and composition have to be reported to distinguish a possibly existing gravity anomaly from prosaic explanations and artefacts. This of course applies to all experiments of group 1 or 2.

In the course of trying to measure signals from potentially minuscule (if existent) forces, constant vigilance for more prosaic explanations must be maintained. This is true when measuring not only the type of force (*e.g.,* magnetic, electrostatic, mechanical, gravitational), but also the origin of the force. In even the most carefully designed experiments, it is easy to be fooled by a litany of potential effects resulting from experimental procedures that may have been poorly conceived or carried out. This section provides a brief overview of some of those considerations.

For example, the direct determination of the mass of a test sample cooled to cryogenic temperatures typically requires that the test sample be attached to a measurement device (a balance for example). If the balance is not cooled, the connection between test sample and instrument must accommodate a very great temperature difference, and problems may arise due to condensation and buoyancy as discussed above. In order to shield the test sample from the effects of condensation, the sample can be enveloped either in a vacuum or a non-condensable (at the operating temperature) gas. Placing the sample in a vacuum mitigates both buoyancy and condensation concerns, but may preclude effective cooling of the sample and its precise temperature measurement. Hence a compromise must be struck between the need to isolate the sample from the environment and the speed at which cooling is expected. Balances which operate in vacuum are useful in this regard.

Determination of the weight of a test sample which is near a superconductor, such as that in Podkletnov's experiment [3], requires that the test sample be isolated from the environment as much as possible, to avoid being shaken, buoyed by air or gas currents, or otherwise perturbed. However, since the sample itself may not need to be cooled, the requirement for strict temperature isolation can be somewhat relaxed.

Exacerbating the problems inherent in cryogenic experiments are those introduced by nearby electric and magnetic fields. As the experimenter endeavors to increase the mechanical isolation between the test sample and the local laboratory environment, the sample's susceptibility to electromagnetic and electrostatic influences becomes greater.

The following summary of the some of the more important experimental concerns is organized into general categories, and there will inevitably be some overlap. It is by no means exhaustive.

6.1. Mechanical Effects

6.1.1. Thermal Effects

This includes thermally-driven air, gas or liquid currents, vapor condensation, and thermal expansion issues when a difference in temperature is encountered. Thermally driven currents act to artificially alter the weight of test samples they encounter. It is often exceedingly difficult to calculate the resulting forces, which tend to randomly fluctuate both temporally and spatially. Condensates can alter the mass of the test article, with the most problematic condensate in cryogenic experiments being water vapor from the air, and, to a lesser extent, oxygen and nitrogen condensate when working at LHe temperatures. Experiments must be designed to avoid any part of a cryogenically-cooled apparatus being in contact with condensable gasses. When dealing with severe temperature differences, structural apparatus and other test articles must not be subjected to thermal expansion or contraction, unless this effect can be accurately analyzed and factored into the resulting measurements.

6.1.2. Buoyancy Effects

Materials display different buoyancy in different media. If a test sample is weighed in air, then weighed in LN_2, its apparent weight will clearly differ and be shape-dependent. More subtly, if the sample is cooled to cryogenic temperatures, its buoyancy cross-section changes as a function of its temperature, producing a time-varying weight change that could be misinterpreted as a gravitational interaction. Buoyancy differences may also arise in a single medium (such as air) due to stratification and temperature differences in the medium.

6.1.3. Seismic/Vibration Effects

Scales, balances and force sensors generally operate by detecting or displaying the relative distances between fixed and movable masses, and it is necessary to avoid introducing artificial mechanical motion/noise into the measuring system. Vibrations can occur from laboratory and other sources, such as compressors, fans, and local traffic (both pedestrian and vehicular), and are usually transmitted into the measuring apparatus by structural elements. Mitigation is typically achieved either by loosening the structural design through the installation of dampers, or tightening the design by anchoring the support structures to larger, relatively immobile masses. If these are not easily or completely achievable, altering the frequency of the chief modes of vibration by adding or subtracting masses is often helpful. Wires and cables carrying signals and power may appear negligible as structural vibration carriers, but for experiments at cryogenic temperatures the wires may freeze and become rigid. In addition, the pulsing of current through any portion of a wire with curvature will produce a flexure in the wire, leading to displacement or, for alternating currents, oscillatory vibrations. The possibility also exists that the measuring system is non-linear over some or all of its range and may suffer from hysteresis or viscous effects which can lead to apparently steady but erroneous readings due to partial rectification of external vibrations. For example it is well-established that vibration applied to a standard analytical chemical balance will cause its steady readings to drift because the air dashpot or similar damping device is non-symmetrical and can apply more force in one direction than the other.

6.1.4. Vacuum Effects

All materials will to more or less extent outgas (release vapor) in vacuum, hence the use of low-outgassing materials for any experiment performed in a vacuum or low ambient pressure environment is desirable. Even a low outgassing material may have minute amounts of contaminant (oils, fingerprints, *etc.*) on its surface, the release of which can be detected by sensitive balances. Other contaminant material on the inside walls of the vacuum chamber housing the sample may be released and coat or otherwise impinge upon the sample. Of more subtle nature is the slight movement of the vacuum chamber and encompassed test apparatus as the vacuum chamber is pumped down. Such small movements can offset previously aligned balances, producing an apparent weight change in a test sample simply due to a slight shifting in the support structures. It is also worth noting the high degree of variation in outgassing between materials which may be used in test apparatus or support structures involved in such experiments. For instance, brass has a high outgassing rate due to the presence of high vapor pressure zinc; if a brass test mass is used in a vacuum environment, it's actual weight loss could change over the course of the experiment and mask other experimentally-induced effects.

6.1.5. Liquid Effects

The evaporation or boiling of liquid (usually the cryogen) around a test sample or the suspension system can clearly contribute noise into the measurement system. This is particularly true when the test sample is suspended in, or is in contact with, a liquid cryogen used to cool it below its transition temperature. Boiling of the cryogen when inserting a sample will cause spurious readings and potentially mask effects near the critical temperature until sample and cryogen are in thermal equilibrium. Similarly, lifting the sample from a cryogen produces high rates of evaporation (often in addition to dripping cryogenic fluid), and will cause an apparent weight change as the temperature of the test sample rises back to ambient thermal conditions.

6.2. Temporal Effects

Certain experiments require a test sample to be cooled in cryogenic gas or liquid while being suspended from the movable arm of a balance. The precise determination of sample temperature as a function of time is critical to ascribing an observed weight change to a superconductive attribute. However, the attachment of a thermocouple to the superconducting sample necessitates feeding wire leads from the movable sample to the fixed laboratory frame on which the temperature readout sits. Movable slip rings are too stiff and unpredictable to be used as rotating/movable joints in most delicate experiments. Testing for the attainment of critical transition temperature *via* Meissner flux expulsion is often too cumbersome, and would of necessity disturb the balance. If direct temperature determination proves exceedingly difficult, the best path is for the experimenter to simply allow sufficient time for the test sample and any supporting mass to reach the required equilibrium temperature.

Additionally, if the expected gravity effects are correlated with a narrow transition around T_{crit}, the caveats in Section 3 should be noted.

6.3. Electromagnetic Effects

6.3.1. Magnetic Coupling

The incorporation of superconductors in gravity experiments necessarily involves magnetic fields. Potential interactions with the magnetic fields arising from current-carrying and signal wires must be taken into account in the vicinity of the experimental apparatus. Residual magnetism can be found in certain "non-magnetic" metals and alloys, which may interact with the measurement system. The assumption of perfect magnetic shielding by high-mu magnetic shielding alloys is usually not valid, and care must be taken that the proper level of electromagnetic shielding required by the experiment is obtained. Time-varying electromagnetic fields generated by test equipment, power sources, leads, and contacts can interact with nearby conducting bodies and cause often subtle but nevertheless spurious measurements. More dramatically, the sudden release of trapped magnetic fields in superconductors raised above transition temperature can affect, and be affected by, nearby magnetic or conductive structures.

6.3.2. Electric Coupling

As in the use of high-mu metals to shield magnetic fields, an over-reliance on a Faraday Cage for complete exclusion of DC or quasi-static electric fields is often not realistic, and extreme care should be taken to account for and mitigate any spurious electric fields that may interfere with the test specimen or measurement apparatus.

6.3.3. Electromagnetic Coupling

Improper shielding or conductor pairs that are not tightly twisted can cause unwanted crosstalk on nearby signal cables, even shielded ones. Switching transients in high-power circuits can easily couple into signal sensing circuits, causing artificial or spurious readings; this is particularly important for circuits incorporated into electronic balances. Using RF power with improperly-matched loads or inappropriately sized matching networks can cause intense electromagnetic interference in nearby circuits.

6.3.4. Grounding Issues

The physical layout of signal grounds should be carefully considered to avoid nearby magnetic interference. Shield grounds should be checked for potential ground loop problems by incorporating a single-point grounding connection. Contacts should be checked for loose, corroded, or bi-metal connections which can introduce unwanted contact potentials.

6.4. Electrostatic and Related Effects

6.4.1. Gradient Effects

The operation of an apparatus at high DC voltages relative to a nearby ground will usually produce spatial gradients in the electric field, unless the experiment is specifically designed to produce uniform fields in the vicinity of the test apparatus. Such gradients may induce subtle motion in nearby free bodies whether conductive or not, such as samples suspended from a balance.

6.4.2. Charge Pooling and Induced Charges

Even under high vacuum, without sufficient surface preparation and thermal baking, the interior surfaces of vacuum chambers are covered with invisible pools of surface charges, usually in the form of films of polar water molecules. In very sensitive measurement systems, or where the sample mass being weighed is in close proximity to a wall, the electrostatic forces between such "patch charges" on the wall and test mass may interfere with proper weight measurements. This applies whether or not the surfaces are conductive.

6.4.3. Ion and Molecular Effects

In high-voltage DC or AC experiments, even in vacuum, strong electric fields can accelerate outgassed or residual ions in the vacuum chamber, which in turn can impart momentum to a suspended test mass. This may occur either because the ions are streaming off of or on to a sample held at high voltage, or because the sample may be in the "ion wind" produced by a nearby source of ions held at high voltage. Such ion wind effects are difficult to calculate, and so generally need to be avoided by adhering to strict chamber cleanliness procedures, proper choice of vacuum compatible materials (including wire insulation), and operation at chamber pressures below those needed to produce significant ion wind effects. High voltages can cause ablation or sputtering of even the best insulation.

6.4.4. Charge Leakage

Charges can leak from insulated conductors, either as a quasi-DC corona or (more likely) in bursts called Trichel Pulses in high-voltage experiments. These may be negligible to the force-measuring apparatus at relatively low voltages, but at some threshold value they can suddenly become problematic, and often under generally unpredictable conditions. These can cause direct effects, for instance by impinging on or interfering with balances, or indirect effects through, for example, interference with nearby signal wires. Most commonly, they cause weak conduction paths to nearby grounds which can interfere with high-gain amplifiers, *etc.*

6.5. Experimental Expectations

The above list is but a small sampling of the experimental artifacts that must be considered when designing and performing any sensitive laboratory experiment, and particularly when searching for minuscule effects in support of controversial theories. In addition, there are psychological pitfalls that investigators designing experiments are often not aware of. For instance, *confirmation bias* is the tendency to trust data that supports one's hypothesis, and to distrust (or ignore) data that does not support it. After all, if my data is exactly as my theory predicts, then the experiment is a good one and I made no errors! Another is *errors in logic*. An example of one such incorrect syllogism might be "If A is true, then we will observe B. We do observe B, therefore we have proven A is true". This false logic overlooks other explanations for observing B. Put more correctly the logical statement is: "We do observe B, therefore one possible explanation is that A is true, and there may be other explanations." The all too common *neglect of prior probability* flouts Bayes' theorem, which shows that if the probability of a theory being true before the experiment is low, then a stronger standard of proof is required to support its correctness. These and other considerations must be recognized and added to the usual repertoire of experimental design and protocol which makes up the bulk of the scientific method.

7. APPLICATION OF DATA ANALYSIS

Anomalies and correlations between intrinsic and extrinsic variables can be demonstrated using standard statistical methods like Gaussian fits to scattered data, calculation of correlation coefficients, χ^2-analysis or confidence intervals. Only a very few papers presented on gravity anomalies include error estimates that are necessary to clearly separate an anomaly from an artefact. Application of χ^2-analysis to an appropriate hypothesis without error bars is useless. Error analysis including, *e.g.,* full balance specifications, requires understanding the chain by which individual experimental errors propagate and this implies at least an attempt at a physical model to explain the putative gravity anomaly. Errors introduced by subtle influences such as outlined in the previous sections have to be incorporated. At the very least, these have to be reasonably estimated, if not provided by separate calibration or manufacturer's specification. The same applies to temperature control. Although superconductivity is essentially a cryogenic temperature phenomenon, little attention is paid in many papers to noise levels resulting from the attainment of thermal equilibrium, how to appropriately measure temperature gradient fields, the estimation of parasitic heat flow and the impacts of radiation fields.

In order to safely separate anomalies from prosaic phenomena, a sharp criterion like a 6σ-distance between the corresponding mean values seems appropriate. Overlap of standard deviations, σ (the "noise"), then will be excluded with very high statistical significance.

8. CONCLUSIONS

Experiments to detect mass or gravity field anomalies using weight measurements of HTSC samples should ideally employ magnetic suspension balances that house the samples in a vacuum and use strict temperature monitoring and control, such as used by Tajmar [5]. All experiments including suspension balance measurements have to be designed in such a way as to exclude conflicts between different relaxation rates of thermal transport mechanism during cool-down or warm-up periods, and the response of the electronic system of the superconductor material to the same disturbances. Detection of such anomalies using a non-HTSC test mass in the vicinity of an HTSC should ensure that the test mass is suitably isolated from the ambient environment. Provisions have to be made to switch the superconductor between superconducting and non-superconducting states by applied magnetic fields. Knowledge of the likely prosaic explanations for the observations, and thorough error analysis are indispensable.

REFERENCES

[1] Reiss H. Weight Anomalies Observed During Cool-down of High Temperature Superconductors. Phys Essays 2003; 16: 236-253.

[2] Tajmar M, Hense K, Marhold K, de Matos C. Weight Measurements of High Temperature Superconductors During Phase Transition in Stationary, Non-stationary Conditions and under ELF Radiation. In: M.S. El-Genk, Ed. Space Technology and Applications International Forum (STAIF-05). American Institute of Physics 2005.

[3] Podkletnov E, Nieminen R. A Possibility of Gravitational Force Shielding by Bulk $YBa_2Cu_3O_{7-x}$ Superconductor. Physica C 1992; 203: 441-444.

[4] Hathaway G, Cleveland B, Bao Y. Gravity Modification Experiment Using a Rotating Superconductor Disk and Radio Frequency Fields. Physica C 2003; 385: 488-500.

[5] Tajmar M, Plesescu F, Seifert B. Measuring the Dependence of Weight on Temperature in the Low-Temperature Regime Using a Magnetic Suspension Balance. Meas Sci Tech 2010; 21: 015111.

[6] Reiss H, Troitsky O Yu. The Meissner-Ochsenfeld effect as a possible tool to control anisotropy of thermal conductivity and pinning strength of type II superconductors. Cryogenics 2009; 49: 433-448.

[7] Reiss H. A numerical analysis of shield temperatures, heat losses and residual gas pressures in an evacuated super-insulation using thermal and fluid networks, Part I: Stationary Conditions. Cryogenics 2004; 44; 259-271.

[8] Sandner S. Stoffwerte von reinen Metallen und Metall-Legierungen. In: VDI Heat Atlas, 10th Ed. Springer 2006; Chapters Dea1 - Dea2.

[9] Touloukian Y, DeWitt D. Thermal radiative properties-Metallic elements and alloys. In: Thermophys. Properties of matter, Vol. 7. IFI/Plenum 1970; p. 137.

[10] Kabelac S, Vortmeyer D. Strahlung technischer Oberflächen. In: VDI Heat Atlas, 10th Ed. Springer 2006; Chapters Ka1-Ka11.

[11] Kaganer M. Thermal insulation in cryogenic engineering. Israel Progr Sci Transl 1969; 33-34.

[12] Klan H. VDI Heat Atlas, 10th Ed. Springer 2006; Chapters Fe1 - Fe3.

[13] Gray K. Steady State Measurements of the Quasiparticle Lifetime in Superconducting Aluminum. J Phys F Metal Phys 1971; 1: 290-308.

[14] Epperlein P, Lassmann K, Eisenmenger W. Quasiparticle recombination time in superconducting tin and normal electronic density of states at the Fermi surface from tunnel junction experiments. Z Physik B 1978; 31: 377-384.

[15] Carslaw H, Jaeger J. Conduction of Heat in Solids. Oxford Science 1959; p. 94.

[16] Blatt J M. Theory of Superconductivity. Academic Press 1964; p. 243.

[17] Takayanagi K. Preparations of Gold Nano Wire with Helical Multi-Shell. MEXT press release. Tokyo Inst Tech, July 21, 2000.

CHAPTER 11

Emerging Physics for Gravity-Like Fields

Walter Dröscher[1] and Jochem Hauser[2,*]

[1]*Institut für Grenzgebiete der Wissenschaft, Maximilianstr. 8, 6010 Innsbruck, Austria and* [2]*Faculty H, Ostfalia Univ. of Applied Sciences, Campus Suderburg, Germany*

Abstract: Based on theoretical ideas under development since 2002, termed Extended Heim Theory (EHT), as well as experiments performed at AIT Seibersdorf, Austria since 2006, it is argued that there is evidence for the *existence of novel gravity-like fields* and thus also *different types of matter*. These gravity-like fields are not described by conventional Newtonian (Einsteinian) gravitation, *i.e.*, by the accumulation of mass. Instead, under certain conditions, they should be producible in the laboratory by small ring or disk shaped masses rotating at cryogenic temperatures. *EHT*, in describing these novel fields, postulates six fundamental physical interactions, three of them of gravitational nature. The two additional gravity-like fields may be both attractive and repulsive. It is further argued, based on both *EHT* and experiments, that these gravity-like fields are outside the known four physical fundamental forces, and may result from the conversion of electromagnetic into gravitational fields. The gravitomagnetic effect of these fields is found to be some 18 orders of magnitude larger than classical frame dragging of General Relativity. This fact seems to be in accordance with recent experiments performed at AIT Seibersdorf. A non relativistic semiclassical model will be presented as an attempt to explain the physical nature of the novel gravity-like fields. There seems to be a special phase transition, triggered at cryogenic temperatures, responsible for the conversion of electromagnetic into gravitational fields. The features of the six fundamental physical interactions are utilized to investigate the potential of the novel gravity-like fields for propulsion purposes as well as energy generation.

Keywords: Gravitational forces, theories of gravitation, superconductors, general relativity, gravitomagnetism, novel gravity-like fields, fundamental physical interactions, extended heim theory, phase transitions, classical frame dragging, geometrization of physics.

1. PHYSICS OF GRAVITY-LIKE FIELDS

Physics, as we know it, is based on the belief of the existence of exactly four fundamental forces. There are two long range forces (interactions) electromagnetism and gravitation. Gravitation is believed to be always attractive, while electromagnetism can be both attractive and repulsive. The bosons that mediate the force between the charged particles that is, particles having mass and/or electric charge, are the hypothetical graviton and the photon, respectively. The other two interactions are the weak force (β decay) and the strong force (atomic nuclei), which are of short range, *i.e.,* their range is about 10^{-15} m. The two cornerstones of modern physics, general relativity and quantum theory, are predicting highly different magnitudes of the energy of the vacuum, whose ratio is about 10^{120}, which means the error is in the exponent [51]. This might be an indication that physics as it is understood at present is not complete and has to be complemented by novel concepts in the form of additional interactions.

Based on ideas, first proposed by Heim in the late 1950s, a so called poly-metric tensor has been constructed by the authors. In this article the main physical concepts leading to the proposed six fundamental forces are presented; the details can be found in the references (see Section 2). These theoretical concepts are termed *Extended Heim Theory* (*EHT*), though *EHT* does not have reached the status of a theory, but should be conceived as a *phenomenological model* to explain the existence of the six fundamental forces. These concepts demand two additional gravity-like fields that may be both attractive and repulsive originating from an interaction of electromagnetism and gravitation as well as new types of matter. Therefore, there should exist six fundamental interactions, three of them of gravitational type. It was

***Address correspondence to Jochem Hauser:** Faculty H, Ostfalia Univ. of Applied Sciences, Campus Suderburg, Germany; E-mail: jh@hpcc-space.de quite unexpected when in 2006 experiments were published on the existence of gravity-like fields by Tajmar *et al.* performed at AIT (Austrian Institute of Technology) and in 2007 by Graham *et al.* (Section 1.3). Also, the NASA-Stanford Gravity-Probe B experiment might have been subjected to these gravity-like fields, see below. These novel fields, *if confirmed*, may have the potential to serve as a basis for a completely different technology on transportation, as well as energy generation.

Hence, it is of great interest to obtain at least a heuristic understanding of the underlying physical phenomena and to establishing rules for the scaling of the gravitational effects that are of non Newtonian type. In this article we consider possible applications to propulsion, but the concepts can be directly employed to any type of transportation and might also aid in the stabilization of fusion plasmas. The physics of space flight is based on the century old rocket equation that is an embodiment of the conservation of linear momentum [52]. As a consequence, chemical rockets, which are the most powerful propulsion systems of today, are flying fuel tanks. In the following paragraph, a brief discussion of the advanced concepts for space propulsion is presented. It should be noted, however, that all of the physical principles presented have been known since the 1950s, but no significant progress regarding their technical realization was made [1-5].

1.1. Recent History of Gravity-Like Field Experiments

During the last two decades, numerous experiments related to gravity shielding or gravitomagnetic interaction (coupling between electromagnetism and gravitation) were carried out. In the 1990s a Russian scientist claimed to have measured gravitational shielding. However, Woods *et al.* [6] have delivered experimental evidence that these two claims cannot be substantiated. This kind of gravitational shielding does not seem to exist. In 1997, a Japanese free fall experiment using an encapsulated spinning gyro [7] reported differences in free fall time depending on the direction of rotation. The authors concluded that an asymmetry (parity violation) caused the generation of anti-gravity. But the equations of motion for a free falling body in an atmosphere are highly complex, because of aerodynamic lift and drag (pressure drag, friction drag) that also may be time dependent. Although the spinning gyro was encapsulated, it might have transferred angular momentum, for instance, by friction effects, to the free falling body causing a rotation of the body. This rotation would have changed the free fall time. If the gyro had a small geometrical asymmetry, a rotation in one direction (18, 000 rpm) might cause a transition from laminar to turbulent flow, and thus may lead to a different transfer in angular momentum, depending on the direction of rotation. Since the authors do not address these possible sources of error, it cannot be decided whether or not this set of experiments is reliable. The experiments were conducted at room temperature.

Hence, the conclusion concerning the above experiments is that they must be considered as incorrect or, at least, inconclusive, and therefore are not considered in our experimental analysis.

1.2. Theoretical Concepts for Geometrization of Physics

On the theoretical side, most advanced physical theories like string theory, grand unification theory, or super-symmetry under development for several decades and aiming at unifying the four known fundamental forces, do not predict any additional physical interactions. In addition, they require a higher dimensional space, but so far there is only evidence for the existence of four-dimensional spacetime. On the other hand, the geometrical concept of Einstein [8], trying to geometrize physics by associating a metric tensor with each known interaction, has been abandoned long ago by theoretical physicists, since all attempts by Einstein and others in constructing a metric which encompasses all physical interactions proved to be unsuccessful. Since then, this concept is believed to be valid for gravitation only.

However, in the 50s of the last century the German physicist Heim and the Italian mathematician Finzi presented novel ideas on how to construct a polymetric describing all physical interactions in order to revive the powerful geometric idea of Einstein. Heim introduced the concept of a six-dimensional space, interpreted and utilized by us as an internal space, possessing three subspaces and constructed several metric tensors that he associated with physical interactions. In Extended Heim Theory *EHT* as proposed by the authors [9, 10], these ideas were extended to an eight-dimensional internal space by adding the subspace I^2 that enables the exchange of information among physical systems. Consequently a total of 15 metric tensors (termed Hermetry form which stands for physical meaning of geometry) are constructed. It should be noted that the name Extended Heim Theory is not really appropriate in that *EHT*, at least at present, is only a collection of ideas on the geometrization of physics for determining the number and type of all fundamental physical interactions. The name of Heim is used, since he was the first one to constructing a poly-metric or multiple metric by employing the idea of structured (internal) space.

The most important aspect of *EHT* is that it postulates the existence of six fundamental interactions. The additional two gravity-like fields predicted by *EHT*, might serve as the basis for a completely novel propulsion principle based on force fields. In other words, according to *EHT*, it should be possible for a space vehicle to generate its proper gravity-like field capable of accelerating the vehicle against its reference frame. Such a propulsion scheme would work without propellant. Even if such an acceleration field would be weak compared to the acceleration on the surface of the Earth, but it may be large when compared with today's state of the art ion propulsion that delivers a thrust of about 20 mN, and, if sustained over a long enough time, the $\Delta \mathbf{v}$ would be substantial.

The question, of course is, does there exist any experimental evidence of such fields? Surprisingly, the answer seems to be *affirmative*, and the recent experiments presented in the following section, might be a first sign for the existence of novel physical interactions. If this turns out to be true, a propulsion scheme based on the concept of a gravity-like force field might be realizable, since the technology, derived from the novel gravity-like fields, should be *substantially easier to realize than chemical propulsion*, once the underlying physics is completely understood.

1.3. Existence of Additional Gravitational Force Fields

Up to 2006, to the knowledge of these authors, no credible experiments on novel physical forces existed. However, on 23 March 2006, the European Space Agency (ESA), on their webpage, announced novel experimental results, reporting on the generation of gravitomagnetic and gravity-like fields (acceleration or artificial gravitational field in the laboratory [11-13]). So far several different interpretations of this phenomenon have been given by the experimenters; the latest one indicates that the effect is due to the action of the liquid He itself. A different physical interpretation of these experimental results, based on the existence of two additional gravity-like fields, for instance, was published in [10].

Since 2006, Tajmar *et al.* [11-13] have refined their experimental setup, repeating their experiments. In July 2007, Graham *et al.* [14] published a paper seeing a similar effect for a superconducting lead disk, but using a completely different measurement technique. However, the sensitivity of their laser-ring gyro was about two orders of magntiude less than the equipment of Tajmar, so their results are *not conclusive*. In September 2007 Tajmar *et al.* [13] published a comparison between the two sets of experiments. The most recent results by Tajmar *et al.* show varying signal strengths depending on the layout of the geometry as well as the materials utilized in the experimental setups A, B, and C employed. *EHT* was used to analyze these experiments and to provide an explanation for the different strengths of the measured gravitomagnetic fields. Moreover, in 2007 results of the Stanford-NASA Gravity Probe B (GP-B) [15, 45] experiment became available. GP-B employed four small, extremely precise gyroscopes in the form of Nb coated quartz spheres at liquid He temperature as its main tool for detection of geodetic (spacetime deformation by a static mass) and frame dragging (spacetime twisting by a rotating mass) effects. Since in GP-B there are two pairs of counter-rotating superconducting niobium coated quartz spheres, each of the two gyros in a pair should feel the presence of the gravitomagnetic field generated by its partner. This situation bears a certain similarity to Tajmar's and Graham's experiments (see Section IA [10]). According to *EHT*, an extreme gravitomagnetic field should be observed, that is completely different from the gravitomagnetic field predicted by *GR*, leading to substantial *gyro misalignment (i.e.,* spacetime *twisting acting as a force that pushes on the gyro axis and forces it out of alignment as it circles around the Earth).* Second, a tangential gravity-like field (acceleration field) should be observed, generated by the rotation of a gyro in the spatially non-homogenous gravitomagnetic field of its partner, where one of the gyros is *accelerated* and the other one is *decelerated.* Therefore, the misalignment of the axes of the GP-B gyroscopes over time should not be completely explainable by conventional assumptions like electrostatic forces between the housing of the gyro and the gyro surface *etc.* The same holds true for the second gyro anomaly, namely the measured gyro frequency shifts in GP-B, meaning that one of the gyros in each of the two gyro pairs was accelerated over the measuring period of 10 months, while the other one was decelerated. This behavior was analyzed in detail [16-20]. It has been shown that there is room for this so called *Tajmar effect*, but, because of the other sources of misalignment present, it cannot be concluded that the extreme gravitomagnetic fields have been generated in GP-B.

The generation of gravitomagnetic and/or gravity-like field(s) in all three sets of experiments, namely by Tajmar, Graham and GP-B seems to follow a similar mechanism.

- First, an *extreme gravitomagnetic field* is observed for a rotating Nb or Pb ring, sphere, or disk at a specified cryogenic temperature (above the critical temperature, T_C, for superconducting and therefore *different* from the mechanism of the Cooper pairs), which is many orders of magnitude larger than predicted by *GR*.

- Second, a *gravity-like field* (gravitational acceleration field) seems to be generated by producing a time-dependent gravitomagnetic field. The ensuing gravitational force observed is more in the range of the electromagnetic force. In other words, the coupling constant for the gravity-like field must be completely different from common gravitational coupling. A coupling between electromagnetism and gravitation seems to take place, however, not with Newtonian gravitation.

Hence, we are in the fortunate situation to have at least some experimental data for the validation of these entirely novel physical concepts and models, in contrast to current theoretical physics (*e.g.*, string theory). Arguably, there exist three different types of experiments, employing very different measurement techniques, performed both on *Earth* and in space, but reporting similar physical phenomena.

Regarding the reliability of the experimental results, investigated in [10], it should be mentioned that Tajmar *et al.* [11-13] at the AIT Seibersdorf, Austria (a certified ESA test center) had carried out these experiments over a period of about three years prior to publication. The authors report that a rotating Nb ring of some 15 cm diameter generated a *gravitomagnetic field* below temperatures of about 30 K, *i.e.*, the temperature where the phase transition takes place. In addition, every time the cryogenic Nb ring was subjected to angular acceleration at around 4 K, a *gravity-like field* was measured in the plane of the ring in circumferential direction. The induced acceleration field was opposite to the angular acceleration, following some kind of gravitational Lenz rule. In addition, an acceleration field was also observed when the Nb ring was rotating with constant angular velocity, but undergoing a phase change that is, from the normal to the (superconducting) state. This was achieved by reducing the temperature below 9.2 K, the critical temperature for Nb. Moreover, no acceleration was measured when the Nb ring was in normal conducting state.

GR predicts that any rotating massive body (*Earth*) drags its local spacetime around, called the *frame dragging* effect, generating the so-called *gravitomagnetic field.* Since spacetime itself gets twisted, a gyro whose axis of rotation remains invariant, should exhibit a misalignment, since the local coordinate system is no longer inertial. This effect, predicted by Lense-Thirring in 1918, however, is far too small to be seen in any laboratory on Earth. For this reason the GP-B experiment was launched in 2004 after more than 40 years of preparation (see Section IA in [10]). On the other hand, the values measured by Tajmar *et al.* [11-13] were *about 18 orders of magnitude higher* than predicted by *GR*, and therefore are outside *GR*. They *cannot be explained by the classical frame dragging effect of GR* and would represent a *new kind of physical phenomenon.* In other words, the cryogenic rotating Nb ring, with a mass of about 400 grams, causes approximately the same frame dragging effect as a white dwarf, as was calculated [20].

1.4. Novel Experiment for Vertical Gravity-Like Field

When analyzing the experiments by Tajmar *et al.* [11-13], it became obvious that an experiment could be devised, demonstrating the generation of a gravity-like field pointing in the *vertical direction* (along the axis of rotation) that might be capable of lifting a vehicle from the surface of the Earth [10] if scaled accordingly.

Since this effect only occurs at very low temperatures it is surmised that a *phase change* takes place. In *EHT* the concept of matter is extended to both ordinary and non-ordinary matter. This means that *particles of real and imaginary mass* should exist under spacial conditions. Consequently, the concept of virtual particle is extended also to virtual particles of imaginary mass and real charge. This type of virtual particle is deemed to be responsible for the conversion of the electromagnetic into a gravitational force, which do not necessarily have to form Cooper pairs, but instead might form some kind of Bose-Einstein condensate (although the temperature is much higher), causing these novel physical interactions. Based on calculations of *EHT*, the technical requirements for a vertical gravity-like field, for instance, magnetic induction field strength, current density, supply power *etc. should easily be met with present technology.*

Required values for such an experiment are substantially lower than for the previously proposed experiments that assumed *fermion coupling to achieve vacuum polarization* as discussed in [17].

Extreme caution is needed when announcing novel physical interactions. Several years ago, a novel physical interaction (fifth force) was announced by a group of physicists at Purdue University and Brookhaven National Laboratory (Fischbach, 1986) who claimed to have found experimental evidence for a distinct deviation from Newton's law at intermediate distances between 1 m and 1 km. After four years of elaboration and 12 more experiments, agreement was reached that there was no fifth or sixth force. This time, however, the situation is different in that no geological structure or chemical composition of the environment is of importance. The physical effects measured by Tajmar and Graham are large, and should be clearly reproducible, independent on location and geological formation.

What does theoretical physics have to say about these experiments and physical phenomena? The answer is very short, namely there is no place for these experimental results. According to currently favored string *theory* or any other *supersymmetry theory* nothing should have been measured. What about alternative theories? At present, there are no theories available that are capable of completely explaining these experiments. Using the geometrization approach of *EHT*, the so called doubled coordinate transformation (for instance [26]) that is currently reformulated using the mathematical framework of nonlinear σ – theory [27] in order to obtain a connection between geometry (specific metric tensors also termed Hermetry forms), results in postulating two additional fundamental gravitational like interactions, which are subsequently utilized to analyze all three types of experiments (see Section I in [10]). However, experiments are producing numerous new results, and no final quantitative conclusions are available at present.

2. THEORETICAL CONCEPTS FOR GRAVITOMAGNETIC AND GRAVITY-LIKE FIELDS

In this section a discussion of the theoretical background for recent experimental results by Tajmar *et al.* [11-13], Graham *et al.* [14], and the Stanford-NASA Gravity-Probe experiments is presented, which shows that there is evidence for additional gravity-like fields. Before, however, the physical fundamentals of *EHT* are discussed, which aims at the geometrization of all physical forces, additional evidence for the existence of gravity-like fields will be evoked stemming from astrophysics observations. These extraordinary phenomena could be related to the recent experiments by Tajmar *et al.* The hitherto unexplained gravitational phenomena in the form of dark matter and dark energy clearly exist, and have been verified beyond doubt by cosmological observations.

2.1. Hints from Cosmology for Additional Gravitational Fields

In the following, cosmological observations will be discussed that provide evidence for additional gravitational interactions. As will be shown below, *EHT* predicts six fundamental interactions. Three of these fundamental forces are of gravitational nature, which can be both attractive and repulsive. From astrophysical observations it is known that about 4% of the matter in our *Universe* is ordinary (visible) matter made of neutrons and protons. This baryonic matter is described by known physics, *i.e.*, the standard model, which now needs 26 free parameters. All the fundamental properties of matter have to be provided to the model. The coupling constants of the four fundamental forces need to be supplied and cannot be calculated from the model. However, galaxy formation requires additional mass beyond known baryonic matter. The current status can be summarized such that a new type of invisible matter (*dark matter*), dominating ordinary matter (about 26%), exists in the *Universe*, but current physical theory cannot explain its nature. Moreover, observations of very distant Type I supernovae require the accelerated expansion of the *Universe*. This phenomenon is attributed to the so called *dark energy*, which comprises about 70% of the total mass of the *Universe*. Dark energy causes a repulsive force, whose density has changed in the course of the evolution of the *Universe*. The physical properties of this unusual energy are virtually unknown. Both, dark matter and dark energy exhibit gravitational interaction. In quantum theory, gravitation, because of its weakness, is not accounted for. Quantum field theory (QFT) only works because it neglects Newtonian gravity. The situation would become even more severe if two additional gravitational interactions existed, that also cannot be accounted for in QFT. On the other hand, *GR* only explains the gravitational behavior of ordinary matter, *i.e.,* those 4% of the total matter of the *Universe* that are assumed to be known. However, for the remaining 96% of matter present physics does not have a satisfactory explanation.

However, if dark matter and dark energy each are associated with their own proper gravitational interaction, then there exist three gravitational forces that can be both *attractive* and *repulsive*. In addition, there are three non-gravitational forces described by quantum theory (electromagnetism, weak, and strong force). Thus, cosmological considerations suggest that there are six fundamental forces. If this is the case, *novel field quanta* must also exist. The messenger particles for these interactions, as predicted in *EHT*, are identified as gravitophoton (ν_{gp}^{0}), which can decay into attractive ν_{gp}^{+} and repulsive, ν_{gp}^{-} gravitophotons or, *via* a second decay mode, into a graviton (ν_g) and a quintessence particle (ν_q, repulsive), for instance see the description in [9, 10, 23-25]. The latter decay mode seems to take place in the experiments by Tajmar *et al.* [11-13]. There is the possibility that one of these unknown gravitational fields might be usable as novel propulsion principle. Furthermore, it might be necessary to *extend the principle of momentum and energy conservation* to both ordinary and non-ordinary matter and including the interaction of the local spacetime filed with these two types of matter [9, 10, 24]. For instance, $\mathbf{m}+\mathbf{m}'=0$ where $\mathbf{m}$ and $\mathbf{m}'$ denote momentum of ordinary and dark matter, respectively. If only $\mathbf{m}$ is measured this would be seen as violation of the momentum conservation principle. As will be shown in the next section, *EHT* predicts that photons can be converted into gravitophotons (enabling an interaction between electromagnetism and gravitation), but this type of gravitation is not mediated by the grviton, which only is the messenger particle for Newtonian gravitation. The combined energy densities of attractive ν_{gp}^{+} (negative energy density) and repulsive ν_{gp}^{-}) (positive energy density) gravitophotons should be zero. It is currently believed that the combined energy density of baryon mass and gravitation add up to zero. However, ν_{gp}^{+} interacts with ordinary matter and thus generates a force. The repulsive gravitophoton ν_{gp}^{-} should not interact with ordinary matter since otherwise Newton's law could not be valid; instead, it seems to drive the *Universe* apart at the present time. The total amount of energy extracted from the vacuum would be zero. However, there would be a net effect on ordinary matter the spacecraft is made from. In general, *conservation principles need to be applied to a closed physical system, which includes both ordinary and non-ordinary matter and the local* spacetime *field.*

2.1.1. Geometry and Physical Interactions

In this section the fundamental ideas of *EHT* are presented with regard to gravitomagnetic phenomena. In Dröscher and Hauser [20] these predictions are compared further with recent experiments. Presenting theory only without experimental validation may lead to the loss of physical reality. Experiment without theoretical insight might lack any deeper understanding, and necessary guidelines for advanced propulsion systems might not be obtainable.

In a series of articles, starting 2002, the existence of additional gravity-like fields was proposed by the authors. In the following an overview of the fundamental assumptions on the underlying physical ideas of *EHT* is presented, which is based on concepts from the German physicist Heim [28, 30] and the Italian mathematician Finzi [32].

2.1.2. Theories with Higher Dimensions

The approach taken by *EHT* is different from supergravity, supersymmetry or superstring theories, for instance see [33]. These theories increase the number of real physical spatial dimensions to construct a single $N \times N$ metric tensor to include Maxwell's equations (*i.e.,* the gauge boson for electromagnetic interaction) or the Young-Mills field, in order to describe electromagnetism and the weak and strong interactions. Already in the 1920s additional spatial dimensions were introduced to unify gravitation and electromagnetism. Supergravity increases the number of dimensions further to incorporate matter in the form of quarks and leptons. For instance, in supergravity 11 dimensions are used. Hence the metric tensor would have 11^2 elements, but not all of them are independent. Since the extra dimensions have not shown up in any experiment, they are considered to be sufficiently compactified (Planck length $l_P \approx 10^{-35}\ m$), so that present energy probes cannot see any of these additional dimensions. The metric tensor that characterizes *GR* is given on the left of Equation (2). In the theory by Kaluza [34, 37], in order *to unify gravitation and electrodynamics*, the metric tensor is a 5×5 matrix, which is depicted on the right hand side. That is, at each point in spacetime a circle is attached, meaning that this dimension is curled up, and thus is not directly visible.

$$\begin{pmatrix} g_{11} & g_{12} & g_{13} & g_{14} \\ g_{21} & g_{22} & g_{23} & g_{24} \\ g_{31} & g_{32} & g_{33} & g_{34} \\ g_{41} & g_{42} & g_{43} & g_{44} \end{pmatrix}; \qquad (1)$$

$$\left(\begin{array}{cccc|c} g_{11} & g_{12} & g_{13} & g_{14} & A_1 \\ g_{21} & g_{22} & g_{23} & g_{24} & A_2 \\ g_{31} & g_{32} & g_{33} & g_{34} & A_3 \\ g_{41} & g_{42} & g_{43} & g_{44} & A_4 \\ \hline A_1 & A_2 & A_3 & A_4 & \end{array}\right), \qquad (2)$$

where $A = (\Phi, \mathbf{A})$ denotes the four potential of electrodynamics. Adding six or seven curled up spatial dimensions (hypersphere) leads to a single 10 or 11 dimensional metric tensor that describes gravitation, electrodynamics, strong and weak forces as well as quarks and leptons (matter). However, the gravitomagnetic experiments (as is shown in Section II of [10]) would have no place in this higher dimensional metric tensor. Furthermore, there is no evidence for higher spatial dimensions. There is also *no explanation for dark matter and dark energy through this single metric tensor*. This type of metric is termed *monometric*.

Therefore, physics takes place in a space $M^4 \otimes K$ where M^4 is the four-dimensional spacetime manifold and K denotes the compactified additional space. For example, the 5×5 metric tensor of Kaluza and Klein, 1926, to unify gravity and electromagnetism has 15 independent components, while Einstein's original metric tensor in our four-dimensional spacetime accounting for gravity only, has 10 independent quantities.

EHT differs from this approach in that there is only four-dimensional spacetime, but multiple 4×4 metric tensors are constructed, termed Hermetry forms.

EHT is described by a *poly-metric*, based on the double coordinate transformation that is derived in the subsequent section. The word *Hermetry* is a combination of *hermeneutics and geometry* that is, a Hermetry form stands for the physical meaning of geometry. Each Hermetry form has a direct and unique physical meaning (for more details see [17, 18, 26]). Each Hermetry is represented by its respective metric tensor (matrix).

No additional spatial dimensions exist in *EHT*. However, as will be outlined in Section 2.3, at each point in spacetime an internal eight-dimensional tangent space, termed Heim space H^8, is attached. This picture leads to the geometrization of physics as foreseen by Einstein [8], and provides a different approach toward unification than string theory.

2.2. Internal Space, Hermetry Forms and Fundamental Interactions

EHT is based on the following four fundamental rules, termed the *GODQ* principle. These rules cannot be derived from more basic principles or mathematical assumptions. Briefly stated: *God quantizes.* Again, these four principles cannot be proved mathematically, but their formulation is based on generally accepted observations and intends to reflect the workings of *Nature*. They simply express the underlying principles that seem to apply in the construction of the *Universe* as least, as this process is understood at the moment. Once accepted, the major physical laws and evolution of the *Universe* should unfold. Needless to say, that *EHT* has not reached this state of development, but the rules of *GODQ* have been employed as the basic guidelines.

i. *Geometrization principle for all physical interactions* that is, constructing the proper metric tensor for each physical interaction.

ii. *Optimization (Nature employs an extremum principle)* that is, physical action expressed by its proper path integral.

iii. *Dualization (duality, symmetry) principle (Nature dualizes or is symmetric and antisymmetric, bits)* that is, determining symmetry groups of both the external space, which is the spacetime field itself (M^4 manifold), and the associated internal Heim space H^8. Conservation principles follow from symmetries of both spaces.

iv. *Quantization principle (Nature uses integers only, discrete quantities)* that is, discreteness and finiteness of physical quantities as well as discrete nature of spacetime (Smolin: "atoms of spacetime" [54]).

The duality principle holds, in particular, for the *dual structure* of *universal physical* space, namely four-dimensional external spacetime (the stage for physical events) and internal Heim space H^8 (physical entities, actors, marked by their symmetry groups). Furthermore, the set of 15 Hermetry forms itself can be separated into two classes termed ordinary and non-ordinary matter, see below. Only the graviton belongs to both classes. Experimental physics has revealed that the so called vacuum of space possesses significant physical features, for instance it exhibits the *Casimir* force. Moreover, material particles can be generated out of this vacuum, so there is evidence for the existence of an underlying internal space, and all physical phenomena seem to be based on the existence of these two spaces. Is this were not the case, only geometry would exist. According to B. Heim this was the case in the early *Universe*, until the minimum area (*metron* area) reached a certain smallness, which triggered a phase transition and matter was produced. A short discussion can be found in Dröscher and Hauser [16] in the section on *Speculative Cosmology* and, of course, from Heim Vol I [29] and II [30]. This does not necessarily exclude the big bang, which might be just another name for the phase transition.

2.2.1. Internal Space H^8

For a complete physical description, the internal dynamical physical properties of spacetime need to be taken into account that is, if a general geometrodynamics approach is to be realized. To this end, in *EHT*, an internal eight-dimensional space, *called Heim* space H^8, comprising four subspaces, see Fig. **1**, is attached to each point in spacetime. The subspace structure is introduced by physical arguments. In other words, the symmetry of H^8 is broken due to this structure. This is the only time an explicit assumption on the structure of H^8 is made. This generalized concept of universal physical space S, namely the combination of four-dimensional (external) spacetime, manifold M^4 with eight-dimensional internal space H^8 , is considered to account for both physical phenomena (particles, waves, interactions) and the stage on which these phenomena take place. This picture is fundamentally different from the view expressed by Veltmann [38]. From the duality principle the existence of additional internal symmetries in *Nature* is deduced, and thus such a higher dimensional internal symmetry space should actually exist, whose structure will now be presented. However, as discussed in Section 2.3, the universal physical space, termed S, external and internal, is *not* the direct sum (or Cartesian product) of $M^4 \otimes H^8$, but trough the *entanglement of internal and external coordinates*, *instead* leads to the construction of a *poly-metric tensor* as well as a set of *four symmetry groups*.

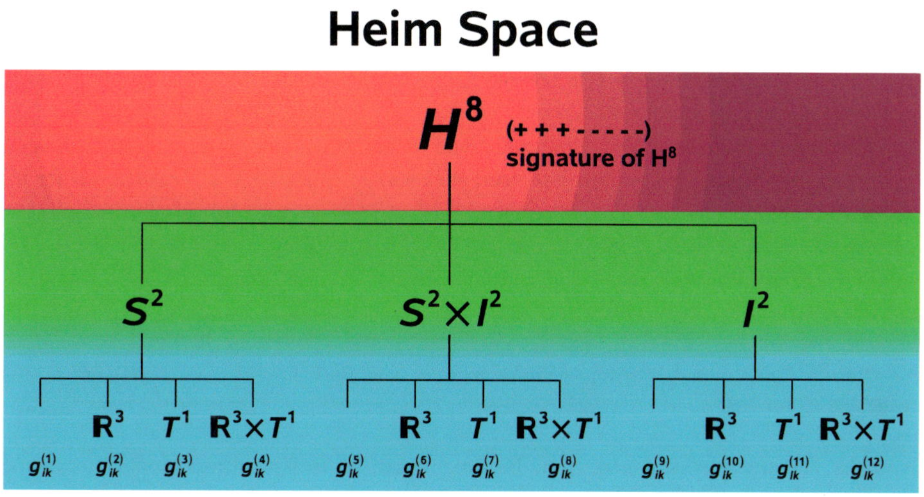

Figure 1: Each point in spacetime is associated with an additional *internal* space, termed Heim H^8 that has eight internal coordinates $\xi^a, a = 1 \ldots 8$. These coordinates are associated with a general 8 component quaternionic scalar field $\varphi = (\varphi^a), \varphi^a \in \mathbb{H}$, $a = 1,.,8$, and $\mathbb{H}$ denoting the set of quaternions. We refer here to the non-linear $\sigma-$ model of Jost [27], which seems to comprise the mathematical description of this physical process. This topic needs to be investigated further. The picture shows the complete set of metric-subspaces that can be constructed from the poly-metric tensor, Equation (4).

Heim space H^8 comprises four subspaces, namely manifolds R^3 (not to be confused with Euclidean space R^3), T^1, S^2, and I^2. The 8 coordinates, ξ^a, of H^8 themselves are functions of the curvilinear coordinates η^μ that is, $\xi^a = \xi^a(\eta^\mu)$ of four-dimensional spacetime, characterized by manifold M^4. Using index selection rules for the poly-metric tensor, a set of submetric tensors can be constructed. Each submetric is denoted as Hermetry form, which has, because of the selection rules, direct physical meaning (see Table **2** in [10, 26]). In order to *construct a Hermetry form, the selection rule requires that internal* space S^2 *or* I^2 *coordinates must be present*. Following this rule, directly delivers a set of 12 Hermetry forms. Internal space H^8 is a factored space that is, it is represented as $H^8 = R^3 \times T^1 \times S^2 \times I^2$. This factorization of H^8 into the internal space-like manifold R^3 and three internal time-like manifolds (signature), namely T^1, S^2 and I^2 comprises the inner structure of H^8. The four subspaces have the following physical meaning: R^3 *is responsible for ponderable mass* (*energy*), the presence of T^1 stands for *charge in general* (*electric, gravitational, weak and color*), S^2 is representing *organization and structure of physical phenomena*, and I^2 is *accomplishing the exchange of information* among the constituents of a physical system.

Postulating the subspace structure of H^8 is equivalent to an explicit symmetry breaking. This is, however, mandatory, since the resulting subspace structure, through their symmetry group structure, is responsible for the physical events to happen. In other words, the evolution of the *Universe* starts at the very instant this subspace structure is realized, which is the beginning of physical time. As long as H^8 retains its symmetry, characterized by group $O(8,q)$, only the geometry of external spacetime exists, and thus there exists the stage only, but not yet the actors. By means of symmetry breaking, perhaps achieved through the mechanism conceived by Heim, see remarks above, the *geometric Universe turns into the physical Universe*. However, for the following discussion, the process of how symmetery breaking is actually achieved and whether a purel geometric Universe pre-existed, is not of concern. Only the fact that this process has occurred is important, *i.e.*, breaking the symmetry of $O(8,q)$ into four subgroups.

In order to geometrize physical interactions, the two principles of *SR* need to be extended, which state that

- Spacetime *is both homogeneous and isotropic* (this same principle applied to the existence of worlds hospitable to life would directly show that they should be most common), and, second,

- *Physical laws are covariant when moving from one inertial system to another.*

To summarize the discussion of above, in *EHT*, the following additional concepts are introduced:

- The third principle to be added, first conceived by Heim (for instance [28] and independently formulated by Finzi [32], states that *each physical interaction is characterized by its own Hermetry form* (metric tensor). Consequently, in order to account for all physical phenomena, instead of a mono-metric, a *poly-metric* has to be constructed. This is a *necessary, but not a sufficient condition.*

- In addition, as was already stated, four-dimensional spacetime is complemented by an *internal 8D* space, H^8 *(causing physical objects to appear on stage), comprising a four-fold sub*space *structure.*

- In order for any physical object to become manifest in four-dimensional spacetime, the Hermetry form of this object must include *coordinates of sub*space S^2 *(organization, structure) or* I^2 *(exchange of information).*

- In H^8, there exist three internal spatial-like coordinates (lower Latin characters *a, b, c* are used for summation in H^8, ranging in general from 1 to 8. Greek indices μ, ν, .. indicate summation from 1 to 4 in four-dimensional spacetime. Characters *i, j, ...* indicate summation in physical space from 1 to 3.

- The first four internal coordinates ξ^1, ξ^2, ξ^3 are for space R^3, and the internal coordinate ξ^4 is used for space T^1.

- The remaining four coordinates are utilized to describing the degree of *organization* and *information* exchange as observed in *Nature*. Coordinates ξ^5, ξ^6, space S^2, denote *entelechial* and *aeonic* coordinates (organization and structure), respectively, and ξ^7, ξ^8 denote coordinates for the exchange of information in space I^2.

- With the introduction of four different types of coordinates, the space of fundamental symmetries of internal space H^8 is fixed and each subspace of H^8 is associated with a specific symmetry group. That is, the set of four symmetry groups describes all particles and fields that can exist, as will be discussed in the following section.

- In summary, the set of 15 Hermetry forms serves as a *classification scheme* for all physical objects. Hermetry form 16, *deemed to be responsible for the inertia field, is special in that it does not contain organization or information coordinates*, but lives directly in four-dimensional spacetime, and its Hermetry form is given by $H_{16}(\mathbb{R}^3 \times T)$ (it should be noted that $\mathbb{R}^3$ stands for a manifold in physical space).

2.2.2. Group Structure of Space H^8 *and Hermetry Forms*

In physics symmetries are represented by mathematical groups and through *Noether's* theorem, with each symmetry transformation a conservation law is associated. Hence, the group structure of any physical theory is of fundamental importance.

Internal space H^8 has eight internal coordinates ξ^a, $a = 1\dots8$ which are associated with a general 8 component quaternionic scalar field $\varphi = (\varphi^a)$, $\varphi^a \in \mathbb{H}$, $a = 1,\dots,8$, and $\mathbb{H}$ denoting the set of quaternions (the reader not interested in the mathematical details can skip this section). It should be noted that the set of quaternions is used instead of the conventional set of real or complex numbers, since $\mathbb{H}$ provides the simplest non-commutative algebra, and will also allow using tensors of third rank as connections. The reason for this will be given shortly.

The subspace structure of H^8 comprising subspaces $R^3 \times T^1 \times S^2 \times I^2$ gives rise to four corresponding quaternionic symmetry groups, namely

$$O(3,q), O(2,q), O(2,q), O(1,q) \text{ with } q \in \mathbb{H}.$$

In other words, each internal coordinate ξ^a corresponds to a scalar function $\varphi(x)$ with $x = (x^\mu(\eta))$ that is $\xi^a(x) \leftrightarrow \varphi^a(x)$. It is now required that $\varphi(x)$ transforms according to $Q\varphi(x)$ with $Q \in O(n,q)$ and $n = 1, 2, 3$, similar to *QFT* as discussed, for instance, by Kaku [34], Zee [39], or Lawrie [40]. This means that internal space R^3 possesses local symmetry under $O(3,q)$, T^1 is locally invariant under $O(1,q)$, and S^2 and I^2 are locally symmetric under $O(2,q)$. The number of generators of each of these groups is 15, 1, 6, and 6, respectively. This local symmetry is underlying the principle of local gauge invariance.

The set of quaternions $\mathbb{H}$ is used in place of the set of complex numbers C, since gravitation is to be included in the Yang-Mills formalism. Complex numbers lead to gauge potentials A_μ, which are first rank tensors. Quaternions, however, result in third rank tensors for the gauge fields.

Furthermore, the 15 generators (gauge fields) of group $O(3,q)$ associated with internal space R^3 (gauge fields), stand for the 15 Hermetry forms of H^8, and thus are responsible for the physics that can occur in the *Universe*. In turn, each Hermetry form or family of Hermetry forms possess their own specific symmetry group.

The four symmetry groups describing the 15 gauge fields of the four interactions are $SU(3)$ for the 8 gluon fields of the strong interaction, $U(1)$ for the electromagnetic field, $SU(2)$ for the three fields of the weak interaction, and $SO(4)$ for the 6 fields of the gravitational interaction. The single generator of $O(1,q)$, internal space T^1 is related to the vacuum field.

Inertia is not described by any Hermetry form in H^8, but is a feature of pure spacetime and thus is described by $H(\mathbb{R}^3 \times T)$. Here spaces $\mathbb{R}^3 \times T$ denote spacetime and not the subspaces of H^8. The physical meaning of these spaces should be clear from the context. Obviously, because of the Minkowski metric, the symmetry group belonging to (also called) Hermetry form, H_{16}, is $SO(3,1)$. In other words, two types of gravitons seem to exist. The one for gravitational mass belongs to the symmetry group $SO(4)$, the other; one responsible for inertia is represented by group $SO(3,1)$. It should be noted that this fact may have major consequences, since the no-go theorem does not hold for $SO(4)$. The 6 generators of $O(2,q)$ for S^2 give rise to 6 Higgs fields (bosons) through which fermions are coming into existence (mass and charge) by the Higgs mechanism as well as the massive bosons, for instance of the weak interaction $W^\pm$, Z^0. Finally, the 6 anti-Higgs bosons of $O(2,q)$ from subspace I^2 procure mass and charge, including imaginary mass for the particles representing non-ordinary matter.

This formulation also answers the important question: what is the *origin of physical fields and particles*? This question cannot be answered by *QFT*. In the context of H^8, it is straightforward to see that it is the *structure of the 8 internal (generalized) coordinates that give rise to physical fields*. Moreover, all fields are of geometrical nature, *i.e.,* possess their proper metric.

The 15 Hermetry forms (gauge fields) represent all aspects of matter that is, *fermions and bosons as well as the six fundamental physical interactions*. This includes the two novel gravitational field quanta, termed *neutral gravitophoton*, ν_{gp}^0, (attractive and repulsive), and a repulsive particle named quintessence, ν_q, that might explain dark energy, which, together with the graviton, give rise to the three gravitational fields predicted by *EHT*. The meaning of the 15 Hermetry forms is discussed below.

1. Hermetry forms H_1 to H_8 describe ordinary matter (OM), both fermions and bosons.

2. Hermetry forms H_1 and H_9 to H_{15} are describing non-ordinary matter (NOM). For instance, neutral leptons, e^0 (believed to be in the range of 40-45 GeV mass?), should exist.

3. Hermetry forms of ponderable mass (possessing a rest mass different from zero) have to contain subspace R^3 in their metric tensor.

4. Hermetry forms carrying any charge include subspace T^1 in their metric.

5. Only the photon can interact with NOM by being converted into a neutral gravitophoton, *i.e.*, $\gamma \to \nu_{gp}^0$.

6. Hermetry forms representing field quanta (messenger particles (bosons) that are mediating forces) have to contain subspace I^2 in their Hermetry form.

7. There is one exception, namely the graviton whose Hermetry form is $H_1 = H_1(\mathrm{S}^2)$.

8. NOM particles *do not contain sub*space R^3 in their Hermetry form, and are thus believed to possess imaginary mass. They may be generated in slids, but, because of their interaction with the lattice, do not behave like tachyons.

9. There is one exception, namely the neutral leptons, which belong to NOM, but have ponderable mass.

10. For NOM, the concept of virtual particle is extended to *virtual particles possessing imaginary mass but real charge*, which are assumed to be responsible for the conversion of photons into gravitophotons, thus establishing the interaction among electromagnetism and gravitation.

11. Each Hermetry form represents a *class of physical phenomena* or *particle family*, for instance, the six quarks or the eight gluons.

12. Dark matter, $H_{14}(\mathrm{R}^3 \times \mathrm{S}^2)$, might consist of neutral leptons, presumably the neutral tau particle τ^0, that are lacking subspace T^1, which is present in the Hermetry form of the charged leptons.

13. The set of quarks is represented by Hermetry form $H(\mathrm{R}^3 \times \mathrm{T}^1 \times \mathrm{I}^2)$.

14. Gluons are given by Hermetry form $H(\mathrm{R}^3 \times \mathrm{I}^2)$.

Hermetry forms contain numerous partial terms in their metric, and therefore, represent families of particles or fields. For example, the quark Hermetry form contains 36 partial metric terms. In how far these partial structures are able to represent the entire range of physical properties of quarks and gluons, including their quantum numbers, such as mass, charge, color charge, weak charge, flavor *etc.* is not known at present. The spin of a particle does not seem to be directly contained in a Hermetry form.

This picture of symmetry groups is different from the current approach utilized in the standard model, since it results in a hierarchical group structure. In this regard, a single group describing all elementary particles like $E(8)$, as, for instance, postulated recently by Lisi [43, 44], *should not exist* in *Nature*.

The extremal principle in physics is characterized by the Lagrange density, and, as a simple example, we consider a charged particle in an electric potential whose Lagrangian is given by $\mathcal{L}=\frac{1}{2}mv^2-e\Phi$. Replacing classical coordinates q_i by scalar fields φ^i gives the general formulation of the Lagrangian $\mathcal{L}=\partial_\mu\phi^\dagger\partial_\mu\phi-V(\varphi^\dagger\varphi)$. The writing of this formula deserves further explanation. In general, a Hermetry form is written as $H_\ell=H_\ell(\xi^a)$, where indices a are determined from the subspaces that occur in the metric of this form. The Lagrangian $\mathcal{L}_\ell$ associated with this H_ℓ, can, however, only depend on the three component scalar function $\phi=(\varphi^1,\varphi^2,\varphi^3)$, since its corresponding $O(3,q)$ group belongs to R^3. On the other hand, the potential function V is locally invariant as long as it depends on the product $\varphi^\dagger\varphi$. Moreover, a Hermetry form causes curvature in four-dimensional spacetime, which in turn determines the equations of motion. At the same time, the equations of motion for any kind of particle can be determined from its Lagrangian, and thus this Lagrangian should reflect its dependence on φ. We believe that a description of the Lagrangians, constructed from their associate Hermetry form, in the framework of the *nonlinear* σ *-model*, as given in the next section, should provide the hitherto missing link between geometry and physics.

2.3. Double Coordinate Transformation and the Non-Linear σ – Model

Einstein introduced the principle of geometrization in physics, which successfully identified the gravitational potential by the metric tensor of spacetime. However, his attempts to geometrize the other force fields, in particular the electromagnetic field, remained unsuccessful. An extension of the Einstein mono-metric tensor to a poly-metric tensor is needed to comprise the other force fields. Since physical fields were identified with Heim space H^8, it is straightforward to surmise that the coordinates ξ^a should be involved in the construction of the poly-metric. This should be done in such a way that if space H^8 is not present, the mono-metric of ξ^a *GR* is obtained. This is achieved by the so called double-coordinate transformation as demonstrated in Section 2.3.2. In addition, there is one more step, namely to find a physical principle that associates geometry with physics that is, $\xi^a\leftrightarrow\phi^a(\eta)$, finding a relationship between internal coordinates of H^8 and a set of scalar functions that will enter into the Lagrangian. Einstein, as is discussed at the end of this section, did not have such a principle and *simply postulated curvature equals energy*. In other words, matter curves spacetime.

In the following the mathematical details of constructing such a poly-metric tensor are presented, whose roots are going back to the ideas of Heim. However, it will turn out, that it is most important to be specific about the type and dimension of the internal space being introduced as well as the inherent symmetries (group structure) this space possesses. Moreover, it is of special consideration whether the Lie algebra associated with these groups' lives in a real or complex vector space, or a different mathematical field should be employed. As the discussion in Section 2.2.2 revealed, the mathematical field of quaternions, $\mathbb{H}$, should be applied instead.

As will be shown below, the poly-metric tensor that is obtained from the introduction of Heim space H^8, gives rise to a set of 15 metric subtensors or Hermetry forms that possess physical meaning, see Section 2.3.2. The concept of an internal 8D space, comprising four subspaces, as discussed above, leads to a modification of the general transformation employed in *GR*. In *GR* there are two sets of coordinates, Cartesian coordinates x and curvilinear coordinates η, linked by a relation between their corresponding coordinate differentials, Equations (21) and (22). If Heim space were not to exist, the poly-metric collapsed to the mono-metric of *GR*.

Having constructed a poly-metric tensor and its associated set of Hermetry forms, however, is only half of the task, providing, so to speak, the left hand side (curvature tensor) of the Einstein field equations. In order to establish a connection between geometry and physics, Einstein used the energy-momentum tensor $T_{\mu\nu}$ as the right hand side of his field equations. In the same way one could add the energy-momentum tensor of electromagnetism and arrive at a set of equations that describe both gravitational and electromagnetic phenomena. This ad hoc approach does not have any deeper physical interpretation, and adding the electromagnetic energy-momentum tensor does not even have a geometrical interpretation. Therefore, Weyl, in 1918, made a first attempt to construct a unified field theory by proposing that the length of a vector changes when parallel displaced. The change in length was dependent on a four-potential, identified with the vector potential of electromagnetism. However, this gave the unphysical result in that the natural frequencies of atoms changed according to their position in spacetime.

The construction process using H^8 results in a genuine poly-metric, but still the relationship between Hermetry forms and physical fields needs to be established. In Hauser and Dröscher [24], the real part of internal coordinates ξ^a in H^8 was already identified with scalar fields ϕ^a with $a = 1,\ldots,8$. However, there is a mathematical formulation termed non-linear σ-model. Comparing the double coordinate transformation with this non-linear σ - model as described by Jost in [27], and identifying coordinates ξ^a with scalar fields ϕ^a, eventually leads to a Lagrange density in the action integral S of the non-linear σ model that can be expressed through the formula of the double coordinate transformation Equation (8). In this way, each Hermetry form is endowed with its proper Lagrangian.

2.3.1. Mono-Metric Tensor of GR

In *GR* there exists four-dimensional spacetime only, comprising three spatial coordinates with positive signature (+) and the time coordinate with negative signature (-). It should be remembered that the Lorentzian metric of R^4 has three spatial (+ signature) and one time-like coordinate (- signature). Signatures are not unique and may be reversed. Numbering of coordinates was chosen such that coordinates of positive signature are numbered first. The corresponding metric is called Minkowski metric and the spacetime associated with this metric is the Minkowski space. The plus and minus signs refer to the (local) Minkowski metric (diagonal metric tensor, see Equation (21). Therefore, the squared proper time interval is taken to be positive if the separation of two events is less than their spatial distance divided by c^2. Let coordinates x^μ with $\mu = 1,\ldots,4$ denote Cartesian coordinates $x^1 = x$, $x^2 = y$, $x^3 = z$, and $x^4 = ct$. A general coordinate system for a spacetime manifold, M, needs to be described by curvilinear coordinates η^ν with $\nu = 1,..,4$ and $\eta = (\eta^\nu) \in \mathsf{M}$. In *GR* the equations relating the two systems of coordinates are given by $x^\mu = x^\mu(\eta^\nu)$ or $\eta^\nu = \eta^\nu(x^\mu)$ and the distance between two neighboring events with coordinates η^ν and $\eta^\nu + d\eta^\nu$ is given by the square of the line element $ds^2 = g_{\nu\mu} d\eta^\nu d\eta^\mu, \nu, \mu = 1,\ldots 4$ where the metric tensor is of the form

$$g_{\nu\mu} = \mathbf{e}_\nu \cdot \mathbf{e}_\mu = \frac{\partial x^\alpha}{\partial \eta^\nu} \frac{\partial x^\alpha}{\partial \eta^\mu} \tag{3}$$

and $\mathbf{e}_\nu = \partial \mathbf{x} / \partial \eta^\nu$ with $\mathbf{x} = x^\mu \hat{\mathbf{e}}_\mu$. The vectors $\mathbf{e}_\mu$ are the curvilinear (covariant) base vectors and $\hat{\mathbf{e}}_\mu$ denote the Cartesian unit vectors.

2.3.2. Poly-metric Tensor and Double Coordinate Transformation

In this section, the set of metric subtensors obtained from the ξ^a coordinates of H^8 is constructed. In turn, each of subtensor is describing a *class* of physical phenomena (physical interaction or particles). This leads to the concept of Hermetry form, which was introduced in Section 2.1. Thus, the connection between physical space and physics (symmetries) might be established in a way foreseen by Einstein, namely by the geometrical properties of spacetime, provided an additional principle can be found that allows to *calculate the energy-momentum tensor* $T_{\mu\nu}$ *from external* spacetime *and internal* space *coordinates*. This should be achievable by employing the nonlinear σ-model that was mentioned in Section 1.4 and will be further discussed below. However, in order to reach this objective, spacetime had to be complemented by an internal space H^8 to *model its intrinsic physical properties*. Once the internal space with its set of coordinates has been determined, everything else is fixed, and Equation (4) is a direct consequence of H^8. In contrast to *GR*, now the relation between the coordinate systems (x^μ) and (η^ν) is *via* the internal space with coordinates ξ^a that is $x^\mu = x^\mu(\xi^a(\eta^\nu))$ or $\eta^\nu = \eta^\nu(\xi^a(x^\mu))$. The mathematical and physical consequences, which are substantial, are discussed in the following two sections. The approach in *EHT* is fundamentally different, since a set of 15 different 4 × 4 metric tensors is constructed that all live in four-dimensional spacetime. The existence of internal space H^8 demands a more general coordinate transformation from a spacetime manifold In the concrete case of *GR* spacetime manifold M^4 would be used M to a manifold N *via* the mapping M (locally

R^4) $\to$ H^8 $\to$ N (locally R^4). In *EHT*, therefore, a double transformation, Equation (4), involving Heim space H^8 occurs. The global metric tensor is of the form

$$g_{\mu\nu} = \frac{\partial x^\alpha}{\partial \xi^a}\frac{\partial \xi^a}{\partial \eta^\mu}\frac{\partial x^\alpha}{\partial \xi^b}\frac{\partial \xi^b}{\partial \eta^\nu} = \frac{\partial x^\alpha}{\partial \xi^a}\frac{\partial x^\alpha}{\partial \xi^b}\frac{\partial \xi^a}{\partial \eta^\mu}\frac{\partial \xi^b}{\partial \eta^\nu} \tag{4}$$

and

$$g_{ab}(\xi(\eta)) := \frac{\partial x^\alpha}{\partial \xi^a}\frac{\partial x^\alpha}{\partial \xi^b} \tag{5}$$

where indices $a,b = 1,...,8$ and $\mu,\nu,\alpha = 1,...,4$, and thus $g_{\mu\nu}$ comprises 64 components. Length being geometric is invariant under such a re-parametrization, and thus Equations (3) and (4) describe exactly the same geometric object. So it seems that nothing has been achieved by this double coordinate transformation, since, obviously, all other geometrical features of the manifold remain also invariant.

However, the associated complete metric tensor (Equation (4)) with its total of 64 terms, Equation (6), *does not have any physical meaning by itself*. The construction process for the set of the Hermetry forms is *accomplished* as follows.

Extracting a certain number of terms from the global metric described by Equation (4) employing the selection rules mentioned above, the complete set of 15 different Hermetry forms is eventually obtained.

A single component of the metric tensor belonging to one of the four subspaces is given by Equation (6). Only special combinations of the $g_{ik}^{\alpha\beta}$ reflect physical quantities, *i.e.,* Hermetry forms. Because of the double transformation, each physically meaningful metric does comprise a different subset of the 64 components. In other words, depending on the Hermetry form, *via* index selection, only specified components from the complete metric tensor in spacetime, Equation (4), are chosen. Hence, each Hermetry form is marked by the fact that *only a subset of the 64 components* is present. This subset is different for each Hermetry form. Therefore each Hermetry form leads to a different metric in the spacetime manifold, and thus describes a different physical phenomenon. In other words, this approach is equivalent to the solidarity principle of Finzi, namely each class of physical phenomena (Hermetry form) determines its proper curvature in four-dimensional spacetime [32]. This is why Equation (4) is termed the *poly-metric tensor*. It serves as a *repository for the 15 Hermetry forms*. This construction principle is different from Einstein's approach. Only in the special case of vanishing space H^8, *EHT* reduces to *GR*,

$$g_{\mu\nu}^{ab} = \frac{\partial x^\alpha}{\partial \xi^{(a)}}\frac{\partial \xi^{(a)}}{\partial \eta^\mu}\frac{\partial x^\alpha}{\partial \xi^{(b)}}\frac{\partial \xi^{(b)}}{\partial \eta^\nu}. \tag{6}$$

The poly-metric tensor can be written as

$$g_{\mu\nu} = \sum_{a,b=1}^{8} g_{\mu\nu}^{a,b}. \tag{7}$$

A single Hermetry form is given by

$$g_{\mu\nu}(H_\ell) := \sum_{a,b\in H_\ell} g_{\mu\nu}^{a,b} =: \sum_{a,b\in H_\ell} (a,b). \tag{8}$$

Any Hermetry form can be written as the sum of its symmetric and anti-symmetric part, where indices S and A denote the splitting of the partial metric terms into their symmetric and anti-symmetric parts

$$(a,b)_S := \frac{1}{2}[(a,b)+(b,a)] \quad \textbf{(9)}$$

and

$$((a,b)_A := \frac{1}{2}[(a,b)-(ba)]\,. \quad \textbf{(10)}$$

For instance, the Hermetry form of the first neutral gravitophoton field which decays into a graviton and a quintessence particle, $\nu_{gp}^{0} \rightarrow \nu_g + \nu_q$, is represented as

$$H(\nu_g + \nu_q) = H(\nu_g) + H(\nu_q) \quad \textbf{(11)}$$

where the Hermetry forms of the graviton and the quintessence particle can be written in the form

$$H(\nu_g) = (55)_S + (56)_S + (66)_S \quad \textbf{(12)}$$

and

$$H(\nu_q) = (77)_S + (78)_A + (88)_S \quad \textbf{(13)}$$

The metric tensor representing any Hermetry form can therefore be written in general form

$$g_{\mu\nu}(H_\ell) = \sum_{a,b\in H_\ell} (a,b)_{S,A} \quad \textbf{(14)}$$

The analysis of the recent experiments by Tajmar [13] led to the conclusion that four different partial gravitomagnetic fields should have been generated, whose strengths were calculated from those metric components describing both the graviton and quintessence particles. Further details can be found in [10].

2.3.3. Asymmetric Hermetry Forms

As the experiments by Tajmar *et al.* [11-13], Graham *et al.* [14] and also the anomalous spin drift observed at Gravity-Probe B have shown, the *magnitude of the gravitomagnetic field strongly depends on the direction of rotation* that is, the *gravitomagnetic field in clockwise direction differs substantially from the counter-clockwise field* leading to some kind of parity violation.

By assigning a spin or parity value to the respective Hermetry forms, a heuristic explanation is given for the experimentally detected asymmetry in the magnitude of the clockwise and counter-clockwise gravitomagnetic fields denoted as $\mathbf{B}_{gp}^{CW}$ and $\mathbf{B}_{gp}^{CCW}$.

To this end, we associate a value of $+1$ with each symmetric term, and 0 with each antisymmetric term in Equation (12). Next, the so called *strength* value, S (not to be confused with the symmetric subscript), for each Hermetry form is introduced. It should be noted that the various terms comprising a Hermetry form are representing physical potentials. The S value of a Hermetry form is therefore a measure of the strength of this potential, and hence determines the magnitude of the gravitomagnetic fields $\mathbf{B}_{gp}^{CW}$ and $\mathbf{B}_{gp}^{CCW}$. With respect to Equation (12) one finds

$$S(H(\nu_g)) = 3 \quad \text{and} \quad S(H(\nu_q)) = 2 \quad \textbf{(15)}$$

Now the strength values $S^{CW}(H(\nu_g + \nu_q))$ and $S^{CCW}(H(\nu_g + \nu_q))$ need to be determined. The argument of S reflects that, according to *EHT*, in the experiments by Tajmar *et al.* [11-13] as well as Graham *et al.* [14], the predicted neutral gravitophoton is decaying into a graviton and quintessence particle, *i.e.*, $\nu_{gp}^{02} \rightarrow \nu_g + \nu_q$.

In addition, two parity factors are introduced, namely $P_x = \pm 1$ and $P_t = \pm 1$. P_x is associated with space inversion $\mathbf{x} \rightarrow -\mathbf{x}$ that is, $P_x^{CW} = +1$ and $P_x^{CCW} = -1$. P_t denotes the time inversion parity parameter. Values $P_t^{CW} = +1$ and $P_t^{CCW} = \pm 1$ are postulated. We do this in order to reproduce the four ratios of CW and CCW gravitomagnetic fields, measured by Tajmar *et al.* [11-13] for different combinations of materials. However, the quantitative ratio of the CW and CCW gravitomagnetic fields is *calculated from the properties of the Hermetry forms* of ν_g and ν_q. For the CCW rotation P_t^{CCW} seems to depend on the material, but at present this dependence is not known.

The strength in CW direction is given by

$$S^{CW}(H(\nu_g + \nu_q)) = P_x^{CW} P_t^{CW} \left(S(H(\nu_{.g}) + S(H(\nu_q)) \right) = 3 + 2 = +5 \tag{16}$$

where the arguments of the parity parameters are the same as for the S values. In a similar way the strength in the CCW direction is calculated

$$S^{CCW}(H(\nu_g + \nu_q)) = P_x^{CCW} P_t^{CCW} S(H(\nu_g) + P_x^{CCW} P_t^{CCW} S(H(\nu_q)) = (\pm 3) + (\pm 2)\,. \tag{17}$$

The four numerical values are.

$$S^{CCW}(H(\nu_g + \nu_q)) = (+5, +1, -1, -5)\,; \tag{18}$$

forming the ratio of Equations (16) and (18), gives four relative signal strengths for the gravitomagnetic fields in the CW and CCW direction of rotation.

In the latest measurements by Tajmar *et al.* [13] three experimental configurations A, B, and C are employed. Configuration C is not considered in this analysis, since the measured uncertainties are large and sometimes of the same magnitude as the measured quantities. Configuration C is, however, discussed in [10]. Analyzing the measured magnitudes of the gravitomagnetic field and taking the measurement uncertainties into account, the above determined four values can actually reproduce the measured ratios. However, the influence of the material on the gravitomagnetic field cannot be directly deduced at present.

If the gravity-like field is considered, the factor of 3/5 needs to be used in Equation (19), since only the graviton interaction is seen, while the quintessence particle does not contribute to the acceleration.

$$-\frac{1}{2}\int \frac{\partial \mathbf{B}_{gp}}{\partial t} \cdot d\mathbf{S} = \alpha_g^{-1} \oint \mathbf{E}_{gp} \cdot d\ell\,. \tag{19}$$

The gravity-like force is determined by integrating Equation (19) over the surface $\mathbf{S}$ of the Nb ring where vector $d\ell$ is taken positive in the direction along curve C that encloses surface $\mathbf{S}$.

2.4. Physical Meaning of Hermetry Forms

In this section the physical importance and meaning of Hermetry forms is summarized and their physical consequences are discussed. There are 15 admissible combinations of metric subtensors (termed Hermetry form) to which physical meaning is ascribed. A Hermetry form is denoted by H_ℓ with $\ell = 1,...,15$. It should be noted that for a final theory an internal space H^{12} might be needed, because there should exist an additional subspace, G^4, which is believed to be responsible for the *steering of the interference of probability amplitudes*. Since the present analysis is concerned with physical fields and particles, space H^8 is deemed to be sufficient.

A set of 12 Hermetry forms can be directly constructed using the selection rules that require the presence of coordinates of spaces S^2 or I^2, see Fig. **1**. In addition, there exist three so called *degenerated* Hermetry forms that is, these forms have the subspace combinations of the photon $\left(T^1 \times S^2 \times I^2,\ H_2\right)$, and neutrino $\left(\mathbb{R}^3 \times S^2, H_6\right)$, but several of the partial terms in the metric of these particles are missing. This results in particles that do not belong to ordinary matter, and consequently this type of matter is termed non-ordinary matter (NOM). There is also a degenerated form of the quintessence particle $\left(I^2,\ H_{10}\right)$, which is part of NOM, and represents the exchange particle for a repulsive gravity field, and thus might be attributed to dark energy. Hence three additional groups of Hermetry forms exist, giving a total of 15. The complete tables of OM and NOM particles/fields are given in [10, 24]. For the following discussion only two more NOM particles are mentioned. The first one is needed in the explanation of the gravitomagnetic experiments, namely the *imaginary photon,* $\gamma_I\left(H_{11}\right)$), which is responsible for the conversion of photons into gravitophotons, *i.e.,* interaction between electromagnetism and gravitational fields. The second one is called the neutral ($\approx 1GeV$ mass), which is one of the neutral leptons $\left(e^0, \mu^0, \tau^0, H_{15}\right)$ of NOM. There are also NOM particles of imaginary mass, which are *virtual particles* in the sense that they do not occur in the initial and final states of a process.

The corresponding conversion equations play a major role in the explanation of the experiments by Tajmar *et al.* [11-13], namely converting the electromagnetic into a gravity-like force, as well as in the proposed vertical gravity-like field propulsion experiment.

It is interesting to see that a Hermetry form of space $S^2 \times I^2$ describes neutral gravitophotons $\left(\nu_{gp}^0\right)$, and a Hermetry form constructed from space $S^2 \times I^2 \times T^1$ represents photons (γ). This is an indication that, at least on theoretical arguments, photons can be converted into gravitophotons, if somehow the time dependent part T^1 of the photon metric can be canceled. Fig. **1** shows the possible Hermetry forms in *EHT*.

For instance, the Hermetry form (photon metric) comprises only coordinates from subspaces T^1, S^2, and I^2 and is denoted by $H_2(T^1 \times S^2 \times I^2)$. The neutral gravitophoton Hermetry form is given by $H_9(S^2 \times I^2)$. Since gravitophoton and photon Hermetry forms are described by different coordinates, they lead to different Christoffel symbols, and thus to different geodesic equations, see Equation (26). Furthermore, if there were a physical process to eliminate the T^1 coordinates, *i.e.*, the corresponding Christoffel symbols are 0, the photon would be converted into a gravitophoton. This is how mixing of particles is accomplished in EHT. We believe this to be the case in the experiments by Tajmar *et al.* [11-13]. The fundamental question, naturally, is how to calculate the probability of such a process, and to determine the experimental conditions under which it can take place. Hermetry forms alone only provide the potential for conversion into other Hermetry forms, but nothing is said about physical realization. In any case, if Hermetry forms describe physical interactions and elementary particles, a completely novel scenario unfolds by regarding the relationships between corresponding Hermetry forms. Completely new technologies could be developed from the conversion of Hermetry forms. Fig. **2** depicts the six fundamental forces predicted by *EHT*. From the neutral gravitophoton metric and from the forces measured in the experiments by Tajmar *et al.* it is deduced that the gravitophoton decays into a graviton $\left(H_1\right)$ and a quintessence (H_9, repulsive) particle. Fig. **1** shows the set of metric-subspaces that can be constructed.

In addition, *six Higgs fields* should exist, whose lightest particle mass should have, according to considerations from *EHT*, a mass of 182.7 ± 0.7 GeV.

2.5. Six Fundamental Interactions

In physics a question of great importance concerns the number and type of fundamental physical interactions. So far this question has not been answered theoretically, but *experimentally four physical forces* have been observed. However, with the *advent of extreme gravitomagnetic and gravity-like fields* (as was shown in [20]) additional long range force fields are needed, which should be of gravitational type. The number and type of these novel fields should be derivable from the combined structure of internal space and external spacetime. Naturally, the number and type of these physical interactions depend on the dimension and structure of internal space H^8, since its composition of four subspaces eventually determines the set of the 15 admissible Hermetry forms, Section 2.2.2, and consequently predicts six fundamental forces.

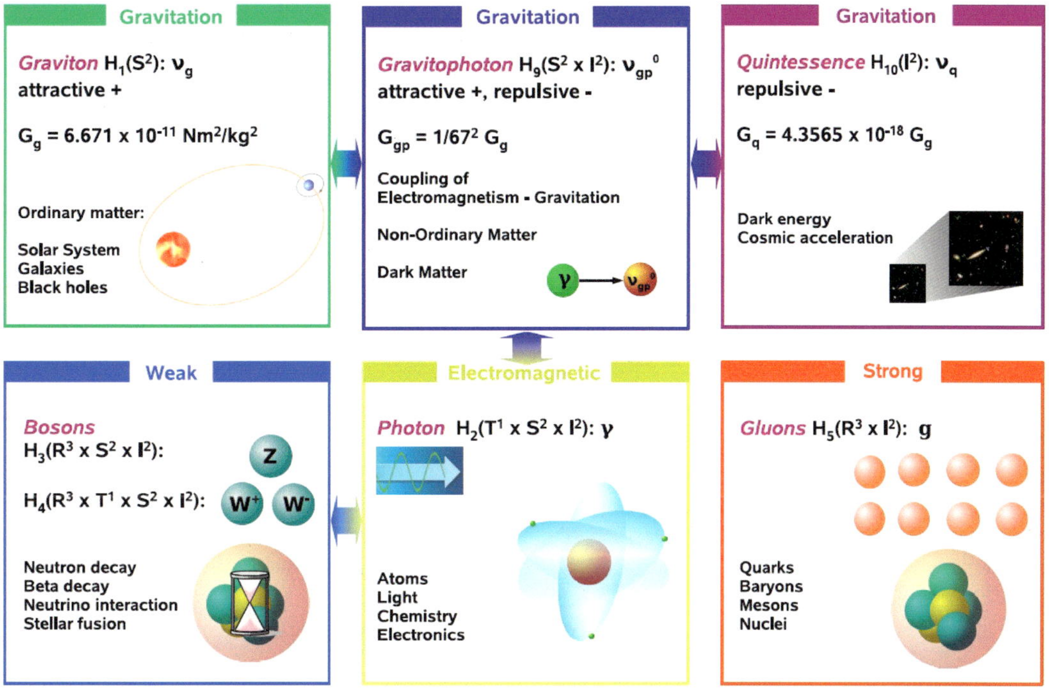

Figure 2: Six fundamental forces are predicted by *EHT*. Three of them are gravity-like fields, mediated by the graviton (attractive, upper left), gravitophotons (attractive and repulsive, upper middle), and the quintessence particle (repulsive, upper right).

Contrary to the ideas employed in *string* theory, see for example [33, 37], spacetime dimensions are not increased in *EHT*. Instead, at each point in spacetime there exists a local internal space H^8 (*internal* space *of eight dimensions*) that is responsible for the existence of all physical objects in spacetime, but is not part of spacetime. Physical spacetime is of *dual nature*, namely comprising external (four-dimensional spacetime) and internal space (Heim space H^8).

The crucial point lies in the construction of the internal space whose subspace composition has to come from basic physical assumptions, which must be generally acceptable [22]. The *coupling strengths* for the three gravitational fields, in the past, were calculated from set theory, *i.e.*, their values are obtained from pure mathematical considerations. However, a moreconvential way has been found, namely the Coleman-Weinberg potential. The arguments of a Hermetry form indicate the subspaces that are forming this metric tensor. The internal symmetry space H^8 comprises four subspaces, namely R^3 responsible for mass (energy), T^1 accounting for charge in general, S^2, for organization and I^2 for information. Hermetry forms can be converted into different forms by experimentally canceling certain subspace coordinates. Since *GR* does not possess any internal structure, it has a very limited geometrical structure, namely that of pure spacetime only. Because of this limitation, *GR* cannot describe other physical interactions than gravity, and consequently needed to be extended.

EHT in its present form without any quantization, *i.e.*, not using a discrete spacetime structure, reduces to the continuum *GR* when this internal space is omitted. The metric tensor, as used in *GR*, has purely geometrical means that is, it is of immaterial character and does not represent any physics. Consequently, the Einsteinian Geometrization Principle (EGP) is equating the Einstein curvature tensor, constructed from the metric tensor, with the stress tensor, representing energy-momentum distribution. In this way, the metric tensor field is related to a physical object whose behavior is governed by an action principle, like that of other physical quantities.

As will be presented in more detail in a forthcoming paper, but was briefly mentioned above, the non-linear σ-model, as given in the general mathematical formulation by Jost [27], can be used to construct the Lagrangian of the corresponding Hermetry form H_ℓ. In the following only the main idea is given. It should be remembered that Hermetry forms are obtained *via* index selection from the poly-metric tensor obtained by the double coordinate transformation incorporating internal space H^8. The internal coordinates ξ^a of subspaces R^3, T^1, S^2, and I^2 have physical meaning and are associated with *mass* (R^3), *general charge* (T^1), *organization* (S^2), and *information* (I^2). However, in spacetime only real physical fields of mass and charge are present. Identifying ξ coordinates with the vectorfield

$$\phi = ((\phi^1,\phi^2,\phi^3),(\phi 4),(\phi^5,\phi^6),(\phi^7,\phi^8)) \tag{20}$$

the function integral $S(\phi)$ for the corresponding Hermetry form representing a physical interaction is set up. The structure of ϕ is determined from the symmetry groups of the subspaces that are present in the respective Hermetry form.

As a consequence of internal Heim space H^8, this geometric approach predicts two further fundamental physical interaction fields, in addition to the four experimentally known ones. The nature of these two novel fields is gravity-like. The six fundamental interactions emerge in our four-dimensional spacetime and represent real physical fields carrying energy. The two additional interaction fields are identified as gravitophoton interaction (*i.e.*, the conversion of photons into a gravity-like field), and quintessence or vacuum interaction (a conversion of photons into a repulsive type of gravity-like field). The gravitophoton interaction is mediated by two massless, so called gravitophoton particles, one which is gravitationally attractive, and the other one is gravitationally repulsive. The massless quintessence or vacuum interaction particle mediates a very weak repulsive gravity-like force that is much smaller in magnitude than the gravitophoton interaction.

Now the important question poses, namely under which experimental circumstances do these interactions manifest themselves as real physical phenomena? The answer is that in the experiments by Tajmar *et al.* [11-13] Graham *et al.* [14] as well as possibly in the Stanford-NASA Gravity Probe-B experiment these novel physical phenomena might have become visible. This will be discussed in more detail in the subsequent section. Moreover, dark matter and dark energy might be interpretable as the result of the novel gravitational interactions.

Hence, according to *EHT*, the gravitational force that we experience is composed of three different fundamental interactions, having six interaction quanta, namely two different neutral gravitophotons that can decay *via* two modes. The first mode gives a decay into a graviton (the usual graviton of quantized *GR* theory) and the repulsive quintessence particle. The second mode results in two gravitophotons - one being attractive, the other one repulsive. The gravitophoton and quintessence interactions should be invariant under an SO(4) group and not SO(3, 1), see [10]. The laws of momentum and energy conservation are strictly obeyed, but both *ordinary and non-ordinary matter* must be accounted for [10].

This means that the gravitational constant G contains contributions of all three gravitational constants, termed G_g, G_{gp} and G_q, respectively. The internal coordinates of H^8 depend on the local (curvilinear) coordinates of spacetime. This is analogous to gauge theory in that a local or gauge transformation is used. In gauge theory it is the particles themselves that are given additional degrees of freedom, expressed by an internal space. Consequently in the geometrization of physics, it is spacetime *instead of elementary particles* that has to be provided with internal degrees of freedom.

The introduction of an internal space has major physical consequences. The structure of H^8 determines the number and type of physical interactions, and subsequently leads to a polymetric. This means that spacetime *comprises both an external and internal structure.* In general, only the external structure is observed, but it has long been known experimentally that matter can be generated out of the vacuum. This is a clear sign that spacetime has additional and surprising physical properties. Therefore, any physical theory that aims at describing physical reality needs to account for this fact. Since *GR* uses pure spacetime only, as a consequence, only part of the physical world is visible in the form of gravitation.

This idea of constructing a polymetric was first conceived by the German physicist B. Heim. A similar principle was mentioned by the Italian mathematician Finzi. The polymetric tensor resulting from this concept is subdivided into a set of sub-tensors, and each element of this set is equivalent to a physical interaction or group of particles, and thus the complete geometrization of physics is achieved. This is, in a nutshell, the strategy chosen to accomplish Einstein's lifelong goal of geometrization of physics. There is of course a second aspect, namely the quantization of the spacetime field. The general theory of Einstein was the first theory to show that fundamental symmetries should have a local character.

It must be noted that this approach is in stark contrast to elementary particle physics, in which particles possess an existence of their own and spacetime is just a background staffage [38]. In *EHT*, considered as the natural extension of *GR*, matter is a consequence of the additional internal physical features of spacetime. These two physical pictures are mutually exclusive, and experiment will determine which view ultimately reflects physical reality. It is, however, well understood that the concept of a point-like elementary particle is highly useful as a working hypothesis in particle physics.

2.6. Geometrization of Physics and the Equations of Motion

Gravity manifests itself as spacetime curvature. If this principle is accepted as the basis for all interactions in physics, a polymetric tensor needs to be constructed. The polymetric tensor describes the complete dynamics of spacetime and leads to a geometrodynamic picture of all physical interactions. In *GR* the geometrical structure of spacetime leads to a single metric that describes gravitational interaction. Einstein's pioneering efforts in the geometrization of physics revealed themselves unsuccessful [8] when more interactions were discovered, and attempts to geometrize physics were abandoned. Einstein did not succeed in constructing a metric tensor that encompassed all physical interactions.

Einsteinian spacetime [37, 42] is indefinitely divisible and can be described by a differentiable manifold. In the following derivation, which relates the metric tensor to physical interactions, this classical picture is used, though spacetime should be treated as a quantized field. The quantization of spacetime seems to play a role in the concept of hyperspace or parallel space [17].

GR can be summarized in a single sentence: *matter curves* spacetime. In curved spacetime the metric is written in the form

$$ds^2 = g_{\mu\nu} d\eta^\mu d\eta^\nu \tag{21}$$

where $g_{\mu\nu}$ is the metric tensor, η^1, η^2, η^3 are the spatial coordinates, and η^4 is the time coordinate. These coordinates can be curvilinear. Einstein summation convention is used, *i.e.*, indices occurring twice are summed over. From the strong equivalence principle it is known (for instance see [41]) that at any point in spacetime a local reference frame can be found for which the metric tensor can be made diagonal, *i.e.*, $g_{\mu\nu} = \eta_{\mu\nu}$ where $\eta_{\mu\nu}$ is the Minkowski tensor. Note that the Minkowski tensor $\eta_{\mu\nu}$ must not be confused with curvilinear coordinates η^μ. Reference coordinates are locally Cartesian $(x^1, x^2, x^3, x^4) = (x, y, z, ct)$. This is equivalent to a transformation between the two sets of coordinates, namely

$$dx^\mu = \Lambda^\mu_\nu \, d\eta^\nu \text{ and } \Lambda^\mu_\nu = \frac{\partial x^\mu}{\partial \eta^\nu} \tag{22}$$

In the free fall frame of the x coordinates the acceleration is 0, and thus the equation of motion simply is

$$\frac{d^2 x^\mu}{d\tau^2} = 0 \text{ and } ds^2 = \eta_{\mu\nu} \, dx^\mu dx^\nu \tag{23}$$

where τ denotes proper time, *i.e.*, the time registered by a clock in its own reference frame. This means the clock is stationary in this frame, and the time measured is the time shown on the clock's dial. In order to obtain the equation of motion for curvilinear coordinates η, one only needs to insert the transformation relations, Equation (22) into Equation (23), which results in the geodesic equation

$$\frac{d^2\eta^\alpha}{d\tau^2} + \Gamma^\alpha_{\mu\nu}\frac{d\eta^\mu}{d\tau}\frac{d\eta^\nu}{d\tau} = 0 \tag{24}$$

where the $\Gamma^\alpha_{\mu\nu}$ are the well known Christoffel symbols or affine connections. Rewriting the geodesic equation (24) in the form

$$\frac{d^2\eta^\alpha}{d\tau^2} = f^\alpha \text{ with } f^\alpha = -\Gamma^\alpha_{\mu\nu}\frac{d\eta^\mu}{d\tau}\frac{d\eta^\nu}{d\tau} \tag{25}$$

and comparing Equation (25) with the equation of motion for a free falling particle Equation (23), the right hand side of Equation (25) can be regarded as a force coming from a physical interaction, which has caused a curvature of the surrounding space, marked by the presence of nonzero Christoffel symbols.

The metric of the proper equation of motion is determined by the Hermetry form obtained from Equation (14). The respective Christoffel symbols denote the physical fields of this Hermetry form. In *GR* mass (matter particles) distributions are determining spacetime geometry, while in electrodynamics the electromagnetic field (waves) is connected to its spacetime geometry. However, in the microscopic world, matter-energydistribution possesses both particle and wave character, which cannot be separated from each other. The existence of the 15 Hermetry forms suggests that the structure of the Einstein field equations provides the framework of the field equations for all physical interactions. This means, Einstein's equations should be representative also for the microscopic world, transcending gravitation and electromagnetism, see Section 2.2. The left hand side of equation (25) can be written as $m_I\mathbf{a}$ where m_I denotes inertial mass and $\mathbf{a}$ is acceleration. Multiplying f^α by its proper charge results in the equation of motion for the respective physical interaction. In the case of gravity, because of the equality of inertial and gravitational mass, charges on the left and right hand sides cancel out. For all other physical interactions this is not the case. In that respect *gravity has a unique role, namely that it curves also the surrounding* space. For all other physical interactions, if pointlike charges are assumed (classical picture), space is curved only at the location of the charge.

One major point of course is that the relation $x = x(\eta)$ as used in *GR* delivers only a single metric, which Einstein associated with gravitation. The *fundamental question is, therefore, how to construct a metric tensor that gives rise to all physical interactions*. The answer lies in the fact that in EHT there exists an internal space H^8. Therefore, in *EHT* the relation between coordinates x and η is $x = x(\xi(\eta))$. In contrast to *GR*, EHT employs a double transformation as specified in Equation (4). From this double transformation a set of different metric tensors can be constructed, which, in turn, lead to a set of individual geodesic equations having the same structure as Equation (25), but depending on the specific charge and Christoffel symbols inherent to this specific physical interaction. An individual equation of motion will have the form

$$m_I\frac{d^2\eta^\mu}{d\tau^2} = e_c f_c^\mu \text{ with } \mu = 1,...,4 \tag{26}$$

where e_c denotes the specific charge of the interaction and quantities f_c^μ are the associated Christoffel symbols. The equation of motion describes a particle of mass m_I with charge e_c, subjected to the respective field of this interaction, represented by its proper Christoffel symbols, *i.e.*, they stand for the curvature of space generated by this interaction. The total number of physical charges e_c is determined by the subspace structure of H^8 in concert with combination rules to constructing a metric that has physical meaning. Equation (26) describes a very difficult physical problem. First, the number of physical charges and their coupling constants need to be determined. There should exist twelve Higgs fields (two $O(2,q)$ groups) that endow physical particles (fields) with their inherent charges. Therefore, all particles of OM or NOM (as described in [9, 24]) are supposed to interact with their respective Higgs particle (field) in order to obtain their corresponding charges (*e.g.,* color, gravitational mass or electric charge *etc.*), but their inertia (energy) should come from group $O(1,q)$, which denotes a special Hermetry form, H_{16} from subspace T^1, related to energy (mass) *via* $\Delta E \Delta t = \hbar$. It is remarkable that the 15 generators of the $O(3,q)$ group define the existence of all possible particle (field) families, *i.e.,* the existence of physical phenomena in material or energetic form is fixed by this symmetry group. It should be noted that the set of 15 families comprises six fermions (quarks, electrons and neutrinos as well as neutral electron, imaginary electron and imaginary quark) in combination with the nine gauge bosons (graviton, photon, vector bosons $W^{\pm}$ and Z^0, gluons as well as imaginary gluon, imaginary photon and the additional gravitational particles, namely the gravitophotons and quintessence particle. The double transformation as given in Equation (4) represents the particle aspect and leads to eigenvalue equations whose eigenvalues have the dimension of an inverse length. In these eigenvalue equations the Christoffel symbols occur. Using the inverse of the Planck length expressed as $m_p c / \hbar$, results in a correspondence between inverse length and mass. Since particles and fields form a unity, the transformation from spacetime into internal space, $\mathsf{M} \rightarrow \mathsf{H}^8$, should represent the field aspect, because derivatives of internal coordinates $\xi^a, a = 1,...,8$ with respect to curvilinear coordinates η^μ lead to an expression $\left(e_c^a / m_I\right) h_{\mu\nu}^a$. The $h_{\mu\nu}^a$ denote deviation from the flat metric, and physically represent the tensor potential of the charge e_c^a with m_I as associated inertial mass. The metric coefficients thus assume energy character. This short discussion implies a comprehensive mathematical program, namely the determination and solution of the above mentioned eigenvalue equations as well as the derivation of the tensor potentials for the interactions. The task is not finished, but this brief discussion should have conveyed an idea how the introduction of an internal symmetry space leads to a correspondence between geometry and physical quantities. The mathematical framework to determine the charges and to obtain the correspondence between geometry and physics is quite involved. It seems that internal coordinates are described by quaternions. There is an interesting question, namely: *What is the Hermetry form of the vacuum field*? If the vacuum has an energy density different from zero, it should not be the case that its Christoffel symbols are 0. However, we feel, in order to answer this question, a quantization procedure for Equation (26) has to be established. *This approach differs substantially from GR, and should lead to the complete geometrization of physical interactions.*

2.7. Fundamental Interactions and their Field Equations

If each physical interaction causes its own spacetime curvature, the structure of the Einstein field equations should provide the mathematical frame for the description of all six forces. In the following it will be shown how, at least in principle, these equations can be constructed. In general, spacetime curvature is small, *i.e.,* except for black holes, neutron stars *etc.*, and thus the metric tensor can be linearized

$$g_{\mu\nu} = \eta_{\mu\nu} + h_{\mu\nu} \tag{27}$$

and Cartesian coordinates

$$(x^\mu) = (x, y, z, ct),\ \mu = 1,...,4. \tag{28}$$

The geometric part of the Einstein equations, that is, the part describing spacetime curvature, can be written as

$$R_{\mu\nu} = \frac{1}{2}\left(\Delta - \frac{1}{c^2}\frac{\partial^2}{\partial t^2}\right); \tag{29}$$

$$h_{\mu\nu} = \frac{1}{2}\Omega\, h_{\mu\nu}, \tag{30}$$

where denotes the d'Alembertian wave operator. In a space free of matter

$$\Omega\, h_{\mu\nu} = 0 \tag{31}$$

and thus gravity waves should be possible. In the case of non-relativistic velocities $v << c$ and neglecting pressure as well as internal energy distributions in the energy-momentum tensor, *i.e.,* there is only one component $T_{44} \approx \rho c^2$ where ρ denotes mass density, the time-independent Einstein field equations lead to

$$\Delta\, h_{44} - \frac{1}{c^2} R\, h_{44} = 2\kappa\, T_{44}\,. \tag{32}$$

However, the *Einstein equations are not self-consistent*, since the energy-momentum tensor is a different physical entity, and has to be supplied from the outside, *i.e.,* does not result from the geometry of spacetime. In order *to make this system self-consistent* it is assumed that in general

$$T_{\mu\nu} = T_{\mu\nu}(g_{\alpha\beta})\,. \tag{33}$$

Regarding a special Hermetry from H_ℓ the energy-momentum tensor in accordance with this Hermetry form can be formally written as

$$T_{\mu\nu} = T_{\mu\nu}(H_\ell)\,. \tag{34}$$

For the above mentioned special case a linear relationship is assumed in the form

$$T_{\ell;44} = k_\ell\, \rho_\ell\, h_{\ell;44} \tag{35}$$

where the index ℓ denotes the Hermetry form and k_ℓ is a specific constant. The quantity $m_\ell(i.e., \rho_\ell)$ describes the type of mass-energy belonging to H_ℓ. If the rest mass is different from 0, it is believed that an eigenvalue equation can be derived from geometrical principles whose spectrum gives the physically admissible particle masses, in a similar sense as already postulated by Heim. Thus a complete geometrodynamic approach would be accomplished.

3. PHYSICS AND MATHEMATICS OF EXTREME GRAVITATIONAL FIELDS

The gravitomagnetic effect as known from *GR* that is, the *Lense-Thirring* effect cannot be utilized to explain the experiments of Tajmar *et al.* [11-13], nor can it serve as propulsion principle, because of its extremely small magnitude. When the Einstein-Maxwell (the linearized Einstein field equations) equations are employed in the explanation of the extreme gravitational fields, it becomes obvious that neither the magnitude nor the sign of the acceleration field measured by Tajmar *et al.* is described correctly. A physical model that aims at providing a phenomenological framework of extreme gravitational fields cannot be based on these equations.

The conclusion is that the laboratory generated extreme gravitomagnetic fields are, because of their enormous field strength, outside *GR* and also the gravity-like fields. It is these gravity-like fields that might be usable as basis for a novel technology of energy generation as well as (space) propulsion principle. In the next section the physical model for the generation of extreme gravitational fields is outlined. One of the key issues is that gravity-like fields are produced by conversion of electromagnetic into gravitational fields, which might explain their strength. Needless to say, this requires the introduction of novel physical concepts.

3.1. Physical Principles of Gravitational Field Generation

At present, theoretical physics does not foresee an interaction between electromagnetic and gravitational fields, *i.e.*, no terms coupling the Maxwell equations with the Einstein-Maxwell equations are known. In the following, a brief discussion of the fundamental physical mechanism deemed to be responsible for the generation of extreme gravitational fields is given.

Some aspects have been discussed in [9, 20, 24], but a more comprehensive discussion will be given in a forthcoming review paper [Preprints are available through the authors]. In this article we present the set of basic equations and physical mechanisms that make up the phenomenological theory for calculating extreme gravitational fields.

First, no explanation within the framework of current physics seems to be possible. Either the recent experiments by Tajmar *et al.* [11-13] and Graham *et al.* [14] somehow misinterpret unknown disturbances as large gravitomagnetic fields, or some novel physics does exist. As a reminder, it should be noted, that despite its name, a gravitomagnetic field is a purely gravitational field. There is no magnetic component or interaction with electrodynamics. The name was derived from the similarity of the Einstein-Maxwell equations with the Maxwell equations. Gravitomagnetism represents the dynamic nature of matter.

A similar situation seems to exist regarding the strong gyro misalignment observed in the GP-B experiment. Although the NASA-Stanford University team claims that there is an electrostatic patch effect (the Nb coated quartz sphere of the gyro is not a perfect equipotential surface) there could be room for the effect seen in Tajmar's experiments. It should be noted that the torque acting on the spinning cryogenic gyros is given by $\mathbf{N}_g = \frac{1}{2}m(\omega \times \mathbf{r}) \times \mathbf{B}_{gp}$. According to *EHT*, at cryogenic temperatures the four gyros used in the GP-B experiment should produce their own $\mathbf{B}_{gp}$ field due to the same phenomenon, which in turn would be the source of a major gyro misalignment by producing a substantial torque in comparison to the small effect of *GR* frame dragging. An analysis for the GP-B experiment was performed in [20].

3.2. Conservation Principles

Physics is governed by symmetry (invariance) where each symmetry stands for a conservation law, termed Noether's principle. Momentum conservation is derived from the homogeneity of space, which means that the Lagrange function is independent of its position in space that is, invariant under translation. Therefore, its variation δL must satisfy $\delta L = 0$ for a translation from $\mathbf{x}$ to $\mathbf{x}+\mathbf{a}$ where $\mathbf{a}$ is a constant vector. Since spacetime is four-dimensional, the homogeneity in time stands for the invariance of the Lagrangian with respect to translation in time. This is expressed by the fact that the Lagrangian does not depend explicitly on time, *i.e.*, $L = L(q_k, \dot{q}_k)$.

There is a stern rule governing all classical physics and, in particular, applies to all propulsion systems: identify the physical system and wrap a closed surface, control surface (CV), around it. Check the amount of momentum going through the CV. What is going on inside the physical system is of absolutely no interest. If this amount is zero, there is no momentum change, and thus no propulsion. Time and again, this principle has been misunderstood, for instance, confusing energy with momentum (all our propulsion is non relativistic). Moreover, the definition of physical system is not always straightforward.

In *EHT*, the spacetime field itself is part of the physical system. The propulsion principle is similar to the Mossbauer effect (the large crystal absorbs the recoil momentum of the particle) that is, the momentum of spacetime is changed almost imperceptible, but the space system is accelerated. The key issue is to find a physical concept that can achieve a sufficiently strong interaction with spacetime, *i.e.*, not through frame-dragging as takes place in *GR* which is far too weak.

The present principle of propellantless propulsion *via* gravity-like fields is different from the above concept in that it involves the spacetime field itself through gravitomagnetic interaction as part of the physical system that is, there is an exchange of momentum and energy between the space vehicle and the local spacetime. The amount of this exchange is determined by the magnitude of the generated $\mathbf{B}_{gp}$ field. The spacetime field that is, electrogravitic and gravitomagnetic fields have just as much claim to reality a the original concept of momentum as electric and magnetic fields together with the charged particles, *i.e.*, gravitational (mass) and electric charge. Momentum is transferred from a charge to the field, through the field, and perhaps to other charges. Therefore, any propulsion system based on gravity-like fields is based on strong interaction, compared to *GR*, with the surrounding spacetime. The experiments at AIT have already demonstrated that these fields are about 10^{18} times stronger than those produced by movable masses. It is conjectured that the accelerated expansion of the *Universe* might be a direct consequence of universal momentum conservation that is, the physical system that comprises all masses in the *Universe* as well as spacetime field [31].

3.3. Ginzburg-Landau Theory of Extreme Gravitational Fields

First, if *one accepts the existence of gravity-like fields from the measurement by Tajmar et al.* [11-13], the question naturally arises: how are these extreme gravitational fields possible? In Section IV of [20] it is shown that none *of the four known fundamental physical interactions* can explain the nature and magnitude of these fields.

It is believed that the physics of extreme gravitational fields is a solid state phenomenon, owing to the formation of a Bose-Einstein condensate *via* a phase transition at a cryogenic temperature T_C, but caused by a new type of matter, which has similarities to superconductivity, but the underlying physical mechanism is believed to be entirely different, representing novel physics.

The analysis that we are presenting here does not aim at explaining the physics of the extreme gravitational fields in that a microscopic quantum theory is presented in the form of the BCS (Bardeen, Cooper, and Schrieffer) theory [21]. BCS theory describes the phenomena within the superconductor, *i.e.,* the physical reasoning why Cooper pairs (phonon coupling of two electrons of opposite momentum and spin) are formed, what their features are, and why they are able to move without friction through the lattice.

Instead, the physics of extreme gravitational fields is treated as an application of the Landau theory of phase transitions to superconductivity combined with the novel concepts of matter of imaginary mass and the possibility of converting imaginary electromagnetic fields (*i.e.,* their origin is from imaginary matter) into gravitational fields.

We accept the pragmatic standpoint of London, by simply saying that superconductivity is characterized by zero friction and thus Ohm's law has to be replaced, leading to a different relation between current density and vector potential. Maxwell's equations are still entirely true, but have to be complemented by the London equations. The same is assumed to hold true for the current of imaginary bosons, denoted e_I^B, formed *via* phase transition by a multiple (even number) of the imaginary electrons e_I in analogy to Cooper pairs. How this mechanism actually takes place is of no concern at the moment; neither is the calculation of the density n_I of the e_I^B bosons, nor their coupling strength or the penetration depth of the gravitomagnetic field. No attempt is made to determine the critical temperature T_C. These questions can be addressed later on, when the phenomenon of extreme gravitational fields is firmly established.

Two fundamental and completely novel physical concepts are introduced, stated below, namely

1. The existence of particles of imaginary mass, *i.e.,* the imaginary electron e_I which has the same charge as its counterpart the real electron, and the imaginary quark q_I. There is only one imaginary quark that should have the elementary charge $+e$. The q_I particle is assumed to be generated within the proton, but then there is also a fourth gluon color termed the *imaginary color* ic, and thus a *total of 15 gluons exists.* Imaginary matter is a direct consequence of the concept of Hermetry form, which is graphically represented by the four-dimensional hypercube as published in [24]. In other words, the implication of the 16 Hermetry forms is the extension of the concept of matter from real to imaginary. Imaginary matter should not be directly visible, but its presence can be deduced from the existence of the extreme gravitational fields.

2. The conversion (phase transition) from electromagnetic into gravitational fields is the key process for generating extreme gravitational fields. In order to achieve this type of hitherto

unknown conversion, the existence of matter of imaginary mass in the form of imaginary electrons e_I and quarks of imaginary mass q_I were instrumental. Imaginary matter is not directly observable, but its presence can be deduced form the existence of the extreme gravity-like fields.

Therefore, in analogy to the successful semi-classical explanation of Bose-Einstein condensates, for instance the phenomena of superconductivity, ferro-magnetism, superfluidity, or spin waves, the phenomenological mathematical description of extreme gravitational fields is based on a similar phenomenological approach in which the ideas of due to London, Ginzburg, and Landau are combined with the concept of mass of imaginary matter and the conversion of electromagnetic into gravitational fields.

That is, the generation of extreme gravitational fields is characterized by the following equations and/or physical processes

- For electrodynamic phenomena the Maxwell equations are utilized, however, modified by the London equations, similar to the case of superconductivity, to account for the phase transition,
- The gravitational field mediated by gravitons ν_g that is, Newtonian gravitation, is not described by the equations below, since it solely depends on mass (stationary or moving) and is governed by the Einstein-Maxwell equations derived from *GR*, its respective coupling constant is G_N,
- The modified Einstein-Maxwell equations of gravity (which are different from the linearized version of the Einstein field equations of *GR*) and
- The London-Einstein equations that is, the London equations employed to the gravitolelectric field $\mathbf{E}_G$ and the gravitomagnetic field $\mathbf{B}_G$, where the index G now stands for the two gravitational fields, mediated by gravitophotons ν_{gp} or the quintessence particle ν_q. The respective coupling constants are G_{gp} and G_q as given in [24].
- The Ginzburg-Landau theory in conjunction with a magnetic induction field $\mathbf{B}$ whose presence leads to the *asymmetric potential* as shown in [24]. In the Heim experiment such a field is generated by the external current in the coil.
- The assumption of a phase transition responsible for the conversion of the imaginary electromagnetic vector potential into a gravitophoton field, *i.e.*, $e0.328\alpha_{gp}\mathbf{v}\cdot\mathbf{A}_{e_I} + m_p\mathbf{v}\cdot\mathbf{A}_{gp} = 0$ with the conversion factor $\alpha_{gp} = 1/212$ calculated from the Coleman-weinberg potential as derived by Kaku [34],
- When the phase transition occurs, no symmetry breaking is taking place and instead imaginary particles are produced, *i.e.*, imaginary electrons e_I and imaginary quarks q_I, because of the asymmetric shape of the potential curve. These particles are not tachyons, because of their electromagnetic interaction with positive charges of the crystal lattice.

The following sets of equations are utilized. The Maxwell equations in differential form are given by

$$\mathbf{E} = -\nabla\Phi \text{ and } \mathbf{B} = \nabla\times\mathbf{A} - \frac{\partial\mathbf{A}}{\partial t} \quad \textbf{(36)}$$

$$\nabla\cdot\mathbf{E} = \frac{1}{\varepsilon_0}\rho \quad \textbf{(37)}$$

$$\nabla\cdot\mathbf{B} = 0 \quad \textbf{(38)}$$

$$\nabla \times \mathbf{B} = \mu_0 \mathbf{j} \tag{39}$$

$$\nabla \times \mathbf{E} = -\frac{\partial \mathbf{B}}{\partial t} \tag{40}$$

where the electric displacement current $\partial \mathbf{E}/\partial t$ was neglected in Ampere's law Equation (39) and $\varepsilon_0 \mu_0 c^2 = 1$, $\mu_0 = 4\pi \times 10^{-7}\ N/A^2 = 1.256 \times 10^{-6}$ Vs/Am, and $\varepsilon_0 = 8.854 \times 10^{-12}$ As/Vm.

Now we consider the Einstein-Maxwell formulation of linearized gravity that possesses a certain similarity to the mathematical form of the electromagnetic Maxwell equations. In analogy to electromagnetism there also exist gravitational scalar and vector potentials, denoted by Φ_G and $\mathbf{A}_G$, respectively [35].

$$\mathbf{E}_G = -\nabla \Phi_G \quad \text{and} \quad \mathbf{B}_G = \nabla \times \mathbf{A}_G . \tag{41}$$

If the original Einstein field equations are expanded up to first order, see for instance Thorne [36], and introducing the corresponding gravitoelectric and gravitomagnetic fields, one obtains the following set of equations for Newtonian gravitation:

$$\nabla \cdot \mathbf{E}_{G_N} = -4\pi G_N \rho \tag{42}$$

$$\nabla \cdot \mathbf{B} G_N = 0 \tag{43}$$

$$\nabla \times \mathbf{B}_{G_N} = -\frac{16\pi G_N}{c^2} \mathbf{j} \tag{44}$$

$$\nabla \times \mathbf{E}_{G_N} = 0 \tag{45}$$

where $\mathbf{j} = \rho \mathbf{v}$ is the mass flux and G_N is Newton's gravitational constant. The field $\mathbf{E}_{G_N}$ describes the gravitational field from a stationary mass distribution, whereas $\mathbf{B}_{G_N}$ describes the gravitomagnetic field produced by moving masses.

The linearized version of Einstein's equations of *GR*, Equations (43) and (45), have been cast in a mathematical form that resembles the Maxwell equations of electrodynamics, but there are two major differences.

First, it can be seen that there is *no equivalent equation to Faraday's induction law*, since the term $-\partial \mathbf{B}_{G_N}/\partial t$ is absent because the expansion is of first order in terms of v/c only. Including second order terms would restore this term, but also *would lead to additional terms* in the other equations, thus clearly changing the Maxwellian character of the Einstein-Maxwell equations.

Second, comparing the signs of the Maxwell and Einstein-Maxwell equations, there is a minus sign in the source term of the first Einstein-Maxwell equation, which causes Newtonian gravity to be always attractive. Equation (45) is the result from *GR* and therefore the index G_N is used. Comparing the stationary Maxwell and Einstein-Maxwell equations, they can be given exactly the same structure, if $\mathbf{E}_{G_N}$ and $\mathbf{B}_{G_N}$ are multiplied by -1 and redefined as new gravitoelectric and gravitomagnetic fields.

However, the Einstein-Maxwell equations *cannot be employed* in describing gravitomagnetic phenomena caused by gravitophotons. To distinguish the gravitomagnetic field generated by gravitophotons it is denoted as $\mathbf{B}_{gp}$. Because of the phase transition that is responsible for the bosons of imaginary mass that are generating an imaginary supercurrent, the source term $\mathbf{j}_{gp}$ for the extreme gravitomagnetic field $\mathbf{B}_{gp}$ is large, and therefore the term $\partial \mathbf{B}_{gp}/\partial t$ from gravitophoton interaction could become large and thus must appear in the equations. In the equation that is analog to Ampere's law, the term $\partial \mathbf{E}_{gp}/\partial t$ might play a role, but for the Heim experiment it is set to 0. Since the masses in both the Heim and Tajmar experiments are exceedingly small, the divergence for the $\mathbf{E}_{gp}$ is set to 0. The *const* in the gravitational Ampere law still needs to be determined. The mass flux from the spinning disk is also negligible that is, the gravitomagnetic field $\mathbf{B}_{G_N}$ generated from *GR* can be safely neglected as well. Therefore, the mathematical form of the equations describing the generation of extreme gravitomagnetic fields, the Einstein-Heim-Maxwell (EHM) equations, expressed as a set of partial differential equations, are assumed to have the form

$$\nabla \cdot \mathbf{E}_{gp} = 0 \tag{46}$$

$$\nabla \times \mathbf{E}_{gp} = -\frac{1}{2}\alpha_g \frac{\partial \mathbf{B}_{gp}}{\partial t} \tag{47}$$

$$\nabla \times \mathbf{B}_{gp} = const\ \mathbf{j}_{gp} + 2\alpha_g^3 \frac{1}{c^2}\frac{\partial \mathbf{E}_{gp}}{\partial t} \tag{48}$$

$$\nabla \times \mathbf{B}_{gp} = const\ \mathbf{j}_{gp} \tag{49}$$

$$\nabla \cdot \mathbf{B}_{gp} = 0 \tag{50}$$

It should be emphasized that the source in the rotation of the gravitomagnetic field $\mathbf{B}_{gp}$ is coming from the current of the imaginary bosons (phase transition), and thus is of electromagnetic origin. As a final remark, because of the assumed coupling of electromagnetism and gravitation, there should be coupling terms occurring in both the Einstein-Heim-Maxwell equations that are of electrodynamic origin, and maybe (not known at present) in the Maxwell equations that are of gravitophoton origin. Since the current density of imaginary bosons formed by the electrons of imaginary mass,

$$\mathbf{j} = -\frac{n_{e_I}^B (e_I^B)^2}{m_{e_I}^B}\mathbf{A}_{e_I}$$

as given in the London equation, is considered to be responsible for the gravitomagnetic field, it seems reasonable to insert this term as current density in the respective Einstein-Heim-Maxwell equation. Since a mass current density is needed for dimensional reasons, the electromagnetic current density has to be multiplied by a factor like m_p / e and also the coupling strength has to be determined.

The London equations are simply derived from the fact that in the superconducting phase the Cooper pairs move without any friction through the lattice of the solid. This means that Ohm's law $\mathbf{j} = \sigma\mathbf{E}$ is no longer valid. In other words, using Newton's law for electrons moving in a lattice

$$m_e \frac{d\mathbf{v}}{dt} = e\mathbf{E} + \frac{\mathbf{v}}{\tau} \tag{51}$$

and omitting the friction term $\mathbf{v}/\tau$, the first London equation is nothing else but the well known Newton law. Introducing particle density n and charge current density $\mathbf{j} = \rho\mathbf{v}$ one obtains the first London equation

$$\mathbf{E} = \frac{m_e}{ne^2}\frac{d\mathbf{j}}{dt} \tag{52}$$

The first London equation is quite different from Ohm's law, since now the electric field $\mathbf{E}$ is proportional to the time derivative of the current density $\mathbf{j}$, and no longer to $\mathbf{j}$ itself. Forming the curl of this equation and inserting this expression into Faraday's induction law, one directly derives the second London equation

$$\nabla \times \mathbf{j} = -\frac{ne^2}{m_e}\mathbf{B}. \quad \textbf{(53)}$$

which reflects the experimentally well known fact that there is no magnetic induction field $\mathbf{B}$ inside a superconductor and is essential to explain the Meissner-Ochsenfeld effect.

Hence, a rotating superconductor with velocity $\mathbf{v} = \omega \times \mathbf{r}$ where ω is the angular velocity of the rotating ring and current density $\mathbf{j} = \rho \mathbf{v}$ as well as electric charge density $\rho = ne$ generates a magnetic induction field, the so called *London moment*, given by

$$\mathbf{B} = -\frac{2m_e}{e}\omega. \quad \textbf{(54)}$$

The Lagrange densities before and after phase transition for the conversion from the imaginary electromagnetic into the real gravitomagnetic potential are assumed to be related in the following way,

$$0.328\, ie\alpha_{gp}\mathbf{v}\cdot\mathbf{A}_{e_I} + m_p\mathbf{v}\cdot\mathbf{A}_{gp} = 0 \quad \textbf{(55)}$$

where the factor 0.328 comes from the radiation correction (Coleman-Weinberg potential), e is the positive charge of the quark of imaginary mass and $\mathbf{v}$ is the velocity of the rotating disk above the coil. The decay of the imaginary photon into the neutral gravitophoton of the first type, *i.e.*, $\gamma_I \rightarrow \nu_{gp}^{01}$ with coupling strength α_{gp} leads to the real gravitophoton potential $\mathbf{A}_{gp}$. From the above equation a constraint on the direction of the resulting is obtained, which is of the form

$$\mathbf{B}_{gp} = \gamma\mathbf{B}_{e_I} + \beta\mathbf{v} \times \mathbf{B}_{e_I}. \quad \textbf{(56)}$$

This is the most general solution. Immediately it can be seen that in the experiments by Tajmar *et al.* [11-13] $\mathbf{B}_{gp}$ is parallel to $\mathbf{B}_{e_I}$, since there is a coupling between velocity $\mathbf{v}$ and the imaginary field $\mathbf{B}_{e_I}$ that is, $\mathbf{B}_{e_I}$ is parallel to the axis of rotation of the disk. Because of this coupling, $\beta = 0$ and this is actually seen in the experiments of the gravity-like field. Therefore the resulting acceleration field always is in circumferential direction in this type of experiment. In the Heim experiment we need to have

$$\mathbf{B}_{gp} \sim \mathbf{v} \times \mathbf{B}_{e_I}, \quad \textbf{(57)}$$

which means that, if the $\mathbf{B}_{e_I}$ of the London equation is directed along the z-axis, the resulting $\mathbf{B}_{gp}$ is pointing in $\hat{e}_r$ direction, if cylindrical coordinates are used.

For the extreme gravitomagnetic field to be generated - two types of decay occur - first the photon is converted into an imaginary photon $\gamma \rightarrow \gamma_I : \alpha$. Then the neutral gravitophoton is produced according to $\gamma_I \rightarrow \nu_{gp}^{01} : \alpha_{gp} = \sqrt{\lambda}$. The coupling constants of these decays are the fine structure constant α and the gravitophon coupling constant α_{gp}, determined from the Coleman-Weinberg potential, respectively. This is an extraordinary fact, since it seems that the process of gravitomagnetic field generation is governed by quantum electrodynamics, *in contradiction to our earlier assumption* where coupling constants were calculated from number theory, see [30]. In other words, once the *existence of fermions with imaginary mass* is accepted, *i.e.*, the concept of matter has been extended; the basic machinery of current physics seems to apply.

In the experiments by Tajmar *et al.* [11-13] cryogenic temperature are needed to produce this spontaneous symmetry breaking, similar to superconductivity. According to the physical model, derived from *EHT*, the symmetry breaking is generating imaginary bosons e_I^B that are moving without interaction through the lattice of the disk that is, there is no friction and the e_I^B do not participate in the rotation of the disk. Therefore, the form of London equation for these bosons is given by, assuming $\nabla\cdot\mathbf{v} = 0$,

$$\mathbf{j} = -\frac{n^2}{m_e}\mathbf{A} \tag{58}$$

which means that the current density $\mathbf{j}$ is obtained from a vector potential $\mathbf{A}$. The question is which type of potential does generate such a current density? The phase transition that occurs at cryogenic temperature leads to a current of imaginary bosons with charge e_I^B that results in an imaginary vector potential $\mathbf{A}_{e_I}$. Utilizing the structure of the London equation, Equation (58), the gravitophoton current density that would be obtained from the imaginary vector field $\mathbf{A}_{e_I}$ is

$$\mathbf{j} = -\frac{n_{e_I}^B (e_I^B)^2}{m_{e_I}^B}\mathbf{A}_{e_I} \tag{59}$$

The combination of London equation and Maxwell equation leads to

$$\left(-\nabla^2 + \frac{1}{\lambda^2}\right)\mathbf{B}_{e_I} = 0 \tag{60}$$

where the wavelength

$$\lambda = \left(\frac{m_{e_I}^B}{(e_{e_I}^B)^2 n_{e_I}}\right)^{1/2}.$$

It should be noted that λ is an imaginary wavelength since the mass $m_{e_I}^B$ is imaginary.

However, since eventually a gravitomagnetic field $\mathbf{B}_{gp}$ is observed, it is the respective gravitomagnetic vector potential $\mathbf{A}_{gp}$ that needs to be used in the calculation of the gravitomagnetic current density $\mathbf{j}_{gp}$. In the experiments of Tajmar *et al.* [11-13], however, a *real gravity-like field is observed* that is, the potential in Equation (59) should be the *gravitational potential.* On the other hand, this gravitational potential cannot be generated by moving masses. The rotating disk has a mass of about 400 g, while, according to *GR*, the *mass of a white dwarf* would be necessary.

It is the coupling between electromagnetism and gravitation triggered by the phase transition that eventually leads to a gravitational field by converting the imaginary vector potential $\mathbf{A}_{e_I}$ into the real gravitophoton potential $\mathbf{A}_{gp}$. In this process, the Lagrange densities before and after the conversion going from the imaginary electromagnetic into the real gravitomagnetic potential are assumed to be related according to Equation (55), since energy needs to be conserved. The corresponding gravitomagnetic field $\mathbf{B}_{gp}$ is given by Equation (56). It should be noted that *these two vector fields may have different directions.* In general, the Heim experiment, which should produce a axial gravity-like field, the relation between the gravitomagnetic and the imaginary vector field is given by Equation (57). We therefore formally write

$$\mathbf{j}_{gp} = -M^2\mathbf{A}_{gp} \tag{61}$$

where M is some kind of generalized or effective mass. The value of λ cannot be used to calculate the effective mass, since it belongs to the virtual charges *i.e.,* the imaginary current. This would mean that the gravitomagnetic field $\mathbf{B}_{gp}$ cannot penetrate into the disk, except for a thin sheet, because of the exponential decay of the field in the interior of the disk. This is true if M is real.

It seems as if the gravitophoton, ν_{gp}^{02}, in the Tajmar experiment, has gained mass. If one does not know about the cryogenic rotating disk, then there must exist a background field interacting with the $\mathbf{B}_{gp}$ field, and the resulting exponential damping virtually is reducing the range of the otherwise infinite range of the gravitophoton interaction. This is the so called Higgs mechanism and the background field that cannot be perceived, is the Higgs field. Instead, one assumes that the mediator boson has gained mass.

In the particle picture, the conversion of the fields is realized by the conversion of the imaginary photon γ_I into the neutral gravitophoton of the first type ν_{gp}^{01} (Heim experiment) or the second type ν_{gp}^{02} (Tajmar experiments) that leads to the real gravitophoton potential $\mathbf{A}_{gp}$. The details for calculating the corresponding extreme gravitomagnetic field $\mathbf{B}_{gp}$ are presented in Section 3.4. In Tajmar *et al.* [11-13] experiments we need to distinguish two different sets of experiments. In one set, Tajmar *et al.* generated extreme gravitomagnetic fields. Any material body that is moving in such a field is subject to the Lense-Thirring effect, except that the Tajmar effect (*i.e.,* the force acting on the moving body) is outside *GR*, but the force law itself remains unchanged. In the second set of experiments, Tajmar *et al.* accelerated or decelerated the rotating disk by changing its angular velocity (the rpm), which led to the circumferential gravity-like field, described above.

According to *QED*, the gauge theory with its hidden symmetry leads to a profound consequence, namely, if the electron is accelerated, then, the gauge field itself is actually emitted as a quantum particle. In other words, if an electron initially has a momentum $\mathbf{p}$ and an acceleration is applied changing the electron into a state with momentum $\tilde{\mathrm{p}}$, a photon of momentum $\mathbf{p}-\tilde{\mathrm{p}}$ is produced. The gauge field, formerly unobservable, has become a physical entity in form of the photon (in the field picture: a combination of electric and magnetic fields) that is observable in spacetime. It seems that an accelerated electron has generated a real photon (gauge particle of *QED*) with its proper momentum and energy. Light is therefore emitted from accelerated charges.

In the following a series of questions is posed and addressed that are crucial in conceiving gravitomagnetic and gravity-like field experiments, trying to work out the essential features of these fields.

- How is imaginary matter being generated, where is it generated, and what eventually happens to it?
- How do the physical conversion mechanisms differ in the Tajmar and the Heim experiments? That is, why is there a gravity-like field acting in the circumferential direction in the experiments by Tajmar and an axial gravity-like field in the Heim experiment.
- What is the exact energy conversion mechanism from electromagnetic into gravitational fields?
- What happens to energy conservation and momentum conservation?
- Where does the energy and momentum come from that a spacecraft needs to travel without propellant?
- What are the governing equations for gravitational fields? Since there exist gravitons, gravitophotons, and quintessence particles, three different types of gravitational fields should exist.

3.4. Evaluation of Gravitational Field Experiments

Tajmar *et al.* [11-13] were the first reporting the generation of extreme gravitomagnetic fields in the laboratory, which are denoted as $\mathbf{B}_{gp}$, since they are assumed to result from gravitophotons and not from gravitons. It should be noted that the existence of these fields was postulated by the authors before these experiments became known, see for instance [17] that is, theory and experiments were developed independently of each other. Even if these experiments could not be confirmed, extreme gravitomagnetic and gravity-like fields should be producible based on theoretical reasons.

It is important to distinguish between the gravitomagnetic field in the experiments by Tajmar *et al.* [11-13] and the proposed Heim experiment. These two fields are different and subsequently lead to different types of forces.

Modanese *et al.* [19] have tried to explain the Tajmar effect by employing the linear Einstein-Maxwell equations, but have come to the same conclusions as the authors, namely that these equations do not even reproduce the correct sign of the gravity-like field (acceleration field) that was observed by Tajmar *et al.* [11-13] when the angular frequency of the cryogenic ring was subject to change, *i.e.,* the ring was accelerated. The other problem of course is that the $\mathbf{B}_{gp}$ field measured is up to 18 or 19 orders of magnitude larger than predicted by *GR*.

Without further discussion, it should be mentioned that the often cited ratio of the gravitational and the electromagnetic force, which for proton and anti-proton is in the range of 10^{-38}, can no longer be used to justify the negligibility of gravitational effects. This value only holds for Newtonian gravitation. Furthermore, there is a most interesting hypothesis found in the recent book by Zee [39] (pp. 516), which states that gravity could be the square of Yang-Mills theory that is, gravity ~ Yang-Mills × Yang-Mills or in more mathematical terms $M_{gravity} \sim \mathrm{M}_{gauge} \times M_{gauge}$. In other words the spin 2 field of gravitation might be comprised of two spin 1 fields of the Yang-Mills type. The coupling constant of Newtonian gravitation is 10^{-38}, therefore the coupling constant of the two Yang-Mills fields is 10^{-19}, which comes close to the Tajmar effect and could be a hint that additional gravitational fields indeed exist. However, this idea so far has not been investigated in how it could be used to provide an explanation for the extreme gravitomagnetic fields.

The observed gravity-like field *follows a Lenz type rule*, *i.e.*, it is opposing its origin. This exhibits an electromagnetic behavior and contradicts the sign of the Einstein-Maxwell equations. The use of the non-linear equations of *GR* cannot change this picture, since the gravitational fields observed are weak enough to fully justify the linear approximation.

The *Tajmar effect* cannot be explained from *GR*, which becomes clear in comparing the GP-B experiment with Tajmar's experiments. In GP-B, which was orbiting the *Earth* for more than 10 months at an altitude of about 640 km, the predicted Lense-Thirring precession of the gyro spin axis (inertial frame dragging by the rotation of the mass of the *Earth*), initially pointing at a guide star (locked by a telescope), is some *42* milli-arc seconds/year. This value is small compared to the already tiny geodetic effect (spacetime curvature caused by the mass of the *Earth*) of *6.6* arc seconds/year. The geodetic precession occurs in the orbital plane of the satellite; while the Lense-Thirring effect causes a precession of the gyro spin axis in the same direction the *Earth* is rotating (the gyro is assumed to be initially in free fall along the axis of rotation of the *Earth*). For the GP-B experiment an inertial frame was required with non-gravitational acceleration less than 10^{-13} *m/s*. Compared to Tajmar's equipment, his gyroscopes definitely are not capable to detect accelerations that small. One of the major challenges of the GP-B experiment was to provide such a drag-free (weightless) satellite. It is therefore impossible that Tajmar has observed any effect related to *GR*. His effect must therefore be outside *GR*, pointing to a new class of gravitational phenomena, provided, of course, *that his measurements are correct*. This were an indication that the standard picture of gravity as manifested in Einstein's 1915 *GR* does need an extension that goes beyond the picture of gravity of simply being the result of the curvature of four-dimensional spacetime. Therefore, the two additional gravitational fields as postulated in *EHT*, represented by gravitophotons and the quintessence particle, are at least qualitatively supported. In other words, the nature of gravity is more complex than represented by *GR*. All predictions of *GR* are correct, but it seems that it is *GR* which is not complete instead of *QM* (quantum mechanics). Moreover, the geodetic and Lense-Thirring effects show that an interaction between spacetime and massive bodies exist. This could mean that the *Tajmar effect, being many orders of magnitude larger, should have a much stronger interaction with its surrounding* spacetime. This is exactly what is needed for propellantless propulsion, which can only work if there is an intense *exchange of energy and momentum among* space vehicle *and* spacetime, see the discussion in Section 3.2.

In order to explain the *Tajmar effect*, an additional assumption has to be made in order to characterize the phase transition that obviously seems to accompany all extreme gravitomagnetic phenomena. As known from superconductivity the heuristic *London equations*, representing the material equations, in combination *with the Maxwell equations* are essential to calculate both the qualitative and quantitative aspects of superconductivity in a heuristic way.

Therefore, from a physical point of view it is clear that the Einstein-Maxwell equations alone cannot describe the gravitomagnetic experiments of Tajmar, in the same way the Maxwell equations cannot account for the phenomenon of superconductivity.

- The magnitude of the extreme gravitomagnetic field points to an electromagnetic origin, being the only other long range field with sufficient coupling strength.

- Therefore, the London equations will be employed to complement the Einstein-Maxwell equations.

- Moreover, there should be a physical mechanism that converts an electromagnetic into a gravitomagnetic (or gravity-like) field modelled by a phase transition of Ginzburg-Landau type.

- Such a mechanism is not conceivable within the framework of the four fundamental interactions which cannot incorporate additional gravitational fields along with their additional interaction bosons. The standard model cannot accommodate these additional particles and thus needs to be extended.

- For energy and momentum to be conserved, the interaction of the matter of the rotating disk (ring) with the surrounding spacetime field must be accounted for.

- In the Tajmar experiments the gravity-like field of the accelerated ring is acting in the plane of the rotating ring, opposing its origin. In the proposed Heim experiment, the gravity-like field of the disk, rotating at constant angular velocity, is calculated to be directed along the axis of rotation. Therefore, these two experiments seem to be based on two different physical mechanisms.

- The first neutral gravitophoton, indicated by ν_{gp}^{01}, which is deemed to be responsible for the Heim effect, should decay into the positive (attractive) ν_{gp}^{+} and negative (repulsive) ν_{gp}^{-} gravitophotons. The resulting gravity-like field is supposed to be pointing in axial direction.

- The second neutral gravitophoton, indicated by ν_{gp}^{02} decays only if the ring is being accelerated, and the resulting gravity-like field is in circumferential direction, and thus this decay route is believed to occur in the experiments by Tajmar *et al.* [11-13], denoted as Tajmar effect.

- From a technological point of view the axial gravity-like field is the one that could provide the enabling technologies for propellantless propulsion and novel air and land transportation systems as well as green energy generation *etc.* At present, it does not seem possible to fully assess the technological consequences of the existence of such a field.

- The two experiments for circumferential and axial gravity-like fields are fundamentally different, but in both cases a conversion from electromagnetic to gravitational fields seems to take place, triggered by the generation of imaginary electrons, see NOM cube in the 4D hypercube of [24].

In the framework of the current paper a full discussion of the implications of imaginary matter cannot be given, but the basic facts of the conversion mechanism will be presented.

It seems that not only particles and their anti-particles, but, under certain conditions, also particles and their ghost or shadow particles (*i.e.,* imaginary mass) particles exist, or, at least, can be created under special experimental conditions.

At temperatures low enough for the respective phase transition to occur, it seems that the imaginary electrons being produced are forming bosons comprising six imaginary electrons e_I. The imaginary current due to these sextets is deemed to result in an imaginary vector potential $\mathbf{A}_I$ whose interaction with the imaginary quarks q_I (protons) in the rotating disk is eventually leading to a real physical interaction which appears in the form of gravity-like fields. The physical mechanism is complex, but, as can be seen from the experimental setup of Tajmar *et al.* [11-13], the generation of the circumferential gravity-like field is surprisingly simple. The same should hold true for the axial gravity-like field experiment.

Any propellantless space propulsion technology therefore would be substantially simpler and efficient than currently used chemical propulsion, and also inherently safer as well as far more economical.

The experiment for the axial field (Fig. **3**) comprises a cryogenic disk comprised of a given material, denoted as M_D having a diameter of about 0.2 m, rotating at circumferential velocity $\mathbf{v}$. Below the disk a superconducting coil is placed, made of material M_C, that comprises N turns. The disk may also reside inside the coil. It should be noted that disk and coil material need to be complementary. In the experiments by Tajmar *et al.* [11-13] a Nb ring and an Al sample holder seem to give the best results. The third part is a simple device to ensure that the current of imaginary electrons is coupled into the coil. In order to achieve this, the wire of the coil is cut through and a non-superconducting disk of about 1 mm thickness is introduced. The Cooper pairs cannot tunnel through this layer, since its thickness far exceeds the 10 Å of the Josephson effect. However, the Compton wave length of the imaginary electrons is much larger, because of the small limit velocity c_I in solids for particles of imaginary mass, and thus the e_I should be capable of tunneling through, leading to the imaginary current I_I that gives rise to the imaginary vector potential $\mathbf{A}_I$ [It is not known if the direct imaginary current is superimposed by a high frequency alternating imaginary current as observed for Cooper pairs in the Josephson effect].

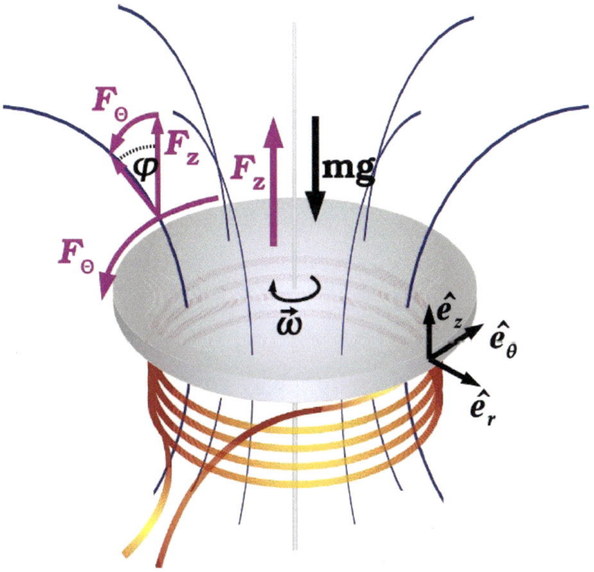

Figure 3: Heim experiment: in this gravity-like field experiment generated should be directed along the axis of rotation. The second component is in the azimuthal direction and should accelerate the ring or disk. Therefore, energy does not need to be supplied to keep the angular velocity of the ring or disk constant. The experimental setup could serve as field propulsion device, if a non-divergence free field were generated (the physical nature of the gravity-like field is not with certainty known at present). It should be noted that in order to produce the gravity-like field the current of the bosons of imaginary mass needs to be induced in the coil, which is supposed to be achieved *via* the Josephson Effect.

The gravitomagnetic mechanism of *GR* clearly is not the mechanism that occurs in the Heim experiment and also in the experiments by Tajmar *et al.* [11-13]. As these experiments demonstrate, the process is a solid state phenomenon, depending on a phase transition, triggered by temperature. Therefore, the generation of the gravitomagnetic field follows a totally different mechanism different than *GR*. Hence, the gravitomagnetic field denoted $\mathbf{B}_{gp}$ must be calculated by a different physical model. According to *EHT*, in the Heim experiment,

- The gravitomagnetic field $\mathbf{B}_{gp}$ is generated by new types of bosons, termed gravitophotons $\nu_{gp}^{\pm}$,
- The origin of the $\mathbf{B}_{gp}$ is the electromagnetic field,
- Conversion from electromagnetism to gravitomagnetism seems to follow the reaction chain starting from photons $\gamma \rightarrow \gamma_{IR} \rightarrow \gamma_I \rightarrow \nu_{gp}^{01} \rightarrow \nu_{gp}^{+} + \nu_{gp}^{-} \rightarrow \nu_g + \nu_q$,
- The ν_g gravitophoton confers the momentum to the space vehicle, the ν_q gravitophoton provides negative momentum to the surrounding spacetime which therefore expands, the total momentum of the physical system remains unchanged, *i.e.*, zero.

As additional material equations for gravitophoton interaction, in analogy to superconductivity, the London equations are employed in determining the magnitude of the $\mathbf{B}_{gp}$ field in conjunction with the conversion mechanism, *i.e.*, its magnitude is determined by the underlying physics of the conversion process. It is well known that in the superconducting case a real super current is generated by electron Cooper pairs, formed by a phase transition at critical temperature T_C, described by the heuristic London equation

$$\mathbf{B} = -\frac{2m_e}{e}\omega \tag{62}$$

where $\mathbf{B}$ is the magnetic induction field caused by the Cooper pairs. In the experiments by Tajmar *et al.* [11-13], as discussed in Section 1, an extreme gravitomagnetic field is generated. For the explanation of these experimental results as well as for the Heim experiment, it is assumed that the current of the superconducting electrons (Cooper pairs) causes a current of imaginary electrons. Imaginary particles are formed *via* the Higgs mechanism, for instance, as described by Kaku, [34] Chapter 10, further details are also given in [10]. Due to the interaction of the imaginary particles with OM in the crystal lattice, they should not behave like tachyons. For the Heim experiment, the imaginary current *needs to be coupled into the superconducting coil by some kind of tunnel effect* as stated above. The Cooper pair current is not important by itself, it only acts as the source for the accompanying imaginary current that is

$$\mathbf{B}_{e_I} = -i\,\alpha\frac{6m_e}{e}\omega_I \tag{63}$$

where ω_I denotes the angular frequency of the imaginary bosons formed by the coupling of the e_I. The magnitude of the imaginary mass m_{e_I} is believed to be $6m_e$ (The charge of the imaginary electron e_I and electron $-e$ are the same). Therefore, this value is used in the above equation. It is important to ensure experimentally that an imaginary current is flowing in the coil, *i.e.*, an *experimental mechanism must be provided to couple this current into the coil*, once the real super-current sets in. It should be mentioned that the chain of formation of the three types of photons $\gamma \rightarrow \gamma_{IR} \rightarrow \gamma_I \rightarrow \nu_{gp}^{01}$ takes only place below a certain critical temperature. The question arises how to couple the electromagnetic energy to the gravitational energy. The value of the coupling constant $\alpha_{gp} = \sqrt{\lambda} \approx 1/221$ is related to the radiative correction of the Higgs field, described by the parameter λ in Kaku [34] (p.353), *via* $\sqrt{\lambda} \approx \alpha_{gp}$. Here, *Kaku* is discussing the use of the Coleman-Weinberg potential to calculate λ and is treating the Higgs field as radiative correction of the electromagnetic field. This has the following meaning. The interaction potential Φ contains a fourth order term with coefficient λ, which was inserted by hand to account for the symmetry breaking, *i.e.,* to model the phase transition process that generates charged particles of imaginary mass.

Therefore, no relation between the two parameters m and λ could be specified. For instance, if λ changes sign, according to the theory Landau, a symmetry breaking will take place. This kind of phase transition that is governed by the quasi symmetry breaking, is supposed to take place in the form of generation of particles of imaginary matter. Kaku then shows (see Figs. **10.4** and **10.5** in [34]) that if the so called radiative correction is used that is, in order to calculate the effective potential by summing up over all one loops in the Feynman diagram, a relationship between the fine structure constant $\alpha = w_{ph}^2 = e^2 / 4\pi\epsilon_0\hbar c$ and the value λ can be established. The following relation was calculated

$$\lambda = \frac{33}{8\pi^2}\alpha^2 = \frac{33}{8\pi^2}\frac{1}{16\pi^2\epsilon_0^2}\frac{e^4}{\hbar^2 c^2}. \tag{64}$$

The coupling constant in quantum electrodynamics is the well known fine structure constant, which has the value $\alpha = 1/137 \approx 7.3\times10^{-3}$, which is small. The factor $6m_{eI}$ is obtained when an imaginary instead of a real mass is considered.

It should be noted that $\mathbf{A}_{e_I}$ is the imaginary vector potential that belongs to Equation (63). Thus the resulting gravitomagnetic field is

$$\mathbf{B}_{gp} = 0.328\,\mathbf{v}\times\alpha\alpha_{gp}\frac{6m_e}{m_p}\omega_\text{I}. \tag{65}$$

As mentioned above, the Lorentz equation also holds for the gravitophoton force (it should be noted that the Maxwell equations and the Einstein-Maxwell equations are similar, and the fully nonlinear equations Einstein field equations are only of interest in the direct neighborhood of black holes or for distances comparable to the diameter of the *Universe*). From the

$$\mathcal{L} = \frac{1}{2}mv^2 + e\Phi + \mathbf{v}\cdot\mathbf{A}_{e_I} \tag{66}$$

it is obvious that the gravity-like force is given by

$$\mathbf{F} = m\,\mathbf{v}\times\mathbf{B}_{gp} \sim m\,\mathbf{v}\times\left(\mathbf{v}\times\mathbf{B}_{e_I}\right), \tag{67}$$

which means that the resulting force is in the direction of the $\mathbf{B}_{e_I}$ field, which, in the *Heim experiment is the axial direction.* In the Heim experiment the neutral gravitophotn is supposed to decay according $\nu_{gp}^{01} \to \nu_{gp}^{+} + \nu_{gp}^{-} \to \nu^{g} + \nu_q : \alpha_g, \alpha_q$, where α_g and α_q are the respective coupling constants that correspond to the graviton, ν_g, and the quintessence particle ν_q. As discussed in the section on conservation principles, Section 3.2, since the spacetime field does exchange energy and momentum in all experiments of extreme gravitomagnetic fields, the force exerted by the gravitons acts on the rotating disk or ring and the force by ν_q is locally pushing against the spacetime field, acting as a repulsive force that in principle leads to an acceleration of the spacetime field, though, most likely, the effect cannot be measured. There are no ideas at present what causes the inertia of the spacetime field and how large it is. The gravitophoton g_{gp}^{+} acceleration acting on the rotating disk or ring then has the form

$$g_{gp}^{+} = 0.328\alpha\alpha_{gp}\alpha_g\frac{6m_e}{m_p}\frac{v^2}{c}\omega_I \tag{68}$$

with

$$\alpha_g = \left(\frac{G_N}{G_{gp}}\right)^{1/2} = 67\,.$$

Since **v** is the circumferential speed of the rotating disk, the average velocity of the particles in the disk is given by

$$v_A^2 = \frac{1}{3} v^2 \quad \textbf{(69)}$$

and therefore the axial acceleration is

$$g_{gp}^{+} = 0.328 \alpha \alpha_{gp} \alpha_g \frac{2m_e}{m_p} \frac{v_A^2}{c} \omega_I . \quad \textbf{(70)}$$

The new variables appearing in the following equation are specified in the numerical example below. The final form for the magnitude of the gravitomagnetic acceleration in z-direction (vertical) is

$$g_g = 0.328 \alpha \alpha_{gp} \alpha_g^2 \frac{2_{m_e}}{m_p} \frac{\rho_D}{\rho_{0D}} \frac{A_D}{A_{0D}} N \frac{A_C}{A_{0C}} \frac{v_A^2}{c} \omega_I \quad \textbf{(71)}$$

As an example for a laboratory experiment to producing a sizable axial field a disk of $d = 0.2m$ diameter together with the following parameters is used:

$$\frac{m_e}{m_p} = \frac{1}{1836}; \; \frac{\rho_D}{\rho_{0D}} = 0.19; \; \frac{A_D}{A_{0D}} = \frac{10^3}{36},$$

where ρ_{0D} and A_{0D} are reference density and reference area of the disk, and $N = 50$ is the number of turns of the coil. A value of $A_C / A_{0C} = 5$ is chosen, where A_C and A_{0C} are the cross section and the so called reference cross section of the coil, respectively. The circumferential speed of the disk is $v = 50 \; ms^{-1}$ and $\omega_I = 10^5 s^{-1}$. Inserting these values results in

$$g_g = 67^2 \times \frac{1}{212} \times \frac{2}{1836} \times 0.19 \times \frac{10^3}{36} \times 50 \times 5 \times \frac{2500}{3 \times 10^{-8}} \times 10^5 \times \frac{1}{9.81} g = \mathbf{1.8 \times 10^{-2}} g \quad \textbf{(72)}$$

where g denotes the acceleration of the *Earth*.

In summary, in the Heim experiment the following conversion takes place: Electromagnetism $\rightarrow$ gravitation + spacetime. Gravitation and spacetime form some kind of unity that is, there is not only the gravitational acceleration from Equation (71) but also spacetime should be locally subject to acceleration mediated by the quintessence particle. With g_g fields matter is accelerated, with g_q field there will be an interaction with spacetime and momentum conservation, in order to be satisfied, need to be applied to both.

For the limit of the real current I_L one finds

$$I < I_L = \frac{2\pi R}{\mu_0} \left(\frac{A_C}{A_{0C}} \right) \frac{6m_e}{e} \omega_I \approx 10 \; A. \quad \textbf{(73)}$$

4. CONCLUSIONS AND FUTURE ACTIVITIES

Since 2002, ideas for a geometric approach of describing physical interactions, termed *EHT* (Extended Heim Theory), have been published. This approach predicts two additional physical interactions in the form of gravity-like fields; see for instance [10, 17, 20, 26, 50]. A popular description of this research may be found in [46-49].

Numerous experiments, carried out by Tajmar *et al.* [11-13] since 2003 and first published in 2006, report on the generation of gravitomagnetic (spacetime twisting) and gravity-like fields (acceleration) in the laboratory. Provided that the experiments of Tajmar and Graham are correct, a similar effect should have been observed in the Stanford-NASA Gravity Probe-B experiment. The comparison of predictions from *EHT* with measurements are given in a recent paper comparing gravity-like field experiments with calculations from *EHT* [20].

Therefore, present situation is characterized by the fact that numerous experiments were performed over a period of more than four years, employing very different measurement techniques, but showing highly similar, but totally unexpected results. These experimental results, if confirmed, are outside *GR*. Since complete electromagnetic shielding took place, none of the four known interactions can generate these extreme gravitomagnetic fields that are of long range. The experimental data, from the completely unrelated experiments by Tajmar *et al.* [11-13] and Graham *et al.* [14], have seen physical effects that are approximately 18 orders of magnitude larger than predicted by *GR*. On the other hand, the Lageos and GP-B experiments (to be finished in fall 2009) have clearly demonstrated that the inertial frame dragging effect, even from celestial bodies, is extremely small and within *GR*. These facts provide evidence for *novel physics in the form of additional fundamental forces*.

Gravity-like fields most likely would lead to novel technologies in the general field of transportation, and thus should be of major interest to the public and, in particular, to industry.

The geometrization concept of *EHT* presented is based on the existence of an eight-dimensional internal symmetry space, which is attached to each point of conventional four-dimensional spacetime. This concept leads to exactly *six fundamental physical interactions*, three of them gravity like, but gravity can now be both attractive and repulsive. The non-linear σ -model has been mentioned as the principle to connect the Hermetry forms resulting from the double coordinate transformation to Lagrangian densities that is, providing the link between geometry and physics. From the subspace structure of internal space H^8 the symmetry breaking of the associated group $O(8,q)$ follows, which gives rise to the assumption of six fundamental forces. Because of the twelve Higgs fields, 12 charges are proposed; four of them should be of gravitational nature. In addition, there seems to be second group $O(8,q)$ that describes the 15 particle families, which result from the $O(3,q)$ subgroup associated with space R^3. This leads to the assumption that there are three colors from quarks and a fourth color from the imaginary quark. This short discussion shows that there are numerous open questions and many more topics to discuss, but, on the other hand, for instance, the predicted existence of 15 gluons and 4 gravitational as well as 2 electric charges shows that physics might be much richer than currently anticipated.

From the ideas presented here, *admittedly speculative* and supported only by the (not yet confirmed) existence of extreme gravitomagnetic and gravity-like fields as measured by Tajmar *et al.* [11-13], it is surmised that these additional gravitational interactions have the potential to provide the physical principle for *propellantless propulsion* and also allow for green energy generation. For instance, in [24] it was suggested to investigate whether the the fusion plasma in a magnetic mirror can be stabilized by axial gravity-like fields and the mirror be closed.

Needless to say, control of gravity would lead to completely new technologies, comparable to the advent of electricity and magnetism in the 19th century.

Concerning the development of advanced space propulsion concepts, the ones described in Section 1 need to be pursued further, but breakthrough space propulsion does need to be based on a *research program to look for fundamentally different propulsion principles* based on the physical ideas presented in this article that is, the existence of novel gravity-like fields.

In how far the concept of geometrization of physical interactions, if it turns out to be true, might also lead to a modified *Weltbild* of physics [53] cannot be decided at present.

5. ACKNOWLEDGEMENTS

This article is dedicated to the eminent *Dr. William Berry* (ret.), head of Propulsion and Aerothermodynamics Division, ESTEC, ESA with whom the second author had the privilege to co-operate for several years.

The assistance by M.Sc. O. Rybatzki, Computer Center, Ostfalia Univ. Applied Sciences, Germany in preparing the figures is gratefully acknowledged.

The authors are particularly grateful to Prof. M. Tajmar, KAIST, Seoul, Korea (formerly at AIT Seibersdorf, Austria) for providing general information, measured data as well as for numerous discussions and hints that helped in the comparisons between *EHT* results and gravity-like experiments.

The authors are grateful to an anonymous reader for suggesting utilizing a white dwarf instead of a neutron star, in order to obtain a more meaningful comparison between the magnitude of gravitomagnetic fields generated by the cryogenic Nb ring and astronomical objects.

The authors are most grateful to Prof. P. Dr. Dr. A. Resch, director of the Institut für Grenzgebiete der Wissenschaft (IGW), Innsbruck, Austria for his support in writing this paper.

The authors are indebted to Dr. A. Müller Eichenau, Germany for reading the manuscript and his comments on the content of this article.

The second author was partly funded by Arbeitsgruppe Innovative Projekte (AGIP) and by the Ministry of Science and Education, Hannover, Germany.

REFERENCES

[1] Kakalios J. The Amazing Story of Quantum Mechanics. Gotham Books 2010.

[2] Choueiri EY. New Dawn for Electric Rockets. Scientific American 2009 February; pp.58-65.

[3] Braun EM, Lu FK, Wilson DR. Experimental Research in Aerodynamic Control with Electric and Electromagnetic Fields. Prog Aero Sci 2009; 45: 30-49.

[4] Kaku M. Physics of the Impossible. Doubleday 2008.

[5] Ketsdever A, Young M. An Overview of Advanced Concepts for Space Access. In: Proc. Of 44th AIAA/ASME/SAE/ASE, Joint Propulsion Conference & Exhibit. Hartford, CT, 20-23 July, 2008. Paper AIAA 2008-5121.

[6] Woods RC, Cooke SG, Helme J, Caldwell CH. Gravity Modification by High-Temperature Superconductors. In: Proc. 37th AIAA/ASME/SAE/ASEE Joint Propulsion Conf.; 2001; Salt Lake City, USA. Amer Inst of Aeronautics & Astronautics 2001.

[7] Haysaka H *et al.* Possibility for the existence of anti-gravity: evidence from a free-fall experiment using a spinning gyro. Spec Sci Tech 1997; 20: 173-181.

[8] Einstein A. On the Generalized Theory of Gravitation. Scientific American 1950 April; Vol. 182, Nr. 4.

[9] Dröscher W, Hauser J. Emerging Physics for Novel Field Propulsion. In: Proc. Of 46th AIAA/ASME/SAE/ASE Joint Propulsion Conference and Exhibit, 26-28 July 2010, Nashville, TN. Amer. Inst. of Aeronautics & Astronautics 2010.

[10] Dröscher W, Hauser J. Gravitational Field Propulsion. Paper AIAA 2009-5069. In: Proc. of 45th AIAA/ASME/SAE/ASE Joint Propulsion Conference and Exhibit, 2-5 August 2009, Denver, Colorado. Amer Inst of Aeronautics & Astronautics 2009.

[11] Tajmar M, Plesescu F, Seifert B, Marhold K. Measurement of Gravitomagnetic and Acceleration Fields Around Rotating Superconductors. In: Proceedings of the STAIF-2007 Conference. AIP Conference Proceedings 2007; 880: 1071-1082.

[12] Tajmar M, Plesescu P, Seifert B, Schnitzer R, Vasiljevich I. Search for Frame-Dragging in the Vicinity of Spinning Superconductors. In: Loughlin MJ Ed. Proceedings of the 2nd International Conference on Time and Matter, Nova Gorica, 2008.

[13] Tajmar M, Plesescu F, Seifert B, Schnitzer R, Vasiljevich I. Investigation of Frame-Dragging-Like Signals from Spinning Superconductors using Laser Gyroscopes. AIP Conference Proceedings 2008; 969: 1080-1090. Tajmar M, Plesescu F, Seifert B. Anomalous Fiber Optic Gyroscope Signals Observed above Spinning Rings at Low Temperature. arXiv 2008: 0806.2271. Available from http://arxiv.org
[14] Graham RD, Hurst RB, Thirkettle RJ, Rowe CH and Butler, PH. Experiment to detect frame dragging in a lead superconductor. Physica C 2008; 468: 383.
[15] Stanford University. Gravity Probe B in a Nutshell. Available from http://einstein.stanford.edu
[16] Dröscher W, Hauser J. Physical Principles of Advanced Space Based on Heim's Field Theory. Paper AIAA 2002-4094. In: Proc. Of 38th AIAA/ASME/SAE/ASE, Joint Propulsion Conference & Exhibit, Indianapolis, Indiana, 7-10 July, 2002. Amer. Inst. of Aeronautics & Astronautics 2002. Available from http://www.hpcc-space.com
[17] Dröscher W, Hauser J. Guidelines For a Space propulsion Device Based on Heim's Quantum Theory. Paper AIAA 2004-3700. In: Proceedings of the 40th AIAA/ASME/SAE/ASE, Joint Propulsion Conference & Exhibit, Fort Lauderdale, FL, 11-14 July, 2004. Amer Inst of Aeronautics & Astronautics 2004.
[18] Dröscher W, Hauser J. Heim Quantum Theory for Space propulsion Physics. In: Proceedings of the STAIF-2005 Conference. American Inst. of Physics 2005.
[19] Modanese G. Personal communication, 30 May 2010.
[20] Dröscher W, Hauser J. Gravity-Like Fields and Space propulsion Concepts. Paper AIAA 2008-5124. Proceedings of the 44th AIAA/ASME/SAE/ASE, Joint Propulsion Conference & Exhibit, Hardford, Connecticut, 20-23 July, 2008. Amer Instof Aeronautics & Astronautics 2008.
[21] Bardeen J, Cooper N, Schrieffer JN. Theory of Superconductivity. Phys Rev 1957; 108: 1175–1204.
[22] Dröscher W, Hauser J. Current Research in Gravito-Magnetic Space propulsion, Paper O-42. In: Proc. of 7th International Symposium on Launcher Technologies, 2-5 April 2007, Barcelona, Spain.
[23] Dröscher W, Hauser J. Heim Quantum Theory for Space propulsion Physics. In: Proceedings of STAIF 2005. American Institute of Physics 2005.
[24] Hauser J, Dröscher W. Emerging Physics for Space propulsion Science. In: Proceedings of Space propulsion & Energy Sciences International Forum SPESIF-2010. American Institute of Physics 2010.
[25] Hauser J, Dröscher W. Physics of Extreme Gravitomagnetic and Gravity-Like Fields for Novel Space propulsion and Energy Generation. In: Proc. Of Space propulsion & Energy Sciences International Forum SPESIF-2011. Elsevier 2011.
[26] Dröscher W, Hauser J. Spacetime Physics and Advanced Propulsion Concepts. Paper AIAA 2006-4608. In: Proceedings of the 42nd AIAA/ASME/SAE/ASE, Joint Propulsion Conference & Exhibit, Sacramento, CA, 9-12 July, 2006. Amer. Inst. of Aeronautics & Astronautics 2006. Available as revised and extended version from http://www.hpcc-space.de
[27] Jost, J. Geometry and Physics. Springer 2009.
[28] Heim, B. Vorschlag eines Weges einer einheitlichen Beschreibung der Elementarteilchen, Zeitschrift Naturforschung 1977; 32A: 233-243.
[29] Heim, B.: Elementarstrukturen der Materie, Band 1, 3. Auflage. Innsbruck: Resch Verlag 1998.
[30] Heim, B.: Elementarstrukturen der Materie, Band 2, 2. Auflage. Innsbruck: Resch Verlag 1996.
[31] Keller JM. Newton's Third Law and Electrodynamics, Am J Phys 1942; 10: 302-307.
[32] Cardone F, Mignani R. Energy and Geometry, World Scientific 2004.
[33] Zwiebach, R. Introduction to String Theory Cambridge Univ. Press 2004.
[34] Kaku, M. Quantum Field Theory, Oxford University Press 1993.
[35] Hobson MP, Efstathiou G, Lasenby AN. General Relativity. Cambridge Univ. Press 2006.
[36] Thorne K. Gravitomagnetism, Jets in Quasars, and the Stanford Gyroscope Experiment. In: Fairbank JD, Deaver BS Jr., Everitt CWF, Michelson PF Eds. Near Zero: New Frontiers of Physics. Freeman 1988.
[37] Witten, E. Reflection on the Fate of Spacetime. Phys Today 1996; 96: 24-30.
[38] Veltman, C. Facts and Mysteries in Elementary Particle Physics. World Scientific 2003.
[39] Zee, A. Quantum Field Theory in a Nutshell. Princeton University Press 2003.
[40] Lawrie ID. Unified Grand Tour of Theoretical Physics. IoP 2002.
[41] Bergstroem L, Goobar A. Cosmology and Particle Astrophysics Physics. Springer 2004.
[42] Liddle A. An Introduction to Modern Cosmology. Wiley 2003.
[43] Lisi AG. An Exceptionally Simple Theory of Everything arXiv 2007: 0711.0770. Available from http://arxiv.org
[44] Collins GP. Wipeout ? Sci Am 2008; 298: 30-32.

[45] Ciufolini I, Pavlis EC, Peron R. Determination of frame-dragging using earth gravity models from CHAMP and GRACE. New Astronomy 2006; 11: 527-550.

[46] Lietz, H. 11 Lichtjahre in 80 Tagen. Von der Feldtheorie Burkhard Heims und ihrer Anwendung fuer einen Raumfahrtantrieb. Telepolis Special 1/2005. Hannover: Heise Verlag; pp. 47-51.

[47] Lietz H. Hyperdrive. New Scientist, 7-13 January 2006.

[48] Lietz H. Plus Vite que la Lumiere. Scilogic 2006; pp. 89-94.

[49] Moton-Nkhata T, Donahue B, Hauser J. Nuclear and Future Flight Propulsion. Aerospace America, December 2007, pp. 60-61.

[50] Dröscher W, Hauser J. Advanced Propulsion Systems from Artificial Gravitational Fields. Paper AIAA 2007-5595. In: Proc. of 43th AIAA/ASME/SAE/ASE, Joint Propulsion Conference & Exhibit, Cincinnati, OH, 8-11 July, 2007. Amer. Inst. of Aeronautics & Astronautics 2008.

[51] Schiller C. Motion Mountain, Part III: General Relativity versus Quantum Mechanics. Google Books 2009.

[52] Czysz PA, Bruno C. Future Spacecraft Propulsion Systems: Enabling Technologies for Space Exploration. Springer, 2009.

[53] Lam C-S. The Zen in Modern Cosmology. World Scientific, 2008.

[54] Smolin, L. Atoms of Space and Time. Scientific American, January 2004.

CHAPTER 12

Gravity-Superconductor Interactions from a Hierarchy of Density Dependent Scalar Fields: A Dark Matter Connection

Glen A. Robertson*

Institute for Advanced Research in the Space, Propulsion & Energy Sciences, Madison, AL 35757, USA

Abstract: This paper discusses the connection between dark matter and gravity-superconductor interactions through a hierarchy of density dependent scalar fields from the far Universe field down to the laboratory scale. A hierarchy of scalar fields in the universe suggests a relationship with properties varying with background conditions or a hierarchy from the far background field to the local background field. Such a theory, called the Chameleon theory (presented in 2004 by Khoury and Weltman), is a dark matter model that is fundamentally dependent on the background density and involves mass coupling to the background field. In this paper, a new hierarchy interpretation of the gravitational constant $G_{\cdot}$ is produced by combining the Chameleon Theory with newly observed astronomical data, where $G_{\cdot}$ is given in terms of the dark matter particle mass and the hierarchy of the background fields-represented by a hierarchy of coupling constants. Whereby, the data seen by gyroscopes in the Tajmar *et al.* rotating superconductor experiments can be attributed to dark matter – like the formation of dark matter rings outside the confines of a galaxy. This relationship provides a starting point for using new galactic structures as guidance for defining new gravity-superconductor phenomena in experiments on the laboratory scale.

Keywords: Gravitational forces, theories of gravitation, superconductors, type-II superconductors, general relativity, quantum gravity, vacuum energy density, gravity-like fields, dark matter, dark matter rings, fields hierarchy, chameleon theory, cosmological scalar fields, gravitational constant G.

1. INTRODUCTION

The appearance of experimental gravity-like phenomena from superconductors has been appearing in literature since the early 1990s. However, in the twenty or so years since, not one sustainable theory has arisen. This fact remains even though a gravity-superconductor connection was present some thirty years earlier by DeWitt [1]. The author notes that it was about eight years after the first superconductor gravity experiments arose before the belief of dark matter in the Universe was confirmed by teams observing supernova. Since then, it has been noted that dark matter could come naturally from quantum gravity [2]; noting quantum gravity theories are abundant in literature. Given that superconductors are quantum systems, it should not be surprising to find evidence of quantum gravity-superconductor interactions. Although Droescher and Hauser [3] make reference to dark matter/energy in relationship to their Extended Heim Theory, to the author's knowledge, no clear theory has been presented connecting gravity-superconductor interactions to dark matter directly. Therefore, it seems appropriate to end this eBook with such a discussion.

1.1. Background

The overwhelming preponderance of matter in the universe is believed to be dark. Therefore, one should expect dark matter to play a fundamental role in the attraction of masses even on the laboratory scale and suggest that there is a hierarchy of scalar fields in the universe with properties varying with background conditions. In this paper, scalar field concepts based on astronomical objects are applied to the laboratory scale by connecting the dark matter particle mass to the gravitational constant. This is accomplished by using both theory and observational data to connect the dark matter particle mass to the reduced Planck constant $\hbar$ and then the Gravitational constant G.

The Khoury and Weltman [4, 5] Chameleon theory, which address a scalar field coupled directly to matter with respect to the surrounding matter density, is such a theory. Whereby, galactic structures attributed to dark matter can be compared with the universe background matter density to provide guidance for laboratory experiments in our local background matter density.

***Address correspondence to Glen A. Robertson:** Institute for Advanced Research in the Space, Propulsion & Energy Sciences, Madison, AL 35757, USA; Tel: 256-694-7941; E-mail: gar@ias-spes.org

Giovanni Modanese and Glen A. Robertson (Eds)

Complementary, to the Chameleon theory is an example of a scalar field based on observational data developed in the context of Cold Dark Matter (CDM) by Silverman and Mallett [6]. The structures they described are alleged CDM vortices in the Andromeda Galaxy, composed of scalar bosons held together by scalar forces. The Silverman and Mallett analytical model pre-dates the observation of a dark matter ring observed by the Hubble Space Telescope [7]. Dark as the ring cannot be seen, although scientists have mapped out the ring's shape using distortions the ring produces in light coming from distant objects behind them; noting that dark matter rings appear to be CDM extending beyond galactic boundaries, even though they have yet to be connected to a scalar field vortex.

The author notes that scalar fields in the Silverman and Mallett [6] CDM vortices and the Khoury and Weltman [4, 5] Chameleon theory are ascribed with somewhat different properties. For example, the Khoury and Weltman model blends astronomical and Planck scale physics through the effective potential $V_{eff}(\phi)$, whereas, the Silverman and Mallett model is based on observed parameters such as the 7 light year radius of CDM vortex in the Andromeda Galaxy. Further, Silverman and Mallett says the boson density varies inside and outside a galaxy, while Khoury and Weltman say the amplitude and mass of the scalar field vary. However, neither addresses the observation of dark matter rings outside the confines of a galaxy.

It is the purpose of the present paper to deduce properties of scalar fields from theory and astronomical data, which could guide laboratory experiments across various physical fields of study and provide an understanding of new astronomical data. As a start, a link between the Tajmar *et al.* superconductor experiments and the formation of dark matter rings outside the confines of a galaxy is discussed (Section 6).

2. THE CHAMELEON THEORY

The Chameleon theory assumes that our Universe is composed of a nearly massless scalar field. From a theoretical standpoint, massless scalar fields are abundant in string and supergravity theories. Indeed, generic compactifications of string theory result in a plethora of massless scalars in the low-energy-four dimensional-effective theory. However, these massless fields generally couple directly to matter with gravitational strength, and therefore lead to unacceptably large violations of the equivalence principle. That is, the Universe scalar field is essentially massless on solar system scales, and therefore subject to tight constraints from tests of the equivalence principle (For a review of experimental tests of the equivalence principle and general relativity, see [9, 10]).

In the Chameleon theory, scalar fields that have cosmological effects, such as quintessence, do not result in large violations of the equivalence principle in laboratories of Earth because of its very dense environment. Thus, the main constraint on this model is that the mass of the field be sufficiently large on Earth to evade equivalence principle and fifth force constraints (For a review of fifth force searches, see [11]). This is because the scalar field acquires a mass whose magnitude depends on the local matter density. In a region of high density, such as on Earth, the mass of the field is large, and thus the resulting violations of the equivalence principle are exponentially suppressed. In the solar system, where the density is much lower, the fields are essentially free.

While the idea of a density-dependent mass term is not new [12-17], the novelty of the Chameleon theory lies in the fact that the scalar fields can couple directly to baryons with gravitational strength; suggesting a strong connection between the Chameleon theory and the gravitational constant.

2.1. Units

The Chameleon theory is a cosmology theory where factors of $\hbar$ and c are omitted as they are routinely set to unity. In this paper, these factors are added to give units of kg on the Universe field ϕ_c, whereby the correct units of kg are derived on the minimum Chameleon mass m_c far from any object. Further, units are explicitly expressed in the present paper to track these conversions for the reader.

2.2. Coupling Constants

In physics, a coupling constant is a number that determines the strength of an interaction. Usually the Lagrangian or the Hamiltonian of a system can be separated into a kinetic part and an interaction part. The coupling constant determines the strength of the interaction part with respect to the kinetic part, or between two sectors of the interaction part. For example, the electric charge (-) of a particle is a coupling constant.

A coupling constant plays an important role in dynamics. For example, one often sets up hierarchies of approximation based on the importance of various coupling constants. In the motion of a large lump of magnetized iron, the magnetic forces are more important than the gravitational forces because of the relative magnitudes of the coupling constants. However, in classical mechanics one usually makes these decisions directly by comparing forces.

In the Chameleon theory, the generation of a density-dependent mass for a given field ϕ results from the interplay of two source terms in its equation of motion, which leads to a Lagrangian dominated by an effective potential $V_{eff}(\phi)$ containing these terms [4, 5].

The first term arises from self interactions, described by a monotonically decreasing potential $V(\phi)$ which is of the runaway form. In particular, the fact that the potential need not have a minimum; rather, it must be monotonic (*i.e.*, never increasing as the values of the independent variable or subscript terms increase accordingly). The second term arises from the conformal coupling to matter fields, of the form $e^{\beta\phi/M_{Pl}}$. The dimensionless coupling constant β need not be small, but >0 with values of order unity or greater allowed.

Unlike the Khoury and Weltman Chameleon theory [4, 5], which basically ignored (with reason) the effects of the coupling constants by setting them to unity (*i.e*, =1), in this analysis, several coupling constants are defined that are allowed to change with the hierarchy of the background fields. These coupling constants are shown to be a multiplier of the local matter density and associated to the far-field though a dark matter hierarchy coupling constant that must be added as a multiplier of the local density in the minimum potential equation established in the Chameleon Theory.

For simplicity, the word-background-in respect to the background field or density is hence forth omitted.

2.3. The Far Field

The coupling constant β in the Khoury and Weltman paper [4, 5] is set to 1, *i.e.*, unity. This is assumed done so as objects are immersed in a near uniform atmosphere, whereby objects couple directly with it. In the Universe field far from other objects, the dark matter particle masses exist in a similar state. Therefore the obvious choice for β in the far field would be 1. However, the author has suggested that when objects are in motion with high enough acceleration, a second motion coupling factor comes into play that can reduce $\beta<1$ [23-25]. Since the universe is expanding, it is reasonable to assume that the far field is in motion. Therefore, the far field ϕ_c coupling constant β_c is expected to be in the interval $0<\beta_c<1$.

In this paper, the universe field ϕ_c far from any object is chosen as the minimum field with units of kg having a (critical) density $\rho_c \approx 1.11778x10^{-26}\ kg/m^3$ and governed by the Khoury and Weltman [4, 5] effective potential

$$V_{eff}(\phi_c) = V(\phi_c) + \rho_c \exp^{\beta_c\phi_c M_{PL}}, \tag{1}$$

where the reduced Planck mass $M_{PL} = \sqrt{\hbar c/8\pi G} = 4.34\times10^{-9} kg$, $G \sim 6.6728x10^{-11}\ m^3/kg{\cdot}s^2$ is the gravitational constant, $\hbar = 1.0546x10^{-34}\ m^2{\cdot}kg/s$ is the reduced Planck constant, $c = 2.9979x10^8\ m/s$ is the speed of light, and where the potential in the far field is given by

$$V(\phi_c) = \frac{M^{4+n}}{\phi_c^n} \rightarrow \frac{M^6}{\phi_c^2}. \tag{2}$$

The units of the factor M are kg and $n = 2$ is the favorite value, such that, the potential $V(\phi_c)$ has units of kg^4.

The miss-match in units in equation (1) between the potential $V(\phi_c)$ with units of kg^4 and the conformal coupling $\rho_c e^{\beta\phi/M_{Pl}}$ with units of kg/m^3 requires that enough factors of $\hbar$ and c be added to construct an equation with a consistent set of units, whereby the effective potential with $n = 2$ is converted as

$$V_{eff}(\phi_c) = \frac{M_E^6}{\phi_c^2}\left(\frac{\hbar^3}{c}\right) + \rho_c c^2 \exp^{\beta_c \phi_c M_{PL}} \tag{3}$$

with the energy scale factor M replaced with the inverse length factor

$$M_E = \left(\frac{\Lambda}{8\pi l_{PL}^2}\right)^{\frac{1}{4}} \sim 10^4 m^{-1} \approx 2M\left(\frac{c}{\hbar}\right) \tag{4}$$

to convert the effective potential $V_{eff}(\phi_c)$ to proper units of J/m^3, which also removes the dark energy scale interaction implied by M from these equations.

That is, the inverse distance factor M_E implies an effective range of dark matter interactions in the far field and not the range of chameleon-mediated interaction, which implies a range on dark energy interactions that are very large (0.1 to 10^3 pc) in the far field [4], where the factor of 2 in equation (4) is needed to set the energy scale factor $M \sim 1.76x10^{-39} kg \left(\sim 10^{-3} eV/c^2\right)$, which is the upper bound associated with dark energy [8].

Of note is that equation (4) provides a connection between the large scale-cosmological constant Λ, which deals with cosmological expansion (dark energy) and the very small scale-Planck length $l_p \approx 1.616252x10^{-35} m$, which deals with Planck scale physics; implying a hierarchal conversion. Noting the same can be said between the energy scale factor M and the reduced Planck constant $\hbar$.

In steady state, the far field ϕ_c attains a value which minimizes the energy density. The minimum is calculated by setting the derivative $dV/d\phi = 0$, as

$$\frac{dV_{eff}(\phi_c)}{d\phi_c} = -3\frac{M_E^6}{\phi_c^3}\left(\frac{\hbar^3}{c}\right) + \left(\frac{\beta_c \rho_c c^2}{M_{PL}}\right)\exp^{\beta_c \phi_c M_{PL}} = 0, \tag{5}$$

where the unit are in $1/m{\cdot}s^2$.

Now noting that the exponential

$$\exp^{\beta_c \phi_c M_{PL}} = 1 + \beta_c\left(\frac{\phi_c}{M_{PL}}\right) + \frac{1}{2}\left(\beta_c\left(\frac{\phi_c}{M_{PL}}\right)\right)^2 \sim 1 \tag{6}$$

since in the far field $\phi_c \ll M_{PL}$ and $0 < \beta \leq 1$, equation (5) yields the far field

$$\phi_c = M_E^2\left(\frac{3}{\beta_c}\frac{M_{PL}}{\rho_c}\right)^{1/3}\left(\frac{\hbar}{c}\right); \tag{7}$$

noting that the far field ϕ_c in equation (7) has the correct units of kg.

2.4. Far Field Chameleon Mass

With factors of $\hbar$ and c to correct the units, the chameleon matter mass in the far field is given from

$$m_c^2 = \left(\frac{\hbar^3}{c^5}\right)\frac{dV_{eff}^2(\phi_c)}{d\phi_c^2}, \tag{8}$$

which from equation (5) gives

$$m_c^2 = 12\frac{M_E^6}{\phi_c^4}\left(\frac{\hbar}{c}\right)^6 + \frac{\beta_c^2 \rho_c}{M_{PL}^2}\left(\frac{\hbar}{c}\right)^3, \tag{9}$$

whereas before $e^{\beta_c \phi_c / M_{PL}} \sim 1$.

It can be shown for $\phi_c << M_{PL}$ that

$$12\frac{M_E^6}{\phi_c^4}\left(\frac{\hbar}{c}\right)^6 >> \frac{\beta_c^2 \rho_c}{M_{PL}^2}\left(\frac{\hbar}{c}\right)^3 \tag{10}$$

to give from equation (9)

$$\phi_c^4 \approx 12\frac{M_E^6}{m_c^2}\left(\frac{\hbar}{c}\right)^6, \tag{11}$$

which when combined with equation (7) yields

$$12\frac{M_E^6}{m_c^2}\left(\frac{\hbar}{c}\right)^6 \approx M_E^8\left(\frac{3}{\beta_c}\frac{M_{PL}}{\rho_c}\right)^{4/3}\left(\frac{\hbar}{c}\right)^4 \tag{12}$$

to give the far field Chameleon mass

$$m_c \approx \frac{\sqrt{12}}{M_E}\left(\frac{\beta_c}{3}\frac{\rho_c}{M_{PL}}\right)^{2/3}\left(\frac{\hbar}{c}\right); \tag{13}$$

noting that the far field chameleon mass m_c in equation (13) has the correct units of kg.

2.5. Estimate on the Dark Matter Particle Mass and the Far Field Mass Coupling Constant

As noted in the introduction, an example of a scalar field based on observational data was developed in the context of Cold Dark Matter (CDM) by Silverman and Mallett [6], whom presented the case for coherent degenerate dark matter specific to a galactic superfluid where the equilibrium $dE/d\xi = 0$ between quantum pressure and gravitational attraction leads to a minimum size (the galactic coherence length)

$$\xi_G = \left(\frac{\hbar^2}{m_{SM}^2}\cdot\frac{3\pi}{G\bar{\rho}_G}\right)^{1/4} \tag{14}$$

determined by the boson mass and condensate mass irrespective of the radial variation in density.

Equation (14) then yields the Sliverman and Mallett dark matter particle mass

$$m_{SM} = \sqrt{\frac{3\pi}{G\bar{\rho}_G} \cdot \left(\frac{\hbar}{\xi_G^2}\right)}. \quad \textbf{(15)}$$

The estimate for the mean mass density of Andromeda Atlas was given as $\bar{\rho}_G \sim 2.0 \times 10^{-24}\ kg/m^3$; noting that it is several orders greater than the critical density ρ_c. Sliverman and Mallett states that if the dark matter in M31 is due primarily to scalar bosons with a coherence length of the order of the size of the M31 luminous core $\xi_G \sim 30kpc\ (\sim 926x10^{20}\,m)$, then equation (15) yields a dark matter particle mass $m_{SM} \sim 1.8340x10^{-23}\ eV / c^2$, which converts to $3.27002x10^{-59}\ kg$ as needed in this paper.

Inputting the Sliverman and Mallett dark matter particle mass into equation (13), where $m_c \equiv m_{SM}$, yields a value for the far field coupling constant $\beta_c \approx 0.1653$, which is within the assumption that $0 < \beta_c \le 1$ for an accelerating far field.

Further, using these values gives the far field $\phi_c \approx 6.930x10^{-29} kg$, using equation (7), which validates the assumptions that in the far field $\phi_c << M_{PL}$.

3. PLANCK AND GRAVITATIONAL CONSTANTS

To better understand the interplay of dark matter from a cosmological prospective, it is noted that the critical density is given in terms of the Hubble constant $H_0 \approx 2.5x10^{-18}s^{-1}$ as

$$\rho_c = \frac{3H_0^2}{8\pi G}. \quad \textbf{(16)}$$

Now defining the dark matter particle mass $m_{DM} \equiv$ the Sliverman and Mallett dark matter particle mass $m_{SM} \equiv$ the far field Chameleon mass m_c of equation (13) and noting $8\pi G = \hbar c / M_{PL}^2$ to remove the gravitational constant, yields the dark matter particle mass in terms of cosmological constants as

$$m_c \equiv m_{SM} \equiv m_{DM} = \left(\frac{\sqrt{12}}{M_E}\right)\left(\beta_c^2 H_0^4 M_{Pl}^2 \left(\frac{\hbar}{c^5}\right)\right)^{1/3} \sim 3.27002x10^{-59}\ kg, \quad \textbf{(17)}$$

for $\beta_c \approx 0.1653$, to yield the reduced Planck constant in terms of the dark matter particle mass m_{DM} as

$$\hbar = \frac{m_{DM}^3}{\beta_c^2} \cdot R\, c^5, \quad \textbf{(18)}$$

where for simplicity

$$R = \left(\frac{1}{12\sqrt{12}}\right)\left(\frac{M_E^{\ 3}}{H_0^{\ 4} M_{Pl}^{\ 2}}\right) = 3.40x10^{97}\, s^4 / m^3 kg^2. \quad \textbf{(19)}$$

To recover the gravitational constant and remove the reduced Planck constant, it is noted that the Planck time

$$T_{PL} = \sqrt{\frac{\hbar G}{c^5}} \sim 5.4x10^{-44} \text{s}, \quad \textbf{(20)}$$

which when combined with equation (18) yields the gravitational constant as

$$G = \frac{\beta_c^2}{m_{DM}^3} \cdot \frac{T_{PL}^2}{R}. \quad \textbf{(21)}$$

4. SCALAR FIELD HIERARCHY

On the cosmological scale, the Planck and Gravitational constants are constant. Then under normal conditions at least down to the atomic scale (*i.e.*, subatomic particles have additional rules), the ratio of the cubed dark matter particle mass m_{DM}^3 to the squared coupling constant β_c^2 must also be constant in all fields ϕ_m. This implies a change to equation (13) as,

$$m_{DM}^3 \equiv m_c^3 = \left(\frac{4}{3}\frac{\sqrt{12}}{M_E^3\, M_{PL}^2}\frac{\hbar^3}{c^3}\right)\beta_c' \beta_c^2 \rho_c^2, \tag{22}$$

where β' is defined as the dark matter hierarchy coupling constant. The assumption is then that the dark matter particulate mass is the same in all fields

$$m_{DM}^3 = \left(\frac{4}{3}\frac{\sqrt{12}}{M_E^3\, M_{PL}^2}\frac{\hbar^3}{c^3}\right)\beta_m' \beta_m^2 \rho_m^2, \tag{23}$$

where from equations (22) and (23), it can be shown that the dark matter hierarchy coupling constant

$$\beta'_m = \frac{\beta_c^2\ \rho_c^2}{\beta_m^2\ \rho_m^2}, \tag{24}$$

such that, the Planck and Gravitational constant are the same in all matter density fields ϕ_m, noting in the far field, the subscript $m = c$, such that, $\beta_c' = 1$.

This implies that the hierarchial form of the minimum potential, equation (1), in any density dependent field ϕ_m be given as

$$V_{eff}\left(\phi_m\right) = V\left(\phi_m\right) + \beta_m' \rho_m exp^{\beta_m \phi_m / M_{PL}}, \tag{25}$$

where the density dependent field

$$\begin{aligned}
\phi_m &= M_E^2\left(\frac{3}{\beta_m'\beta_m}\frac{M_{PL}}{\rho_m}\right)^{1/3}\left(\frac{\hbar}{c}\right); \\
\phi_m &= M_E^2\left(\frac{3}{\beta_c}\frac{M_{PL}}{\rho_c}\right)^{1/3}\left(\frac{\hbar}{c}\right)\cdot\left(\frac{\beta_m}{\beta_c}\frac{\rho_m}{\rho_c}\right)^{1/3}; \\
\phi_m &= \phi_c\left(\frac{\beta_m}{\beta_c}\frac{\rho_m}{\rho_c}\right)^{1/3} = \frac{\phi_c}{\left(\beta_m'\right)^{1/6}},
\end{aligned} \tag{26}$$

which provides an equation for the hierarchy of density dependent scalar fields in the universe.

5. APPLICATION TO THE GRAVITATIONAL FORCE

A hierarchy of density dependent fields implies that additional gravitational forces are induced by gradients or changes in the fields about an object. In the Chameleon theory, these additional field forces F_ϕ on a small object of mass m and density ρ_m placed near a larger object of mass M, radius R, and density ρ_M, with factors of c added to correct the units is given with respect to the field gradient $\nabla\phi$ and the radial distance r as

$$F_{\phi} = m\left(-\beta_m \frac{c^2}{M_{PL}} \nabla\phi(r)_m\right). \tag{27}$$

This is in contrast to Newtonian physics, where the dominate gravitational forces are given by

$$F_g = m\left(\frac{GM}{R^2}\right). \tag{28}$$

By ignoring the spacial derivation of the density about the small object and taking a linearized prospective, where the small object must sample the field above and below, the field about the small object can be estimated from the outward (OW) side field $\frac{1}{2}\phi_m$ given by the object itself plus the field $\frac{1}{2}(\phi_M - \phi_m)$ between the two objects (or inward (IW) of the small object). That is, the linearized field change about the small object can be estimate as

$$\Delta\phi_m \sim \left(\frac{1}{2}\phi_m\right)_{OW} + \left(\frac{1}{2}\phi_M - \frac{1}{2}\phi_m\right)_{IW} = \frac{1}{2}\phi_M, \tag{29}$$

such that, the field gradient $\nabla\phi$ about the small object with respect to the radial distance r is

$$\nabla\phi(r)_m \sim \frac{1}{2}\nabla\phi(R)_M. \tag{30}$$

Then using the form of the density dependent field from equation (26)

$$\nabla\phi(R)_M = \nabla\left(\phi_c\left(\frac{\beta_M}{\beta_c}\frac{\rho_M}{\rho_c}\right)^{1/3}\right);$$
$$\nabla\phi(R)_M = \phi_c\left(\frac{3}{4\pi}\frac{\beta_M}{\beta_c}\frac{M}{\rho_c}\right)^{1/3}\nabla\left(R^{-1}\right); \tag{31}$$
$$\nabla\phi(R)_M = -2\frac{\phi_c}{R^2}\left(\frac{3}{4\pi}\frac{\beta_M}{\beta_c}\frac{M}{\rho_c}\right)^{1/3},$$

which when combined with equations (27) and (28) yields the field force F_ϕ in terms of the normal Newtonian force F_g as

$$F_\phi = \frac{m}{R^2}\left(\frac{2\beta_m\phi_c c^2}{M_{PL}}\left(\frac{3}{4\pi}\frac{\beta_M}{\beta_c}\frac{M}{\rho_c}\right)^{1/3}\right) = 2\left(\frac{c^2}{G}\right)\left(\frac{\beta_m\phi_c}{M_{PL}}\right)\left(\frac{3}{4\pi}\frac{\beta_M}{\beta_c}\frac{1}{M^2\rho_c}\right)^{1/3} F_g. \tag{32}$$

5.1. Fifth Force Analogy

Near the earth $\beta_m \sim \beta_M \sim 1$, whereby from equation (24) with $\rho_m = 5520\, kg/m^3$, the earth's dark matter coupling constant $\beta'_M \sim 10^{-61}$ indicating a very weak coupling to dark matter particulate mass near the earth. Such a very small dark matter coupling near the earth does not mean that there is a weak coupling to the matter density field near the earth, only that coupling to local matter

$$\leq \left(\frac{m_{DM}^3}{\beta'_M}\right)^{1/3} \sim 6.782x10^{-39}\, kg \tag{33}$$

becomes insignificant (*i.e.*, the order of quantum energies, such that quantum effects are suppressed in the earth's local field).

On the other hand, coupling to local matter in the far field

$$\leq \left(\frac{m_{DM}^3}{\beta_c'}\right)^{1/3} \sim m_{DM} \sim 3.270x10^{-59}\,kg \qquad \textbf{(34)}$$

becomes insignificant (*i.e.*, quantum effects and smaller dominate the far field in the absent of normal matter).

That is, one should suspect matter coupling in experiments outside the earth's matter density field to be somewhat different as β_m' may change (noting G should remain somewhat constant under normal conditions). A similar theme was noted by Khoury and Weltman in their paper Chameleon Fields: Awaiting Surprises for Tests of Gravity in Space [5].

On the other hand, since the Gravitational constant is known to slightly vary in experimentation [18], variations in the dark matter coupling may come into play even on the earth. This can be visualized by combining equations (21) and (24) to yield

$$G = \left(\frac{\beta_m'}{m_{DM}^3}\right)\left(\frac{\beta_m^2 \rho_m^2}{\rho_c^2}\right)\frac{T_{PL}^2}{R}, \qquad \textbf{(35)}$$

where by holding everything but the dark matter coupling constant β_m'-constant leads to changes in the Gravitational constant. That is, it may be possible even for experiments on the earth to change the coupling to local dark matter.

For example in fifth force searches, where the force on an object is given by $F = (1+\theta)F_g$, the fifth force factor θ for the Chameleon theory is then given from equation (32) as

$$\theta = 2\left(\frac{c^2}{G}\right)\left(\frac{\beta_m \phi_c}{M_{PL}}\right)\left(\frac{3}{4\pi}\frac{\beta_M}{\beta_c}\frac{1}{M^2 \rho_c}\right)^{1/3}; \qquad \textbf{(36)}$$

noting that the fifth force factor θ is subject to changes (under yet undefined conditions) through the dark matter coupling constant β_m' in the gravitational constant G per equation (35).

The question is "How can the dark matter coupling constant β_m' be changed in the Earth environment?" The answer to this may be found in comparing the formation of dark matter rings outside the confines of a galaxy to current laboratory scale data of anomalous reading from superconductor experiments. These experiments are discussed in the following.

6. ROTATING SUPERCONDUCTOR/LOW TEMPERATURE EXPERIMENTS

In superconductor theory, a scalar field is known to exist $\sim 10^{-6}$ m about a superconductor. This phenomenon is associated with proximity effects as the Josephson Effect. Could the superconductor dark matter coupling factor β_{sc}', of this field be extended to produce disturbance in the local field?

In recent rotating superconductor experiments by Tajmar *et al.* [19-22], a change in G up to 10^{20} is needed to explain the results. Here the field disturbance is explained in terms of a decoupling from the earth's dark matter coupling factor β_M' to be replaced with the superconductor dark matter coupling factor β_{sc}', which does so as the superconductor rotates.

To start, using equation (35), let

$$G^* \approx \left(\frac{\beta'_{sc}}{m_{DM}^3}\right)\left(\frac{\beta_M^2 \rho_M^2}{\rho_c^2}\right)\frac{T_{PL}^2}{R_{sc}} \approx \left(\frac{\beta'_{sc}}{\beta'_M}\right) G \sim 10^{20} G, \quad (37)$$

where the superconductor hierarchy coupling factor β'_{sc} replaces the earth's hierarchy coupling factor β'_M, such that,

$$\beta'_{sc} = \left[\frac{\beta_c^2 \rho_c^2}{\beta_{sc}^2 \rho_{sc}^2}\right] \sim 10^{20} \beta'_M . \quad (38)$$

To give a new cut off mass

$$\leq \left(\frac{m_{DM}^3}{\beta'_{sc}}\right)^{1/3} \approx \left(\frac{m_{DM}^3}{10^{20}\beta'_M}\right)^{1/3} = \left(\frac{m_{DM}^3}{10^{-41}}\right)^{1/3} \approx 10^{-45} kg . \quad (39)$$

The cutoff mass in equation (39) infers that with respect to motion, the test instruments (accelerometers or gyroscopes) feel a 10^{-16} difference in the cutoff mass. For example, in the direction of motion the cutoff mass is as noted in equation (24), where in the opposite direction it raises to that in equation (39). That is, the differences in the cutoff masses produce a gradient $\nabla\phi$ in the local Chameleon field about the superconductor (assumed to be associated with the direction of rotation), subjecting the test instrument mass m_{ti} to a force per equation (27) of the form

$$F_{\phi sc} \approx m_{ti}\left(-\beta_{sc}\frac{c^2}{M_{PL}}\nabla\phi(r)_{sc}\right). \quad (40)$$

It must be noted that in follow-on experiments, Tajmar *et al.* [20-22] noted that the effects on the gyros appeared to be more connected with the liquid helium than to the superconductor test specimen, which suggest that condensed matter (*i.e.*, cold helium) or possibly more coherent atomic or sub-atomic matter (*i.e.*, electron pairing or other atomic condensing phenomena) produces an external field differing from matter tested at normal temperatures.

However one should suspect this effect to be weak at liquid helium temperature. Whereby, it is enhance by rotation, *i.e.*, inducing superconductor like behavior. Such as, the formation London moments-a quantum-mechanical phenomenon whereby a spinning superconductor generates a magnetic field whose axis lines up exactly with the spin axis. Whereby, the formation of London moments in the liquid helium induces superconductor behavior in the liquid helium.

6.1. Time Dilation

One can assert that the coupling between the Cooper pairs in the superconductor and the external field changes during the phase transition, which follows conformal field theory (CFT), when the source is moving (*i.e.*, rotating superconductor), the behavior of forces associated with field gradients changes in a non-intuitive way. For example, a test mass must sample a scalar field at various points in the vicinity, to figure out which way to accelerate. The scalar field can be different at various points both because of geometry, and because the source for the field is not at the same position for different observation points. This interpretation suggest time retardation, where in textbooks the time retardation is expressed using delta functions, or retarded Green's functions; inferring that the field gradient $\nabla\phi(r)_{sc}$ about a rotating superconductor can be complex. Further, one needs to account for local environmental effects (*e.g.*, Coriolis force) as well as the superconductor rotational direction.

A simple way to handle this is to let the superconductor's field gradient $\nabla\phi(r)_{sc} = \nabla\phi(R)_M$ (*i.e.*, that of the earth; this should be intuitive as the superconductor is submersed in the earth's density field). Doing so then requires the definition of a time dilation β^*_{sc} of the superconductor's mass coupling factor, where using equation (40) the time dilation force is simply defined as

$$F^*_{\phi sc} = m_{ti}\left(-\beta^*_{sc}\frac{c^2}{M_{PL}}\nabla\phi(R)_M\right) \quad \textbf{(41)}$$

with $\nabla\phi(R)_M$ given by equation (31) and where for the Tajmar *et al.* [19-22] experiments $\beta^*_{sc} = \beta_{sc}$ as defined by equation (38). That is, the rotating superconductor changes the dark matter coupling of the test instrument mass m_{ti} in the local environment established by the rotating superconductor.

It is noted that the coupling constants β_M, β_{sc} are functions of the densities ρ_M, ρ_{sc}, where the dark matter coupling β'_{sc} about the superconductor is a sole function in the time dilation density ρ^*_{sc} of the superconductor-which in this case is due to superconductor effects under rotation. Noting that the time dilation ρ^*_{sc} of the superconductor's density is not a true density change, but includes the change in the dark matter field density in the superconductor, where under normal conditions, the dark matter field density of an object is equivalent to the mass density of the object.

6.2. Discussion

In the general sense, the Tarmar *et al.* [19-22] experiments indicate that quantum systems, like superconductors, under rotation (or possible just in motion) provides a method to change the gravitational constant; measurable in laboratory experiments; noting that further experiments are needed to validate this claim.

The author also suggests that the rotation of a large superconductor at some speed (yet undeterminable) should produce an optical effect about its rotation perimeter in like to that found with the observations of galactic dark matter rings (discussed further in the next section). Further, there should also be some gyroscope effect differing from normal matter – possibly making the rotating superconductor to appear either smaller or larger in comparison. To pull a rabbit – should a rotating superconductor appear to weigh less, this could account for some of the missing mass in the Universe as the Universe has many superconductor qualities.

7. GENERALIZED INTERPRETATION TO GALAXY DARK MATTER RINGS

The Tajmar *et al.* [19-22] experiments provide for a generalized interpretation of the formation of dark matter rings outside the confines of a galaxy, as follows:

Like the rotating superconductor/helium, the rotating galaxy density field (containing both matter and dark matter) provides a medium (containing quantum systems in motion) to induce a time dilation force on the external density field (*e.g.*, cosmic dust m_{cd} plus dark matter) about a galaxy. This force would then have the same form as equation (41), *i.e.*,

$$F^*_{\phi galaxy} = m_{cd}\left(-\beta^*_{galaxy}\frac{c^2}{M_{PL}}\nabla\phi(R)_{galaxy}\right). \quad \textbf{(42)}$$

Whereby, the time dilation of the rotational force exerted on the external density field about a galaxy shows up as dark matter rings outside the confines of a galaxy. In a sense, the dark matter rings are analogous to the anomalous rotation seen by the gyros in the Tajmar *et al.* [19-22] experiments.

7.1. Other Phenomena in the Universe

The author notes that although equations (40-42) are generic forms of a more complex force equation involving mass rotation, it could be applied to other phenomena in the Universe where rotating masses exists. As an example, these equations also suggest the opposite effect. That is, given rotating gases or matter cloud, one should expect a high density field to be formed at the center, *i.e.*, black holes or jets due to the compression of the central density field.

8. MOTION, TIME DILATION AND THE COUPLING CONSTANT

The author [23-25] has shown that the coupling factor β_m is given as

$$\beta_m \approx \left(1/3\left(\frac{M_E^2}{\hat{\beta}_{C_m}\bar{\rho}_m\bar{R}_m\sqrt{l_p\bar{R}_m}}\right)\left(\frac{2M_{PL}^4}{\bar{\rho}_0}\right)^{1/3}\right)^{1/2}, \tag{43}$$

where $\hat{\beta}_{C_m}$ is the Chameleon field motion coupling factor, $\bar{\rho}_m$ is the Chameleon field density of an object, $\bar{R}_m$ is the Chameleon field radius of an object, $\bar{\rho}_0$ is the surrounding or ambient Chameleon field density.

8.1. Chameleon Field density

The following presents some general information for an object of mass m, specific to the change in the object's Chameleon Field density. There are many Chameleon Field density conditions and each condition takes some thought or the appropriate condition will not be surmised; given invalid results.

Specifically, when calculating the force on a object, effects on the ambient Chameleon fields must also be taken into consideration. For further discussion see [23-25].

8.1.1. Stationary Object

When $\hat{\beta}_{C_m} = 1$ (*i.e.*, stationary Chameleon field densities), $\bar{\rho}_m = \rho_m = 3m/4\pi R_m^3$ the normal density of the object, $\bar{R}_m = R_m$ the normal radius of the object and $\bar{\rho}_0 = \rho_0$ the normal surrounding or ambient density. For example, in Khoury and Weltman all the coupling factors are set to one. Therefore, equation (43) should give a value of $\beta_{earth} = 1$ for the earth. As a check, for the earth, with a density $\rho_m \approx 5520\ kg/m^3$, $R_m = 6.37\times10^6\ m$ and atmosphere density $\rho_{0_{atm}} \approx 1.2\ kg/m^3$ with $\hat{\beta}_{C_m} = 1$, equation (43) yield $\beta_m \approx 0.98$, which is ≈ 1. It is noted that the earth is in motion where one would think $\hat{\beta}_{C_m} \neq 1$. Here however, the coupling under consideration here are to objects near the earth that are moving with the earth. On the other hand, the slight difference, *i.e.*, $\beta_m \approx 0.98$, maybe due to the actual density of the earth being masked by its motion or particulate motion internal to it.

8.1.2. Object in Motion

The Chameleon field density of the object in motion is given as

$$\bar{\rho}_m \approx \rho_m + \left(\frac{F_m}{F_{N_m}}\right)\rho_m \approx \frac{3m}{4\pi\bar{R}_m^3}, \tag{44}$$

where $F_m = ma_m$ with a_m being the total acceleration from the stationary position and $F_{N_m} = mg_N$ with g_N being the Newtonian gravity constant of the nearest large mass, such that for $a_m \gg g_N$;

$$\bar{R}_m^3 \approx \frac{g_N}{g_N + a_m}R^3 \to \frac{g_N}{a_m}R^3 \tag{45}$$

and

$$\bar{\rho}_m \to \left(\frac{a_m}{g_N}\right)\rho_m. \tag{46}$$

8.1.2.1. Motion coupling factor

It is noted that for objects in motion, the determination of the motion coupling factor $\hat{\beta}_{C_m}$ gets messy as the coupling factors β_m must be experimentally determined. However, in most Newtonian cases $(F = ma)$, the Chameleon Field force $(F_\phi << ma)$ is insignificant. Whereby, the Chameleon Field Theory in general is insignificant for calculating additions to Newtonian forces for objects having no Chameleon density change due to internal particulate motion.

8.1.3. Exhausted Mass

The Chameleon field density of the object with internal particulates being exhausted (for example a rocket; see [25]) is given as

$$\bar{\rho}_m \approx \partial\rho_m + \left(\frac{F_m}{F_{N_m}}\right)\rho_m \approx \frac{3\partial m}{4\pi \bar{R}_m^3}, \tag{47}$$

such that

$$\bar{R}_m^3 \approx \left(\frac{F_{N_m}}{F_{N_m} + \left(\frac{m}{\partial m}\right)F_m}\right)R_m^3 = \left(\frac{g_N}{g_N + \left(\frac{m}{\partial m}\right)a_m}\right)R_m^3 \to \left(\frac{g_N}{a_m}\right)\left(\frac{\partial m}{m}\right)R_m^3, \tag{48}$$

where combining equations(47) and (48) yields equation (46).

Note: When calculating the force on the object exhausting mass, the local Chameleon field density change produced by the exhausted mass and effects on the ambient Chameleon fields must also be taken into consideration. This is not as straight forward as it seems here (*e.g.*, see the rocket example; [25]).

8.1.3.1. Time dilation

The exhausted mass system is viewed as a single body system with phased fifth force coefficients about the object due to time dilation and retardation associated with the exhausted mass motion. This implies a phase factor φ,

$$\varphi = \frac{\tau}{\tau + \Delta\tau} << 1; \tag{49}$$

where the changing mass coupling factors can be given simply by

$$\bar{\beta}_m \approx \bar{\beta}_{exhaust} \approx -\frac{1}{6\varphi}\left|\frac{F_m}{F_{N_m}}\right| = -\frac{1}{6\varphi}\left(\frac{a_m}{g_N}\right). \tag{50}$$

The exhaust mass relaxation time

$$\tau \approx \frac{\bar{R}_{exhaust}}{v_{exhaust}}; \tag{51}$$

the ratio of the effective exhaust mass radius $\bar{R}_{exhaust}$ and the exhaust mass velocity $v_{exhaust}$, and the retardation time

$$\Delta\tau \approx \frac{m_{exhaust}}{\dot{m}} > \tau, \tag{52}$$

where $\dot{m}$ is the exhausted mass flow rate t and the exhausted mass

$$m_{exhaust} \approx m - \partial m \; ; \tag{53}$$

m is the initial object mass and $\partial m \approx m - m_{exhaust}$ is the final object mass.

8.1.3.2. Motion coupling factor

The motion coupling factor for the exhausted mass case is then calculated by setting equations (43) and (50) equivalent and using equations (46) and (48).

8.1.4. Internal Particulate Mass Motion

The Chameleon field density of an object with internal particulates being accelerated but not exhausted is given as

$$\bar{\rho}_m \approx \rho_m + \left(\frac{F_i}{F_{N_i}}\right)\rho_i \approx \frac{3m}{4\pi \bar{R}_m^3}, \tag{54}$$

such that,

$$\bar{R}_m^3 \approx \left(\frac{mg_N}{mg_N + m_i a_i}\right) R_m^3 \rightarrow \left(\frac{m}{m_i}\right)\left(\frac{g_N}{a_i}\right) R_m^3 = \left(\frac{F_N}{F_i}\right) R_m^3, \tag{55}$$

where combining equations(54) and (55) yields

$$\bar{\rho}_m \approx \left(\frac{a_i}{g_N}\right)\rho_i, \tag{56}$$

which is similar to equation (46) with only the acceleration of the internal particulates taken into consideration.

Note: Here the condition of the internal acceleration $a_i >> g_N$ is always meet as the particulate speed $v \rightarrow c$ must be met, *i.e.*, the particulates must be at or below the atomic scale.

Also of note, as the object accelerates, the Chameleon density changes accordingly, that is, equation (56) changes to

$$\bar{\rho}_m \approx \left(\frac{a_i}{g_N}\right)\rho_i + \left(\frac{a_m}{g_N}\right)\rho_m = \left(\left(\frac{m_i}{m}\right)a_i + a_m\right)\frac{\rho_m}{g_N}, \tag{57}$$

and equation (54) to

$$\bar{R}_m^3 \approx \frac{g_N}{\left(\frac{m_i}{m}\right)a_i + a_m} R_m^3. \tag{58}$$

8.1.4.1. Time dilation

As with the exhausted mass system, this system is also viewed as a single body system with phased fifth force coefficients about the object due to time dilation and retardation associated with the exhausted mass motion. This implies a phase factor φ,

$$\varphi = \frac{\tau}{\tau + \Delta\tau} << 1 \, ; \tag{59}$$

where the changing mass coupling factors can be given simply by

$$\bar{\beta}_m \approx \bar{\beta}_i \approx -\frac{1}{6\varphi}\left|\frac{F_i}{F_{N_i}}\right| = -\frac{1}{6\varphi}\left(\frac{a_i}{g_N}\right). \tag{60}$$

The particulate mass relaxation time

$$\tau \approx \frac{\bar{R}_m}{v_i - v_r} \, ; \tag{61}$$

the ratio of the effective object radius $\bar{R}_m$ and the particulate mass velocity v_i minus the relaxation velocity v_r, and the retardation time

$$\Delta\tau \approx \frac{m_i}{\dot{m}_i - \dot{m}_r} > \tau \, , \tag{62}$$

where $\dot{m}_r$ is the particulate mass flow rate and $\dot{m}_r$ is the relaxation mass flow rate.

Note: If the relaxation is randomized $v_r \approx 0; \ \dot{m}_r \approx 0$.

8.1.4.2. Motion coupling factor

The motion coupling factor comes is given by equating equations (43) and (60) to yield

$$\hat{\beta}_{C_m} \approx -12\varphi^2 \left(\frac{g_N}{a_i}\right)^3 \left(\frac{M_E^2}{\rho_i \bar{R}_m \sqrt{l_p \bar{R}_m}}\right)\left(\frac{2M_{PL}^4}{\rho_0}\right)^{1/3}, \tag{63}$$

where each variable is reasonably known or can be approximately calculated.

8.1.4.3. Quantum momentum transfer

In the internal time dilation case, the Chameleon Field force $F_\phi \to ma$ and represent a quantum theory of momentum transfer, whereby from equations (41) and (60)

$$F_\phi = m\left(\frac{a_i}{g_N}\right)\left(\frac{1}{6\varphi}\frac{c^2}{M_{PL}}\nabla\phi(r)_m\right) \to ma_m \tag{64}$$

to give the object acceleration

$$a_m \to \left(\frac{1}{6\varphi}\right)\left(\frac{a_i}{g_N}\right)\left(\frac{c^2}{M_{PL}}\nabla\phi(r)_m\right). \tag{65}$$

8.2. Tajmar Experiment

Equation (64) can then be used to determent the time dilation force $F_{\phi_{sc}}$ or quantum momentum transfer to the Chameleon field about the rotating cooled mass or superconductor in the Tajmar experiments through the Chameleon field acceleration

$$a_{\phi_{sc}} \to \frac{1}{6}\left(\frac{a_i}{\varphi_{sc}}\right)\left(\frac{c^2}{g_N M_{PL}}\nabla\phi(R)_M\right), \tag{66}$$

where $\nabla\phi(R)_M$ is given by equation (31) with the exception that the earth coupling factor may be effected by the rotating mass as was noted in the rocket example [25], *which would need to be accounted for as a two body problem, i.e.,* $\left(F_{\phi_m} \pm F_{\phi_{sc}}\right)$ to include acceleration effects on the external Chameleon field. The determination of the particulate acceleration a_i in the rotating cooled mass and the phase factor φ_{sc} are left for future analysis.

9. GENERAL DISCUSSION

In this paper, the Khoury and Weltman Chameleon (scalar) field model [4, 5], which describes causes for an additional force to gravity, is expanded to explain forces at a microscopic level, where objects sense gradients in the Chameleon field produced by other objects. The picture differs from classical field theory, in which distant sources produce varying potential fields.

Since these (Chameleon) fields occupy the same vacuum, gravitation, scalar and other forces all have their origins in some basic structure of the universe; providing similarities down to the laboratory scale. To some degree, the Chameleon model basically involves higher dimensions related to the 11 dimensions in the early universe. This is apparent in the original Khoury and Weltman Chameleon (scalar) field model [4, 5] were the moduli field is remnants of dimensions 5-11, which curled up a few attoseconds after the Big Bang. Whereby, parameters from various structures, such as vortices in the Andromeda galaxy, provide a bridge for connecting other parameters such as the Newton gravitational constant, Planck's constant, Hubble constant and *etc.* for studying new astrological phenomena in the laboratory.

Such studies can be accomplished as the process by which field gradients are sensed by nearby objects allows for local coupling constants β_m that vary with the local field ϕ_m through its local Chameleon density $\bar{\rho}_m$. These local coupling constants being rooted in a hierarchical coupling constant β'_m that provides linkage to the Universe far-field ϕ_c and density $\bar{\rho}_c$; equation (24). With the experimental rules for determining the coupling parameter(s) to include the possibility that the local hierarchical coupling constant $\beta'_m << 1$ can approach the far-field hierarchical coupling constant $\beta'_c = 1$ to allow 20 orders of magnitude enhancement of the gravitational constant G about super-fluids, *i.e.*, superconductors; equation (37).

10. FINAL REMARKS

Connections to physical constants evolving new physics as dark matter and the derivation of a density dependent hierarchy of scalar fields in the universe provides a starting point for extension from cosmology and astrophysics to other physical fields as particle and solid state physics within the earth's density field.

In conclusion, understanding processes that can change the dark matter coupling could account for experimental anomalies and lead to new force producing devices outside known physics while being based on old or new astrological observations.

11. DISCLOSURE

Part of the information included in this chapter has been previously published by the author in a preprint available on the author's organization website at the address

http://ias-spes.org/IASSPES_New/Papers/HDDSF.pdf

12. ACKNOWLEDGEMENT

The author would like to thank retired Penn State particle physics Dr. Raymond Lewis who introduced the author to the Chameleon theory, has greatly helped in its interpretation, and for his helpful comments and suggestions to this paper.

REFERENCES

[1] DeWitt, BS. Superconductors and gravitational drag. Phys Rev Lett 1966; 16: 1092–1093.

[2] Cartwright J. Dark matter could come naturally from quantum gravity. http://physicsworld.com/cws/article/news/38765.

[3] Hauser J, Droescher W. Extended Heim Theory (EHT): Emerging Physics for Gravity-Like Fields. In: Modanese G, Robertson GA, Eds. Gravity-Superconductors Interaction: Theory and Experiment. Bentham 2012; Chapter 11.

[4] Khoury J, Weltman A. Chameleon cosmology. Phys Rev D 2004; 69: 044026.

[5] Khoury J, Weltman A. Chameleon Fields: Awaiting Surprises for Tests of Gravity in Space. Phys Rev Lett 2004; 93: 171104.

[6] Silverman MP, Mallett RL. Coherent degenerate dark matter: a galactic superfluid? Class. Quantum Grav 2001; 18: L103-8.

[7] BBC. Astronomers discover 'dark matter ring'. http://www.abc.net.au/news/newsitems/200705/s1924778.htm.

[8] Brax P, van de Bruck C, Davis A-C, Khoury J, Weltman A. Detecting dark energy in orbit: The cosmological chameleon. Phys Rev D 2004; 70: 123518.

[9] Will CM. Theory and Experiment in Gravitational Physics. 2nd ed. New York: Basic Books/Perseus Group 1993.

[10] Will CM. The confrontation between general relativity and experiment. Living Rev. Relativity 2001; 4: 4.

[11] Fischbach E, Talmadge C. The Search for Non-Newtonian Gravity. New York: Springer-Verlag 1999.

[12] Damour T, Polyakov AM. The string dilaton and a least action principle. Nucl Phys B 1994; 423: 532.

[13] Damour T, Polyakov AM. String Theory and Gravity. Gen Relativ Gravit 1994; 26: 1171.

[14] Anderson GW, Carroll SM. Dark Matter with Time-Dependent Mass. astro-ph/9711288 1997.

[15] Huey G, Steinhardt PJ, Ovrut BA, Waldram D. A Cosmological mechanism for stabilizing moduli. Phys Lett B 2000; 476: 379.

[16] Hill CT, Ross GC. Models and new phenomenological implications of a class of pseudo-Goldstone bosons. Nucl Phys 1988; B311: 253.

[17] Ellis J, Kalara S, Olive KA, Wetterich C. Density-dependent couplings and astrophysical bounds on light scalar particles. Phys Lett B 1989; 228: 264.

[18] Gillies GT. The Newtonian gravitational constant: recent measurements and related studies. Rep Prog Phys 1997; 60: 151-225.

[19] Tajmar M, Plesescu F, Seifert B, Marhold K. Measurement of Gravitomagnetic and Acceleration Fields around Rotating Superconductors. In: El-Genk MS, Ed. Proceedings of the Space Technology and Applications International Forum; 2007: AIP Conf Proc 880; pp. 1071-1082.

[20] Tajmar M, Plesescu F, Seifert B, Schnitzer R, Vasiljevich I. Investigation of Frame-Dragging-Like Signals from Spinning Superconductors using Laser Gyroscopes. In: El-Genk MS, Ed. Proceedings of the Space Technology and Applications International Forum; 2008: AIP Conf Proc 969, p. 1080-1090.

[21] Tajmar M, Plesescu F, Seifert B. Anomalous fiber optic gyroscope signals observed above spinning rings at low temperature. J Phys.: Conf Ser 2009; 150: 032101.

[22] Tajmar M, Plesescu F. Fiber-Optic-Gyroscope Measurements Close to Rotating Liquid Helium. In: Robertson GA, Ed. Proceedings of the Space, Propulsion & Energy Sciences International Forum; 2010: AIP Conf Proc 1208, pp. 220- 226.

[23] Robertson GA. Engineering Dynamics of a Scalar Universe Part I: Theory & Static Density Models. In: Robertson GA, Ed. Proceedings of the Space, Propulsion & Energy Sciences International Forum; 2009: AIP Conf Proc 1103, pp. 687-705.

[24] Robertson GA. Engineering Dynamics of a Scalar Universe Part II: Time-Varying Density Model & Propulsion. In: Robertson GA, Ed. Proceedings of the Space, Propulsion & Energy Sciences International Forum; 2009: AIP Conf Proc 1103, pp. 706-721.

[25] Robertson GA. The Chameleon Solid Rocket Propulsion Model. In: Robertson GA, Ed. Proceedings of the Space, Propulsion & Energy Sciences International Forum; 2010: AIP Conf Proc 1208, pp. 207-219.

CHAPTER 13

Impact of Lifetime and Decay Rates of Thermally Excited States in Superconductors on a Gravity Experiment

Harald Reiss*

Department of Physics, University of Wuerzburg, Am Hubland, D-97074 Wuerzburg, Germany

Abstract: A numerical analysis is presented of a gravity/superconductor interaction experiment in a suspension balance recently reported in the literature. A central question of the analysis is whether thermal design of the experiment fully exploited the high resolution provided by the balance. For this purpose, we calculate sample temperature evolution, T(x,y,z,t), and the number $N_{SC}(t)$ of volume elements that have completed phase change, under strongly differing heat transfer conditions (radiation, convection and boiling). Experimental equilibrium generation or decay rates, $G_{exp}(t)$, of electron pairs initiated by propagation of a thermal perturbation, the *phonon* aspect, are estimated from $dN_{SC}(t)/dt$ using the electron pair equilibrium density, $\rho_{EP}(T)$, of this material. The counterpart to the phonon aspect, *e.g.*, non-equilibrium decay rates, $G_{Exc}(t)$, the response of the *electron* system describing the decay of the "product" of the same thermal disturbance, is analyzed using (i) a diffusion model of their propagation in the solid, and (ii) a sequential model with lifetimes estimated from the uncertainty principle and in analogy to the nucleon-nucleon, pion-mediated Yukawa interaction. Items (i) and (ii) describe decay of the disturbance in space and time, respectively. A discrepancy appears between $G_{exp}(t)$ and $G_{Exc}(t)$ if the sample is warmed-up solely by radiation, like in the suspension balance experiment. The discrepancy creates dead time intervals large enough to spread out weight *vs.* time signals that under other experimental conditions (*e.g.*, using large Biot numbers) could be sharply identified, provided a correlation between gravity and superconductivity really exists.

Keywords: Gravitational forces, superconductors, high-Tc superconductors, low-Tc superconductors, Type-II superconductors, normal/superconductor phase transition, thermal perturbations in superconductors, yukawa-like interaction, Heat transfer in superconductors, equilibrium decay rates, non-equilibrium decay rates, phonons and electrons in superconductors, gravity-like fields.

1. INTRODUCTION

Normal/superconductor phase transition during warm-up or cool-down periods is usually considered to occur at exactly the instant when solid temperature, T(t), coincides with critical temperature, T_{Crit}, at a time, t = t'. In almost all experiments performed to observe phase transition, solid temperature is detected from interaction, by solid/solid contact, of a thermometer with lattice vibrations. But it is not clear that at t = t', a previously normal conducting electron system of a superconductor sample has already completed re-structuring to a mixture of normal conducting and superconducting components, in a two-fluid model.

At very low temperature, the superconducting electron system is decoupled from propagation of thermal waves but reflects its own dynamic response to this or other specific excitations by corresponding relaxation times, τ_{El}. Thermal diffusivity, on the other hand, determines a characteristic time, τ_{Ph}, for propagation of thermal (phonon) waves in solids after a thermal perturbation. Both relaxation times, τ_{El} and τ_{Ph}, after the same disturbance are not necessarily identical; the lattice, if excited, behaves quite differently from the electron system.

A similar situation (two or more different relaxation times) arises in multi-filamentary superconductors, again after a thermal perturbation: Time constant, τ_B, for propagation of magnetic flux density, B, in the superconducting filaments is relatively small while thermal relaxation time, τ_{Ph}, is much larger, by orders of magnitude, and the inverse of this relationship in the matrix material is of enormous importance for

***Address correspondence to Harald Reiss:** Department of Physics, University of Wuerzburg, Am Hubland, D-97074 Wuerzburg, Germany; E-mail: harald.reiss@physik.uni-wuerzburg.de

Giovanni Modanese and Glen A. Robertson (Eds)

obtaining stability against quench in multi-filament superconductors for high field applications. For example, a thermal perturbation emitted from a filament 1 arrives after a time t^{th}_{12} at its immediate neighbour, filament 2. Because of the good conduction properties of the matrix material, development of a localised hot spot in filament 1 is avoided. Magnetic flux density, after a corresponding flux jump in filament 1, needs the time $t^{B}_{12} >> t^{th}_{12}$ to travel the same distance. During the interval $\Delta t = t^{B}_{12}-t^{th}_{12}$, critical current density J_{Crit} in filament 2 will respond, and partly be reduced, solely due to its dependency on temperature, but a total break-down of J_{Crit} in this filament due to full penetration of the magnetic field will not be observed. The time interval, Δt, during which the filament is blind for variations of J_{Crit} with B accordingly can be interpreted as a dead time interval. Numerous examples for dead time intervals exist. They are recognized by experimenters when physical or chemical systems are embedded in a medium with strongly different transport properties for different observables (thermal energy, magnetic field, current) or if there are different decay channels of the excitation, in other words, if the corresponding relaxation times are of strongly different values.

In this paper, we will investigate a possible mismatch between response of lattice and electron systems in terms of lifetime of disturbances. As a special case, thermal perturbations arising during cool-down and warm-up periods and their decay will be discussed as deviations from dynamical equilibrium of the electron system. The results will be used to tentatively explain the reason why clearly observable signals could not be found in a gravity/superconductor interaction experiment reported in the literature that was performed under radiative warm-up of samples in a magnetic suspension balance. A numerical, thermal analysis of this experiment has been performed. First results were reported in the contribution by G. Hathaway and H. Reiss to the present volume. The present paper explains details of the analysis.

The paper is organised as follows: In Section 1.1, a provisional comparison is made between thermal relaxation time, τ_{Ph}, and its electronic counterpart, τ_{El}. Section 1.2 discusses decay of excited electron states initiated by a thermal perturbation. Section 1.3 serves for determination of experimental decay rates from cool-down and warm-up periods (the *phonon* aspect of the decay), and Sections 2 and 3 are concerned with decay in space and time of the *electronic* excitations resulting from the same thermal perturbations (the counterpart of the phonon aspect). The final Section 4 describes dead time intervals resulting from the difference between τ_{Ph} and τ_{El}, and their impact on interpretation of the above mentioned gravity/superconductor interaction experiment.

1.1. Overview: Electron Pairs, Quasi-Particles and Relaxation Times

Below critical temperature, single electrons in superconductors condense to electron pairs, as bound states at energies close to Fermi energy. At constant temperature, breaking of electron pairs and recombination can be described as a statistical process that establishes a dynamic equilibrium of the density of pairs and their decay products. Pair breaking and recombination proceed above a "Fermi sea" of single electrons that is unstable to even the weakest attractive particle/particle interactions.

Because of the weak binding energy (some meV, correlated with an energy gap, $2\Delta E$, in the single electron energy spectrum), electron pairs, besides statistical decay and recombination at constant temperature, are subject also to thermal excitations at any non-zero temperature. When $kT << 2\Delta E$ (k the Boltzmann constant), only few excitations will occur, and the electronic state is highly degenerate, otherwise the number of excitations may become large. After a sudden temperature increase, pair breaking will be observed before a new dynamic equilibrium is established at the increased temperature, with the new equilibrium obtained from recombination of decay products to electron pairs after a certain lifetime of the disturbed electron system.

Pair breaking and recombination proceed under conservation of energy and momentum; recombination is illustrated in Fig. **1a**. Pair breaking is possible only if the corresponding scattering process provides energy of at least the amount given by the width, $2\Delta E$, *e.g.*, by absorption of a phonon, $(h/2\pi)\ \omega \geq 2\Delta E$, from lattice waves. Besides absorption of phonons, there are other pair breaking mechanisms like absorption of photons, tunnelling between superconductors, and supercritical currents.

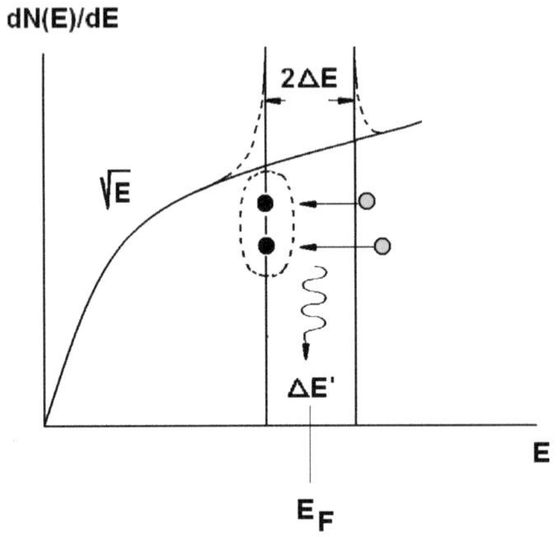

Figure 1a: Decay of excited electron states (recombination to electron pairs) in a superconductor below T_{Crit}, under conservation of energy, after end of a thermal perturbation (schematic, not to scale). The figure shows density, $\rho(E) = dN(E)/dE$, of single particles states (solid curve, proportional to $E^{1/2}$) *vs.* energy, E, the energy gap, $2\Delta E$, at the Fermi energy, E_F, and an escaping phonon of energy $\Delta E'$ of at least $2\Delta E$. The dashed curve around the full circles indicates an electron pair. Each shaded circle denotes an elementary electron excitation.

We shall investigate an aspect that may be important for interpretation of a variety of experiments with superconductors and that apparently has not been dealt with previously. This aspect is not focussed only on perhaps existing but still hypothetical and undiscovered gravity/superconductor interactions. It results from a mismatch between relaxation times of

1) Propagation of a thermal perturbation by conduction in a superconductor during cool-down or warm-up periods,

2) Response of the electron system of the superconductor to the same disturbance, at temperature below T_{Crit}, by recombination of excited electron states to pairs

As will be shown in this paper, mismatch could occur in particular near the phase transition, and if it occurs will have significant impacts on results also of experiments like levitation, measurement of observables like electronic part of specific heat or of thermal conductivity, or of onset or decay of persistent currents; such experiments could be performed to verify existence of dead time intervals.

We will in the following very generally speak of *elementary excitations* of the electron states if the superconductor is below critical temperature but not in the ground state, of *electrons* if single particle aspects in the superconductor are in the foreground (or if decay products from electron pairs are in normal conducting state), and of *quasi-particles* when collective electron aspects are considered.

[Note-Also (Bogoliubov) quasi-particles are elementary excitations in a superconductor. Properties of quasi-particles are similar to those of solid electrons if they reside near the Fermi level, which means they can be assigned mass and momentum. They consist of a moving electron together with a surrounding exchange correlation hole, which means when an electron fights its way through a solid, other electrons must move out from its way, because of the Pauli exclusion principle, and to minimize the repulsive Coulomb energy. If the interaction between electrons is weak, all the interactions that a single electron in a solid experiences can be renormalized into a self-energy that defines the effective mass of a new particle, the quasi-particle. The original *many*-body problem among interacting particles that cannot be solved is by renormalization reduced to a *one-body* problem of free particles (consult standard volumes on superconductivity)].

The special case of quasi-particles has to be taken into account when comparing the lifetimes of disturbances initiated in current injection experiments with lifetimes of thermal perturbations. Though decay of excitations in both cases (current injection, thermal perturbations) proceeds by completely different transport mechanisms, comparison between both kinds of disturbances and their lifetime is important: In current injection experiments, the decay of electron excitations is studied after an artificial increase of the number of quasi-particles. In the present case we have an increase of the number of excitations if sample temperature, T, increases (but $T < T_{Crit}$), with the total number of particles conserved. In current injection experiments reported *e.g.,* by Gray [1] or Epperlein *et al.* [2] in low temperature superconductors (LTSC; Al, Sn), thin films are used, with the injected electrical charge being collected in very thin surface layers, and data are taken at temperature far below T_{Crit}. Here we investigate solid samples (pellets of YBaCuO) with dimensions of millimetres and perform the analysis at temperatures very close to T_{Crit}.

For a provisional estimate of τ_{Ph} and τ_{El}, under these different experimental conditions, consider a thin film of d = 100 nm thickness onto the surface of which a very short thermal pulse (a "Dirac pulse" from a laser) is delivered. Sample thickness in current injection experiments is even smaller than this value of d, but then it is doubtful whether the following series expansion of rear sample surface temperature, T(t), based on solely thermal conduction transport would still be valid. Al coatings evaporated onto thin 6 μm Mylar foils below a thickness of 50 nm become transparent to radiation, and the question is whether it is possible to generate a laser pulse short enough not to interfere with the thermal wave that it initiates in the sample, and during its detection. Assuming the thermal diffusivity, D_{th}, of the film is constant (independent of temperature), a series expansion of T(t), see [3], Chapter 3, reads

$$T(t) = \sum b_n \sin[(n\,\pi\,x)/d]\,\exp(-D_{th}\,n^2\,\pi^2\,t/d^2) \quad \textbf{(1)}$$

with expansion coefficients, b_n, and the summation in Equation (1) over the index n taken from n = 1 to $n \to \infty$. If retaining only the n = 1 term, and using for the diffusivity D_{th} of the film a value in the order of 5 m^2/s, typical for very pure (99.999%) Al or Sn in their superconducting state, the thermal relaxation time, $\tau_{Ph} = d^2/(D_{th}\,\pi^2)$, would be several orders below 1 ns, and if film thickness is below 0.1 μm, as is the case in many injection experiments, thermal relaxation would be completed even after much shorter periods of time. In any case, this value of τ_{Ph} is definitely smaller than τ_{El}, the lifetime of disturbed electron states, on the order of tens of μs, as observed in a variety of reported current injection experiments using Al or Sn films of about the same thickness or other materials like W, or Ta under optical pulses; see below for citations.

Measurement of the quasi-particle recombination time (lifetime of the disturbed electron system) in superconducting Sn after injection of a current pulse was performed in [2] with current i_P = 10.1 mA and t_p = 1 μs the pulse duration. There were about 6.3 10^{10} unit charges injected into a (total) volume of at least 3 10^{-14} m^3, which means an average of the injected particle density of about 2.1 10^{24} $1/m^3$ (see [2], Fig. **3**).

A similar, provisional estimate can be made for high temperature superconductors (HTSC) like YBaCuO. During a thermal perturbation, the probability for exciting the electron system is proportional to $\exp[-\Delta E(T)/kT]$, under equilibrium conditions. Assuming that a thin film sample of the same thickness as before (0.1μm) was heated locally from 90.5 to 91.999 K (just below phase transition, T_{Crit} = 92 K), and using the electron density at temperatures close to absolute zero of about 6 10^{27} $1/m^3$ of which about (½) (1/10) is available for pair formation, with the same (total) volume in which current was injected in [2] (see above), and a standard (BSC) temperature dependence of $\Delta E(T)$, (see later, Equation 8b), this yields the number of thermally excited electron states (over the statistically generated number) of about 4.7 10^{26} $1/m^3$, by a factor M of about two orders of magnitude larger than in the current injection experiment (and still to be considered as a lower limit).

A naïve, order of magnitude estimate of the relaxation time τ_{El} in YBaCuO then can be made by assuming $\tau_{El,therm\ excit} \approx M\,\tau_{El,Current\ inject}$, with M the above constant and with $\tau_{El,Current\ inject}$ again on the order of 0.1 to 1 μs. This would result in a τ_{El} of at least 10 to 100 μs after a thermal perturbation in the HTSC, if temperature is not too close to T_{Crit}. The result is again much larger than the corresponding τ_{Ph}, here about 0.3 ns at T = 90 K using for YBaCuO the value of the diffusivity D_{th} = 3.4 10^{-6} m^2/s in Equation (1), if again a Dirac pulse is applied to the film surface.

Diffusion length of quasi-particles is comparable to diffusion length of electrons, in LTSC in the order of 10 to 100 µm at cryogenic temperature (*e.g.,* 12 µm in Al at 14 K, see Table **1** in [4]). Diffusion length in HTSC is smaller, thus we can safely assume it can be neglected against sample dimensions in the experiments to be described herein (YBaCuO pellets). There will be only little escape of phonons.

Accordingly, decay of excited electron states generated under thermal perturbation, not only after current injection, should be analysed. Like a variety of other excitations in superconductors (absorption of micro-waves, steady-state or pulsed tunnel diode experiments, absorption of incident phonons by the lattice), decay of thermal perturbations of the electronic system can be described as relaxation of strongly populated, non-equilibrium states.

1.2. Decay of Non-Equilibrium States

1.2.1. Dynamic Equilibrium

The superconducting state, according to Section 1.1, consists of two components that under (a) dynamic equilibrium or (b) non-equilibrium conditions, respectively, can be specified as follows:

(a) After a purely statistical fluctuation, under strictly thermal equilibrium:

A super-fluid of electron pairs, and a fluid of excited electron states, electrons or quasi-particles that exhibit some coherence behaviour both existing solely under statistical decay and recombination.

(b) After a temperature variation (a thermal perturbation) and before a new thermal equilibrium is established:

Again a super-fluid of electron pairs, and a fluid of excitations that now results solely from the temperature variation.

In both cases (a) and (b), the total number of solid particles is conserved.

The situation is different from ordinary evaporation/condensation of conventional fluids, for example, a liquid with well defined evaporation temperature and saturation vapour pressure (see Appendix A1): In superconductors and under a temperature variation well below critical temperature, contrary to evaporation/condensation in classical liquids, neither is there a measurable "background" (*e.g.,* idle spectators consisting of excited electron states populated from decaying pairs and de-populated by recombination, both initiated spontaneously), nor will it be possible to simultaneously measure a dynamic "foreground", with a net decay rate of pairs after the end of a temperature increase, or a net condensation rate after completion of a temperature decrease. Following a thermal perturbation, dynamic equilibrium state (a) exists only when the non-equilibrium state (b) has decayed completely until the next disturbance, and before this is accomplished, there is only one state, namely the predominant, non-equilibrium state (b) that can be observed in experiments.

Generally, summation over the inverse of average lifetimes, τ_i, of the ith excited electron state, at a given constant temperature ($T > 0$), with the summation index, i, taken over all $1 \leq i \leq N_{Exc}$ (with N_{Exc} the number of elementary electron excitations), describes the decay rate (negative of the generation rate, G_{Exc}) at time t

$$G_{Exc}[T(t)] = (1/V) \sum \{[1/\tau_{i,Exc}[T(t)]\} \quad \textbf{(2a)}$$

per unit volume, V; dimension of $G_{Exc}[T(t)]$ accordingly is $[1/(m^3\ s)]$.

Dynamic equilibrium between pairs and excitations exists, and is described by Equation (2a), if recombination of excited electron states to electron pairs on the statistical average compensates decay, and *vice versa.* Equation (2a) essentially is identical to Equation (3) in [1]; only a dependence on time of temperature, T(t), has been added. At constant temperature, the rate $G_{Exc}[T(t)]$, too, is constant.

Equation (2a) can be written also in terms of decay widths, Γ,

$$G_{Exc}[T(t)] = (1/V) \sum \{\Gamma_{i,Exc}[T(t)]\}/(h/2\pi) \qquad \textbf{(2b)}$$

of statistically or of thermally excited states, with the usual definition, $\Gamma_i = (h/2\pi)/\tau_i$, and the summation index $1 \leq i \leq N_{Exc}$ of the excited states. Use of the decay widths will be explained later.

1.2.2. Disturbance of the Equilibrium

Thermal equilibrium is displaced if, at a time t_0, temperature is suddenly increased to $T_1 = T + \Delta T$, still with $T_1 < T_{Crit}$. Assume that the temperature increase is initiated by local absorption of a short heat pulse. More electron pairs will decay (more than observed under dynamic equilibrium), and the positive generation rate, $G_{Exc}[T(t)]$, of excited electron states accordingly increases. After the end of the pulse, at time t_1, now the *disturbed* system decays, with negative generation rate $G_{El}[T(t)] = -G_{Exc}[T(t)]$, and after a time t_2 returns to a new dynamic equilibrium where we observe, at the new temperature, $T(t_2)$, new constant decay and generation rates of electron pairs.

If under dynamic equilibrium a total wave function, $\psi(t_0)$, or a set of individual wave functions, $\varphi_i(t_0)$, of which $\psi(t_0)$ is appropriately composed, describes the superconducting quantum state at the original time t_0, a thermal perturbation requires the *whole* set of the $\varphi_i(t_0)$, not only part of them, to be re-arranged to a new total wave function, $\psi(t_1)$, at the end of the disturbance, which again is to be re-arranged to $\psi(t_2)$ now because of decay of the excited state $\psi(t_1)$. The difference, $t_2 - t_1$, then is the lifetime, τ_{Total}, of the disturbed state.

In an ensemble consisting of an arbitrary number N of any kind of particles that have to be re-ordered, after a disturbance from their corresponding equilibrium state, re-arrangement of the total wave function cannot proceed instantaneously (for the superconductor, this will become obvious in the following), but requires a time interval, the total lifetime $\tau_{Total} = t_2 - t_1 > 0$, of the disturbed system, before it has completed return to dynamic equilibrium. Regardless in which manner re-ordering proceeds, the re-arrangement rate, on the average, will be proportional to the ratio N/τ_{Total}. We will consider re-arrangement (decay) of the excited state under two different points of view: Decay proceeds in space and in time.

Assume that a local disturbance (a temperature increase) occurs at a particular position $\mathbf{x}'$ in the superconductor solid (bold symbols denote vector quantities). We accordingly have an increased concentration of excited electron states over the previous equilibrium value at this position. Because of propagation of the thermal wave, there is

1) Decay in *space*, which means that the increased concentration, $c(\mathbf{x}',t_0)$, from any arbitrary position, $\mathbf{x}'$, of excited states is distributed by a transport process to positions $\mathbf{x} \neq \mathbf{x}'$; a particular simple transport mechanism is diffusion.

Application of the usual diffusion approach (solution of Fourier's differential equation) requires definition of a corresponding mean free path of a stepwise (in space) propagating process that must be small compared to the sample dimensions. This is explained in Section 2. The final (stagnation) result will be $c(\mathbf{x},t_\infty) > c(\mathbf{x}',t_0)$ for *all* positions $\mathbf{x}$ (including $\mathbf{x}'$) after a sufficiently long period of time, t_∞.

2) Decay in *time* means that the disturbed total wave function, $\psi(\mathbf{x},t > t_0)$, that describes all electron states, returns to equilibrium shape by recombination processes, with the stagnation result $\psi(\mathbf{x},t_2)$. Thus decay in time means re-arrangement of the total wave function by an operation that models recombination and transforms the total wave function from $\psi(\mathbf{x},t > t_0)$ to its stagnation value, $\psi(t_2)$.

The question then is how contributions from items (1) and (2) sum up to lifetime, τ_{Total}, of the disturbed system (this will be discussed below, and in more detail in Sections 2 and 3).

Re-arrangement (item 2) of the total wave function of the superconducting state cannot be completed before the disturbance (surplus number of excited state) has decayed completely to the concentration $c(\mathbf{x},t_2)$.

Otherwise, re-arrangement would lead to local surplus concentrations that again would have to be depopulated by subsequent diffusion steps. This is because of two inter-related diffusion mechanisms (decay channels) that are responsible for distribution of a thermal perturbation *and* its impacts on the concentration of excited states, as will be explained in more detail in Section 2, Equation (6a,b).

A general method to calculate lifetimes is provided by elements of perturbation theory, with operators describing recombination of single particles to pairs, but this procedure would require a large number of repeated applications. An alternative is time-dependent Ginzburg-Landau theory of the order parameter. In both cases, the computational effort is enormous due to the very complicated structure of the wave functions each comprising components of a very large number of particles involved. Instead, we will use in Section 3 for item (2) an aspect of the Yukawa model of nucleon-nucleon interaction: A "time of flight"-concept with a mediating Boson, considered as an analogue that can be used (with some caution, compare Section 3.2) to describe binding of two electrons to a pair in a superconductor (Fig. **1b**).

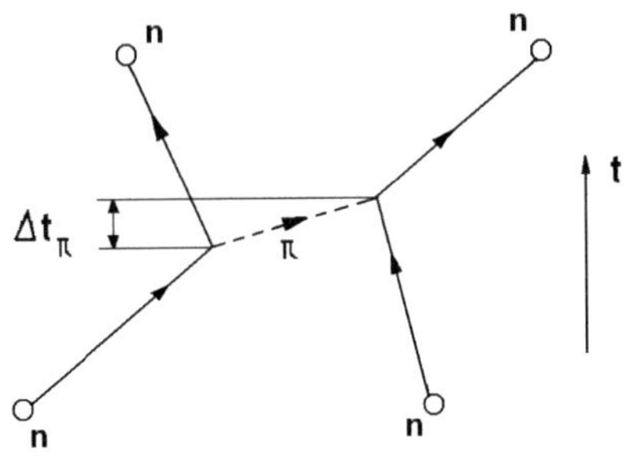

Figure 1b: Yukawa interaction between two nucleons (n) mediated by exchange of a pion (π, dotted line), as a particle system to which reference will be made in the text. The arrow on the right side denotes direction of time, t.

While statistical fluctuations like those around the initial dynamic equilibrium concentrations are superimposed on the proper decay process (initiated by the thermal perturbation), they cannot substantially alter decay of disturbed states. This is explained by the mere existence of persistent currents in superconductors (see Appendix A2). The tiny probability for excitation of electrons after a statistical thermal fluctuation is reflected by the corresponding tiny probability for a necessity to re-arrange the total wave function, $\psi(t)$, describing the superconducting state in the total volume of the sample. In turn, in view of the existence of persistent currents in superconductors, under constant experimental conditions, a possible impact of statistical temperature fluctuations (in addition to thermal perturbations) can entirely be neglected.

For item (2), decay in time, and for practical reasons, recombination to electron pairs is divided into two contributions (Sections 3.1 and 3.2) when calculating intrinsic decay widths Γ and lifetimes τ:

(2i) Contributions Γ_{ij} for two arbitrary particles i and j, that determine an "intrinsic lifetime" of the non-equilibrium state. This contribution results from "correlation", a first step (of two) to be taken for re-combination of particles i and j of a very large number, N. It requires a finite time interval needed *before* the proper (thermal) condensation (or recombination) event to a pair can take place. "Correlation" in the sense used here in a superconductor strictly speaking means exchange of information between two quantum states; in reality it is the exchange of phonons that mediate binding interaction between single particles and that have to travel a non-zero distance between the particles concerned,

(2ii) Contributions $\Gamma_{Rec} = (h/2\pi)/\tau$ resulting from the uncertainty principle, for the proper condensation (or recombination) event, once the particles i and j are correlated in step (2i), that generates one pair from recombination of two correlated particles i and j.

For recombination to *one* electron pair, each of the individual decay widths, Γ_{ij}, and Γ_{Rec}, of the excited electron state contributes by about 10^{-12} and 10^{-14} s, respectively, to lifetime τ_{Total}, for an energy gap of some tens of meV.

But there is a very large number of individual Γ_{ij}, and Γ_{Rec} that have to be taken into account so that the lifetime, τ_{Total}, of the total disturbed state (the time needed to re-arrange the total wave function) may become large.

Because of the Pauli exclusion principle, calculation of lifetime, τ_{Total}, from the two contributions (2i) and (2ii) has to proceed in a step-wise manner; we will later call this procedure a "sequential model" (Section 3.3).

For quasi-particle injection (tunnelling of electrons between two superconductors, under pulsed currents), Gray indicates a set of two differential equations (see Equations 9 and 10 in [1]): For very small injection rate, I_0,

$$\partial(\Delta N)/\partial t = I_0 + \beta\,\Delta N_\omega - 2\Delta N/\tau$$

$$\partial(\Delta N_\omega)/\partial t = \Delta N/\tau - \tfrac{1}{2}\,\beta\,\Delta N_\omega - \gamma\,\Delta N_\omega$$

with τ the lifetime of the excited electronic state, and ΔN and ΔN_ω are deviations from equilibrium of the density (numbers per unit volume) of quasi-particles (N) and phonons (N_ω), respectively. Without going into details of source and sink terms like $\beta\,\Delta N_\omega$, $-\tfrac{1}{2}\,\beta\,\Delta N_\omega$ and $-\gamma\,\Delta N_\omega$ of both species in the two equations, it appears the same balances in principle could be used for the estimate of the lifetime electron states after a thermal perturbation. But application of Gray's equations to the present case is not advisable.

[Note-Injection, absorption and loss terms (or source and sink terms) are well known from other disciplines, like radiative transfer: Variation di(x)/dx of the intensity, i(x), is given by a series of absorption, emission and scattering contributions that in total describes decay, here per unit of optical path length, x, of intensity in a corresponding medium. Another example is radio-active decay. Both describe weakening of an observable (intensity at a position, x', or number of particles, at a time, t', respectively) in relation to the corresponding *original* intensity, i(x=0), or the *original* number of particles, N(t=0). This is not the case described by the variation d(ΔN)/dt because it deals with the decay of a *surplus* number ΔN of quasi-particles from current injection, not with the total number N].

A differential change, d(ΔN)/dt, of the *surplus* number, ΔN, cannot be applied to the thermal perturbation problem in superconductors, whether with respect to thin films or pellets, for reasons already obvious from Section 1.1: For determination of the lifetime (the time needed to re-arrange the total wave function) and calculation of the decay rates $G_{Exc}(t)$, we have to consider the *whole* number N, *e.g.,* of bound electrons plus quasi-particles (or other species), not only an additional quantity, ΔN(t). Otherwise we would separate the solution into an idle spectator contribution, with a corresponding number of constituents, N(t), plus an additional term, ΔN(t), as if the undisturbed number N(t) was not involved at all with re-arranging the total wave function from $\psi(t_1)$ to $\psi(t_2)$; as already mentioned, this situation is quite different from a normal fluid. From a pure *particle* perspective, that approach is sufficient to investigate current injection in *e.g.,* semi-conductors, but in a superconductor we have to re-arrange a very large number of single particle wave functions to a unique, coherent total wave function of same amplitude and phase.

Therefore, we will not apply relations like $\partial[\Delta N(t)]/\partial t$. Instead, we will calculate decay or formation rates (wave function re-arrangement rates), G, as mean values, $G = N/\tau_{Total}$, with τ_{Total} indicating the time needed to re-arrange the total wave function of the disturbed electron system at a new dynamic equilibrium (not the time needed for just decay of a surplus number of particles).

Further, Gray's equations make no reference to position since they speak of quasi-particles, which means: of *quasi-free* virtual particles, a collective state of electronic excitations that in this context not necessarily needs specification as to position. Yet Gray's equations implicitly assume that the response to recombination and condensation to pairs would *immediately* be completed by a corresponding re-arrangement of the remaining quasi-particles and emission of phonons. This is not covered by renormalization of a many-body to a one-body problem.

The basic question is whether *in principle* an immediate response of the electron system to a thermal perturbation is possible. With finite distances between particles to be correlated, this is certainly not fulfilled (exchange of phonons proceeds with rather limited speed), apart from conclusions from the uncertainty relation.

So what did Gray [1] or Epperlein *et al.* [2] measure? Certainly and absolutely correct, the de-population of the surplus number of quasi-particles by decay to electron pairs and to other (escape) mechanisms (a description entirely confined to a particle picture). Is their result, an effective lifetime, τ_{eff}, or an intrinsic recombination time (the lifetime, τ_R, of a quasi-particle) identical with lifetime of the disturbed state, τ_{Total}, needed for re-arrangement of the total *wave function* after a disturbance? Certainly not. Even if the number of surplus particles has disappeared, this does not mean the total wave function is completely restored to a new equilibrium state. Instead, decay of the surplus particles is the *condition* to be fulfilled in order that re-arrangement of the total wave function can be completed within the lifetime τ_{Total}. Of course, determination of τ_{Total} was neither the aim of the reported experiments in [1, 2] nor had the experimental setup allowed measurement of a re-arrangement process. The results obtained in both references thus have to be interpreted rather from the point of view of conservation of electrical charge, under exchange of phonons with the lattice, escape events and current flow.

That lifetime of a surplus of quasi-particles in current injection experiments is different from lifetime τ_{Total} is also obvious from [5]: The authors report that the lifetime they observed did not change when the sample was shortly heated above critical temperature, clearly a reference to a particle related picture.

Next we will extract equilibrium recombination and decay rates from numerical analysis of experiments. In Section 3, estimates of lifetimes, τ_{Total}, of excited electron states will be presented.

1.3. Extraction of Equilibrium Recombination and Decay Rates from Experiments

Recently [7], we reported on the time dependence of the Meissner-Ochsenfeld effect. A numerical, finite element (FE) analysis shows that it takes a superconductor pellet when immersed into a coolant (an YBaCuO pellet in LN_2) a time interval, Δt, in the order of tens of seconds, depending on the anisotropy of its thermal diffusivity, to get a sufficient number of volume elements, N_{SC}, superconducting; otherwise there is no stable levitation position where total levitation force balances gravity.

Another FE analysis has been performed in [8] to obtain sample temperature evolution, T(x,y,z,t), and the number $N_{SC}(t)$ of superconducting volume elements, but now under strongly differing heat transfer conditions (reference [8] is a contribution to the present volume). In terms of thermal resistances to the coolant (at 77 K), to ambient air or to a radiative environment (both at RT), temperature excursion of the pellets was analyzed with respect to the four different situations (boundary conditions, chains of thermal resistances) that are schematically described in Fig. **2**:

1) **Cool-down**, with only liquid boiling/liquid convection at the pellet/coolant interface; heat transfer is modelled with temperature dependent (thus time-dependent) boiling and convection heat transfer coefficients at the interface to the coolant (resistance R_1); there is no further thermal resistance between (single) pellet and coolant,

2) **Warm-up**, with no liquid phase present but with a convection resistance (R_{21}) at the interface between a cylindrical sample holder and ambient air and, in series, a conductive resistance (R_{22}) of sample holder walls between interface and a number of up to 18 thin pellets. The walls of the sample holder are made of PTFE, with 7.5 mm wall thickness. A weak radiation contribution is included in R_{21},

3) **Cool-down**, with conductive resistance between pellets and sample holder walls (R_{31}) and, in series, a liquid boiling/convection resistance (R_{32}) at the interface to the coolant; the sample holder is the same as in item 2,

4) **Warm-up**, solely radiative heat transfer between sample and a cylindrical radiative environment (metallic), with solely a thermal radiation resistance (R_4 only; no other

resistance between environment and single pellet), and with the pellet exposed under high vacuum. This part of the diagram schematically describes inner heat transfer flow directions in a magnetic suspension balance experiment reported by [9].

Total thermal resistances (external to the pellets) increase from items (1) to (4). Internal conductive thermal resistances of the pellets must be added to the external chains of resistances.

The experiments corresponding to items 3 and 4 were performed to investigate whether a change of sample weight would occur at the phase transition. The results reported in [9] cannot confirm a weight change, but there are doubts whether the experiment was designed appropriately to fully exploit the potential of the suspension balance, see description and impacts on results of the suspension balance experiment by dead time intervals in Section 4 of the present paper.

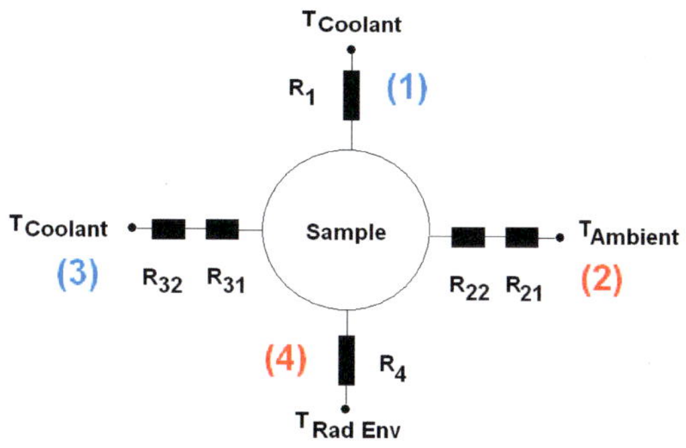

Figure 2: Thermal resistances network (schematic) corresponding to the heat flow paths (radiative, conductive and convective heat transfer), items 1 to 4. Indices of the individual resistances, R_{kl}, are assigned according to direction of heat flow (hot to cold).

The following sections describe numerical experiments/simulations designed to help determine decay rates and lifetimes. Temperature fields, T(x,y,z,t) obtained from the detailed numerical (FE) analysis have already been reported in [8]. From the T(x,y,z,t), the number of superconducting volume elements, $N_{SC}(t)$, is condensed in Fig. **3** of the present paper.

Time t' at which $N_{SC}(t') = 0$ in the suspension balance (item 4, warm-up experiment) is at about t' = 8240 s (obtained with $\alpha = 0.1$ W/(m^2 K), see [8] for explanation). The corresponding t'' in the dipping experiment (item 3, cool-down experiment), where $N_{SC}(t'') = \max(N_{SC})$, is about t'' = 720 s, roughly 1/10 of t'.

The origin of this difference is easily explained: A temperature interval, ΔT, between start temperature in the experiments and phase transition exists between 77 K and T_{Crit} in the warm-up experiments (ΔT only about 15 K), while it is between RT and T_{Crit} during cool-down (ΔT almost 210 K), respectively. Together with the thermal resistances in the pellets and those at the interface to coolant or to radiative environment, this explains the location (in time) of the phase transition (saturation of the curves at maximum $N_{SC}(t)$ during cool-down, or at $N_{SC}(t) = 0$ during warm-up periods, respectively). The shift observed from curve (1) to curve (3) results from the additional conductive resistance in (3), and the corresponding shift from (2) to (4) is due to a very large radiative resistance in the suspension balance experiment.

From $dN_{SC}(t)/dt$, we calculate the corresponding generation or decay rates, $G_{exp}(t)$ (absolute values, per unit volume, V), of electron pairs (EP), under quasi-equilibrium conditions,

$$G_{exp}(t) = [dN_{EP}(t)/dt]/V = \rho_{EP}\, dN_{SC}(t)/dt \quad \textbf{(3)}$$

that arise at temperatures close to but below transition at T_{Crit}. Note that applicability of Equation (3) also to cool-down experiments, from room temperature to below T_{Crit}, is not trivial and will be addressed in

Section 3.6. The equilibrium density, ρ_{EP}, of electron pairs is estimated using as an approximation a fraction of about *(½) (1/10)* of the density of electrons in *high temperature* superconductors ($\rho_{El} = 6\ 10^{27}$ 1/m^3). This fraction is large in comparison to the small fraction in *low temperature* superconductors, in the order of *(½) (1/1000)*, of the density of electrons in these materials (but closer to $\rho_{El} = 8.5\ 10^{28}$ 1/m^3); for the densities, see standard volumes on LTSC and HTSC.

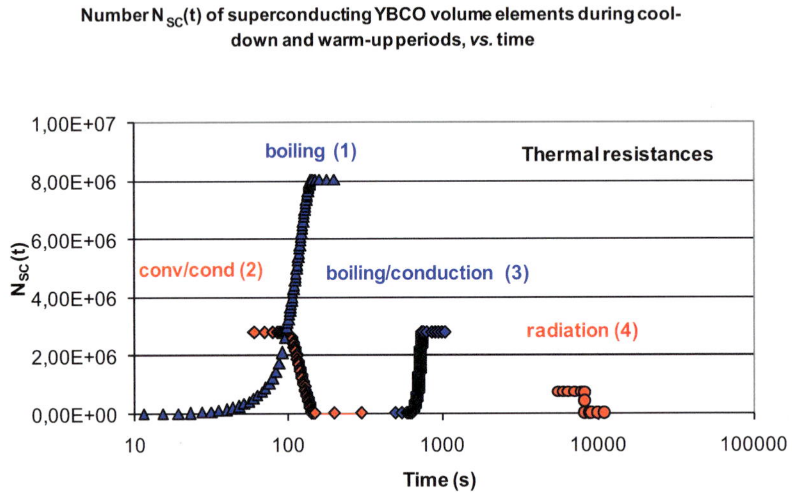

Figure 3: Number $N_{SC}(t)$ of superconducting (SC) elements (width and length each 200 µm, thickness 50 µm) in the cool-down and warm-up experiments reported in [7], *vs.* time, to identify the location (in time) of the phase transition.

The *(quasi-) equilibrium* $G_{exp}(t) = [dN_{EP}(t)/dt]/V$ rates for the 4 experiments are shown in Fig. **4**. Since the $\rho_{EP}(T)$ itself are dynamical equilibrium densities, so are the rates $G_{exp}(t)$, per unit of sample volume (alternatives to determine the $G_{exp}(t)$ could be measurements of enthalpy and entropy differences, but with rather insufficient local resolution). If instead of sample volume, V, we take identical differential (constant) volume elements, dV_{SC}, that have become superconducting or normal conducting, respectively, the rates of items (1) to (4) of course would be identical.

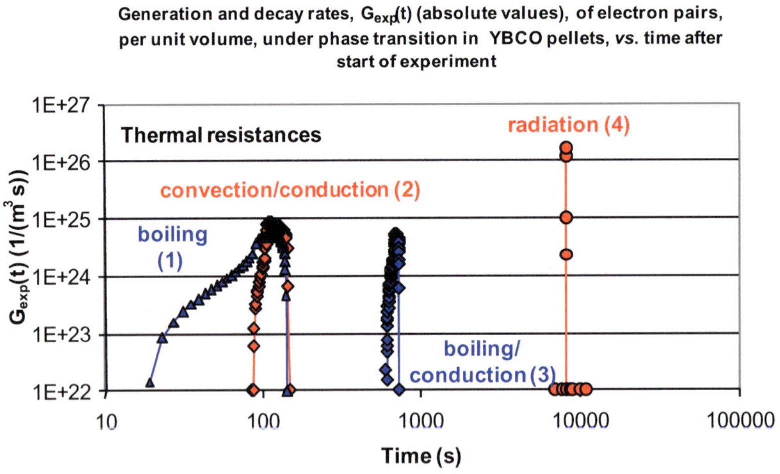

Figure 4: Experimental equilibrium formation and decay rates, $G_{exp}(t) = [dN_{EP}(t)/dt/V]$, per unit volume (absolute values, assuming sample volume-averaged conversion) of electron pairs (indexed EP) during phase transition in the YBaCuO pellets under cool-down and warm-up in the experiments with the thermal resistances (1) to (4) listed in Section 1.3, *vs.* time. The figure has already been shown in [8].

Lifetime, τ_{Total}, is also reflected by the time interval, τ, needed to establish screening super-currents over the sample volume. Lifetime τ_{Total} of excited electron states, or, equivalently, time $\tau = \tau_{Total}$ must be short, otherwise the average N/τ_{Total}l cannot satisfy the experimental rates, $G_{exp}(t)$.

In summary, the central question whether variations with time of the superconducting, dynamical equilibrium states can keep step with the phonon aspect of propagation of the same thermal perturbation, can be put into a simple relation involving the time integrals, with G_{Exc} defined in Equation (2a),

$$\int G_{Exc}[T(t)]\, dt \geq \int G_{exp}[T(t)]\, dt \qquad \textbf{(4)}$$

taken over the short time interval where the sample experiences phase transition. For this purpose, the $G_{exp}[T(t)]$ and $G_{Exc}[T(t)]$ are calculated at temperatures very close to, but below T_{Crit}. Results will be shown later (Section 3.4, Table **1**).

Summation $\sum V_j\ \rho_{EP}(T_j)\ dN_{SC}(t)/dt$ over all V_j consists of a series of *dynamic equilibrium* contributions (each term constant, at constant temperature, T_j) but indicates a total transfer rate (condensation/decay/recombination) in the whole sample volume, V, initiated by a temperature variation, ΔT. Accordingly, the rates $G_{exp}(t)$ obtained in this way are quasi-equilibrium rates.

2. DECAY IN SPACE: LIFETIME FROM A DIFFUSION MODEL APPROACH

Non-equilibrium thermal states, during warm-up or cool-down periods, always arise *locally* in a superconductor from local temperature gradients, from different thermal boundary conditions or from internal heat sources because of its finite conductivity. Temperature evolution in real solid samples never develops perfectly uniformly.

The situation in superconductors when exposed to absorption of a heat pulse is analogous to standard laser flash experiments in which at a time t_0 a very short radiation pulse is directed onto a thin film sample. Assume that the sample initially is at uniform temperature. Absorption of the pulse locally increases sample front surface temperature to a maximum obtained at a time, $t_1 > t_0$ that is overlaid onto the previous uniform temperature (background) level. The laser flash experiment then involves observation of front or rear surface temperature to obtain the thermal diffusivity of the sample material; in the most simple procedure, the diffusivity is obtained at a time, $t_{1/2}$, when rear surface temperature has increased to half its maximum value.

Determination of the lifetime of thermally excited states of superconductors is modelled in the following in a quite analogous way, with the exception that we do not look for the time $t_{1/2}$ when the concentration of surplus excited states has decayed to half its initial value after end of the disturbance. Instead we look for the time t_∞ when the disturbance (the concentration of excited electron states) has decayed completely, $c(\mathbf{x},t_\infty)$ = const. This concept has successfully been applied in [10] where an instantaneous, internal pulse method is used.

In the tunnel (current injection) experiment described in Fig. **3** of [2], delivery of a short *electrical current* pulse to the active contact area of the upper film (the source being a very small volume, V_J, about 10^{-24} m^3) causes the previous dynamic equilibrium concentration of electron excitations to increase to transient values. These excitations decay according to electrical conduction properties of the tunnel junction into electronic states that can be occupied in the lower film and by loss of phonons that cannot contribute to re-absorption/generation of new quasi-particles.

Delivery of a short *heat* pulse, *e.g.,* from a laser directed onto the surface of a thin film or of a pellet, with the heat pulse absorbed in a small volume, $V_{th} >> V_J$, near the surface that is part of the total volume, V, of the sample causes local temperature in V_{th} to increase above equilibrium (start) temperature. As a consequence, there is propagation of (a) a thermal wave and (b) of a concentration wave of excited electron states from V_{th}, because of thermal conductive properties of the solid material and because of a concentration gradient, respectively. It is tempting to assume not only propagation of the thermal wave but also redistribution of the concentration of excited electron states can be described as a diffusion process.

These two decay channels have to be distinguished as follows:

a) Propagation of the thermal wave from V_{th} into the solid increases temperature over local start values, $T = T(\mathbf{x},t_0)$. This also initiates increase of concentration of excited states at the same positions outside V_{th} over local start values, $c = c(\mathbf{x},t_0)$, because of coupling between temperature and temperature dependent density of excitations (there are accordingly "source terms" initiated by the temperature field for increase of concentration).

b) Propagation of the concentration wave from V_{th} into the solid increases concentration of excitations over local start values, $c = c(\mathbf{x},t_0)$, now merely by diffusion under a concentration gradient but without source terms (source and sink terms, if any, would cancel each other).

This means there is a diffusion "front" of the concentration of excitations driven by two different mechanisms one of which is closely coupled to propagation of the thermal wave. Both mechanisms propagate in parallel into the solid material, but with different speed.

With this separation, de-excitation of the over-concentration of excited electron states in V_{th} originating from absorption of the heat pulse can be described by evolution of the total wave function $\psi(\mathbf{x},t)$: If $\mathbf{x} = \mathbf{x}(t)$ indicates the diffusion front of the order parameter, $c = |\psi(\mathbf{x},t)|^2$, we have, in one dimension,

$$dc(x,t)/dt = (\partial c/\partial x)\,(\partial x/\partial t) + \partial c/\partial t \qquad \textbf{(5)}$$

The first factor in Equation (5), the concentration gradient $\partial c/\partial x$, describes de-excitation (decay of the disturbance) by diffusion (or decay "in space"), the second factor, $\partial x/\partial t$, the speed at which the diffusion front propagates, and the term $\partial c/\partial t$ indicates the decay "in time" of the disturbance. The first contribution, $(\partial c/\partial x)$ $(\partial x/\partial t)$, accordingly comprises channels (a) and (b) mentioned above that both will be modelled using diffusion methods. The contribution $\partial c/\partial t$ is investigated in Section 3.

An analogue is an expanding cloud of radio-active particles; in the present case, excited electron states and electron pairs are the analogues of mother and daughter terms, respectively.

Since a superconductor solid is opaque, radiation from the laser source (the incident heat pulse) will penetrate to only a very small distance, d_{Rad}, into the solid volume. Together with the diameter of the target spot, this defines a source volume. For simplicity, we will assume that the sample is kept at adiabatic conditions, at least during a time interval that is long compared to the lifetime of the disturbance.

For modelling the propagation of excited electron states in the solid by diffusion, *e.g.,* the decay channel (b) of the above, the following questions have to be answered:

1) Which kind of interaction is observed that in a diffusion manner (if any) has to proceed stepwise?

2) How large is the mean free path, in relation to sample dimensions, between two such interactions?

3) How fast will the interaction propagate into the sample, in particular compared to the thermal wave?

4) Is it possible to assign a diffusivity to the propagation of the interaction?

At positions x inside the source volume, V_{th}, and during decay of the disturbance the order parameter increases with time because the number of excitations decreases. Outside this volume, under solely thermal conduction condition, the order parameter decreases (distribution of the thermal perturbation at these positions leads to local increase of the number of excitations). It is thus the order parameter that diffuses within the solid volume. What is the driving mechanism that would cause the order parameter to diffuse?

Consider inside the source volume, V_{th}, a number of N excited electron states that result from previous decay of N/2 electron pairs in the same volume. The excited states, or concentrations thereof, can be assigned an enthalpy, g_{Exc}, that is larger than the corresponding enthalpy, g_{EP}, of electron pairs, and it is the enthalpy difference that drives excited states to recombine to electron pairs (consistent with temperature evolution in this volume).

Ohm's law for transport of electrical current of density, **J**, under an electrical field, **E**, is a classical diffusion law: In metals, $\mathbf{J} = [(n\, e^2\, \tau)/2m]\, \mathbf{E}$ (with n the concentration of electrons, e and m the unit electrical charge and mass of electrons, and τ the time of flight between two scattering events). The electrons under the electrical field move with the drift velocity, $\mathbf{v}_D = -e\, \mathbf{E}\, \tau/(2m)$, in Cu at RT with a mean free path of only about 5 nm. A simple expression like Ohm's diffusion law of the movement of single charge carriers (electrons) is distinct from correlation between pairs: A current density $\mathbf{J} > 0$ exists only if the coherent state of *all* electron pairs moves with a common, non-zero centre-of-mass velocity. Neither is it possible to define a mean free path of *one* electron pair nor of the cloud of all coherently coupled pairs. Therefore the main pre-condition for a diffusion approach (drift of individuals or of the cloud, under a potential difference and with interactions imposed by obstacles) for both species is not available in a superconductor.

However, the problem how to describe decay in space of the excited electron states by diffusion can be removed if we consider propagation of *concentrations* of pairs and excitations. In this case, it is no longer necessary to specify the behaviour of individuals, and the mean free path can be defined with respect to decrease of concentration over a path the length of which can be estimated from exponential decay of the concentration, as in radiative heat transfer where the mean free path indicates exponential decay of the radiation intensity, not of a single but of multiple photons, on the statistical average. The mean free path of the concentration of excited electron states under diffusion is given by the Ginzburg-Landau (GL) coherence length, ξ_{GL}. With ξ_{GL} in the order of 1 nm in YBaCuO, this is very small in comparison to pellet sample dimension, at least 20 mm, which means the precondition for "diffusion" to be a stepwise process would be fulfilled.

The velocity by which the diffusion propagates can be approximated by the phonon velocity, v_{Ph}, because v_{Ph} is the velocity of the Boson that mediates the binding interaction between two electrons. As a rough estimate, we take a value v_{Ph} of about 1/100 of the Fermi velocity (the latter describes the velocity of electron pairs carrying a super-current).

Calculation of the evolution with time of the *concentration*, $c_{Exc}(t)$, of excited states propagating by the two above mentioned channels (a) and (b) then is performed by use of Fourier's differential equation. For description of $c_{Exc}(t)$, again in one dimension, we have

$$\partial c/\partial t = D_c\, \partial^2 c(t)/\partial x^2 \text{ with Source terms} \qquad \textbf{(6a)}$$

at all times $t > t_l$ where t_l indicates the end of the thermal perturbation (duration of the heat pulse). For description of the temperature evolution, we have

$$\partial T/\partial t = D_{th}\, \partial^2 T(t)/\partial x^2 \text{ without Source terms} \qquad \textbf{(6b)}$$

Both Equations (6a) and (6b) are coupled by temperature dependence of the concentration $c_{Exc}(t)$, see below.

The thermal diffusivity, $D_{th} = D_{th}(T)$, in Equation (6b) is calculated from temperature dependent values of conductivity and specific heat of YBaCuO applied already in [7] and with density of 5968 kg/m^3.

The concentration diffusivity, D_c, in Equation (6a) in analogy to thermal transport can be determined from

$$D_c = (1 + \delta)\,(5/2)\, v_{Ph}\, \xi_{GL} \qquad \textbf{(7)}$$

(see [10] and references cited therein). Equation (7) applies the conductivity of a two-atomic gas in the hard core approximation. The quantity, δ, is small, in the order of 1/100. Numerical values of D_c, for v_{Ph} constant at about 9.4 10^3 m/s, and in a broad interval $0.1 \leq \xi_{GL} \leq 10$ nm are between 7.9 10^{-7} and 7.9 10^{-5} m^2/s. For $\xi_{GL} = 3$ nm, D_c is about one order of magnitude below the diffusivity (2.5 10^{-4} m^2/s) reported in [11] for quasi-particle diffusion in a W/Al bi-layer diffusion film. Such a decrease of the diffusivity is qualitatively to be expected for a high temperature superconductor.

The source terms in Equation (6a) arise from the temperature dependence of $c_{Exc}(T)$ and its coupling to the temperature field. These terms describe excitations at positions, x, within the total volume as soon as the thermal wave arrives at these positions. For simplicity, we use the equilibrium expression for the temperature dependence of the concentration. In a time interval dt, we have

$$(dc_{Exc}/dT)\,(dT/dt)\,dt = c_{Exc,0}\,\exp[-\Delta E(T)/(kT)]\,dt \qquad \textbf{(8a)}$$

with binding energy, ΔE, taken as temperature dependent using the BSC-result

$$\Delta E(T) = 1.74\,\Delta E(T{=}0)\,(1 - \gamma)^{1/2} \qquad \textbf{(8b)}$$

indicated in Blatt [12], (p. 243). In Equation (8a,b), $c_{Exc,0} = 6\ 10^{26}$ 1/m^3, $\Delta E(T{=}0) = 60$ meV for high temperature superconductors and $\gamma = T/T_{crit}$. Equation (8b) results from an exact thermo-dynamical solution at non-zero temperature provided by the theory of Bogoliubov, Zubarev and Tserkovnikov (as [12], p. 227, points out, one of the very few exact theories in statistical mechanics).

In case there are single electrons near the Fermi energy, the Pauli exclusion principle blocks electron states that otherwise could be populated by decay products from pairs, which means binding energy of electron pairs, or width of the energy gap, decreases when temperature approaches T_{Crit}. This aspect is reflected by the temperature dependence of ΔE. It also provides an infinite slope of $\Delta E(T)$ at $T = T_{Crit}$, with a second order transition, which is consistent with the jump of the electron part of the specific heat at T_{Crit}.

Because of $T = T(x,y,z,t)$, all quantities γ, ΔE and dc_{Exc}/dt in Equation (8a,b) depend on positions (x,y,z) and time.

Calculations in the following have been made for a sample pellet with radius, r = 16 mm, and thickness, d = 20 mm (this is of identical size as the sample used in the cool-down experiment, Section 1.3, item 1), and for comparison, for a thin film (thickness d = 100 nm). Start temperature is 90.5 K in both cases. Length of the heat pulse is 1 μs and 1 ns, respectively. Intensity of the pulse has been adjusted to increase surface temperature of the samples within these periods of time to $T_{Crit} = 92$ K. For the pellet, it is assumed the heat pulse is generated in tiny volume elements located at the centre of the sample. For the thin film, Equation (6a,b) is solved with target volume given by a thin layer of thickness d_{Rad} covering the whole front surface into which radiation has penetrated (the calculation thus is 1D). In both cases, we have cylindrical symmetry.

Solution of Equation (6b) for pellet and thin film using the diffusivity, D_{th}, is shown in Fig. **5**. In the thin film, thermal equilibrium (decay of the thermal perturbation indicated by coincidence of front and rear surface temperature) is reached after 10^{-8} s, while in the pellet, decay is completed not before about 100 s after start of the disturbance. Thermal equilibrium satisfies stagnation temperature in thin film and pellet (90.770 and 90.504 K, respectively), in both cases a very small uniform temperature increase under these conditions.

Solution of Equation (6a) for the thin film with the diffusivity D_c yields evolution of (relative) concentrations, $c_{Exc}(t)$, of excited states in the whole sample. Results are shown in Fig. **6** for different values, ξ_m, of the mean free path (since the mean free path is on the order of the Ginzburg-Landau coherence length, variations of ξ_m assumed in the calculations then also account for uncertainties of the phonon velocity). This calculation was performed without inclusion of the source terms, dc_{Exc}/dt, in Equation (6a).

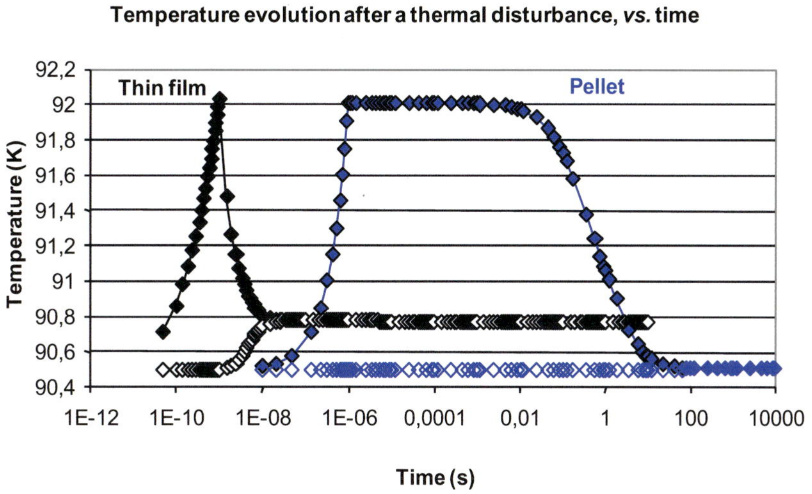

Figure 5: Temperature evaluation calculated within target source volumes (solid symbols) and rear (thin film) or periphery (pellet) YBaCuO sample surfaces (open symbols) under a thermal perturbation generated by a laser pulse absorbed on a thin film (d = 100 nm) and of an arbitrary heat pulse absorbed at the centre of a sample pellet (d = 20 mm), with duration of the pulse of 1 ns and 1 μs, respectively. For the thin film, the laser pulse is applied to a target volume given by a thin layer covering the whole sample surface whereas for the pellet, a heat pulse is generated at the sample centre. Temperature dependent values of thermal diffusivity have been taken from [7].

Fig. **6** shows that equilibrium is obtained at the latest after 10^{-8} s as before, indicated by coincidence of the front and rear surface concentrations. This means that the lifetime of the excited states, if they were to decay solely by diffusion of the local surplus concentration of excitations from the source (target) to the whole sample volume followed by decay "in time" (Section 3), would be at least of this value. Stagnation concentration is 1.282 10^{26} 1/m^3 for the thin film (or a relative concentration of 1.181 over the start value at 90.5 K, as confirmed by Fig **6**). Almost no differences are observed, at times shorter than 10^{-7} s, if the source terms were included. For the pellet (not shown in the figure), the corresponding values are 1.089 10^{26} 1/m^3 and 1.002).

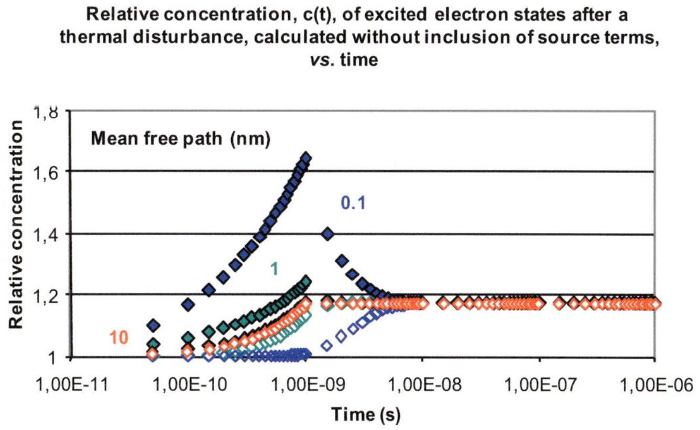

Figure 6: Relative concentration, c(t), of excited electron states at front and rear surface (solid and open symbols, respectively) of the YBaCuO *thin film* sample under a thermal perturbation of 1 ns duration, calculated for different values of the mean free path with the diffusion model using the diffusivity, D_c, in Equation (6a), without inclusion of source terms initiated by propagation of the thermal wave outside the target volume. For values of D_c compare text after Equation (7).

From the relative concentrations, $c_{Exc}(t)$, obtained for the thin film sample from application of the diffusion model and with the diffusivity, D_c, we calculate local (absolute) *decay* rates (full diamonds in Fig. **7**; the open circles are explained below) using for the decay rate the expression ρ_{Exc} $dc_{Exc}(t)/dt$, with ρ_{Exc} the equilibrium density of excitations at start temperature (90.5 K). Results are given at rear surface of the thin film. With or without inclusion of source terms, and for the different values of the mean free path of the interactions, the electron system relaxation time τ_{El} of diffusion in space to total lifetime in this presentation is estimated from the decrease of the decay rate to (relatively) small values. The contribution, τ_{El}, of the surplus concentration of excited states thus is between 5 10^{-9} and 10^{-7} s, for mean free paths decreasing from 10 to 0.1 nm.

The results indicated by the solid and open diamonds in Figs. **6** and **7** for estimation of the lifetime contribution, τ_{El}, have been obtained solely from the diffusion approximation. It yields the propagation of the surplus concentration of excited states generated in the source volume using Equation (6a) and the diffusivity, D_c, into the whole solid.

Alternatively, the decay rates can be calculated from propagation of the *thermal perturbation* using the thermal diffusivity D_{th}. Application of Equation (8a,b), with T(t) and $\Delta E(T)$ and of Equation (6b) yields the decay rates given by the open blue circles in Fig. **7**. The corresponding estimate of thelifetime, τ_{th}, yields about 7 10^{-8} s, approximately the same value that is obtained from the pure *concentration* diffusion model if there the mean free path would be limited to 0.1 nm (solid yellow diamonds in Fig. **7**). However, since the mean free path must be larger than this value, at least comparable to the Ginzburg-Landau coherence length of about 1 nm, τ_{El} (estimated from the curves with the solid symbols) is definitely smaller, at least an order of magnitude, than τ_{th} (open circles).

Lifetimes τ_{El} and τ_{th} estimated by the same method for the pellet are much longer, by orders of magnitude, due to larger sample thickness and increased duration of the heat pulse, compare Fig. **5** (temperature evolution) and Fig. **8a** and **b** (decay rates given at centre and periphery of the pellet). But the overall result is as before: τ_{El} is definitely smaller than τ_{th}, again at least an order of magnitude, except for the case that an unrealistically small mean free path (0.1 nm, Fig. **8b**) of the interaction would be assumed.

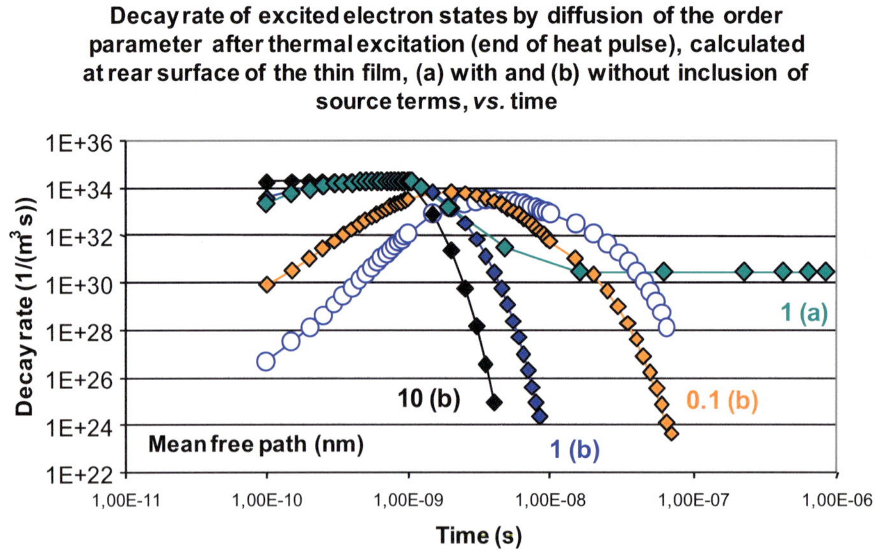

Figure 7: Decay rate of excited electron states (solid diamonds) at rear surface of the YBaCuO *thin film* sample under the thermal perturbation, calculated for different values of the mean free path using the diffusion model and the diffusivity, D_c, in Equation (6a), calculated (a) with and (b) without inclusion of the source terms initiated by propagation of the thermal wave outside the target volume. Open circles indicate decay rates calculated using (i) Equation (6b) for T(t) with the thermal diffusivity, D_{th}, and (ii) the initial density of excited states, $\rho_{Exc,0}$ and the energy gap, $\Delta E(T)$, in Equation (8a,b).

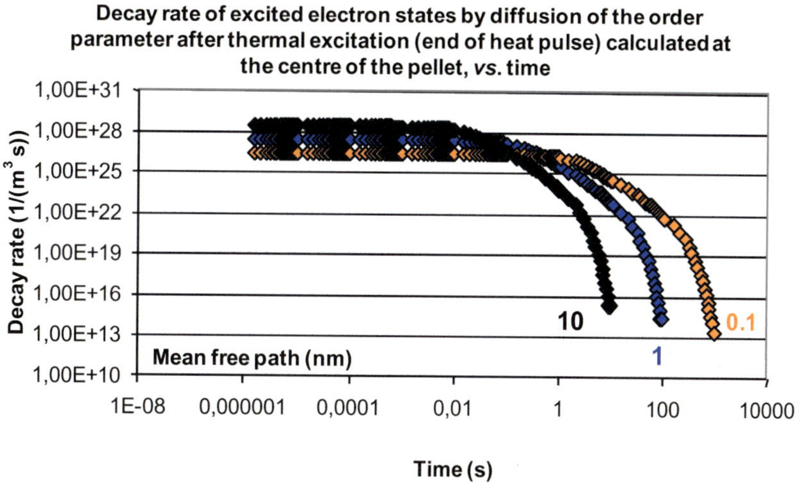

Figure 8a: Decay rate of excited electron states at the centre of the YBaCuO *pellet* under the thermal perturbation, calculated for different values of the mean free path using the diffusion model and the diffusivity, D_c, in Equation (6a) (without inclusion of source terms initiated by propagation of the thermal wave outside the target volume.

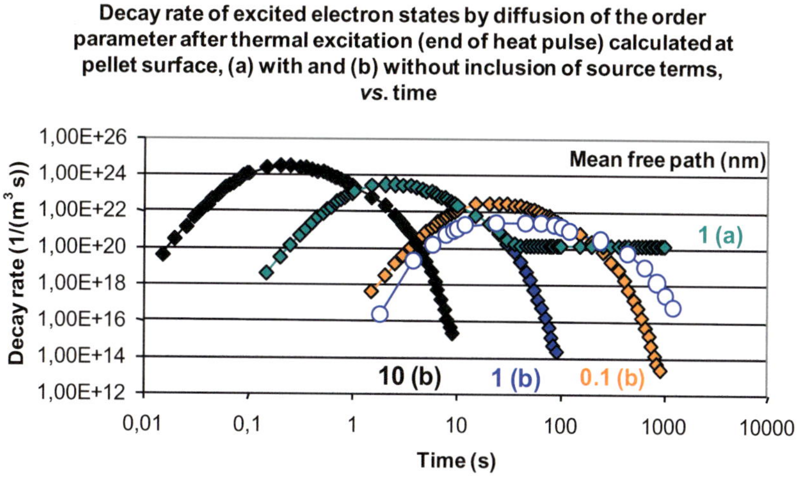

Figure 8b: Decay rate of excited electron states (solid diamonds) at the periphery of the YBaCuO *pellet* sample under the thermal perturbation, calculated for different values of the mean free path using the diffusion model and the diffusivity, D_c, in Equation (6a) and (a) with and (b) without inclusion of the source terms initiated by propagation of the thermal wave outside the target volume. Open circles indicate decay rates calculated as in Fig. 7 using the thermal diffusivity, D_{th}, taken from [7].

These conclusions rely entirely on the assumption that decay "in space" of surplus concentration of excited states would proceed *solely* by diffusion in the sample pellet. Radiation from the source volume to positions outside is not considered separately; a (very small) contribution of this mechanism to distribution of the thermal perturbation in the solid can be integrated, as a radiative conductivity (another diffusion approximation), into the solid thermal conductivity. Though sample dimensions used in the suspension balanc (Section 1.3, item 4) and in the other experiments (items 2 and 3) were different from the present sample pellet, the conclusion is the same: *Decay in space by diffusion proceeds on a very small time scale.*

The total wave function cannot be re-arranged to its final equilibrium shape, by "decay in time" before the surplus concentration of excited states has completely decayed "in space" by the diffusion steps. If this were done, the total wave function obtained in a series of intermediate decay and recombination steps would have to be re-arranged repeatedly until the final result (all recombination steps completed to equilibrium concentrations) is obtained. This is the consequence of the difference between τ_{El} and τ_{th}: If the concentration of excitations, c(**x**'), at a position **x**' outside the source volume according to Equation (6a) increases by diffusion (decay of the original disturbance "in space"), a decrease of c(**x**') due to decay "in time" (Section 3) would subsequently (because of the lower propagation velocity of the thermal wave) be corrected by the diffusion solution now of Equation (6b), again a decay of the disturbance "in space".

Re-arrangement of the total wave function accordingly requires solution of the second step of the lifetime calculations, as is represented by the term $\partial c/\partial t$ in Equation (5), the intrinsic lifetime contribution, or the "decay in time" of the disturbance.

The results obtained for the relaxation time from the diffusion model as the first step to calculating total relaxation time and generation or decay rates are, in a general sense, the reverse of the results achieved from estimates of τ_{Ph} and τ_{El} in Section 1.1: There, comparison was made between propagation of a thermal wave or its phonon lifetime τ_{Ph}, and decay of the electron excitation in a *particle-related* decay model or its electron lifetime τ_{El}. Here the comparison is made between τ_{th} and the first contribution to total relaxation time, τ_{Total}, that is needed to re-arrange the total wave function to its new dynamical equilibrium shape. Contributions from the term $\partial c/\partial t$ in Equation (5), τ_{Total} can become enormous when temperatures approaches T_{Crit}, which means the corresponding decay rate, N/τ_{Total}, on the average finally will become very small.

3. DECAY IN TIME: LIFETIMES FROM A SEQUENTIAL MODEL

As mentioned in Subsection 1.2.2, there are two contributions to the total lifetime, τ_{Total}, or total decay width, Γ, of the electron state after a thermal perturbation: a) contributions Γ_{ij} for two arbitrary particles i and j, that determines an "intrinsic lifetime" of the non-equilibrium state, b) contributions Γ_{1EP} resulting from the uncertainty principle, for the proper condensation or recombination event during formation of one electron pair (indexed below as 1EP).

3.1. Estimate of the Contribution Δt_{1EP}

The following estimate of the contribution Δt_{1EP} for recombination (Section 1.2.2, item 2) is obtained from an analogue to numerous examples reported in atomic, molecular and nuclear physics in which decay of excited states is made by application of the uncertainty principle. The contribution Δt_{1EP} can be estimated from

$$\Delta t_{1EP} = (h/2\pi)/\Delta E_{1EP} \tag{9}$$

with the binding energy (the energy gap), ΔE_{1EP}. The time interval, Δt_{1EP}, holds for decay of *one* electron pair (1 EP) and for recombination, or if an arbitrary large number of electron pairs are formed *instantaneously* from particles contained in the volume of the superconductor. With provisionally ΔE of 60 meV (at very low temperatures in high temperature superconductors), the interval Δt_{1EP} amounts to about 10^{-14} s.

For this approach, averages of the lifetime must be taken over a (virtual) volume V_C at all positions, **x**. Determination of the size of V_C may become one of the critical points of the analysis: On the one hand, we have to confine calculation of the average to a very small volume, otherwise the number of particles concerned would be too large to be handled reasonably with numerical methods, on the other hand an arbitrary dependence of the final results on size of V_C must be avoided. In the following, we will estimate V_C.

3.2. Estimate of Δt_{ij}, the Intrinsic Lifetime Contribution

A sequential model will be used in this Subsection that determines lifetime from the ratio of the distance between two particles before recombination and the velocity by which an exchange Boson mediates binding between both particles.

Two arbitrary electrons cannot have exactly the same wave function, because of the Pauli exclusion principle. If electrons could be considered as purely classical particles (point masses), with no interactions among each other, they all would statistically be independent, and the probability of finding an electron at a position $\mathbf{x}_j$ near an electron at a given position $\mathbf{x}_i$ would be the same as for finding an electron at any position $\mathbf{x}_j$. This is not the case in a superconductor or, generally, if the Pauli exclusion principle is to be followed.

3.2.1. Average Distance between Electrons before Recombination

In the ground state (gs), the total wave function, with particles i and j counted by numbering 1, 2, 3,...N, reads (compare *e.g.,* [13], Chap 6, for an introduction)

$$\Psi_{gs} = C \sum(-1)^P P[\varphi(1,2)\ \varphi(3,4)\ \varphi(5,6)\ldots\varphi(N-1)\ \varphi(N)] \qquad \textbf{(10a)}$$

with the sum extending over all N! permutations, P, of the total number, N, of particles, and zero centre of mass motion. All pair wave functions, $\varphi(i,j)$, must be anti-symmetric, with respect to exchange of particles i and j.

Each of the $\varphi(i,j)$ can be written as a product of a symmetric space dependent, $\varphi_S(i,j)$, and of an anti-symmetric, spin dependent part, $\chi_A(i,j)$.To determine the distance, $d = r_i-r_j$, between two arbitrary electrons i and j, we in principle need expectation values of the space part, $\varphi_S(i,j)$, that describes the charge distribution of particles i and j. In a particularly simple picture, the $\varphi_S(i,j)$ are expanded in terms of Bloch waves each of which is a free electron plane wave,

$$\varphi_S(i,j) = \sum a_k \exp\,[ik\,(x_i - x_j)] \qquad \textbf{(10b)}$$

for description of s-wave states. This can be generalized to functions $f(|x_i - x_j|)\ Y_{lm}(\theta,\eta)$, with inclusion of direction of the vector $(\mathbf{x}_i - \mathbf{x}_j)$, but we need in the following only the distances, $(x_i - x_j)$.

For a description of excited states, instead of a transforming Ψ_{gs} (Equation 10a) to a total wave function, $\Psi_{Exc} = \Psi(t_1)$, the BCS model starts with creation and destruction operators acting on the vacuum and in the following on the ground state. If just one electron pair is broken into two electrons, they have momentum vectors **k'** and **k''**; with uncorrelated electrons, there is no electron with exactly **k''** = **-k'**. Indication of spin-up or spin-down is omitted for simplicity. In the BCS-model (compare *e.g.,* Blatt [12], p. 174 to 180, or [13] Chapter 6), the total wave functions of ground and excited states read

$$\Psi_{gs} = \Psi(t_0) = [\prod\,(u_k + v_k\, b_k^{+})]\ |0> \qquad \textbf{(11a)}$$

$$\Psi_{exc} = [\prod\,(u_k + v_k\, b_k^{+})]\ a_{k',+}^{+}\ a_{k'',-}^{+}|\ 0> \qquad \textbf{(11b)}$$

with probability amplitudes u_k and v_k to find corresponding k-correlated pairs: For each electron pair broken, the k-pairs in the BCS-picture consist of two quasi-particles, one state unoccupied, the other occupied, respectively (the missing electron with momentum **k''** = -**k'** is represented by the "unoccupied" state). Creation and destruction operators are denoted by a_k and b_k, one for each index, k, and the index runs over all $k \neq k'$, k''. The vacuum state is denoted by |0>.

The operators $a_{k',+}^{+}\ a_{k'',-}^{+}$ in Equation (11a,b) do not alter the functional dependence of the $\varphi_S(i,j)$ on particle/particle distance (this does not imply the distances are the same as before the excitation). While wave functions for central potentials are tabulated, centre of mass positions, r_i, r_j, cannot be calculated using the corresponding $|\varphi_S(i,j)|^2$: Because of symmetry of the $\varphi_S(i,j)$, the results would be identical. Wave functions for other than central potentials are difficult to handle. An alternative is to directly estimate distances, without explicit recourse to wave functions, as is made in the following:

A first, direct but very rough estimate for the ground state can be performed using the inter-particle distances, $d = (V/N)^{1/3}$, with N the number of particles filling a sample volume, V. With the inverse of the electron density in high temperature superconductors ($\rho_{El} = 6\ 10^{27}\ /m^3$), this yields a mean distance between any two (among all) electrons of about 0.55 nm, or when taking only that part of the electrons that condense to electron pairs, a fraction of (1/10) of the total number, the mean distance increases to about 1.2 nm. This is about the coherence length of an electron pair in the ab-plane of high temperature superconductors.

A similar result is obtained from consideration of the Coulomb interaction within the cloud of a large number of electrons. For this purpose, we start with an estimate of the number $N_{El}(t_0)$ of electrons that at the initial time, t_0, are available for electron pair formation in a virtual spherical volume, V_C.

The volume, V_C, will be used below for calculation of average decay rates. It has to be small, much smaller than the sample volume, V, in order to limit the number of particles contained in V_C to values that can be handled as integers in the summations carried out in a sequential numerical model. Once the decay rates are calculated, they will be normalized to dimensions $1/(m^3\ s)$ by division of the result by V_C to compare with the experimental rates from Section 1.3. To determine V_C, we need its radius, r_C.

The radius, r_C, of V_C shall be given by the condition that the Coulomb potential energy between two arbitrarily selected electrons i and j, E_C, is minimized and below the binding energy, $2\Delta E$, of a single electron pair. The Coulomb energy amounts to

$$E_C(r_C) = \chi\ e_i\ e_j/(4\pi\ \varepsilon_0\ r_C) \leq 2\Delta E(T) \quad \textbf{(12)}$$

with e_i and e_j denoting electron charge and ε_0 the dielectric constant.

In metals, the effective Coulomb force between quasi-particles is modified by screening. In the solid, the two-quasi-particle electrostatic (repulsive) Coulomb potential consists of (a) the repulsive interaction (interpreted as a mean field), and (b) the (attractive) positive ion charges in the solid. The two contributions are superimposed. In the simplest form, this can provisionally be simulated using a screening factor, χ, in Equation (12) that essentially modifies the dielectric constant. The Thomas-Fermi potential, as the classical example of a screened Coulomb potential, will be considered below. Provisionally, arbitrary values, χ = 0.01, 0.1 and 1 (no screening), are used in the following.

Pairing of electrons in conventional superconductors is by formation of highly symmetric, singlet s-waves of charge distribution ($\mathbf{S} = \mathbf{s}_1 + \mathbf{s}_2 = 0$, from $\mathbf{s}_1 = 1/2$, $\mathbf{s}_2 = -1/2$, and $\mathbf{L} = \mathbf{l}_1 + \mathbf{l}_2 = 0$, from $\mathbf{l}_1 = 0$ and $\mathbf{l}_2 = 0$), and the energy gap is finite and isotropic around the Fermi surface, in the ground state.

Pairing of electrons in unconventional superconductors like YBaCuO, with $\mathbf{S} = 0$ and $\mathbf{L} = 2$, is identified by spins and angular momenta $\mathbf{s}_1 = 1/2$, $\mathbf{s}_2 = -1/2$, $\mathbf{l}_1 = 1$ and $\mathbf{l}_2 = 1$. The corresponding charge distributions can approach each other more closely than in the highly symmetric case $\mathbf{s}_1 = 1/2$, $\mathbf{s}_2 = -1/2$, $\mathbf{l}_1 = 0$ and $\mathbf{l}_2 = 0$ states. Also, the energy gap is non-uniform, with zeros in particular directions.

If we estimate r_C using Equation (12), singlet s-waves of charge distribution cannot approach distances below r_C, which in turn means the virtual sphere, V_C, of the radius, r_C, roughly contains *only* $\mathbf{S} = 0$ and $\mathbf{L} = 2$ electron pairs (states with larger but even angular momentum, $\mathbf{L} = 4, 6,\ldots$, are not very probable because they would indicate too large a rotational energy). Accordingly, in this very rough approximation, almost all electrons, $N_{El}(t_0) = (½)\ (\rho_{El}\ V_C)/10$ contained within the volume V_C do not belong to $\mathbf{S} = 0$, $\mathbf{L} = 0$ but to $\mathbf{S} = 0$, $\mathbf{L} = 2$ spin and angular momentum states. Crystal imperfections and impurities could lead to false, s-wave-*like* charge distributions (see [14]), this will be neglected by assuming perfect crystalline order and very clean materials.

Positions x_i, x_j of particles i and j, in one dimension, at time, $t \geq t_1$, are predicted in the following using random variables RND_i and RND_j, with $0 \leq RND_{i,j} \leq 1$ applied to the r_C which yields the random distance $d(t) = x_i(t)\text{-}x_jt)$ of particles before electron pair formation. Selection of a lower limit of d(t) does not have significant influence on the final results (in the calculations, the coherence length of YBaCuO in c-axis direction, was assumed for this limit).

As an alternative, results can be obtained also using the Thomas-Fermi potential,

$$E_{TF} = E_C(r_C) \exp(-r/r_{TF}) \qquad \textbf{(13)}$$

with $E_C(r_C)$ from Equation (12), without the factor χ, and r_{TF} the scattering length. Literature values of r_{TF} (*e.g.*, [15]) are in the order of 0.5 nm.

As Blatt ([12], p. 212) explains, screening of the Coulomb field in the superconducting state of the sample does not differ much from screening in the normal state; accordingly, we will continue with the familiar, normal state expression, Equation (12), or Equation (13).

3.2.2. Boson Mediated Binding Energy

Distances between electrons were estimated in the previous Subsection from statistics, as the first information needed to calculate Δt_{ij}. The second information needed concerns the Boson that mediates binding of two particles to a pair. As an initial approach, we consider a nucleon-nucleon interaction model (with some caution) as an analogue to binding between two electrons in an electron pair:

Nuclear forces (compare standard volumes on nuclear physics) are short-range, saturation forces. In the Yukawa-model (Fig. **1b**), the pion (π), a Boson with spin zero, needs a time interval (in a rough picture a "time of flight") of about $\Delta t_\pi = 4.7\ 10^{-24}$ s, much smaller than lifetime of charged pions (about 10^{-8} s), to mediate the binding energy between two nucleons. This time interval is estimated from the uncertainty principle using $\Delta E = m_\pi c^2$, with m_π the rest mass of the pion and c the velocity of light. It is not clear that its mass necessarily would be the rest mass of a free solid particle, but the range of the pion mediated nuclear binding force (the "uncertainty of the nucleon radius"), $d = \Delta t_\pi c$, is about $1.4\ 10^{-15}$ m, a value surprisingly close to the radius of the nucleon.

In the Deuteron, the only stable bound, two-particle nucleon system, we have a central binding force (plus a small electrical quadrupole moment) and a comparatively small binding energy so that the inter-particle distance between proton and neutron even exceeds the range of the nucleon/nucleon interaction force. This is in analogy to binding in the BCS-model: It is sufficient that there is a (negative) binding energy that even may be a arbitrarily small.

There are of course differences between the three cases considered (nucleon-nucleon interaction, Deuteron and electron pair): (a) In the Deuteron, proton and neutron couple to a spin triplet ($^3S_1$)-state (parallel spins) and it is a free particle; (b) While the exchange Boson in the nucleon-nucleon interaction interacts between two solid particles, it does so only in the interior of a nucleus (we do not consider p-p or p-n scattering reactions); (c) In electron pairs of a superconductor, the exchange is between electrons, with the lattice vibrations that mediate exchange of energy and momentum.

But the other aspects of the electron pair formation, *e.g.*, a two particle interaction, a Boson as the exchange particle and weakly bond, two-particle states, get the electron pair in superconductors, though only from formal aspects, at least marginally similar to its nucleon/nucleon analogues. An alternative comparison could be made with the two-electron system in the ^{4}He-atom, but this comparison formally suffers from the central potential that the electrons in superconductors do not experience.

Formation of both a nucleonic bound state and of an electron bound state (the electron pair) in this model then would proceed within a time interval (the time of flight, or lifetime, Δt_π or Δt_{ij}, respectively) that the corresponding exchange Boson (π or ω, respectively) needs to mediate the binding interaction (or *correlate* the corresponding single particles before recombination). After the time interval Δt_π, re-arrangement of the two single wave functions of the nucleons to a wave function describing a coupled state of two nucleons is accomplished, and Δt_π accordingly is the lifetime of two *uncoupled* nucleons considered as virtually disturbed states before they couple to a bound two-particle state in a nucleus. In the same way, Δt_{ij} is the time interval that is needed for re-arrangement of the total wave function after recombination of two electrons to a pair.

The present model to estimate the contribution Δt_{ij} in a superconductor accordingly is based on the following analogy:

1) In the Yukawa model, the time of flight determines the uncertainty of the size of the *nucleon* (or the *lifetime* of two uncoupled nucleons before they combine to a nucleon-nucleon pair in a nucleus),

2) In the electron pair, the time of flight of the mediating Boson determines the uncertainty of the size of the *electron pair* (or the *lifetime* of two excited states before they recombine to an electron pair in a superconductor). The uncertainty of the size of the electron pair is the average "distance" between the two particles concerned.

In both cases, dividing this distance by the velocity of the corresponding exchange Boson, $d(t)/v_{Boson}$, gives a measure for the "lifetime of the interaction" and, if summed up over all particles concerned, the intrinsic part of the "lifetime of the *disturbance*", to be compared with the term $\partial c/\partial t$ in Equation (5). This intrinsic lifetime concerns the disturbed state, not the proper existence of a pair; lifetime of a bound, two particle system is of course different from this value (the Deuteron is a stable particle, and without disturbances electron pairs in superconductors on the average exist indefinitely, otherwise there would be no persistent currents). Individual partners i and j in both cases are not coupled forever; other potential partners, i' or j', may take the role of i and j, like particle i coupled to particle j' or i' coupled to j, provided selection rules with respect to momentum or spin are fulfilled.

3.3. Sequential Model

Calculation of total lifetime requires contributions Δt_{1EP} and Δt_{ij} to be weighted by the number of allowed open decay channels. Weighing has to take into account the Pauli exclusion principle: Only when re-arranging of the total wave function after creation of a new (recombined) pair (i,j) is completed will the next re-arranging step be allowed. At a time t, this prevents formation of pairs $\varphi(1,2)$ and corresponding total wave functions, ψ_{gs}, if at a previous time, t', this pair would already have been formed, in any permutation. Formation of the total wave function, ψ, therefore cannot be completed before *each* permutation, $P[\varphi(1,2)\ \varphi(3,4)\ \varphi(5,6)\ldots\varphi(N-1)\ \varphi(N)$, contributing to the total sum is completed.

The maximum number N_{cor} of possible correlations (two potential candidates suitable for building one pair) therefore is to be determined from a total of N_{Exc} particles,

$$N_{cor} = N_{Exc}(t)!/[(N_{Exc}(t)-2)!\ 2!] \qquad \textbf{(14)}$$

While the ratio $d(t)/v_{PH}$ on the average determines the time needed for *one* correlation attempt, total time for mediating the exchange energy between *all* potential candidates i and j, to *one* electron pair (i,j), then is obtained by summation over all N_{cor} open correlation steps to yield $\Delta t_{ij}(t)$, the time needed to re-arrange the total wave function after the recombination to (i,j); the excited electron system will not take just the very first from a sequence of all potential combinations. The summation breaks up when for a particle, i, its appropriate particle partner, $j \neq i$, has been identified, to recombine to the pair (i, j); in the summation taken over a very large value of $N_{Exc}(t)$ particles, we thus have on the average a factor of ½ to be considered in $\Delta t_{ij}(t)$.

Only when by this procedure the two electrons i and j properly are identified (by the conditions $\mathbf{s}_i = 1/2$, $\mathbf{s}_j = -1/2$, $\mathbf{l}_i = 1$, $\mathbf{l}_j = 1$,and $\mathbf{p}_i = -\mathbf{p}_j$ at positions $x_i(t)$, $x_j(t)$ within V_C), the time interval Δt_{1EP} from Equation (9) is added to $\Delta t_{ij}(t)$ to yield $\Delta t_{ij}^{rec}(t) = \Delta t_{ij}(t) + \Delta t_{1EP}$. This time interval incorporates correlation *and* recombination (index "rec") of two electrons to one pair, in terms of the total wave function: after one more electron pair has been added to the number of already existing pairs. After formation of a number N_{EP} of electron pairs, we have $N_{Exc}(t) = N_{Exc}(t_1) - 2N_{EP}$ as the number of excitations to be used in Equation (14) for the next selection of potential electron pairs, now from the reduced number $N_{Exc}(t)$. This process is repeated until all N_{El} candidates (available in ΔE and V_C) are coupled to pairs, which means we have for the total time interval, Δt^{total}, needed to accomplish re-arrangement of the total wave function,

$$\Delta t^{total} = \sum \Delta t_{ij}^{rec}(t) \tag{15}$$

with the summation index running over $2 \leq \Delta k \leq N_{Exc}(t)$ in Equation (15) using $\Delta k = 2$. It is clear this procedure to calculate lifetime strongly differs from a particle picture, the latter would have to include particle motions.

But computation time required to calculate the sum in Equation (15) is enormous: For T = 90.5 K, we have $N_{El}(t_l)$ and N_{cor} in V_C in the order of 10^4 and 10^8, respectively, which means summations over a number of terms in the order of 10^{12} would be the consequence. To simplify the procedure, instead of calculation of random positions, we apply a mean value d_{mean} of the inter-particle distance, d(t).

Mean distances, d_{mean}, at the time, t, are given in Fig. **9** for values $0.01 \leq \chi \leq 1$ of the screening factor and for the Thomas-Fermi potential. When $T \rightarrow T_{Crit}$ (92 K), all d_{mean} plotted in Fig. **9** diverge, due to the dependence of ΔE on T and thus of the radius r_C of the volume V_C in Equation (12). This result is indeed the familiar one: All electron pairs finally decay into single, uncorrelated electrons when $T \rightarrow T_{Crit}$. The divergence of d_{mean} at temperatures close to T_{Crit} ($d_{mean} > 10^{-6}$ m) should cautiously be interpreted as tendency, but it is clear that probability of decay of electron pairs increases as soon as the distance between both particles exceeds coherence length. The figure also shows that with the screening factor $\chi = 0.01$ in Equation (12), the result roughly coincides with the result obtained form the Thomas-Fermi potential.

Note again that we speak of correlations. In the traditional particle picture, particle/particle distances like those in Fig. **9** for $T \rightarrow T_{Crit}$ are much larger than average electron distances in any normal conducting solid. There is no longer any direct interaction between two electrons separated by such distances. In a superconductor and in the correlation picture used here, strongly increased d_{mean} destroys correlation between two potential candidates, and they no longer would be able to combine to an electron pair. Accordingly, electron pairs do no longer exist if $T \rightarrow T_{Crit}$, and superconductivity breaks down.

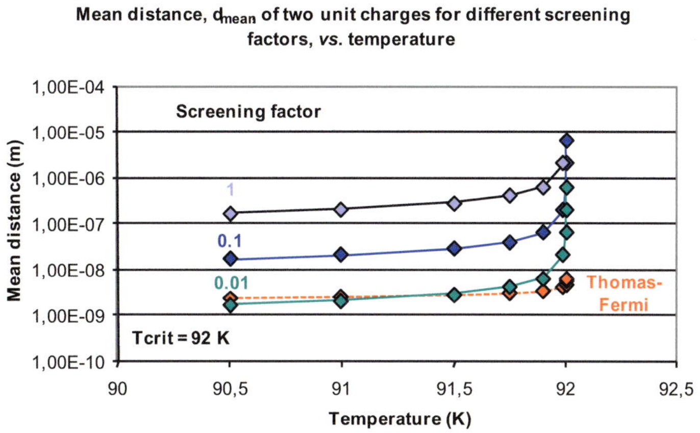

Figure 9: Mean values, d_{mean}, of inter-particle distance, d, calculated at a time, t, between two unit charges (electrons) in the virtual volume V_C before return to equilibrium state (recombination to electron pairs), calculated for different factors, χ, screening the Coulomb potential (Equation 12, see text), and for the Thomas-Fermi potential (Equation 13), *vs. constant* temperature. The divergence of $d(t_l)$ to values much larger than the coherence length (in the order of 1 nm) observed when $T \rightarrow T_{Crit}$ indicates a tendency or probability for eventual break-up of finally all electron pairs.

The sequential model to determine lifetimes accordingly is based on:

a) The uncertainty principle to estimate the time interval needed for the proper (condensation-like) recombination or (evaporation-like) decay of a pair,

b) The total wave function of excited states factorized into space components, $\Phi_S(i,j)$, that depend on the distance between arbitrary particles i and j,

c) A model to estimate the distance between particles i and j before recombination,

d) An analogy between the phonon-mediated binding force between two electrons with the pion-mediated Yukawa nucleon-nucleon force (together with the "time of flight"-concept), and

e) The Pauli principle that is reflected in a sequential model for the calculation of total lifetime of disturbed state.

But the situation might be more complicated (de Gennes, ([16], p. 99)): Ion charges oscillating near the resonance frequency of thermal ionic motion could "over-screen" the negative charge of either electrons 1 or 2 in Equation (12), or the mean field of all other electrons. We have to leave open this problem to another study.

Summarizing the contributions from diffusion and sequential model, we have for the total lifetime near the phase transition $\tau_{Total} \approx \Delta t^{total}$, because the contribution to τ_{Total} from diffusion was found to be very small (Section 2).

3.4. Generation/Decay Rates during Warm-up at $T < T_{Crit}$

Total conversion rate, from $N_{Exc}(t > t_1)$ particles located in the volume V_C to $N_{Exc}(t_1)/2$ electron pairs, by application of the lifetimes, $\tau_{Total} \approx \Delta t^{total}$, estimated in the previous Section,

$$G_{Exc}(t) = dN_{Exc}/dt\ N_{Exc}(t_1)/\tau_{Total} \qquad \textbf{(16)}$$

yields the rates indicated in Fig. **10**, at different constant temperatures, $T < T_{Crit}$, and for a deviation $\Delta T \ll T_{Crit}$. While both d_{mean} and, as a consequence, τ_{Total}, diverge, $\tau_{Total} \to \infty$ as soon as temperature approaches T_{Crit}, the rates $G_{Exc}(t)$ are finite and finally converge to $G_{Exc}(t) = 0$ at $T = T_{Crit}$. At this limit, neither will any quasi-particles be left to recombine to electron pairs nor exist any electron pairs at all at this temperature, apart from statistical fluctuations that may generate electron pairs even under this condition or at temperatures randomly exceeding T_{Crit}.

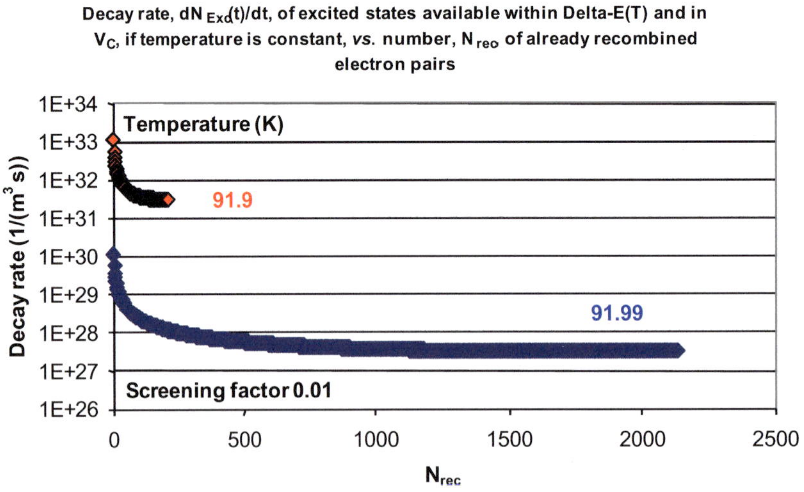

Figure 10: Decay rates, $dN_{Exc}(t)/dt$, of excited electron states contained in the volume V_C. Results are given per unit volume and time for different temperatures, $T < T_{Crit}$, if temperature is *constant* and for the screening factor $\chi = 0.01$. Data are plotted *vs.* number of already existing pairs. For the velocity of the exchange Boson, the value $v_{Ph} = v_F/100$ has been used, to calculate upper limits of the decay rates (see text). "Decay" in this and the subsequent figures is understood as return (re-arrangement) of the disturbed *total* wave function to a new total wave function of a new dynamical equilibrium state at the new temperature.

For temperature clearly below T_{Crit}, the total volume of the solid contains a very large number of single cells, V_C, with the G_{Exc} in each cell being identical. This means the decay rates G_{Exc} calculated in Equation (16) per unit cell volume yield also the decay rates of the whole solid because decay proceeds in parallel in each cell. The decay rates initially depend on the volume V_C because Δt^{total} is calculated by means of the number of particles contained within this volume. The procedure at this temperature thus relies strongly on the size of the cells, V_C, but the situation relaxes if temperature during warm-up converges to T_{Crit}. The distances, d(t) or d_{mean}, and thus the volume, V_C, increase strongly which means V_C finally fills the total sample volume. At a certain time, t', the distances d(t) or d_{mcan} exceed sample dimensions. Then $G_{Exc}(t) \rightarrow 0$ as before and dependence $G_{Exc}(t)$ on size of V_C has disappeared.

In Fig. **10**, results are given as a function of the number $N_{rec}(t)$ of already existing pairs that during decay of the disturbance have recombined (indexed "rec"). The smaller N_{rec}, the more correlation "channels" are open for the remaining candidates in the volume V_C to re-arrange to the total wave function to an equilibrium state. The curves accordingly decrease with increasing N_{rec}, and they terminate as soon as all pairs have been formed (all particles available from ΔE and V_C for recombination are consumed).

Termination of the curves shown in Fig. **10** occurs after a time, t_f, needed for completion of the recombination process (during decay of the thermal perturbation), *e.g.*, $t_f = \Delta t^{total}$, when finally all $N_{El}(t_1)$ particles have recombined, if temperature is kept constant. Times t_f accordingly are the required intrinsic lifetime contributions.

These contributions are plotted in Fig. **11** *vs.* temperature. The curves calculated for $\chi = 0.01$ and 0.1 approach very large values when temperature $T \rightarrow T_{Crit}$. The divergence of t_f simply indicates the tendency of the system to approach dynamic thermal equilibrium, which at constant temperature is of quasi-infinite lifetime. The τ_{Total} in the temperature region close to T_{Crit} accordingly should be interpreted as probabilities for the time needed to complete decay of the excited electron system.

It is not possible to extend these calculation to exactly T (constant) = T_{Crit}: Divergence resulting from the dependence on temperature of ΔE, see Equation (8b), is too strong to be handled numerically already if the temperature difference is below 10^{-6} K. In the extreme case $[T_{Crit} - T(\text{constant})] \rightarrow 0$, T, *with T constant*, it would take the superconductor indefinitely long time to allow recombination of all available excitations to pairs, again a natural consequence of the uncertainty principle: The case $T = T_{Crit}$ (T constant) simply is the dynamic equilibrium of the electron state, with no longer disturbances to be compensated but only statistical fluctuations of the density of pairs and decay products.

This result, increase of τ_{Total} to indefinitely large values if $[T_{Crit} - T(\text{constant})] \rightarrow 0$, cannot be achieved if summations over contributions $d(t)/v_{Ph}$ or d_{mean}/v_{Ph} would be omitted and in total be replaced by a constant. The same conclusion applies to the decay widths (see below, Fig. **15**) that when $T \rightarrow T_{Crit}$ would not increase strongly without the summations over the N_{cor} potential decay channels.

The first impression is that this finding apparently agrees with Gray [1] who reports there are indications that lifetime may increase strongly with temperature very near T_{Crit}. The temperature range very close to T_{Crit} even has been excluded from the experiments reported in [17], to avoid divergences of the order parameter relaxation time. But lifetime, τ_{eff}, in both references is different from and necessarily smaller than t_f. Note also that in [17], lifetime τ_{eff} is modelled in terms of sample dimensions and as a transport (transmission) process which certainly would not appropriate for modelling lifetime t_f, a quantity not related to a particle picture.

Fig. **11** also shows the influence of the screening factor, χ, on t_f. All curves approach very large values, with the exception of the Thomas-Fermi case. The latter will be replaced in the following by the curve obtained for $\chi = 0.01$, see Fig. **9**, that shows there is not much difference between the Thomas-Fermi and the $\chi = 0.01$ curve. However, more investigations are needed to clarify how t_f really behaves at temperatures even more close to T_{Crit} than presently can be simulated.

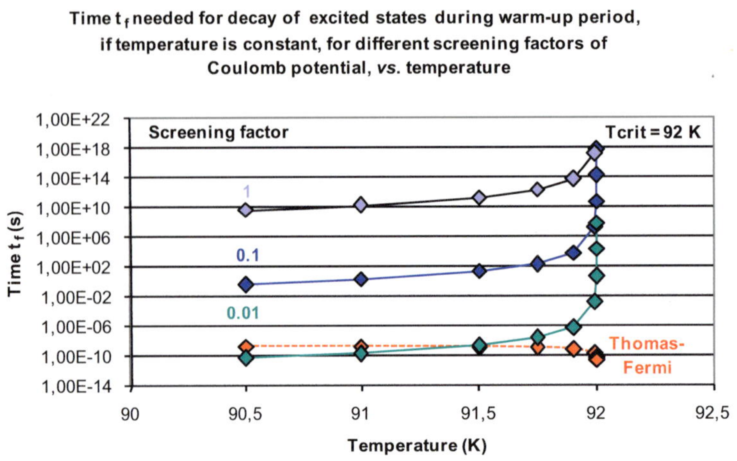

Figure 11: Time t_f (lifetime) needed for return to thermal equilibrium (completion of decay of all excited electron states available within $\Delta E(T)$ and in the virtual volume V_C, *e.g.,* return of the disturbed total wave function to a new total wave function at the new dynamic equilibrium). Results are given *vs. constant* temperature, $T < T_{Crit}$, and for different values of the factor χ screening the Coulomb potential, Equation (12). The divergence of t_f to very large values observed when $T \rightarrow T_{Crit}$ simply indicates tendency or probability to approach dynamic thermal equilibrium, of quasi-infinite lifetime at constant temperature.

But since warm-up continues, temperature, T(t), and all quantities that depend on temperature, accordingly are *not* constant. In the following, we will approach the continuous evolution, T(t), by a series of discrete, short time steps within each of which temperature assumes different values but can be considered as constant in these short intervals. This also explains application of the equilibrium density, $\rho_{El} = 6\ 10^{27}\ 1/m^3$, to calculate the rates $G_{exp}(t)$ and the $G_{Exc}(t)$, which means the calculated rates again are quasi-equilibrium quantities.

If tentatively plotting lifetimes, t_f, *vs.* $\Delta E/(kT)$, the curves do not follow a typical Arrhenius plot, contrary to the results reported from current injection experiments (compare with the figures shown in *e.g.,* [1] or [2]). This is because lifetime in the present situation is not constant but depends on temperature. A linear relationship of decay rate *vs.* $\Delta E/(kT)$ can be obtained only in a log-log-plot. The behaviour of these curves that integrate a very large number of single events can be explained already by the quantum-mechanical (uncertainty) prediction for a transition between two energy states separated by a pairing energy, ΔE, and without caring about the source from which the two constituent particles might come, but with the temperature dependent ΔE.

There is also the previous information (Fig. **10**) that the transfer rates decrease with increasing number of already existing electron pairs (from recombination of the decay products, after the disturbance). We therefore can identify not only final (t_f) but also initial and mean recombination rates, as has been done in Fig. **12**. Initial decay rates at temperatures close to T_{Crit} of course are much larger than the mean rates. Mean rates can be considered as "final" rates in the figures as they are the final result when taking averages over the total number of steps in the calculations.

The dashed red line given in Fig. **12** indicates maximum of the experimental $G_{exp}(t)$ that results from the suspension balance, *warm-up* experiment (Section 1.3, item 4) the thermal analysis of which has been reported previously [6]. The dashed blue horizontal line refers to the maximum of the of the *cool-down* experiment (Section 1.3, item 1).

The initial and mean $G_{Exc}(t)$ are much larger, by orders of magnitude, than the experimental rates (horizontal lines in Fig. **12**), but at temperatures clearly below T_{Crit}. If temperature approaches T_{Crit} very closely, the considerably smaller $G_{Exc}(t)$ (note the logarithmic scale) lead to a discrepancy between the mean $G_{Exc}(t)$ and the $G_{exp}(t)$. In the limit $T \rightarrow T_{Crit}$, the rates $G_{Exc}(t)$ converge to zero.

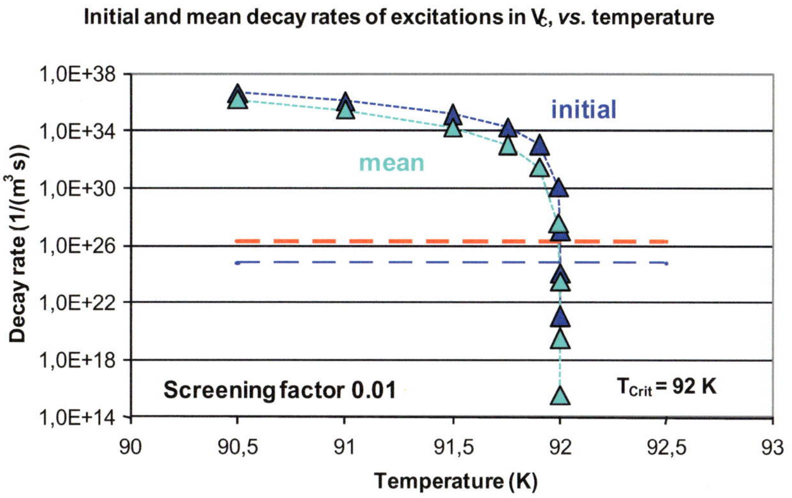

Figure 12: Initial and mean (final) non-equilibrium decay rates, $dN_{Exc}(t)/dt$, of excited electron states (solid triangles) during warm-up period, for the screening factor $\chi = 0.01$ in Equation (12) and if temperature would be *constant*. Also indicated is the maximum of the experimental $dN_{exp}(t)/dt$ (dashed horizontal lines) from suspension balance, warm-up experiment (red) and cool-down experiments (blue), see Section 1.3. The $dN_{exp}(t)/dt$ are averages taken over sample volume, while the $dN_{EXc}(t)/dt$ are calculated as average taken over the volume V_C. The dashed blue line relies on the reversibility of the heat transfer experiment, see Section 3.6 and Appendix A3 for explanation.

The discrepancy becomes obvious from Fig. **13** where the decay rates are plotted *versus* $\Delta E(T)/kT$. The decay of non-equilibrium states arising from a temperature variation ΔT performed at a temperature very close to T_{Crit} ($T < T_{Crit}$) will proceed rather slowly (and must do so), and as a consequence, phase change from normal- to super-conduction cannot not proceed instantaneously.

While comparison between the decay rates $G_{exp}(t)$ and $G_{Exc}(t)$ is made in Figs. **12** and **13** *versus temperature* (or its inverse), in Fig. **14** we have plotted both rates of the suspension balance experiment (Section 1.3, item 4) *versus time* (after start of the warm-up period).

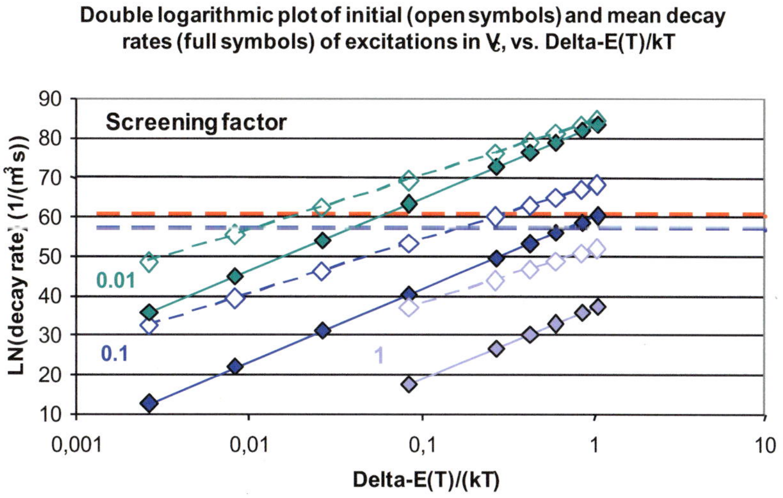

Figure 13: Log-log plot of initial and final (average) non-equilibrium decay rates of excited electron states during warm-up period, $dN_{Exc}(t)/dt$, for screening factors $\chi = 1$, 0.1 and 0.01, at constant temperature, *vs.* $\Delta E(T)/(k\ T)$. The horizontal dashed lines are explained in Caption to Fig. **12**. The $dN_{exp}(t)/dt$ are averages taken over sample volume, while the $dN_{EXc}(t)/dt$ are calculated as average taken over the volume V_C. At T = 91, 91.99 and 91.999999 K, $\Delta E(T)/(k\ T)$ amounts to 1.465, 0.145 and 0.00145, respectively.

The thin vertical, dashed line in Fig. **14** serves to identify dead-time intervals, see Section 4. Data for G_{Exc} are given at constant temperature and thus have to be interpreted as quasi-equilibrium rates in the short time intervals. As mentioned, the re-arrangement rates approximate the experimental rates only at temperatures not too close to T_{Crit} and only in case the screening factor, χ, does not exceed the value $\chi = 0.01$, otherwise there would be a discrepancy by several orders of magnitude between both rates. The volume elements in this case would warm-up at a rate too large to be followed by the decay (recombination) of excited states to electron pairs (again, the time for re-arrangement of the total wave function), which means "overheating" as in an evaporation experiment, could be the consequence. Re-arrangement of the total wave function to a new dynamic equilibrium of the electron system then cannot keep step with the "phonon" (warm-up) behaviour of the solid.

The same conclusion can be drawn from comparison (Equation 4) of the time integrals $\int G_{exp}(t)\, dt$ and $\int G_{Exc}(t)\, dt$:

The $\int G_{exp}(t)\, dt$ are calculated (Table **1**) within the periods of time during which $dN_{EP}(t)/dt/V > 0$ (compare Fig. **4**), with a mean value taken over the 4 experiments of 2.902 $10^{26}/m^3$. Deviation of the individual $\int G_{exp}(t)\, dt$ from the mean is small, below ± 3%, as it should be because pellet substance (YBaCuO) and its critical temperature are identical in the experiments.

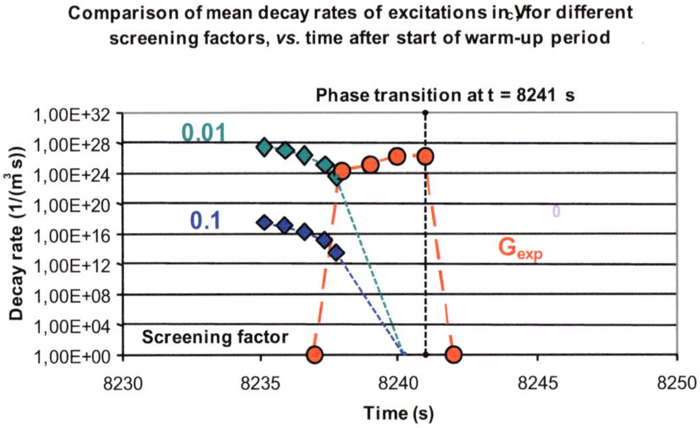

Figure 14: Final (mean) non-equilibrium decay rates, $dN_{Exc}(t)/dt$, of excited electron states (solid diamonds) for screening factors $\chi = 0.01$ and 0.1 in Equation (12) and at constant temperature, for comparison with the experimental decay rates, $dN_{exp}(t)/dt$ (full circles), from the suspension balance experiment during warm-up period (item 4, Section 1.3, temperature not constant), *vs.* time. The $dN_{exp}(t)/dt$ are averages taken over sample volume, while the $dN_{Exc}(t)/dt$ are calculated as average taken over the volume V_C. Phase transition observed in the numerical simulations in this experiment occurs at t = 8241 s (dashed vertical line). The solid diamonds clearly indicate the tendency or probability of the numerically obtained decay rates to decrease to zero when $T(t) \rightarrow T_{Crit}$ (the dashed lines attached to the solid diamonds indicate the corresponding extrapolations). The figure also serves for calculation of the time integrals, Equation (4), within the period $8237 \leq t \leq 8242$ s, *e.g.*, when $dN_{EP}(t)/dt/V > 0$ (see Fig. **4**; results will be given in Table **1**).

Table 1: Values of the integrals $\int G_{exp}(t)\, dt$ ($1/m^3$) obtained using experimental, quasi-equilibrium $G_{exp}(t)$ ($1/(m^3\, s)$) for the thermal resistances (items 1 to 4). The $\int G_{exp}(t)\, dt$ are calculated within the periods of time during which $dN_{EP}(t)/dt/V > 0$ (see Fig. **4**).

Thermal Resistances	**Experimental Generation/Decay Rate $Q_{exp}(10^{26}/m^3)$**
(schematic in Fig. **2**)	
Boiling (1)	2.933
Convection/conduction (2)	2.851
Boiling/conduction (3)	2.837
Radiation (4)	2.988
Mean	**2.902**

Contrary to the $\int G_{Expt}(t)$ dt in Table 1, the $\int G_{Exc}(t)$ dt calculated for the radiation balance experiment (Section 1.3, item 4) are much smaller and in units of $10^{26}/m^3$ amount to 0.53 and 5.3 10^{-11} for screening factors $\chi = 0.01$ and 0.1, respectively, instead of the mean $\int G_{Exc}(t)dt = 2.902\ 10^{26}/m^3$. The deviation of $\int G_{Exc}(t)dt$ from $\int G_{exp}(t)dt$ obviously becomes substantial if the screening factor would be increased, $\chi >$ 0.01.

It is clear from Fig. **14** that the deviation also would become the larger, the more the length of the integration interval is reduced around the phase transition at t = 8241 s.

3.5. Comparison with Classical Exponential Decay Formula

If $N_{Exc}(t)$ denotes the number of excitations still existing after start of decay of a thermal perturbation, adoption of the classical exponential decay law yields

$$N_{Exc}(t)/N_{Exc}(t=0) = \exp[-\lambda(T)\ t] \quad \textbf{(17)}$$

with a decay constant, $\lambda(T)$, that in the present model depends on time and temperature. This can be explained as follows: Using $\Gamma(t) = (h/2\pi)\ \lambda(t)$, decay of the excited state of the electron system, after a thermal perturbation, can be described by the decay width, $\Gamma(t)$, the probability for decay, that should increase when the number, $N_{Exc}(t)$, approaches zero. This is different from classical, *e.g.,* radioactive, decay: There the probability of decay of instable nuclei is always constant, regardless how large the number of still existing nuclei, and so are λ and Γ. Here, the smaller $N_{Exc}(t)$, the faster the decay of this number, and the larger the decay width or probability when time increases until decay is completed.

Using Equation (17), with the number of still existing excitations, after start of decay of the thermal perturbation, we calculate in the usual way, from $LN[N_{Exc}(t)/N_{Exc}(t=0)]$, decay constant $\lambda(t)$ and decay width, $\Gamma(t)$, compare Fig. **15** for T = 91.9 K and a screening factor $\chi = 0.01$ (if using other temperatures or screening factors, the qualitative behaviour $\Gamma(t)$ is the same).

As was expected, the decay width, or the probability of decay, increases with time as soon as $N_{Exc}(t)$ approaches zero (Fig. **15**): If only few particles are left to recombine to pairs, completion of recombination or rearrangement of the total wave function proceeds the faster the smaller this number (particles i find their partner particles j more quickly), and the probability for decay accordingly increases.

This behaviour, strong increase of the decay probability when $T \rightarrow T_{Crit}$, would not be obtained without the correlation step, *e.g.,* the summations described in Section 3.3.

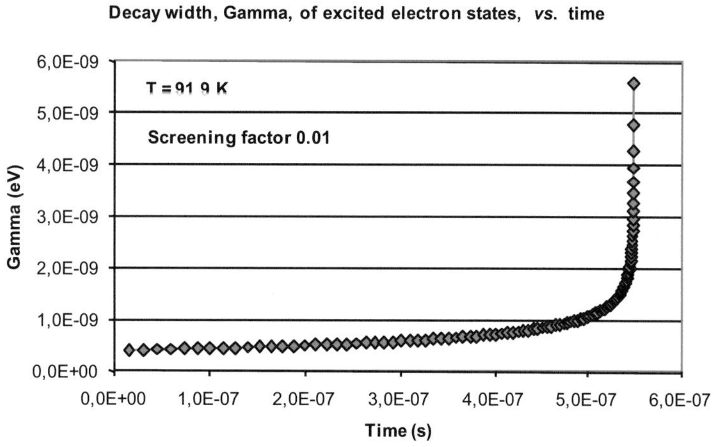

Figure 15: Decay width, $\Gamma(t)$, of excited electron states (probability of recombination to electron pairs) in dependence of time using Equation (17). For explanation of the time-dependence of Γ see Section 3.5.

3.6. Cool-Down from $T > T_{Crit}$

We still have to justify introduction of the dashed, blue horizontal line in the previous figures. In cool-down experiments like items 1 and 3 (Section 1.3), the temperature field T(x,y,z,t) approaches the phase transition from $T > T_{Crit}$. This means the model described in the foregoing Subsections cannot be applied without alterations, since ΔE is not defined at $T > T_{Crit}$, and the $G_{Exc}(t)$ from $T < T_{Crit}$ (or from negative temperature interval $\Delta T = T - T_{Crit}$) cannot simply be shifted to the region $T > T_{Crit}$ (or to the corresponding positive interval $\Delta T = T - T_{Crit}$).

However, instead of shifting $G_{Exc}(t)$, we can shift the experimental rates, $G_{exp}(t)$ under the condition that the temperature fields T(x,y,z,t) for the two cool-down experiments would be identical under a transformation of the time scale, t, into the scale -t, to hopefully obtain an identity like T(x,y,z,-t) = T(x,y,z,t), under time reversal. More discussion is presented in Appendix A3, which indicates the model described in the foregoing Subsections is applicable also to the present case.

4. IMPACT FROM DEAD-TIME INTERVALS ON THE SUSPENSION BALANCE GRAVITY EXPERIMENT

An estimate of the temperature interval, ΔT_D, a "dead *temperature* interval" during which the $G_{Exc}(t)$ cannot satisfy the equilibrium $G_{exp}(t)$ rates, can be obtained from Fig. **12**. The ΔT_D are obtained from the intersection of the data points (solid triangles) with the red and blue, dashed horizontal lines. At temperatures very close to T_{Crit}, the ΔT_D are very small and cannot visually be resolved in the figure. A high order polynomial fit to the data points (triangles) saves to extract ΔT_D = 7.3 and 3.7 mK, for the suspension balance (item 4) and the cool-down experiments (items 1 and 3 in Section 1.3), respectively.

Considering the derivatives, dT(t)/dt, of the temperature evolution (Fig. **16**), with a maximum of about 10^{-4} K/s observed for the suspension balance experiment, the length, Δt_D, of the corresponding "dead *time* interval", $\Delta t_D = \Delta T_D/(dT/dt)$ in the suspension balance experiment is in the order of 10 s. Larger values of χ would yield even strongly increased dead time intervals, compare the case $\chi = 0.1$ in Fig. **14**. During this period of time, the experimental setup is "blind" against measurement of any observable that could be correlated with variations of the excited electron system, *e.g.,* its decay to a new dynamic equilibrium.

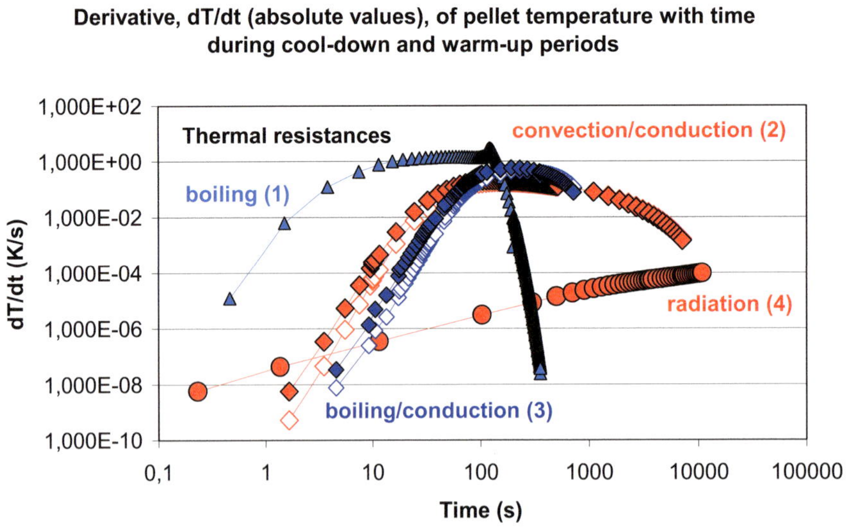

Figure 16: Derivatives, dT/dt (absolute values), of temperature at central pellet positions during warm-up and cool-down periods. Results are given for the dipping experiments (items 1 to 3 in Section 1.3, triangles and diamonds) and suspension balance (item 4, circles), *vs.* time. Full and open diamonds indicate bottom and top vertical positions in the piles, respectively. Part of the results are from [8].

A trivial example for the importance of dead time intervals is behaviour of critical current density under temperature variation: Assume a dead time interval, Δt_D, of arbitrary length but with the same origin as before (a warm-up experiment, in the present case provided by a short heat pulse). A stationary value of J_{Crit} during Δt_D cannot be detected because the number of superconducting charge carriers (the electron pairs) under the heat pulse first decreases and in the following, after the end of the pulse (the period when decay of the excited electron system has started) continuously increases until a new dynamic equilibrium and, correspondingly, a new stationary J_{Crit} will be established. This is the analogue to the example given in the introduction (Section 1): Detectors positioned at filament 2 would indicate a temperature increase but nothing else, in particular no response of J_{Crit} to the magnetic field emitted during the flux jump at filament 1, before end of the dead time interval.

The situation is different in the warm-up period in the suspension balance experiment. Here the temperature does not decrease but continuously increases, which means an increase of J_{Crit} due to a temperature decrease will not be observed. Instead, J_{Crit} continuously decreases (more and more electron pairs will decay). While a stationary thermal state cannot be achieved during the warm-up period, this does not imply a continuous decay process of the excited electron states would not exist. To which extent it exists, and whether decay of the excited electron states can follow or, in an extreme case, even compensate decay of pairs, depends on the decay rate, dQ_{Exc}/dt.

Temperature evaluation in the solid during the simulated warm-up periods can be divided into stepwise processes. If dQ_{Exc}/dt is very large, it at least reduces in each step the decay rate, dQ_{EP}/dt, of the electron pairs or may even overcompensate dQ_{EP}/dt, which means a decrease of the number of electron pairs due to warm-up would immediately be compensated, and in each step a new equilibrium state be established. If on the other hand dQ_{Exc}/dt is definitely smaller than dQ_{EP}/dt, as is illustrated in Figs. **12-14** at temperatures close to T_{Crit}, the density of electron pairs decreases and the density of excited electron states continuously increases, and no equilibrium will be established until superconductivity in the sample breaks down. The density, ρ_{EP}, of electron pairs then is zero.

As mentioned in the introduction, break-down of superconductivity is usually expected at the instant, t', when a thermal detector at this instant indicates that temperature T(t') of the solid (the phonon response recognized by the detector) has reached T_{Crit}. But if the rate dQ_{Exc}/dt is very small, the density ρ_{EP} of electron pairs has not decreased to zero at t = t'. Any observable the behaviour of which under temperature variation is correlated with existence of the superconducting state, like specific heat, c_p, and its jump at $T = T_{Crit}$, will not respond to the temperature increase with a clear signal at t = t': No jump of c_p can be observed at t = t'. Transition from superconducting to normal state proceeds continuously during the dead time interval, Δt_D. This means that the jump of the specific heat will no manifest itself instantaneously at t = t' but develops within Δt_D the length of which depends on the difference between dQ_{EP}/dt and dQ_{Exc}/dt. The signal "jump" of c_p accordingly is broadened over the length Δt_D, and if this period of time is large, the signal will disappear in the noise of the measurement.

For the suspension balance experiment, this means a sharp signal indicating a change of weight at the phase transition, if any, cannot be expected; the signal "change of weight" is the more broadened the larger Δt_D. A change of weight at the phase transition, if any, certainly would be very small. The consequence of Δt_D in the order of 10 s is that the signal will disappear in the noise to an extent that it cannot be identified even with the high resolution of the suspension balance. The high resolution of this device can be exploited only if actions are taken to reduce Δt_D.

Each volume element of a superconducting sample experiences an individual dead time interval, Δt_D. Since temperature evolution of the sample investigated in the suspension balance experiment is very homogeneous (see the analysis in [8]), all volume elements in this experiment almost instantaneously recognize the same dead time interval, because of the weak heat radiation heat transfer (this is reflected in Fig. **4**, the very small width of the curve obtained for item 4). Accordingly, if there is a (presently unknown) correlation between sample weight and superconducting state, no contributions of the electron system to variations of sample weight could be detected within Δt_D in any of the volume elements of the sample applied in the suspension balance experiment.

Conversely, for the cool-down experiments, items 1 and 3, the derivatives dT(t)/dt (see again Fig. **16**) are on the order of 0.1 K/s (peak values), which means Δt_D is in the order of 10 ms, short enough to ensure immediate (within standard experimental resolution) response of an observable like weight to crossing the phase transition *if* such a dependency would exist.

This means the success of measurements of observables apparently depends also on internal and external heat transfer, as is to be expected since temperature evolution in the sample and density, ρ_{Exc}, of excited electron states are correlated, by the temperature dependence $\rho_{Exc} = \rho_{Exc}(T)$. As a consequence, there are experimental situations when no sharp signal, of whatever structure and correlated to whatever observable can safely be detected if it is investigated with a setup and/or with the warm-up rates designed as reported in [9] and if the observable should respond to normal/superconductor phase transitions. Appropriate design of a gravity experiment in a suspension balance, or in any other device yielding comparable high experimental resolution, accordingly should carefully take into account also its *thermal* design aspects.

In summary, if a superconductor/gravity correlation really would exist, with the normal/superconductor phase transition in the suspension balance experiment located at t = 8241 s after start of the warm-up period, sharp signals indicating possible variations of sample weight cannot be observed in the dead temperature or dead time intervals

$$T_{Crit} - \Delta T_D(\chi = 0.01) \leq T(t) \leq T_{Crit}\ (K), \text{ with } \Delta T_D = 7.3\ mK, \text{ or}$$

$$(8241 - \Delta t_D(\chi = 0.01) \leq t \leq 8242\ s, \text{ with } \Delta t_D \approx 10\ s.$$

These intervals result from a mismatch between experimental and theoretical decay rates of the disturbed system, or the relaxation times, τ_{PH} and τ_{El}, respectively. Careful comparison of sample weight safely before and after phase transition probably is a better way to detect variations, *if any*, of the superconductor sample weight under a phase transition.

In Fig. **9**, a screening factor χ = 0.01 and the Thomas-Fermi potential yielded roughly the same mean distance, which means the order of magnitude estimate of the dead time intervals, Δt_D, made above should be representative.

That dead time intervals in general must exist is clear from the different relaxation times. However, high resolution (in time and internal sample temperature) experiments to measure quantities like the electronic part of specific heat or thermal conductivity, or the onset or decay of persistent currents, during phase transition would be highly beneficial to also quantitatively verify existence of dead time intervals in superconductors near their phase transition.

5. SUMMARY AND OUTLOOK

Internal and external heat transfer (thermal diffusivity of the solid superconductor and radiation exchange, convection and boiling) during and after a thermal perturbation are shown in this paper to exert significant influence on relaxation time for return of the disturbed electron system to a new dynamic equilibrium. This is simply because the number of elementary excitations depends on the local temperature field that in turn results from the different heat transfer mechanisms.

The results presented in the foregoing sections indicate no sharp signal of whatever physical observable could be expected if the rates $G_{Exc}(t)$ cannot approximate the required equilibrium $G_{exp}(t)$. Results otherwise could be subject to "overheating" in a warm-up experiment with the consequence that signals indicating a dependency of observables on the phase transition probably are very small, too small even if a suspension balance or other devices with high experimental resolution are applied. Full exploitation of the resolution obtainable with a suspension balance can be achieved only if all its design aspects, in particular thermal design, are taken into account and tailored more specifically.

Further investigations are necessary to confirm the results presented in this paper and the present models, in particular the procedures by which average decay rates were estimated using the "time of flight"-concept (the process exchanging information between two quantum states) with mean distances including the role of the volume, V_C, and the foregoing diffusion model approach for the decay of the disturbance in space. Note that the result, lifetime of an excited electron system measured in terms of the period of time needed to re-arrange the total wave function, cannot be obtained in a particle-related experiment where conservation of charge, mass *etc.* are in the foreground.

Experiments should be performed with very high resolution, in time and of sample temperature, when a superconductor approaches critical temperature, and thermal boundary conditions carefully be tailored to minimize dead time intervals. Candidates to verify dead-time intervals are measurement of specific heat, thermal conductivity, onset of persistent currents, elastic moduli, thermal expansion, speed of sound and their dependence on variables like mechanical pressure or magnetic field. The methods presented in this paper and in our joint contribution (Hathaway and Reiss, to this volume) should be applied also to stability of superconductors against quench.

6. ACKNOWLEDGEMENT

The author wishes to express his gratitude to Mr. George Hathaway, P. Eng., Hathaway Consulting Services, Ontario, Canada, for carefully reading and significantly improving quality of the manuscript.

REFERENCES

[1] Gray KE. Steady state measurement of the quasiparticle lifetime in superconducting aluminium. J Phys F: Metal Phys 1971; 1: 290 – 308.

[2] Epperlein PW, Lassmann K, Eisenmenger W. Quasiparticle recombination time in superconducting tin and normal electron density of states at the Fermi surface from tunnel junction experiments. Z Physik 1978; B 31: 377 – 384.

[3] Carslaw H S, Jaeger J C. Conduction of Heat in Solids. 2nd Ed. Oxford Science Publ., Clarendon Press 1959.

[4] Hellwege KH. Einführung in die Festkörperphysik. 3rd Ed. (in German). Springer 1988.

[5] Barends R, Baselmans JJA, Yates SJC, Gao JR, Hovenier JN, Klapwijk TM. Quasiparticle lifetime and noise in Tantalum High Q superconducting resonators. J Low Temp Phys 2008; 151 518-523.

[6] Wagner W. Grenzflächen und disperse Systeme. In: Weißmantel C, Lenk R, Forker W, Linke D, Eds. Atom- und Kernphysik, Struktur der Materie. Harri Deutsch 1983.

[7] Reiss H, Troitsky O Yu. The Meissner-Ochsenfeld effect as a possible tool to control anisotropy of thermal conductivity and pinning strength of type II superconductors. Cryogenics 2009; 49: 433 – 448.

[8] Hathaway G, Reiss H. Experimental Considerations in Superconductor Gravity Experiments, in: Modanese G, Robertson GA Eds. Gravity-Superconductors Interactions: Theory and Experiment. Bentham 2012; Chapter 10.

[9] Tajmar M, Plesescu F, Seifert B. Measuring the dependence of weight on temperature in the low temperature regime using a magnetic suspension balance. Meas Sci Technol 2010; 21: 015111; 1 – 7.

[10] Reiss H, Troitsky O Yu. A thermophysical model to numerically determine the diffusivity of highly excited nuclear matter with an instantaneous, internal pulse method. J Chem Eng Data 2009; 54: 2483–2497.

[11] Loidl M, Cooper S, Meier O, Pröbst F, Sáfrán G, Seidel W, Sisti M, Stodolsky L, Uchaikin S. Quasiparticle diffusion over several mm in cryogenic detectors. Nucl Instr Methods Phys Res A 2001; 465: 440-446.

[12] Blatt JM. Theory of Superconductivity. Academic Press 1964.

[13] Annett JF. Superconductivity, Superfluids and Condensates. (reprinted Ed.) Oxford University Press 2010.

[14] Pines D. Understanding high temperature superconductivity: Progress and prospects. Manuscript Director's Colloquium. Los Alamos National Laboratory. 1997: 1 – 16

[15] Mannhart J. Changes in the superconducting properties of high-T_c cuprates produced by applied electric fields. Mod Phys Lett 1992; B 6: 555 – 571.

[16] de Gennes P. Superconductivity of metals and alloys. W. A. Benjamin 1966.

[17] Holdik K, Welte M, Eisenmenger W. Quasi-particle recombination time of superconducting tin films in a parallel magnetic field. J Low Temp Phys 1985; 58: 379-388.

APPENDIX A1

Dynamical Equilibrium in Conventional Boiling, and the Existence of Spectators

The following explanation might be helpful to improve understanding of a particular aspect of the difference between a multi-particle quantum state (a superconductor, described by a single wave function) and a conventional multi-particle state (a liquid):

At any non-zero (and constant) temperature, there is dynamical equilibrium flow of vapour molecules from an interfacial layer between conventional liquid and its vapour to the space above liquid level (in a closed vessel), and in the reverse direction from the vapour volume back to the liquid. For example, in water at 20 deg C, the flow rate is about 10^{20} molecules/(cm^2 s), which corresponds to a residence time (lifetime in this layer) of water molecules in the order of some microseconds ([6], p. 400); intermolecular binding energies are strong enough to prevent the surface from having a nebulous appearance. Dynamic equilibrium flow between liquid and vapour phase accordingly consists of the fraction of all liquid molecules contained in the interfacial layer that have gained surplus kinetic energy to overcome inter-molecular binding energies by which they otherwise would be captured in the liquid. The fraction of these molecules is given by the tail of the Maxwell velocity distribution in relation to binding energies, and the corresponding molecules leave the liquid surface by diffusion and convection. Under dynamic thermal equilibrium, the rate of condensing vapour molecules on the average compensates the evaporation rate, to establish constant vapour pressure above the liquid, in a closed vessel (reflowing, condensing vapour molecules experience reflections at the liquid surface and re-absorption by liquid molecules in the interfacial layer).

If liquid temperature is increased (a thermal perturbation), more vapour molecules leave the liquid, because additional heat delivered to the liquid increases kinetic energy of the liquid molecules and thus changes their velocity distribution. Evaporation rate then shortly exceeds condensation rate during a time interval, Δt, after which a new dynamical equilibrium is established, at the higher temperature and at increased saturation vapour pressure above the liquid surface. Kinetic energy and internal states of excitations of molecules leaving the liquid and condensing are the *same*, on a statistical average, and all are definitely *uncorrelated* (apart from collisions among themselves and with walls of a container or if we neglect near-field clusters of water molecules), under thermal equilibrium and within the interval Δt.

The situation in a conventional liquid is quite different from decay of excited electron states or of electron pairs under a thermal perturbation: In the conventional liquid, we have (a) generation/condensation of only *one* sort of particles, (b) there is a large number of idle “spectator” molecules: These are given by the fraction from total number that is not contained in the tail of the Maxwell velocity distribution. The probability that these molecules would ever leave the liquid by evaporation, in a closed vessel, is very small, (c) while the velocity of the molecules follows the Maxwell distribution, the momentum of all electron pairs and thus their centre of mass velocity is identical, within uncertainty principles, (d) no single wave function comprehensively describes the state of all liquid/vapour molecules while the opposite is true for the electron pairs in a superconductor.

APPENDIX A2

Impact of Statistical Fluctuations

The following is a discussion whether statistical fluctuations could have impacts on re-arrangement of the total wave function, after a thermal perturbation:

Once a super-current is induced, it will quasi-indefinitely circulate in the superconductor or on its surface. Assume that there is a superconducting ring (a wire), and in the ring an accidentally occurring thermal fluctuation in a volume element, $V_\xi = \xi^3$, with ξ the coherence length. Let the cross section, ξ^2, of this volume element be very small in relation to the total cross section of the wire. If we assume the fluctuation would get the volume element V_ξ normal conducting, it locally introduces an electrical resistance against the induced current which accordingly has to find a new, zero-resistance path around the disturbance (such paths are available as long as critical current density will not be exceeded). Assume that the equilibrium density of single electrons is about $10^{22}/cm^3$, and roughly $\xi = 1$ nm, in a high temperature superconductor. This means we have about N = 10 electron pairs in V_ξ. A condensation energy for one electron pair, in the order 20 meV, yields a total condensation energy, E_{Cond}, in V_ξ of about 200 meV (for high temperature superconductors, this is a small number). The probability for this single fluctuation event in V_ξ is given by the Boltzmann factor $\exp[-E_{Cond}/(kT)]$. The excitation energy necessary to decompose one electron pair is at least $2\Delta E(T)$. At low temperatures, the probability of finding one of the two electrons excited to a state, **p**, near $\mathbf{p}_F$ then is given by $\exp[-E_{Cond}/2/(kT)]$. For a temperature fluctuation $\Delta T = 1$ K at T = 100 K, the probability amounts to about exp(-23) or $2\ 10^{-11}$ (from the difference of the probabilities of the states at 101 and 100 K), and the event accordingly is highly improbable.

For low temperature superconductors, this probability even would be much smaller, by many orders of magnitude, since ξ is significantly larger, and T in the order of 10 K. Accordingly, the induced current will not decay.

Consider again the decay of just one of the $1 \leq i \leq N$ electron pairs (N = 10, in the volume V_ξ), under the thermal fluctuation: The new total wave function, $\psi(t')$, of the total electron state, with remaining N – 1 pairs at the later time, t', describes a new dynamical equilibrium, once it is established, and $\psi(t')$ has to be anti-symmetrical, as before. Decay of the next of the $1 \leq i \leq N-1$ electron pairs in V_ξ, again requests re-arrangement of $\psi(t')$ to another total anti-symmetric $\psi(t'')$, and this continues until all N = 10 electron pairs in V_ξ have decayed. Re-arrangement of the total wave function, $\psi(t'')$, is not confined to particles contained only in the volume element V_ξ. Instead, its re-arrangement covers the *whole* superconducting electron state of the wire, in its total volume, V_{Total}, otherwise we had, after completion of anti-symmetrization, at least two different total wave functions, which is contrary to experience. In view of the enormous numbers of single wave functions, $\varphi_i(t)$, of which $\psi(t'')$ is composed, re-arrangement of $\psi(t'')$ after a statistical thermal fluctuation, too, is highly improbable, and so for any time, t.

APPENDIX A3

Generation and Decay Rates during Cool-down from $T > T_{Crit}$

For simplicity, assume for the moment that the thermal conduction problem, with radiative boundary conditions and no internal heat sources, for all $T > T_{Crit}$, can be treated as one-dimensional. Fourier's differential equation applied to the pellet volume and under *warm-up* reads

$$\rho\, c_p\, dT(t)/dt = d/dx\, [-\lambda\, dT(t)/dx] + d/dx\, [Q_{Rad}(t)] \qquad \textbf{(A3.1)}$$

with ρ and λ density and thermal conductivity of the pellet, c_p its specific heat and Q_{Rad} [W/m^2] the radiative heat flow received by the pellet. Assume, still under warm-up, that the phase transition is completed and that a temperature field, $T(x,t') > T_{Crit}$, is established in the pellet at a time, t', after start of the experiment. At t – t', the pellet shall have obtained thermal equilibrium, which means its temperature, T(x,t') necessarily is below a stable temperature of the cavity.

Now consider *cool-down*, as the reverse heat transfer process. Assume for this purpose that the temperature field T(x,t') defines the initial condition for a cool-down period applied to the same pellet, with same thermal/radiative environment as before. After a sufficiently long time, t''' > t', the pellet will arrive at a temperature field, $T(x,t''') < T_{Crit}$, with T(x,t''') necessarily higher than cavity temperature. Accordingly, cool-down is described in Equation (A3.1) by reversing the orientation of conduction and radiation heat flow and the results calculated in Sections 2 and 3 hold also for this case.

The temperature field, T(x,t''') is identical with T(x,t') if conductive properties of the pellet are constant and homogeneous, and if radiative boundary conditions (emissivities of pellet and cavity) are the same as during the warm-up period. This means that also the gradients dT(x,t'')/dx and thus the $dN_{SC}(t'')/dt$ will be the same, but time t'' during cool-down is not identical to time, t, during warm-up. But t'' can easily be shifted to the (common) time scale, t, because of reversibility of the heat transfer problem under constant and homogeneous materials properties and emissivities. Instead of the gradient -dT(t)/dx in the divergence of the conduction term, we have the reverse temperature gradient, +dT(t'')/dx, $t \leq t'' \leq t'''$, and the sample also experiences no longer a positive divergence of radiation exchange with its surroundings but the divergence becomes negative (the pellet emits more radiation heat flow than it receives). As a consequence, both conductive and radiative divergences change sign in Equation (A3.1) so that also dT(t'')/dt becomes negative.

Subject Index

Made in the USA
Las Vegas, NV
03 November 2026

57424457R00200